Basic Drafting Using Pencil Sketches and AutoCAD®

JAMES M. KIRKPATRICK
Eastfield College

Upper Saddle River, New Jersey
Columbus, Ohio

Editor in Chief: Stephen Helba
Executive Editor: Debbie Yarnell
Media Development Editor: Michelle Churma
Production Editor: Louise N. Sette
Production Supervision: Karen Fortgang, *bookworks*
Design Coordinator: Diane Ernsberger
Cover Designer: Bryan Huber
Cover art: Bryan Huber
Production Manager: Brian Fox
Marketing Manager: Jimmy Stephens

This book was set in Times Roman by STELLARViSIONs. It was printed and bound by Courier Kendallville, Inc. The cover was printed by Phoenix Color Corp.

Pearson Education Ltd., *London*
Pearson Education Australia Pty. Limited, *Sydney*
Pearson Education Singapore Pte. Ltd.
Pearson Education North Asia Ltd., *Hong Kong*
Pearson Education Canada, Ltd., *Toronto*
Pearson Educación de Mexico, S. A. de C.V.
Pearson Education—Japan, *Tokyo*
Pearson Education Malaysia Pte. Ltd.
Pearson Education, *Upper Saddle River, New Jersey*

Copyright © 2003 by Pearson Education, Inc., Upper Saddle River, New Jersey **07458.** All rights reserved. Printed in the United States of America. This publication is protected by Copyright and permission should be obtained from the publisher prior to any prohibited reproduction, storage in a retrieval system, or transmission in any form or by any means, electronic, mechanical, photocopying, recording, or likewise. For information regarding permission(s), write to: Rights and Permissions Department.

10 9 8 7 6 5 4 3 2 1
ISBN: 0-13-094069-0

Preface

The purpose of this textbook is to teach the fundamentals of drafting through the use of sketches on gridded paper done with triangles, a circle template, and a minimum of other tools and supplies. In addition, the student will be introduced to the AutoCAD program as partially drawn AutoCAD drawings identical to the pencil sketches are completed. These drawings are included in the textbook on a CD-ROM and will be compatible with any version of AutoCAD or AutoCAD LT from Release 14 to the present. Tutorials will not be tied to any specific version of AutoCAD, as commands will be typed from the keyboard.

This approach allows the instructor to cover more subject matter without concentrating on the rigors of manual instrument drawing and mastering skills or purchasing expensive instruments that have little application in today's industries. After the student completes sketches on a topic, one or more tutorials on the AutoCAD program are presented so that the student will also be comfortable with AutoCAD when the course is completed. A variety of exercises are provided in each chapter to challenge students on many levels.

Contents

1 Introduction   1

     Drafting Fundamentals   1
     Necessary Personal Characteristics   1
     Purpose of This Book   2
     How to Use This Book   2
     Types of Drawings to Be Covered in This Book   4
     The Hardware and Software Necessary to Complete the Computer-Aided Tutorials in This Book   12

2 Sketching Tools, Supplies, and Their Uses   17

     Sketching   17
     Sketching Tools   17

3 AutoCAD Fundamentals   32

     Introduction   32
     Making the Setup for the First Drawing Exercise   32
     Tips for Beginning AutoCAD and AutoCAD LT Users   41

4 Lettering for Pencil Sketches   44

     Introduction   44
     The Sketching Alphabet   44
     Developing a Good Lettering Style   46

5 Sketching Line Weights and Drawing Constructions   49

     Lines Used in Technical Drawings   49
     Sketching Good Lines   50
     Construction Terms   51
     Drawing Constructions   53

6 Linetypes and Drawing Constructions Using AutoCAD   69

     Linetypes   69
     Drawing Constructions   69

7 Reading and Sketching Orthographic Views   91

     Orthographic Projection   91
     Identifying Surfaces and Features   94
     Sketching Orthographic Views   101

8 Making Orthographic Views with AutoCAD   114

Two-Dimensional Drawings in AutoCAD   114

9 Sketching Sectional Views   150

Uses of Sectional Drawing   150
Constructing a Sectional View   152
Elements of Sectional Drawing   152

10 Making Sectional Views with AutoCAD   163

Hatching Using AutoCAD or AutoCAD LT   163

11 Sketching Auxiliary Views   177

Uses of Auxiliary Views   177
Drawing a Primary (Also Called Single) Auxiliary View from a Three-Dimensional View   178
Drawing a Secondary (Also Called Double) Auxiliary View from a Three-Dimensional View   179

12 Making Auxiliary Views with AutoCAD   186

Drawing Auxiliary Views Using AutoCAD or AutoCAD LT   186

13 Sketching Pictorial Views   203

Pictorial Drawing Forms   203

14 Making Isometric Views with AutoCAD   223

Isometric Drawing Settings   223

15 Sketching Dimensions   233

Introduction   233
Standard Dimensioning Practices   233

16 Dimensioning with AutoCAD   246

Dimensioning   246

17 Sketching Threads and Fasteners   254

Introduction   254
Thread Specification for English Units   254
Thread Specification for Metric Threads   258
Symbols for Drawing Threads   259
Types of Fasteners   259
Sketching a Thread   262

18 AutoCAD Drawings of Fasteners Using Blocks and Attributes   268

19 Sketching Development Drawings   282

20 AutoCAD Developments with an Introduction to AutoCAD 3D   297

21 The Drawing System   313

Appendix   320

Exercises to Chapters   324

Index   471

1 Introduction

OBJECTIVES

After completing this chapter, you will be able to

- Describe why drafting fundamentals are important in manufacturing and construction.
- List seven personal characteristics needed to become successful in technical graphics occupations.
- Describe the purpose of this book.
- Describe how to use this book.
- Identify the types of drawings to be assigned.
- Identify the hardware and software necessary to complete the computer-aided tutorials in this book.

DRAFTING FUNDAMENTALS

Drafting fundamentals make up a language that is used in manufacturing and construction for communicating ideas and instructions. These fundamentals are used by designers, engineers, and architects to design products, to communicate, and to sell ideas. After these products have been approved for manufacturing or construction, they are refined and built by people who must understand the same language of graphic communication. Without this means of communication, manufacturing of products in large numbers and building construction would be greatly hampered if not impossible.

NECESSARY PERSONAL CHARACTERISTICS

Although there are people of many different personality types in the business of technical graphics, they have a few common traits, namely,

They work closely and for long periods with details.
They tolerate repeated, daily contact with the same people.
They accept changes and corrections to their work.
They sit in the same place for long periods. This can be a health problem for people who do not get enough of the right kind of exercise. If you decide to make a career of this type of work, develop a good exercise program and stay with it.
They visualize objects in two and three dimensions, as shown in Figure 1–1.
They check their own work systematically and carefully. This is difficult for beginners, but it is extremely important. Most companies are very understanding with beginners who are slow, but they are not so patient with employees who are consistently inaccurate. The extra time it takes to check your work carefully and correct it before you turn it in is well worth it. Checking your own work and knowing it is right encourages you to develop the final personal trait.
They develop pride in their work. Students who do not develop strong graphics skills will quickly become discouraged when a deadline approaches. Trying to do

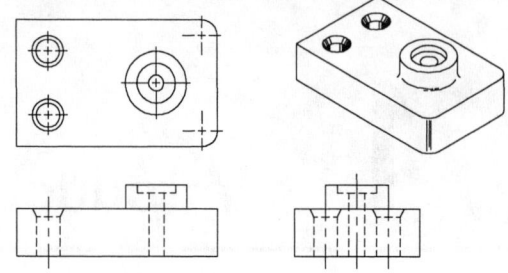

FIGURE 1–1

Drawings in Two and Three Dimensions

good work in a reasonable time with poor skills can be corrected only by spending the time necessary to develop good skills and the resulting pride in your work.

Many students in beginning technical graphics classes either do not have the time or will not spend the time necessary to develop good drawing and visualizing skills and believe that some mysterious trait was left out of their set of abilities. In most cases this is not true. Drafting fundamentals come more easily at first for some people, but those who have to work at it often develop better skills than those who have an extremely easy time of it.

The technical graphic language is foreign to many beginners. Learning to read and draw with this new language is exciting and well worth the time necessary to learn it. Whether or not you make a career of this type of work, the graphic language of drafting provides a set of skills that will allow you to understand and communicate many things in your life that you would not be able to otherwise.

PURPOSE OF THIS BOOK

The purpose of this text is to teach the fundamentals of drafting through the use of sketches on gridded paper done with triangles, a circle template, and a minimum of other tools and supplies. If the instructor wishes, only pencil and eraser are absolutely necessary. In addition, the student will be introduced to the AutoCAD or AutoCAD LT programs as partially drawn AutoCAD drawings that are the same as or similar to the pencil sketches are completed. These drawings are with the text on a floppy disk and are compatible with any version of AutoCAD or AutoCAD LT from Release 14 to the present. Tutorials are not tied to any specific version of AutoCAD, as commands will be typed from the keyboard.

This approach allows the instructor to cover more subject matter without concentrating on the rigors of manual instrument drawing and mastering skills and purchasing expensive instruments that have little application in today's industries. After the student completes sketches on a topic, one or more tutorials on the AutoCAD program are presented so that the student will also be comfortable with AutoCAD when the course is completed. A variety of exercises are provided in each chapter to challenge students on many levels.

HOW TO USE THIS BOOK

Using Sketches to Learn Drafting Fundamentals

Step 1.   Read the assigned chapter.
Step 2.   Remove the sketch assignment page from your book.
Step 3.   Read the assignment carefully.
Step 4.   Make the sketch using pencils, erasers, triangles, scales, and a circle template, or use only pencils and erasers per your instructor.

Step 5. Complete the title block using a good lettering style of a consistent height and slant.
Step 6. Check your drawing and make any necessary corrections.
Step 7. Give the drawing to your instructor for grading.

Using the AutoCAD or AutoCAD LT Tutorials

All the tutorials require you to type commands. Because the many versions of AutoCAD and AutoCAD LT have commands located on toolbars or menus that are labeled differently, typing is the only common means of operating these software packages. If you wish to click commands from the menus or toolbars, you will have to study their locations and find them when you want to use them. In many cases typing is the fastest way to operate AutoCAD software. The specifics of how the tutorials are constructed are listed next.

Drives

This book assumes that the hard drive of your computer is labeled C. It also assumes there are one or two floppy disk drives labeled A and B. If you have a compact disk drive, this book assumes it is labeled D.

Prompt and Response Columns

Throughout the exercises in the book, Prompt and Response columns provide step-by-step instructions for starting and completing a command. The Prompt column text repeats the AutoCAD or AutoCAD LT prompt that appears in the Command: prompt area of the display screen. The text in the Response column shows your response to the prompt and appears as follows:

1. All responses are shown in bold type.
2. ↵ is used to indicate the enter response. Either a button on the pointing device or a key on the keyboard may be used. On keyboards this key is marked ↵, Enter, or Return.
3. A response that is to be typed and entered from the keyboard is preceded by the word Type: and is followed by ↵ (for example, Type: **L↵**).
4. Function keys are the keys marked F1 through F10 on the keyboard (F11 and F12 are available on many keyboards). If the response is to use a function key, the key name will be preceded by the word Press: (for example, Press: **F7**).
5. When you are required to use the click button on your digitizer mouse (the left button) to select an object or to specify a location, the word Click: is used followed by a description (for example, Click: **D1**).
6. Helpful notes, such as (F1 is the flip screen function key for DOS; F2 is the flip screen function key for Windows), are provided in parentheses.

Margin Notes

Helpful notes, tips, and warnings are included in the margin.

Steps in Using the Tutorials

Step 1. Copy all drawings from the floppy disk included with your book to a directory (a folder if you are using Windows) on the hard drive of your computer. If you are using a computer in a school laboratory, be sure to get your instructor's permission before taking this step.
Step 2. Open the assigned drawing.
Step 3. Follow the steps in your book to make the assigned drawing.
Step 4. Carefully check your drawing on the screen, and make any necessary corrections.
Step 5. Save your drawing in at least two places.
Step 6. Print or plot your drawing as instructed.

Step 7. Check the hard copy of your drawing, make corrections, and reprint if necessary.
Step 8. Give the drawing to your instructor for grading.

TYPES OF DRAWINGS TO BE COVERED IN THIS BOOK

Line Weights and Drawing Constructions

Figure 1–2 is an example of the types of drawing constructions for sketching and AutoCAD. These exercises are provided to teach some of the basic skills you will need to make the other drawings in this book.

Orthographic Drawings

Two-dimensional (2D) drawings are called *orthographic drawings* in engineering and drafting occupations. These drawings are the universal language of technical drawing.

FIGURE 1–2
Drawing Constructions in Sketches and AutoCAD

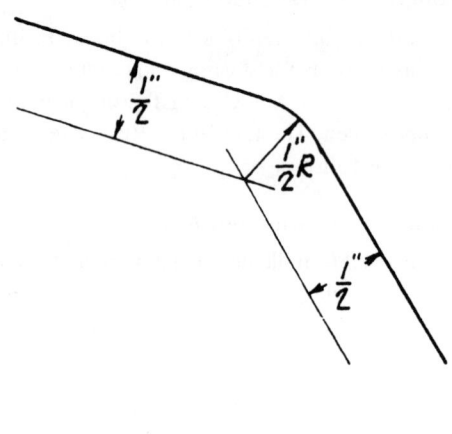
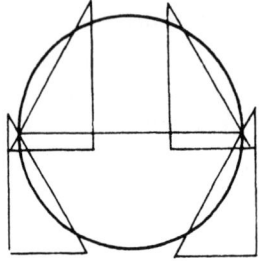
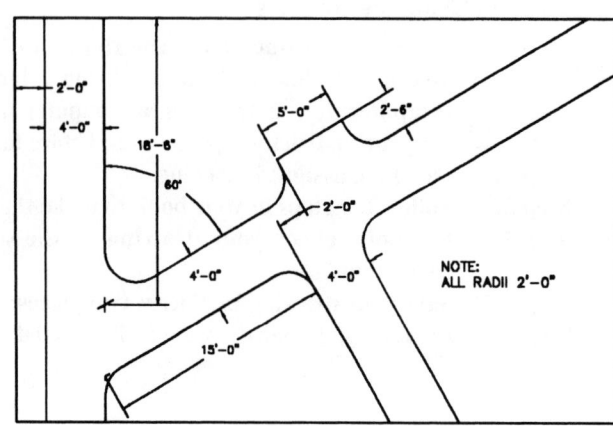

FIGURE 1-3
Orthographic Sketch and AutoCAD Drawing

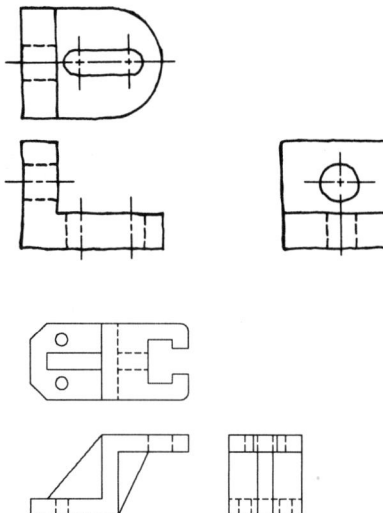

Only two dimensions are seen in any one view. Those dimensions may be height and width, height and depth, or width and depth. Figure 1-3 shows some examples of orthographic drawings.

Sectional Drawings

Sectional drawings are used in many different industries to clarify internal or hidden external construction. Figure 1-4 shows two sectional drawings: one sketch and one made using AutoCAD. This type of drawing can be done with more consistent lines and with much less effort using AutoCAD than drawing it manually. Shading lines are often drawn with a thinner line by using a layer command. Layers are used to separate linetypes or other features that allow the drawing to be used more efficiently.

Auxiliary Drawings

Auxiliary drawings are used to show the true shape of slanted surfaces. Figure 1-5 shows two auxiliary drawings: one sketched and one made using AutoCAD.

FIGURE 1-4
Sectional Sketch and AutoCAD Drawing

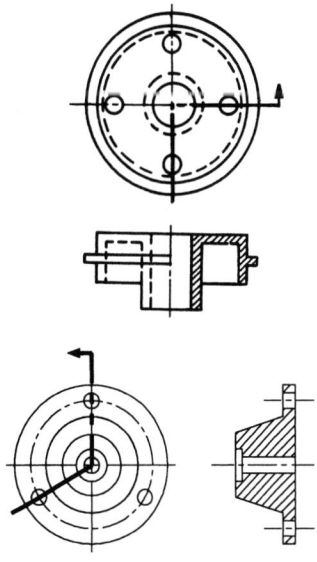

Introduction

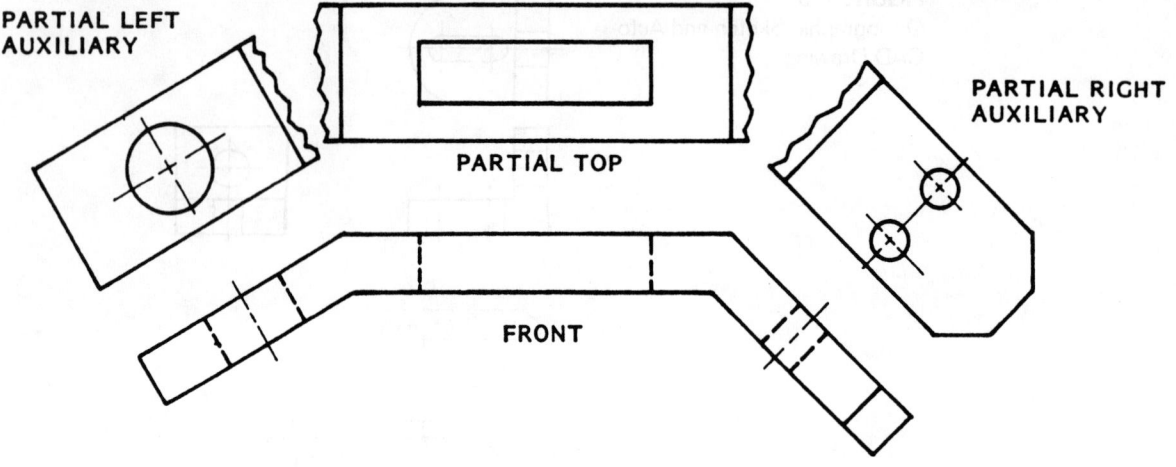

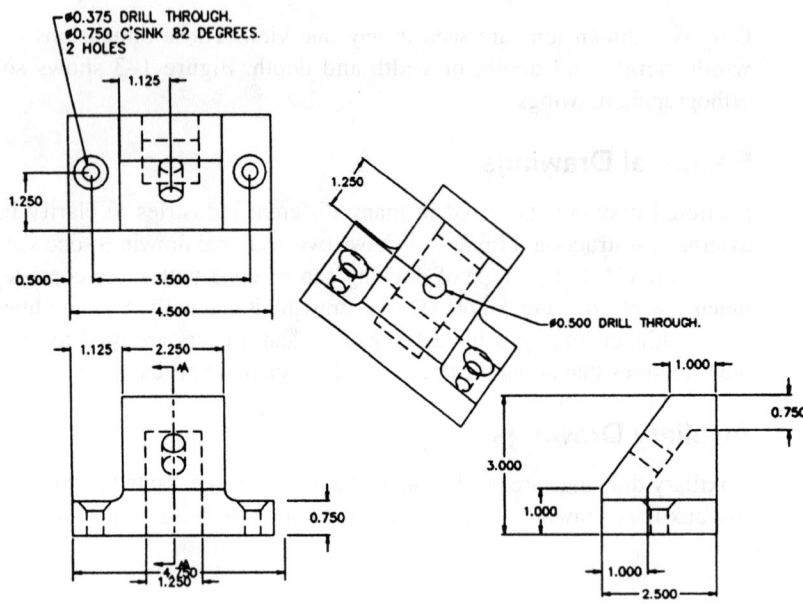

FIGURE 1-5
Auxiliary Sketch and AutoCAD Drawing

Isometric Drawings

Isometric drawings are often used to show pictorial views of objects. Although isometric drawing is not three-dimensional, it is a two-dimensional drawing form that shows what a part looks like in three dimensions. Figure 1–6 shows examples of isometric drawings.

Dimensioned Drawings

A dimensioned drawing is a type of orthographic drawing that gives the size and location of features. AutoCAD is very useful in dimensioning because it can become almost automatic when drawings are made full size. Figure 1–7 shows two drawings: a sketch and one done with AutoCAD.

Toleranced Drawings

All drawings used to build parts or structures have tolerances (limits) that are placed on all dimensions. These tolerances may be very specific, as shown in Figure 1–8. For example, an understood tolerance of $\pm\frac{1}{4}''$ on a 3″ dimension means that a part, when inspected, must measure between $3\frac{1}{4}''$ and $2\frac{3}{4}''$ to pass. A more accurate tolerance for the same 3″ dimension could be $\pm.005''$. The upper limit in this case would be 3.005″, and the lower limit would be 2.995″. AutoCAD allows parts to be drawn with extreme accuracy using certain commands. It also allows parts that do not require accuracy to be drawn faster but with less accuracy. Figure 1–9 also shows a feature called *geometric tolerances* (those symbols, letters, and numbers inside boxes), which are done easily with the tolerance routine on the dimensioning menu or toolbar of the latest versions of AutoCAD and AutoCAD LT.

FIGURE 1–6
Isometric Sketches and AutoCAD Drawings

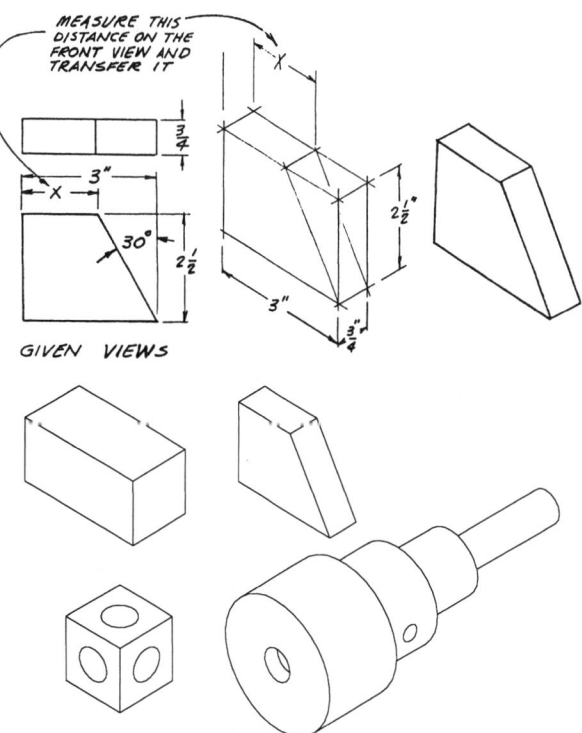

FIGURE 1–7
Dimensioned Sketches and Auto-CAD Drawings

HOLE SCHEDULE					
HOLE	A	B	C	D	E
SIZE	.13	.19	.25	2.00	.38

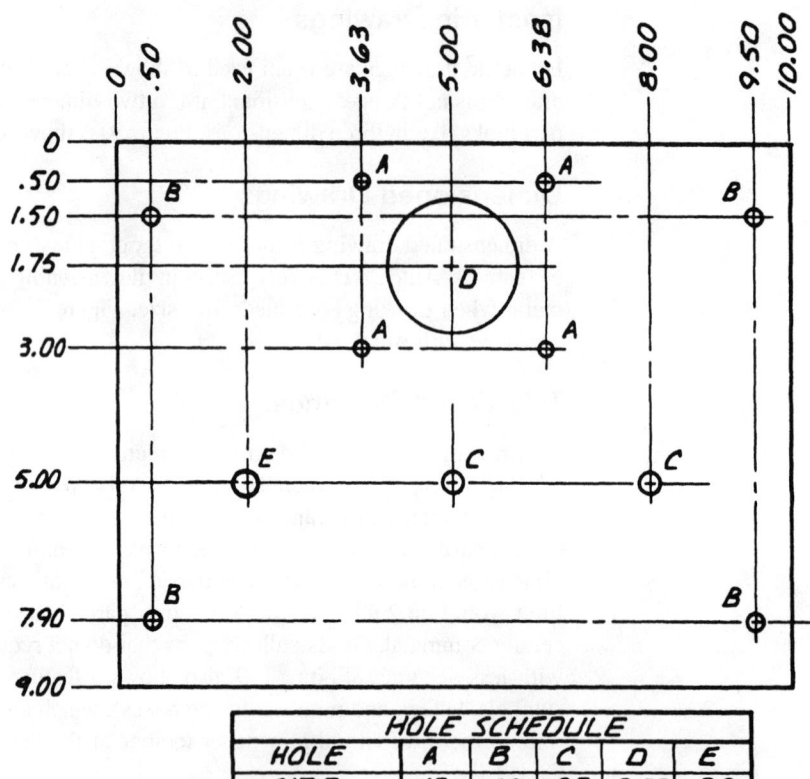

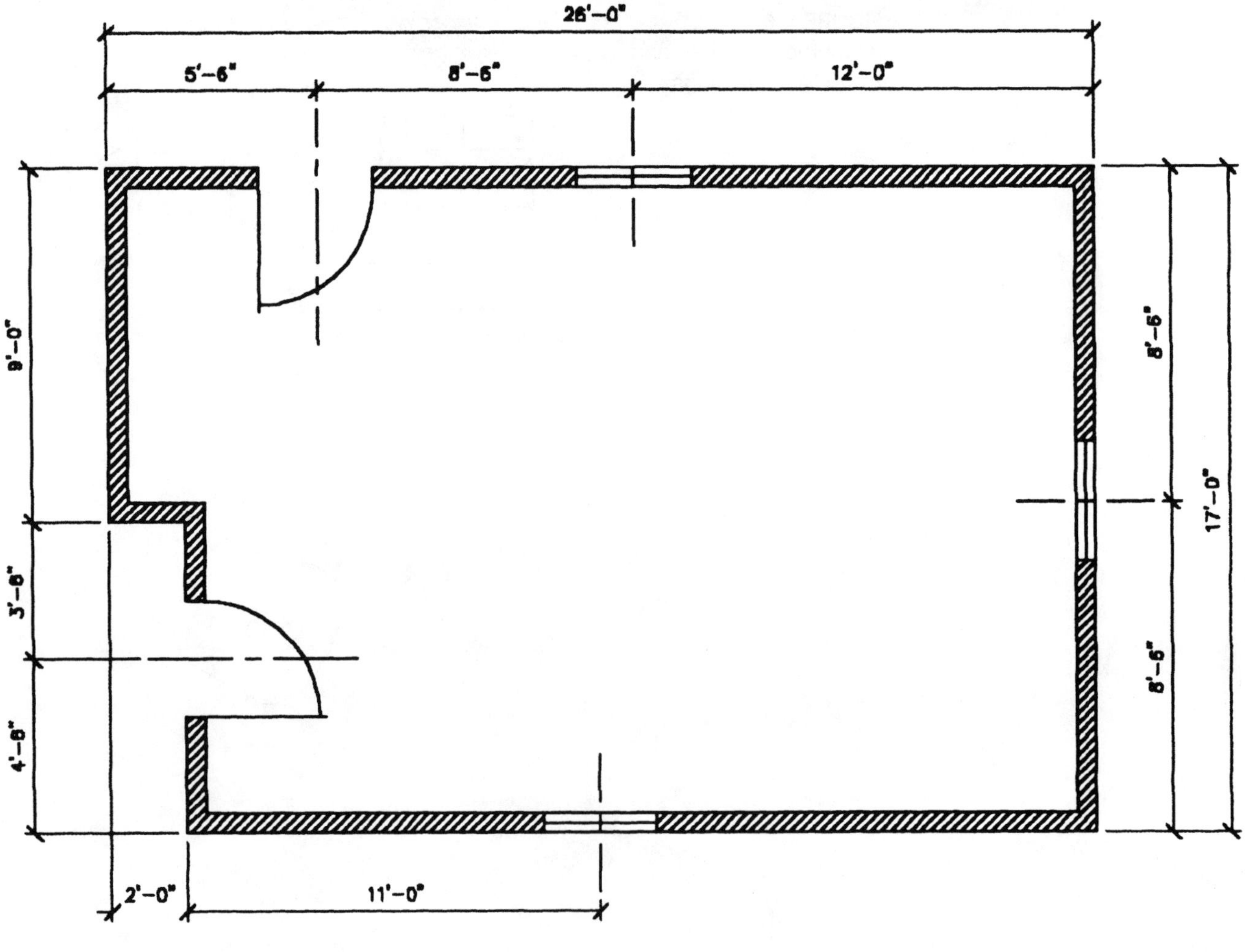

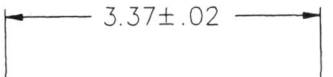

FIGURE 1-8
Tolerances

Drawings Containing Threads and Fasteners

Figure 1–10 shows two drawings containing threads and fasteners. The symbols and the specifications attached to them are used to identify where threads and fasteners are located and precisely what they are.

Development Drawings

Development drawings are used to create flat patterns of sheet metal shapes. These flat patterns are bent on fold lines to create shapes such as those shown in Figure 1–11.

Detail Drawings

The drawings shown in Figure 1–12 are complete drawings of a single part showing all the information needed to manufacture that part.

Assembly Drawings

Assembly drawings are used to show how parts fit together to form an assembly. They also identify all parts of that assembly and list them on a document called a *parts list* or *list of materials*. Figure 1–13 (on p. 12) is a formal AutoCAD assembly drawing.

FIGURE 1-9
Geometric Dimensions and Tolerances

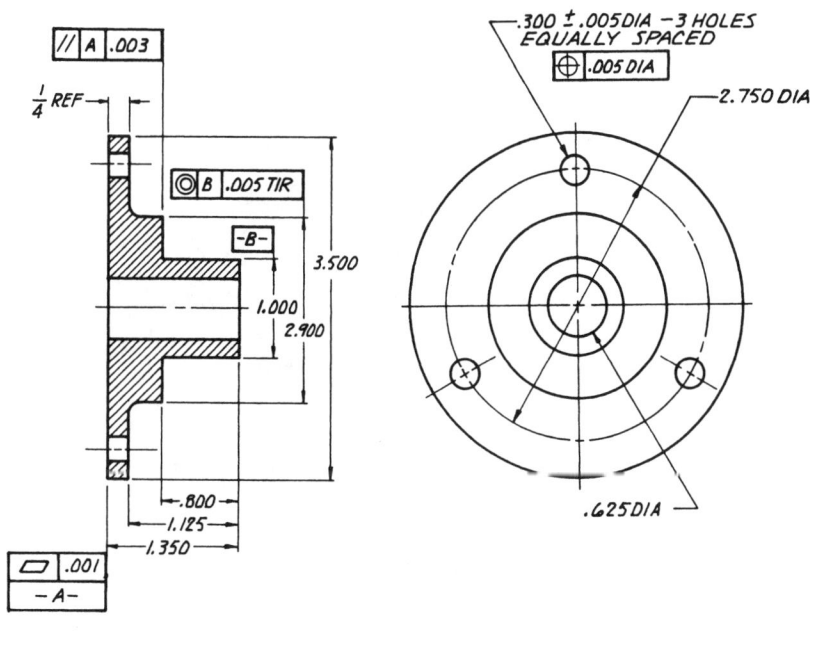

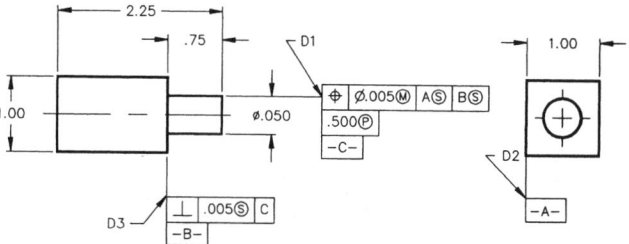

Introduction

FIGURE 1–10
Threads

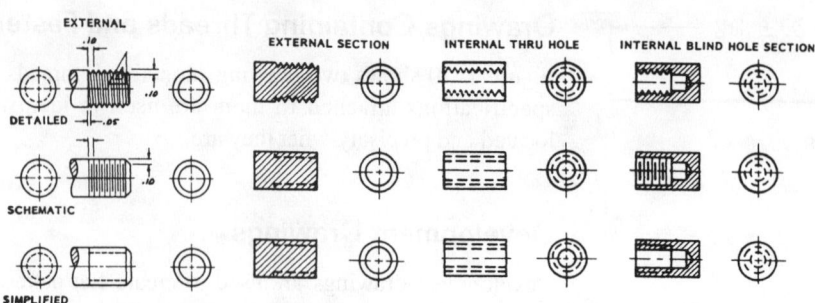

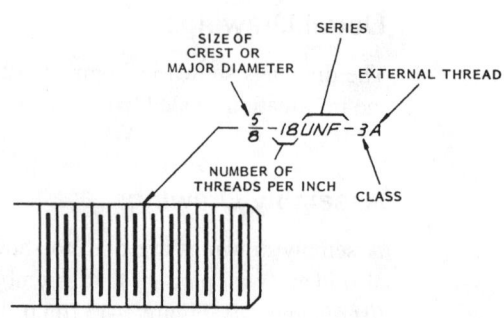

FIGURE 1–11
Development Drawings

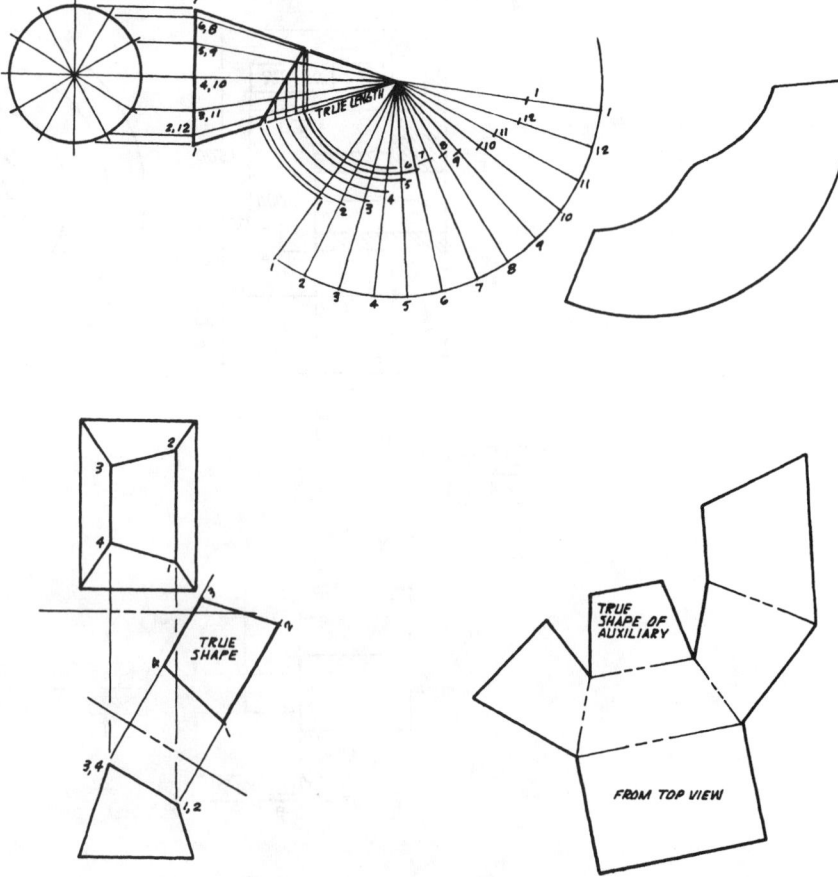

10 Chapter 1

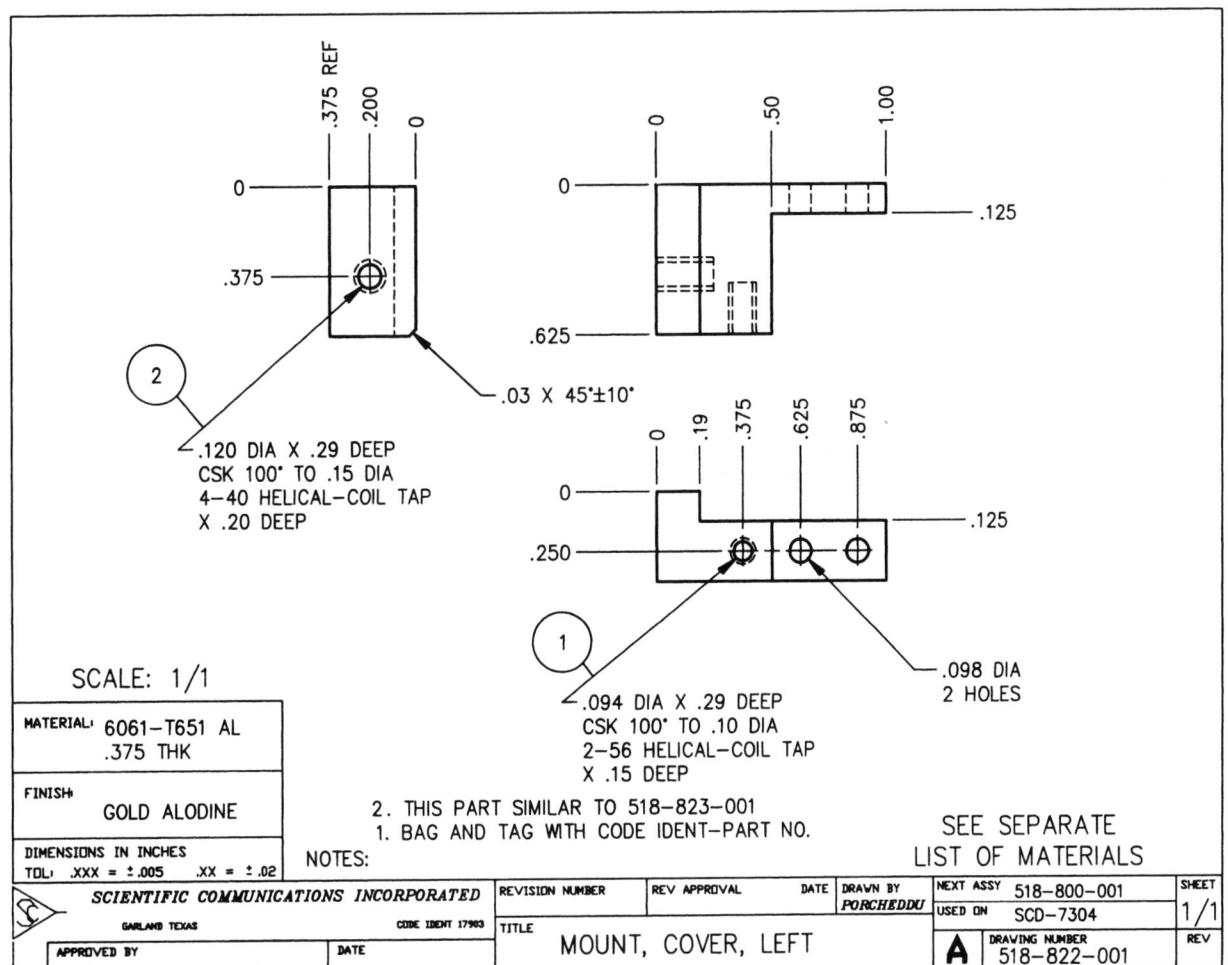

FIGURE 1-12
Detail Drawing

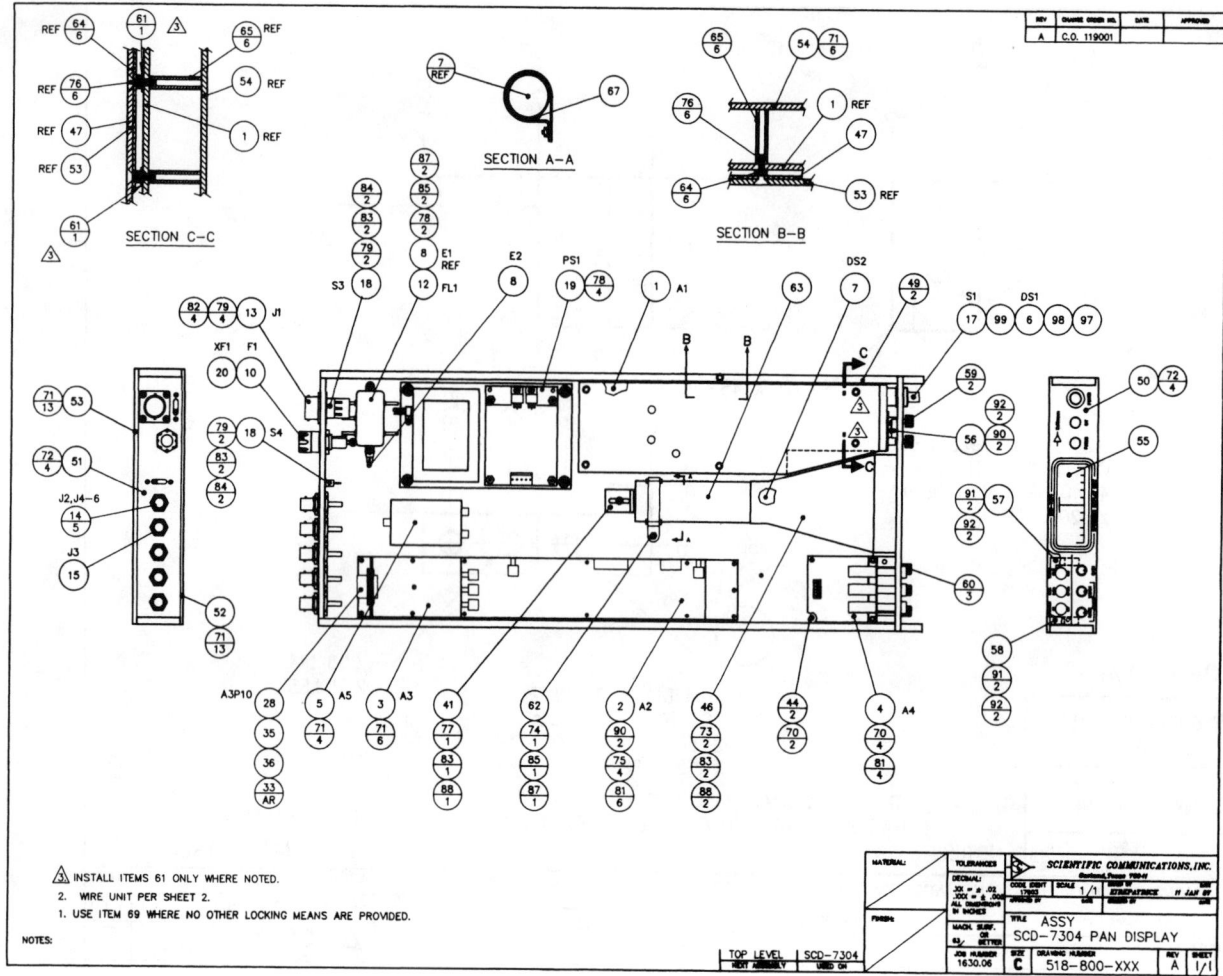

FIGURE 1-13
Assembly Drawing

THE HARDWARE AND SOFTWARE NECESSARY TO COMPLETE THE COMPUTER-AIDED TUTORIALS IN THIS BOOK

Hardware

The following are parts of a typical personal computer system (Figure 1-14) on which AutoCAD or AutoCAD LT software can be used:

Computer
Floppy disk drive
Hard disk drive
Compact disk drive
Video monitor
Keyboard
Printer
Plotter

Computer

The computer should be of sufficient capacity to run AutoCAD or AutoCAD LT easily. The computer should also have a graphics card (a printed circuit board that allows a high-quality display of graphic data) and other features that will allow the display (the screen)

12   Chapter 1

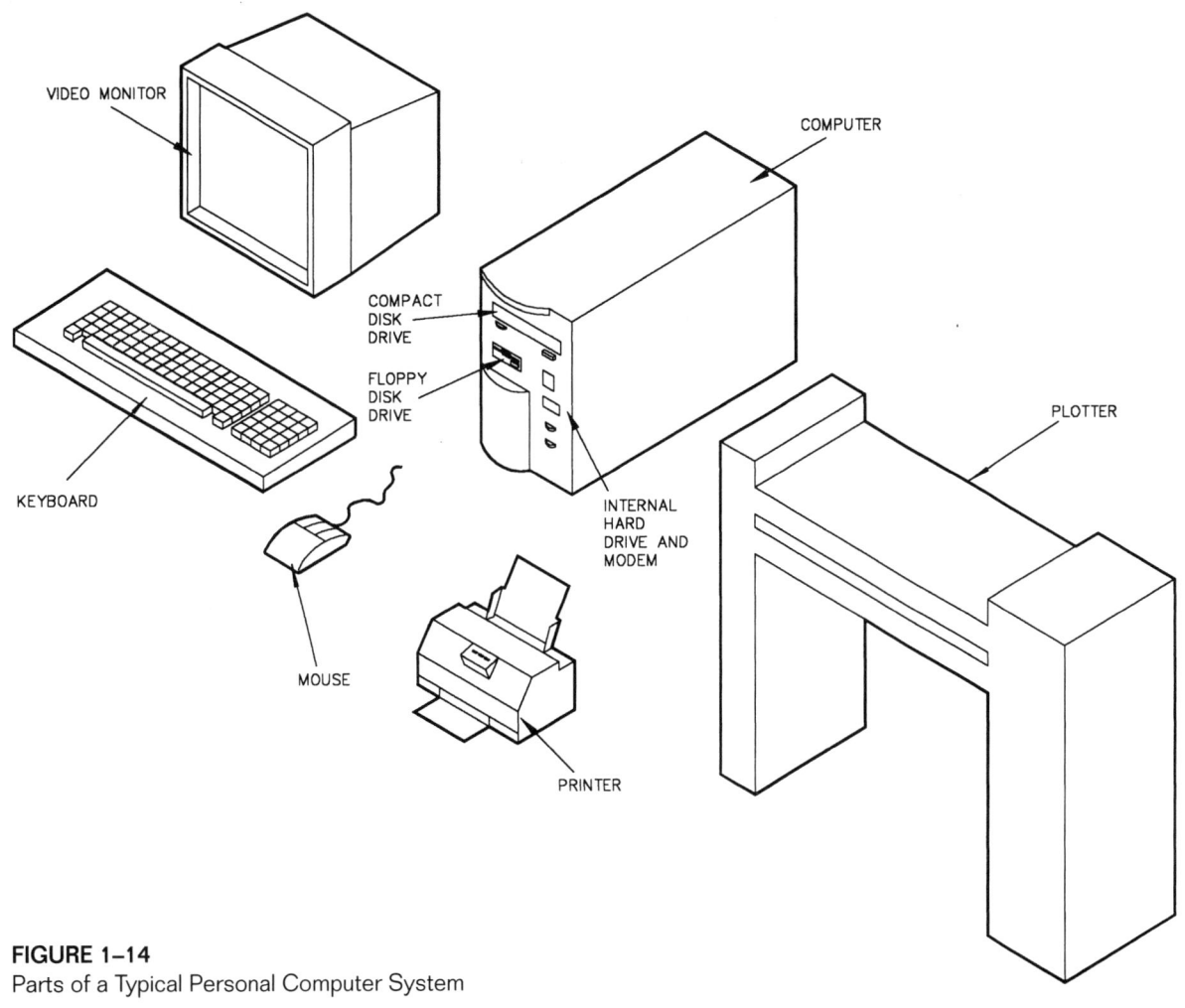

FIGURE 1–14
Parts of a Typical Personal Computer System

to be easily read and quickly generated. In addition, both AutoCAD and AutoCAD LT require a math coprocessor chip. A slow response to commands or a poor display will defeat the advantages of this valuable drawing tool.

Floppy Disk Drive

At least one floppy disk drive is needed to move large blocks of information into and out of the computer. Floppy disks are inserted into a floppy disk drive. Like a record player, the disk drive spins the floppy disk on a spindle to read information from it or write information to it. AutoCAD and AutoCAD LT drawings can be stored on floppy disks. Floppy disk drives are identified by a one-letter name followed by a colon (for example, A:). The $3\frac{1}{2}''$ high-density disk drive uses a $3\frac{1}{2}''$ floppy diskette. This diskette stores approximately 1.44 MB (1,440,000 bytes) of information.

Compact Disk Drive

Compact disk drives are available in a variety of speeds. The faster they are, the higher the price. A double-speed compact disk drive is sufficient to load AutoCAD or AutoCAD LT software, but a quadruple or faster speed drive is nice to have. Because much of the new software is available on compact disks and as a result is more convenient and faster to load, a compact disk drive is a necessity. Read/write compact disk drives allow you to store AutoCAD drawings on a compact disk. One CD stores 700 MB of information.

Introduction

Hard Disk Drive

A hard disk drive, also called a hard drive or hard disk, is usually permanently installed inside the computer. It can store much more information than the removable floppy disks and is used to store the AutoCAD or AutoCAD LT program. AutoCAD drawings may also be stored on the hard disk drive. The hard disk drive is commonly called drive C (C:). A hard drive with adequate storage capacity for your situation is necessary; 6 gigabyte (GB) to 60 GB hard drives are common.

Video Monitor

A video monitor is similar to a television screen. A color video monitor is a necessity for most drawings. The physical size of the screen is not as important as the resolution. The resolution of the video is stated in *pixels*, which is the number of dots arranged in rows and columns to make the visual video on the screen—the finer the resolution, the better. AutoCAD and AutoCAD LT require a video screen of reasonably high resolution.

Keyboard

The keyboard has three parts:

Alphanumeric keys Located in the center of the keyboard, these are used to type the lettering and numbers that will appear on your drawings and occasionally to type commands. The number keys can also be used as a calculator with an AutoCAD LT command.

Function keys Keys labeled F1–F12, often located to the left or above the alphanumeric keys of many keyboards. These keys are used to perform special functions such as turning a grid on or off. These keys are used, and their functions are explained, in later chapters.

Numeric keys Often located to the right of the alphanumeric keys, these keys can be used to type numbers that will appear on your drawings and can also be used as a calculator in combination with an AutoCAD or AutoCAD LT command. The directional arrows, which can be toggled ON or OFF, may be used to move the location of the pointer, although moving the pointer with a mouse is usually much faster.

Mouse

A mouse is used to select commands from the AutoCAD or AutoCAD LT toolbars or menus that appear on the right edge or top of the AutoCAD LT screen video. The mouse allows the eyes of the operator to remain on the screen at all times. It is also used to enter points of a drawing, such as where a line starts or where a circle is located. It is moved across a tabletop or pad and its action is described on the video screen by the movement of a crosshair. The crosshair is positioned to highlight a command or to locate a point. The click button (usually the far left button) on the mouse is pushed to select a command or enter a point on the drawing. The extra buttons on a mouse can be assigned to perform different tasks. Most commonly one extra button is assigned return ↵.

Printer

Laser printers are available that produce hard copies of excellent quality in black and white. Many of the newer color printers are relatively inexpensive and produce high-quality hard copies. The low-cost color printers are slow, however, so take the speed factor into account if you buy one.

Plotter

To make high-quality, usable drawings (hard copies), a plotter is often used. One type of plotter commonly available uses pens similar to technical drawing pens for manual inking. Felt markers are also available for this type of plotter. Both types of pens can be used for multicolor drawings. A good plotter makes drawings with smooth curves, dense lines, and crisp connections. Plotters may have one pen or multiple pens. A plotter may accept only $8\frac{1}{2}'' \times 11''$ paper, or it may accept larger sizes and rolls of paper. Electrostatic and ink-jet plotters that produce excellent color copies are also available.

Software

Any release of AutoCAD or AutoCAD LT from Release 14 to the current version can be used to make the drawings in this book.

The drawings contained on the disk that is included are Release 14 drawings. These drawings can be opened in any of the releases described above. If, however, you work on one of them in Release 2002 of AutoCAD or AutoCAD LT, that drawing becomes a Release 2002 file and cannot be used in earlier versions unless you save it as a Release 14 drawing.

REVIEW QUESTIONS

Circle the best answer.

1. Checking your own work is unnecessary because someone else can check it better.
   a. True
   b. False
2. Which of the following is *not* used to make pencil sketches in this book?
   a. Triangles
   b. Circle template
   c. Compass
   d. Eraser
   e. Scales
3. When the words Click: **D1** are shown in the Response column, which button on your mouse do you press?
   a. The left button
   b. The right button
   c. The middle button
4. Which of the following is the flip screen function key for Windows?
   a. F2
   b. F3
   c. F5
   d. F7
   e. F9
5. Orthographic drawings show how many dimensions in any one view?
   a. One
   b. Two
   c. Three
   d. Four
   e. None
6. Dimensioning in AutoCAD can be almost automatic.
   a. True
   b. False
7. A tolerance of $\pm.005''$ on a dimension of 4″ means that the feature, when inspected, must fall in which of the following ranges?
   a. 4.500–3.950
   b. 4.005–3.995
   c. 4.050–3.950
   d. 4.000–4.100
   e. 4.000–4.010
8. Sectional drawings are used to
   a. Make 3D drawings
   b. Show where to place dimensions
   c. Introduce 3D commands
   d. Introduce 2D commands
   e. Clarify internal details
9. Isometric drawing is not the same thing as a 3D model.
   a. True
   b. False

10. This book assumes that the hard drive of your computer is labeled
    a. A
    b. B
    c. C
    d. D
    e. This book does not assume anything about your hard drive.
11. List five personal characteristics necessary to become successful in technical graphics.

    _____

    _____

    _____

    _____

    _____

12. What must be done to use a Release 2002 drawing in Release 14 of AutoCAD?

    _____

13. What must be done to use a Release 14 drawing in Release 2002?

    _____

14. List two items that a video monitor must have to do most of the drawings in AutoCAD.

    _____   _____

15. Which releases of AutoCAD are required to complete the AutoCAD tutorials in this book?

    _____

2 Sketching Tools, Supplies, and Their Uses

OBJECTIVES

After completing this chapter, you will be able to

- Correctly identify the tools used for sketching.
- Describe how these tools are used.
- Read scales accurately.
- Draw lines to scale.

SKETCHING

Sketching well can be very useful for people who are designing and drawing any product. Sketching can be done with nothing more than a pencil, eraser, and a piece of paper. Sketching can also be done with triangles, a circle template, pencils, eraser, scales, drafting powder, and paper. Sketching can even be done on the computer using AutoCAD software. This book is designed so that you may use as elaborate a sketching system as you want.

If you choose not to use triangles and circle template for manual sketches, be aware that technical sketches such as the ones that will be assigned in this course must be accurate, neat, and legible. This book will show you how to make manual sketches with or without the aid of triangles and circle templates. It also will introduce you to the Auto-CAD program by using commands that are common to any version of the program within recent years.

SKETCHING TOOLS

The tools for sketching include pencils, erasers, scales, grid paper, and, if desired, triangles, circle template, and drafting powder. This chapter describes these tools, how to use them, and how to keep them in good working condition.

Pencils

Both thin-lead mechanical pencils and wooden pencils work well for technical sketching.

Mechanical Pencils

Lead for mechanical pencils, Figure 2–1, is made in several degrees of hardness and several widths. The following are those most commonly used for technical sketching:

    0.5 mm diameter—2H or H hardness: used for thinner lines
    0.7 mm diameter—2H or H hardness: used for thicker lines

FIGURE 2–1
Lead Hardness

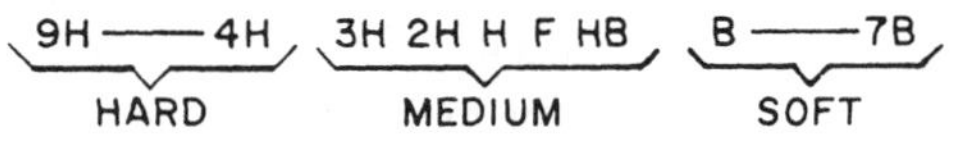

9H —— 4H | 3H 2H H F HB | B —— 7B
HARD | MEDIUM | SOFT

17

FIGURE 2–2
Drawing Lines with a Mechanical Pencil

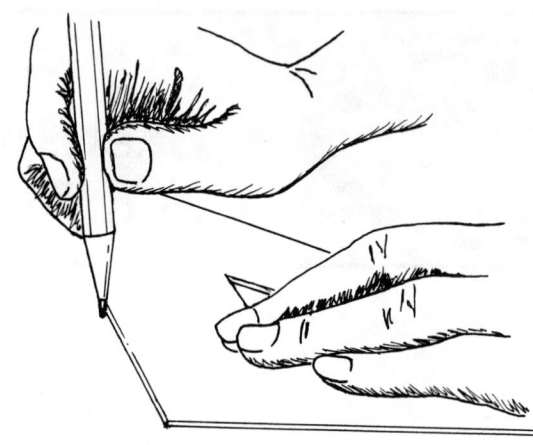

FIGURE 2–3
Crushing the Fine Point

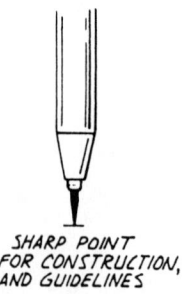

Hold mechanical pencils perpendicular to the paper when using them with triangles or templates so that you draw with the full diameter of the lead and avoid breaking the lead (Figure 2–2).

Wooden Pencils and Lead Holders

Wooden drawing pencils are made in several degrees of hardness, from 9H to 7B. 9H is the hardest, 7B is the softest. H, HB, or 2H is a good hardness for sketching. A standard no. 2 pencil with a soft pink eraser on the end of it is fine for most sketches. Keep wooden pencils fairly sharp. Crush the end of a finely sharpened pencil or lead holder a little so that sketch lines will not be too fine (Figure 2–3).

When using wooden pencils or lead holders with triangles or templates, hold the pencil or lead holder at about a 60° angle to the paper and roll it slightly between the fingers to make lines of uniform width (Figure 2–4).

FIGURE 2–4
Correct Angle for Wooden Pencil

Erasers

A Pink Pearl eraser is best for erasing on bond or vellum papers. You will find this type of eraser on the end of a no. 2 pencil as well as in a round or rectangular form (Figure 2–5).

Grid Paper

Your book contains pages with grids that will aid you in sketching. This paper is also available at most stores that sell architectural, engineering, or drafting supplies. Grids at $\frac{1}{10}''$, $\frac{1}{4}''$, or isometric angles are available at these locations.

Triangles

Both 30-60° and 45° triangles (Figure 2–6) are used for technical drawing. Inexpensive triangles are fine. To draw vertical lines, place the triangle on a grid line or a straight edge such as a T-square, and hold the triangle firmly with one hand as you draw upward with the other hand. You will soon learn to return downward over a line to improve its density. Notice that the pencil is slanted in the direction of the line (Figure 2–7) but is not tilted in relation to the edge of the triangle. Mechanical pencils are held perpendicular to the paper.

FIGURE 2–5
Erasers

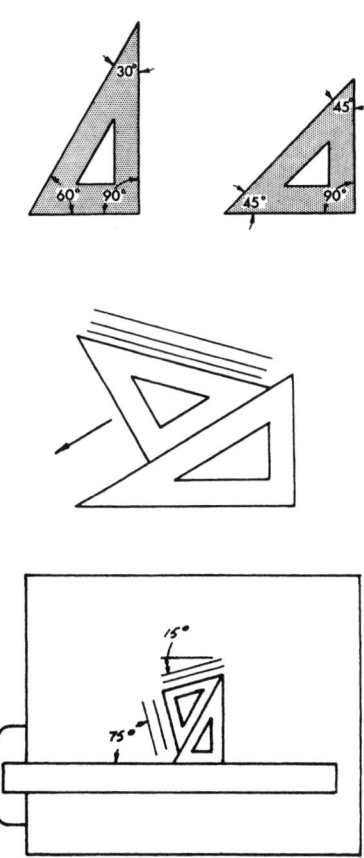

FIGURE 2–6
30-60° and 45° Triangles

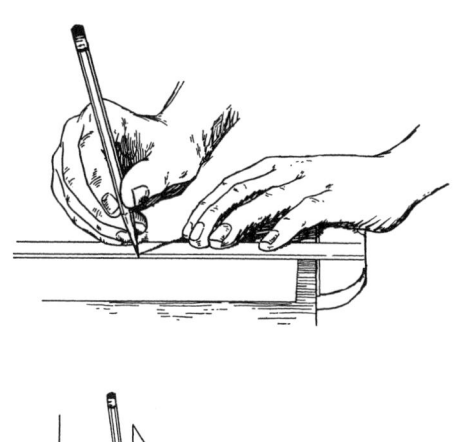
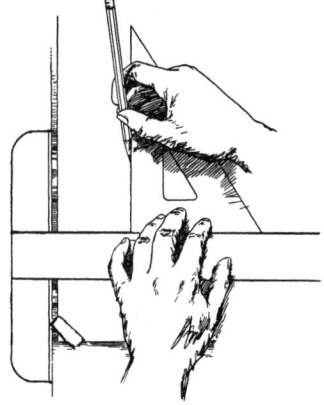

FIGURE 2–7
Drawing Horizontal and Vertical Lines with a Triangle and a Sharpened Lead

Circle Template

Circle templates can be used to draw most of the circles in the assignments in this book. The template should contain circles from $\frac{1}{8}''$ to $1\frac{1}{2}''$ in diameter. To use the circle template, align the crosshair of the template on the centerline of where the circle will be drawn. Draw the circle, holding the pencil perpendicular to the paper (Figure 2–8).

Drafting Powder

An excellent aid for keeping drawings clean is the drafting powder bag or can. The bag is made of a coarsely woven material that allows the coarse powder to fall out of it when the bag is kneaded (Figure 2–9). The powder is sprinkled over the drawing before a drawing is started. When the drawing is completed, the powder is rubbed very lightly over the drawing to pick up the excess graphite. The powder is then brushed off the drawing.

Scales

Scales are made in a variety of styles. All scales are designed to be used by selecting the correct scale for the situation and measuring with it. There is no arithmetic involved. You do not have to multiply, divide, add, or subtract to use a scale—you just read it. For example, the 5 mark on a scale can be only .5, 5, 50, 500 or some other number with a 5 and more zeros, or a decimal point with zeros and then a 5 (.005). It cannot be read as 10, 200, or any other number.

Although there are many different types of scales, let's keep it simple and use only the architect's scale, the civil engineer's scale, a metric ruler, and the steel rule.

Sketching Tools, Supplies, and Their Uses

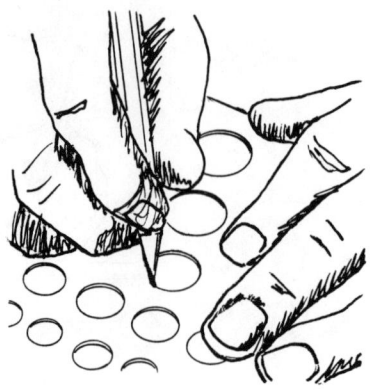

FIGURE 2–8
Drawing a Circle with a Circle Template and a Mechanical Pencil

FIGURE 2–9
Using Drafting Powder

Architect's Scale

The architect's scale is designed to be used to draw architectural structures and is therefore divided into sections for feet and inches. The six-sided triangular architect's scale has a total of 11 different scales marked on its edges.

Using the architect's scale to represent feet and inches:

Figure 2–10 shows the edge of the architect's scale that contains the $\frac{1}{4}$ scale on the right end and the $\frac{1}{8}$ scale on the left end. Notice that the $\frac{1}{8}$ scale uses the smaller divisions marked on the shorter lines and is read from the left 0 on the scale. The $\frac{1}{4}$ scale uses the larger divisions marked on the longer lines and is read from the right 0 on the scale.

The section to the left of the left 0 shows a foot on the $\frac{1}{8}$ scale divided into inches. There are six spaces in this section, and since there are 12 inches in a foot, each mark represents 2 inches. The section to the right of the right 0 shows a foot on the $\frac{1}{4}$ scale divided into inches. There are 12 spaces in this section, so each space represents 1 inch. Notice the labeled measurement on the $\frac{1}{8}$ scale. The measurement showing 6′-8″ starts at the right of the 0 on the 6′ mark (two marks to the right of 4) and stops at the fourth space to the left of the 0. These 4 spaces represent 8 inches, since each mark to the left of 0 is 2 inches.

Study the labeled measurements on the $\frac{1}{4}$ and $\frac{1}{8}$ scales until you are satisfied you know how to read these two scales. After you know how to read these scales, you know how to read all scales on the architect's scale.

Using the architect's scale to represent inches and fractions of an inch:

The architect's scale can also be used to represent inches with the section to the right or left of 0 representing fractions of an inch. Figure 2–11 shows the $\frac{3}{4}$ scale as it represents inches and fractions of an inch. The labeled measurement closest to the $\frac{3}{4}$ scale shows the

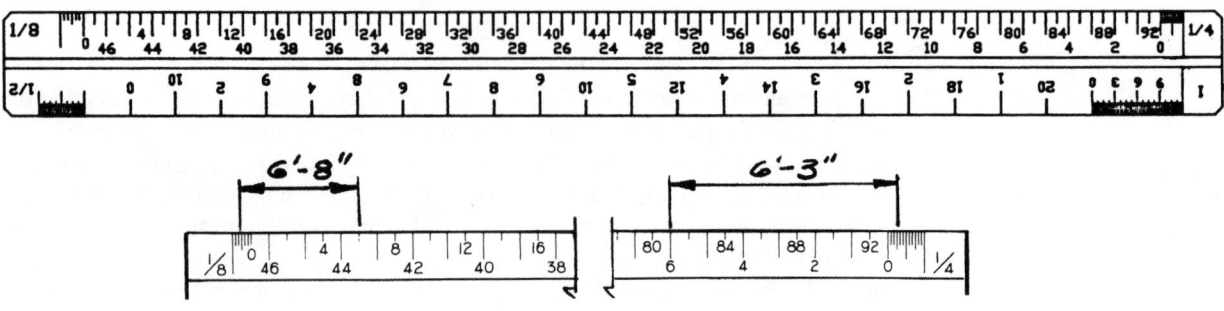

FIGURE 2–10
Architect's Scale

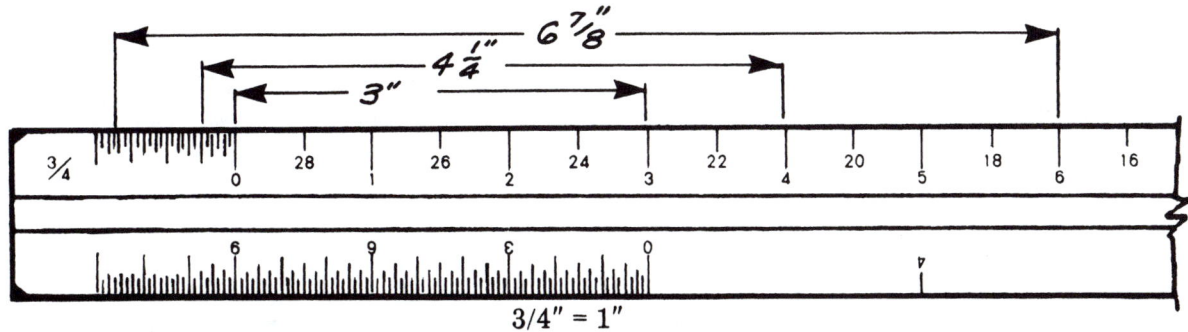

FIGURE 2–11
Reading Inches and Fractions of an Inch

right end of the measurement stopping on the 3″ mark and the left end stopping on 0. The next line up shows the right end on the 4″ mark and the left end $\frac{1}{4}$ of the distance across the section that in this case shows fractions of an inch. The top line shows the left end three marks short of the full inch. Since there are 24 spaces in this inch scale; three marks is $\frac{1}{8}$ $\left(\frac{8}{8} - \frac{1}{8} = \frac{7}{8}\right)$.

Civil Engineer's Scale

The civil engineer's scale divides the inch into a certain number of units. On the scale shown in Figure 2–12 the inch is divided into 20 units on one edge and 40 units on the other edge. The lower part of Figure 2–12 is an enlarged view of the 40 scale. If you needed to draw something to a scale of 1″=400′, you would use the 40 scale, and each of the smallest spaces on the scale would represent 10 feet. On the 40 scale in Figure 2–12, there is a 0, then four small marks, then a longer mark. The longer mark makes it easier to locate measurements on the scale (every 5th mark being long). Notice that the 200th mark is identified by the number 2, the 400th by the number 4, and so on. If you use those numbers, adding zeros after them to represent the correct scale, you will find this scale simple to use.

On a drawing with a scale of 1″=400 miles, for example, you would select the 40 scale and add two zeros after each number on the scale; each of the smallest spaces on the scale would then represent 10 miles.

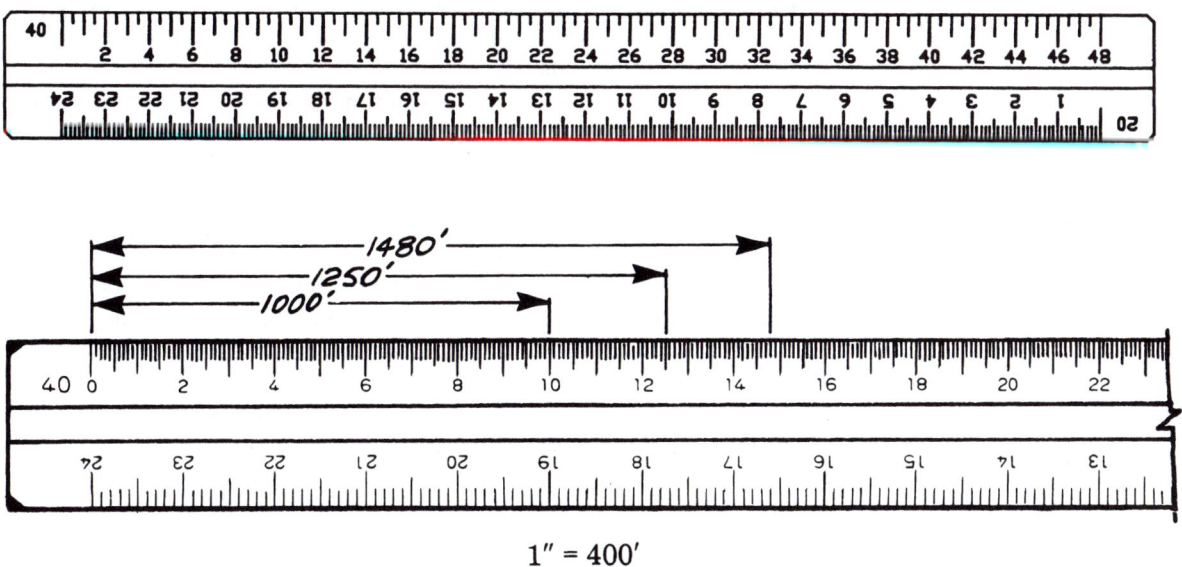

FIGURE 2–12
Civil Engineer's Scale

Sketching Tools, Supplies, and Their Uses

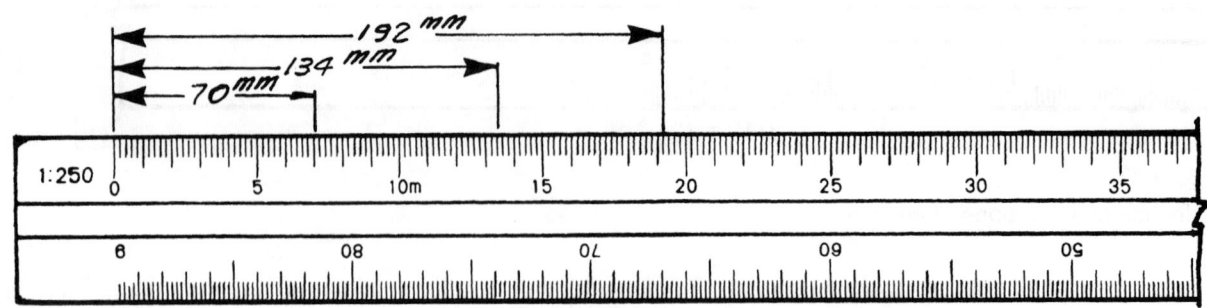

1mm = 2.5mm

FIGURE 2–13
Metric Scale

To draw at a scale of 1″=20′, you should select the 20 scale and add one zero to each of the numbers on the scale; on a scale of 1″=60′, select the 60 scale and add one zero; on a scale of 1″=5000 miles, select the 50 scale and add three zeros to each of the numbers on the scale.

As with other scales, to use the civil engineer's scale you simply select the correct scale and measure with it. **No arithmetic is necessary.**

Metric Scale

Metric scales come in several shapes and have several different scales (Figure 2–13). Some of the most common ones are 1:20, 1:10, 1:100, and 1:2.5. These are all size scales, which means that the drawing is a fractional size of the actual object. For example, 1:20 means that the drawing is $\frac{1}{20}$ the size of the object. This scale would be shown on the drawing as 1 mm = 20 mm (1 millimeter on the drawing represents 20 millimeters on the object). It could also be shown as 5 cm = 1 m (5 centimeters on the drawing represents 1 meter on the object). This may not be readily apparent until you realize that 1 meter divided by 20 is .05 meter.

The scale 1:10 means that the drawing is $\frac{1}{10}$ the size of the object; 1:100 means that the drawing is $\frac{1}{100}$ the size of the object; and the 1:2.5 scale means that the drawing is $\frac{1}{2.5}$ the size of the object or a little less than half as big.

None of the exercises in this book use any other than the full metric scale (a ruler marked in millimeters). Remember that 25.4 mm = 1 in.

Steel Rule

The most common form of the steel rule is 6″ in length and is a very accurate measuring device in many cases. Figure 2–14 shows an enlarged picture of the left end of a steel rule. Notice that the top edge of the rule is the decimal rule and is marked in 100ths of an inch (.01). The bottom edge is the fractional rule and is marked in 64ths of an inch.

Reading the Decimal Edge (Figure 2–15)

1. Look at the largest divisions, those marked with the numbers 1 and 2. These are inches and would be written as 1, 2, 1.0, 2.0, 1.00, 2.00 or 1.000, 2.000.

2. Look at the next smaller divisions, marked with numbers 1 through 9. These are tenths of an inch ($\frac{1}{10}$ or 0.1).

3. Look at the smallest divisions. These are hundredths of an inch ($\frac{1}{100}$ or 0.01).

4. Study the longer marks that have no numbers. These make it easier to locate and count the smallest divisions. For example, the longest mark between the tenth-inch divisions

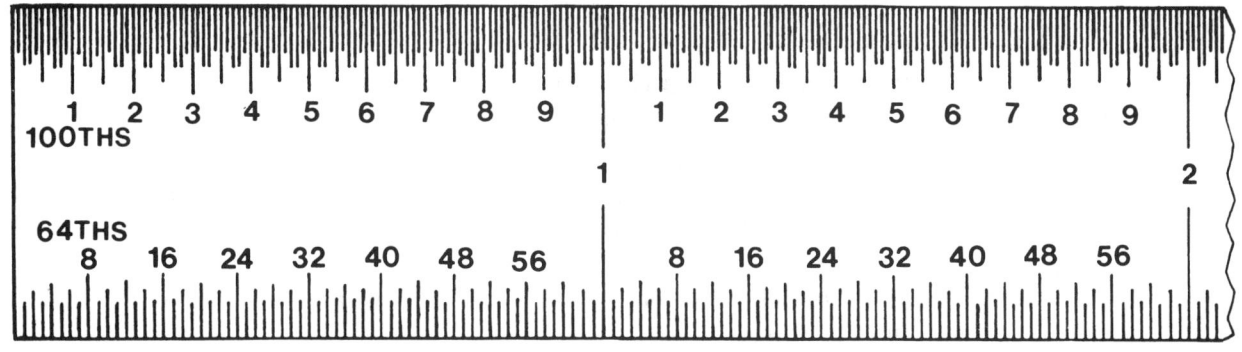

FIGURE 2-14
The Steel Rule

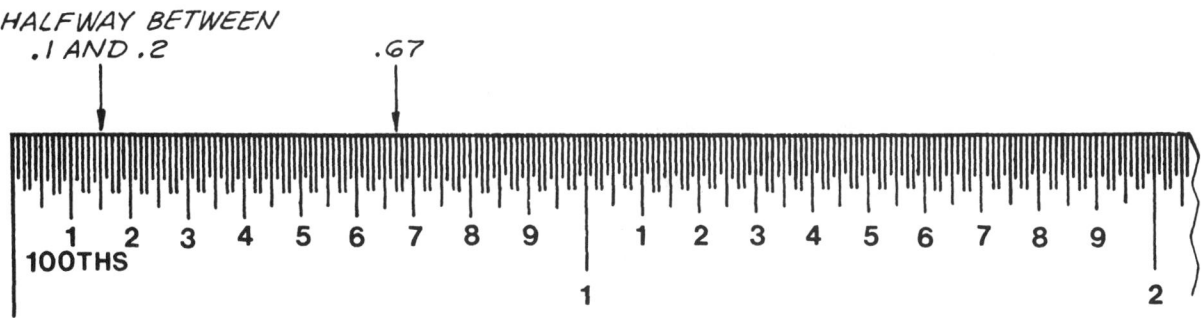

FIGURE 2-15
Reading the Decimal Edge of the Steel Rule

marked 1 and 2 is halfway between the two divisions and is $\frac{15}{100}$ or .15 and is half the distance between 0.1 and 0.2.

5. To locate a dimension of 0.67, for example, find the center point between 6 and 7 on the scale and count two more small marks to the right.

Reading the Fractional Edge (Figure 2-16)

1. Look at the largest divisions, those marked with the numbers 1 and 2. These are inches and would be written as whole numbers 1 and 2.

2. Look at the next smaller divisions, marked with numbers 8, 16, 24, 32, 40, 48, and 56. These are the number of 64ths of an inch and are helpful in locating and counting

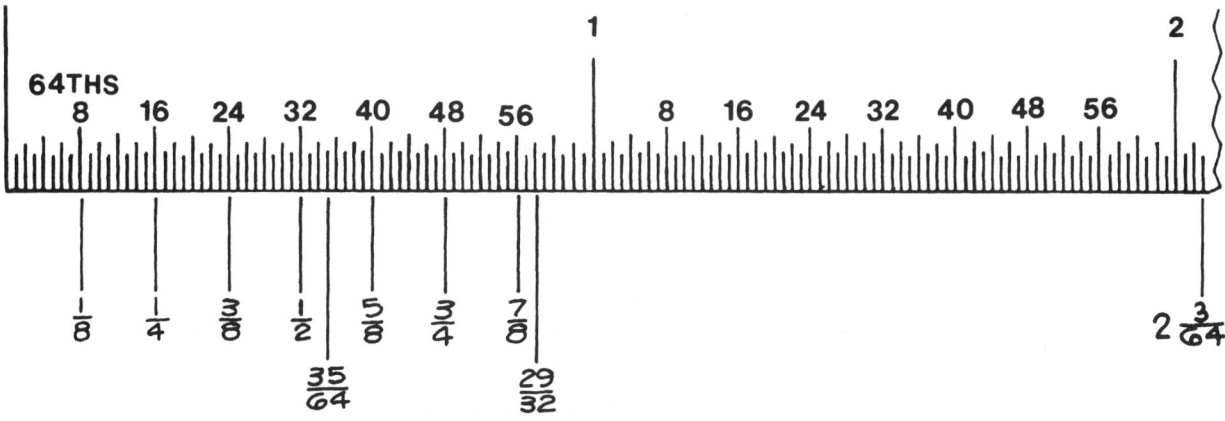

FIGURE 2-16
Reading the Fractional Edge of the Steel Rule

Sketching Tools, Supplies, and Their Uses

fractions such as 64ths and 32nds. These units, however, are written in their reduced forms: $\frac{8}{64} = \frac{1}{8}$; $\frac{16}{64} = \frac{1}{4}$.

3. Study the other marks, which identify 32nds and 16ths.
4. To locate a dimension of $\frac{35}{64}$, find the number of 64ths closest to it (32), and count three more of the smallest spaces to the right.
5. To locate a dimension of $\frac{29}{32}$, change $\frac{29}{32}$ to $\frac{58}{64}$ (multiply the numerator and denominator times 2). Find 56 on the fractional edge, and count two of the smallest spaces to the right ($\frac{2}{64}$ or $\frac{1}{32}$).
6. To locate a dimension of $2\frac{3}{64}$, find the 2-inch mark, and count three of the smallest spaces to the right.

EXERCISE 2–1
Drawing Lines to Scale

Remove the two sheets labeled Exercise 2–1 from your book and use the following instructions to draw lines to scale.

Using the Architect's Scale to Draw Lines to Scale

Step 1. Use the $\frac{3}{4}$ scale to represent $\frac{3''}{4}=1'$ and draw a line 4′-3″ long (Figure 2–17).

1. Determine which scale is required to fit the drawing on the sheet size you are sketching. In this case use the $\frac{3}{4}$ scale.
2. Align the 3″ mark (six of the small spaces to the left of 0) with the start point of the measurement, and make a small mark or dot at that point with your pencil on one of the grid lines of your gridded paper.
3. Locate the 4′ mark and make a small mark or dot at that point with your pencil.
4. Use a triangle to draw a line between the marks.

Step 2. **On your own:**

1. Use the $\frac{3}{4}$ scale to represent $\frac{3''}{4}=1'$ to draw a line 6′-2″ long.
2. Use the $\frac{1}{8}$ scale to represent $\frac{1''}{8}=1'$ to draw a line 15′-6″ long.
3. Use the $\frac{1}{2}$ scale to represent $\frac{1''}{2}=1'$ to draw a line 7′-10″ long.

Step 3. Use the $\frac{3}{4}$ scale to represent $\frac{3''}{4}=1''$ to draw a line $5\frac{3''}{4}$ long (Figure 2–18).

1. Determine which scale is required to fit the drawing on the sheet size you are sketching. In this case use the $\frac{3}{4}$ scale.

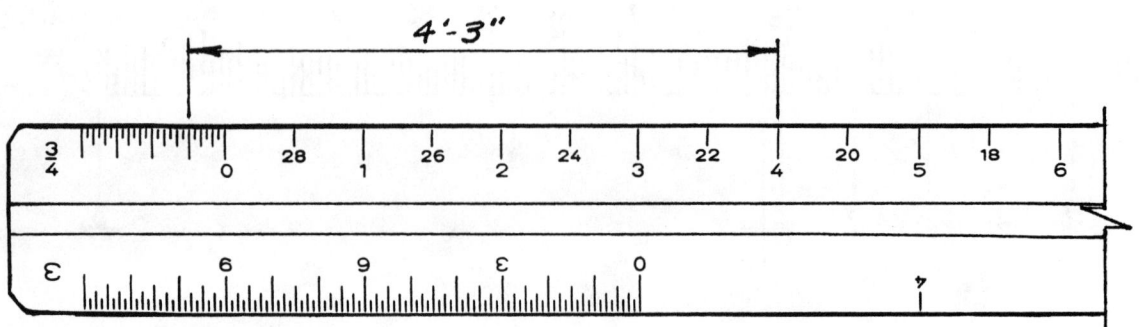

FIGURE 2–17
Drawing a Line 4′-3″ Long on a Scale of $\frac{3''}{4} = 1'$

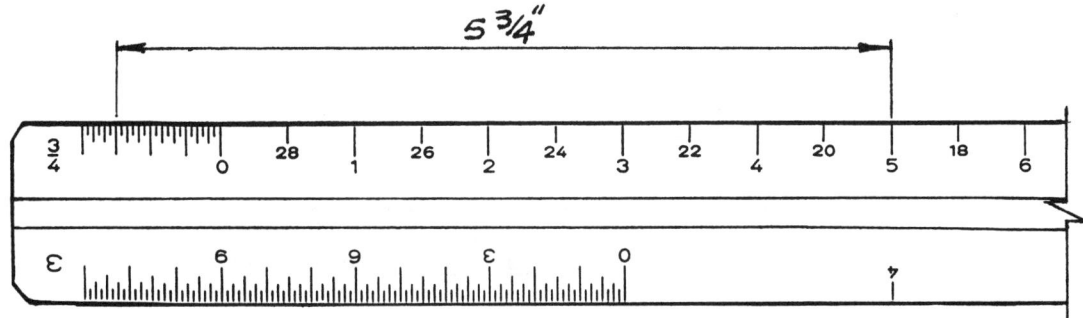

FIGURE 2-18
Drawing a Line $5\frac{3}{4}''$ Long on a Scale of $\frac{3}{4}'' = 1''$

2. Align the $\frac{3}{4}''$ mark to the left of the 0 with the start point of the measurement, and make a small mark or dot at that point with your pencil.
3. Locate the 5" mark to the right of 0 and make a small mark or dot at that point with your pencil.
4. Use a triangle to draw a line between the marks.

Step 4. **On your own:**
1. Use the $\frac{3}{4}$ scale to represent $\frac{3}{4}''=1''$ to draw a line $6\frac{3}{4}''$ long.
2. Use the $\frac{1}{8}$ scale to represent $\frac{1}{8}''=1'$ to draw a line $20\frac{1}{2}''$ long (a very short line).
3. Use the $\frac{1}{2}$ scale to represent $\frac{1}{2}''=1''$ to draw a line $8\frac{1}{4}''$ long.

Using the Civil Engineer's Scale to Draw Lines to Scale

Step 5. **Use the 20 scale to represent 1"=20' to draw a line 84' long (Figure 2–19).**

1. Determine which scale is required to fit the drawing on the sheet size you are sketching. In this case use the 20 scale. To read this scale, place a 0 after each of the numbers on the scale, so that 1 becomes 10, 2 is 20, 10 is 100, etc.
2. Align the 0 on the scale with the start point of the measurement, and make a small mark or dot at that point with your pencil on one of the grid lines of your gridded paper.
3. Locate the number 8 mark and move four of the small marks to the right (8 represents 80, and each of the small spaces is 1 because there are 10 spaces between each one of the larger numbered marks), and make a small mark or dot at that point with your pencil.
4. Use a triangle to draw a line between the marks.

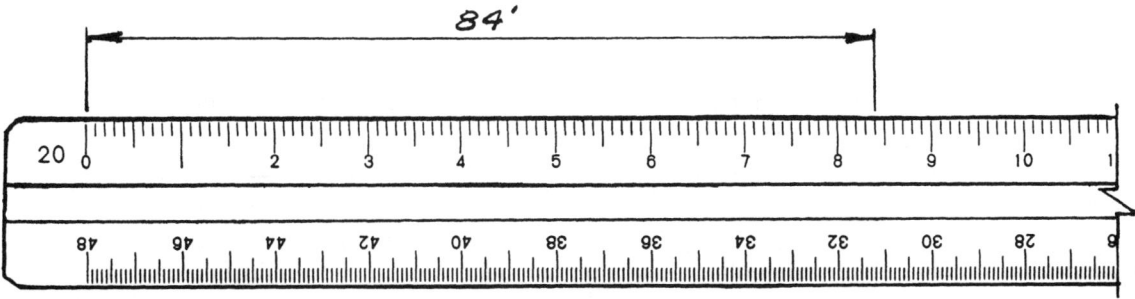

FIGURE 2-19
Drawing a Line 84' Long on a Scale of 1" = 20'

Sketching Tools, Supplies, and Their Uses

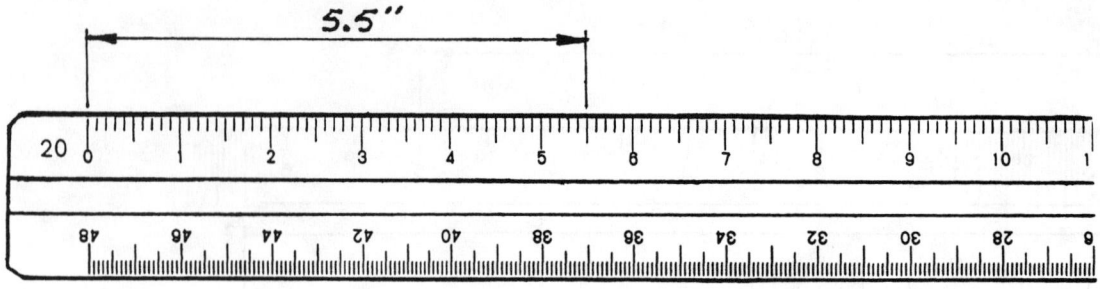

FIGURE 2–20
Drawing a Line 5.5″ Long on a Scale of $\frac{1}{2}″ = 1″$

Step 6. **On your own:**
1. Use the 20 scale to represent 1″=20′ to draw a line 68′ long.
2. Use the 40 scale to represent 1″=400′ to draw a line 1650′ long.
3. Use the 50 scale to represent 1″=50 miles to draw a line 214 miles long.

Step 7. **Use the 20 scale to represent $\frac{1}{2}″=1″$ to draw a line 5.5″ long (Figure 2–20).**
1. Determine which scale is required to fit the drawing on the sheet size you are sketching. In this case use the 20 scale.
2. Align the 0 on the scale with the start point of the measurement, and make a small mark or dot at that point with your pencil on one of the grid lines of your gridded paper.
3. Locate the number 5 mark to the right of 0 and move five small marks to the right (each of the small marks is .1 because there are 10 spaces between each of the larger numbered marks), and make a small mark or dot at that point with your pencil.
4. Use a triangle to draw a line between the marks.

Step 8. **On your own:**
1. Use the 20 scale to represent $\frac{1}{2}″=1″$ to draw a line 6.2″ long.
2. Use the 40 scale to represent $\frac{1}{4}″=1″$ to draw a line 17.5″ long.
3. Use the 40 scale to represent $\frac{1}{4}″=1″$ to draw a line 11.4″ long.

Using the Metric Ruler to Draw Lines Accurately

Step 9. **Use the metric ruler to draw a line 86 mm long (Figure 2–21).**
1. Align the 0 on the ruler with the start point of the measurement, and make a small mark or dot at that point with your pencil on one of the grid lines of your gridded paper.

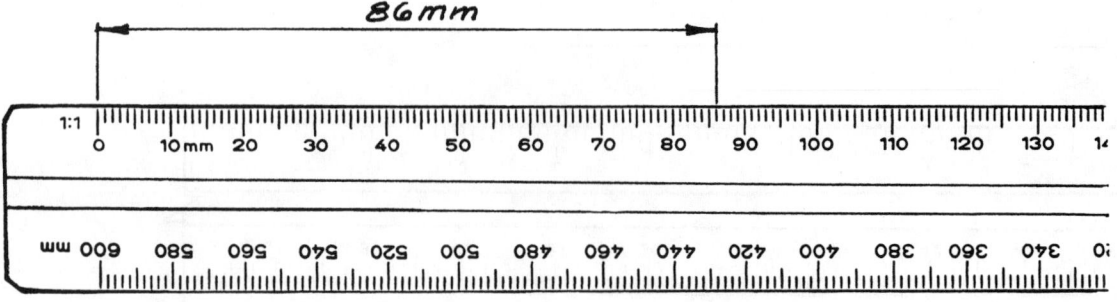

FIGURE 2–21
Drawing a Line 86 mm Long Using a Metric Ruler

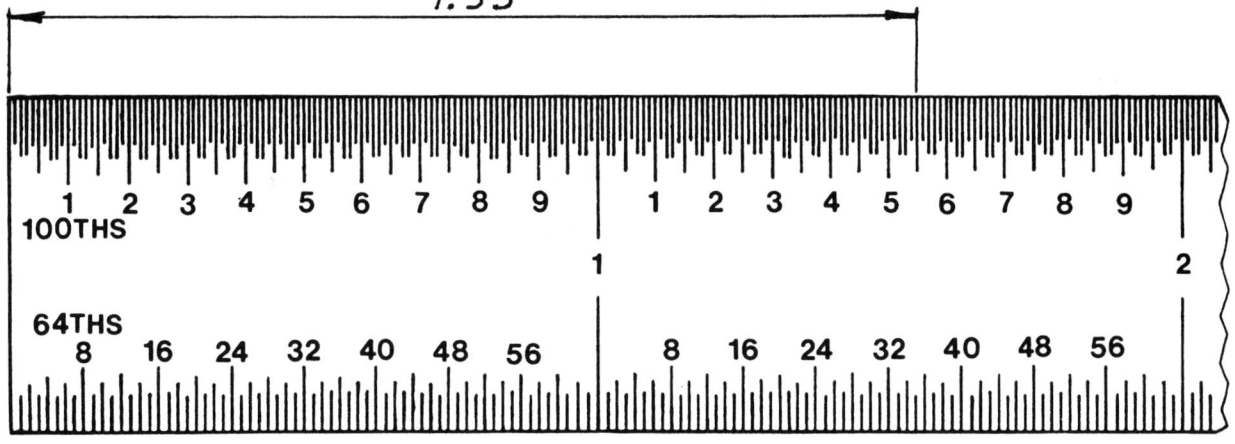

FIGURE 2–22
Drawing a Line 1.55″ Long Using a Steel Rule

    2. Locate the number 80 mark to the right of 0 and move six small marks to the right (each of the small marks is .1 mm because there are 10 spaces between each of the larger numbered marks), and make a small mark or dot at that point with your pencil.

    3. Use a triangle to draw a line between the marks.

Step 10. **On your own:**

    1. Use the metric ruler to draw a line 35 mm long.

    2. Use the metric ruler to draw a line 122 mm long.

    3. Use the metric ruler to draw a line 155 mm long.

Using the Steel Rule to Draw Lines Accurately

Step 11. **Use the decimal edge of the steel rule to draw a line 1.55″ long (Figure 2–22).**

    1. Align the 0 on the ruler with the start point of the measurement, and make a small mark or dot at that point with your pencil on one of the grid lines of your gridded paper.

    2. Locate the number 5 mark to the right of the 1″ mark, and move five of the small marks to the right (each of the small marks is .01″), and make a small mark or dot at that point with your pencil.

    3. Use a triangle to draw a line between the marks.

Step 12. **On your own:**

    1. Use the steel rule to draw a line 2.875″ long.

    2. Use the steel rule to draw a line 4.65″ long.

    3. Use the steel rule to draw a line 5.735″ long.

Step 13. **Use the fractional edge of the steel rule to draw a line $1\frac{3}{8}″$ long (Figure 2–23).**

    1. Align the 0 on the ruler with the start point of the measurement, and make a small mark or dot at that point with your pencil on one of the grid lines of your gridded paper.

    2. Locate the number 1 inch mark and move to the right to the line numbered 24 $\left(\frac{24}{64} = \frac{3}{8}\right)$, and make a small mark or dot at that point with your pencil.

    3. Use a triangle to draw a line between the marks.

Sketching Tools, Supplies, and Their Uses

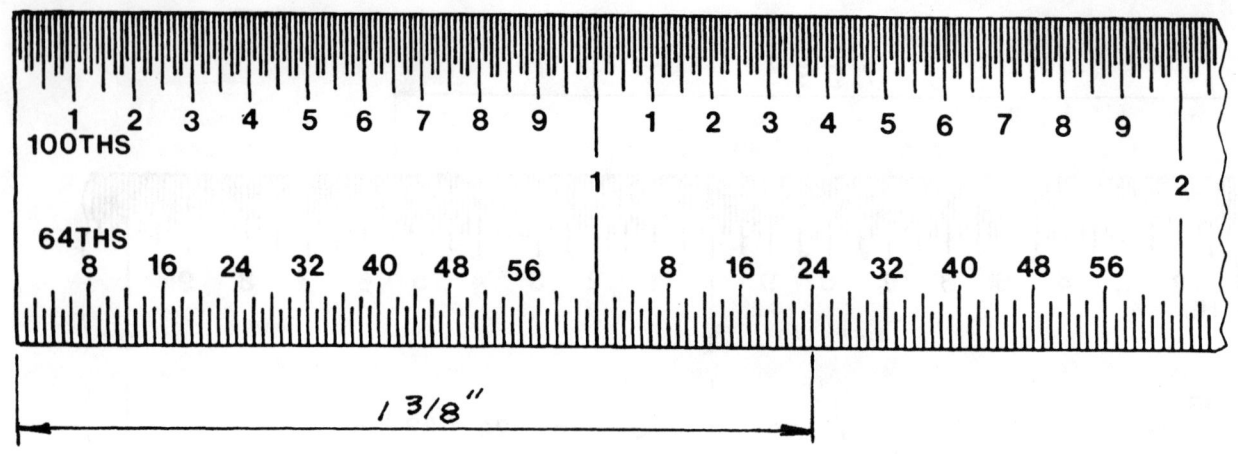

FIGURE 2–23
Drawing a Line 1$\frac{3}{8}$″ Long Using a Steel Rule

Step 14. On your own:
1. Use the steel rule to draw a line 4$\frac{1}{8}$″ long.
2. Use the steel rule to draw a 3$\frac{3}{64}$″ line.
3. Use the steel rule to draw a 2$\frac{5}{32}$″ line.

EXERCISES

EXERCISE 2–1   Complete Exercise 2–1, sheets 1 and 2, using steps 1 through 14 described in this chapter. Fill in the title block with your best lettering. Title the drawing SCALES. If you do not have a civil engineer's scale or a steel rule, use the architect's scale for those measurements.

EXERCISE 2–2   Complete the sheet labeled Exercise 2–2 by drawing lines parallel to the printed lines. Your final drawing should look like Figure 2–24. Use these steps:

Step 1. Make sure lines are parallel by using your 30-60° and 45° triangles in the manner shown in Figure 2–7. Refer to Chapter 5 if necessary for a description of this procedure.

FIGURE 2–24
Lines for Exercise 2–2

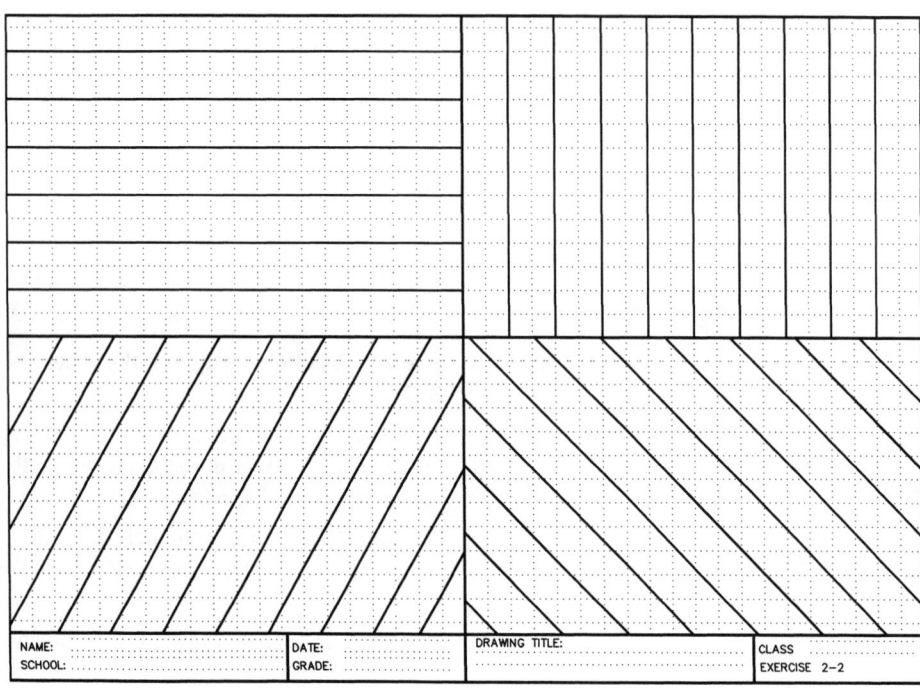

FIGURE 2–25
Dimensions for Exercise 2–3

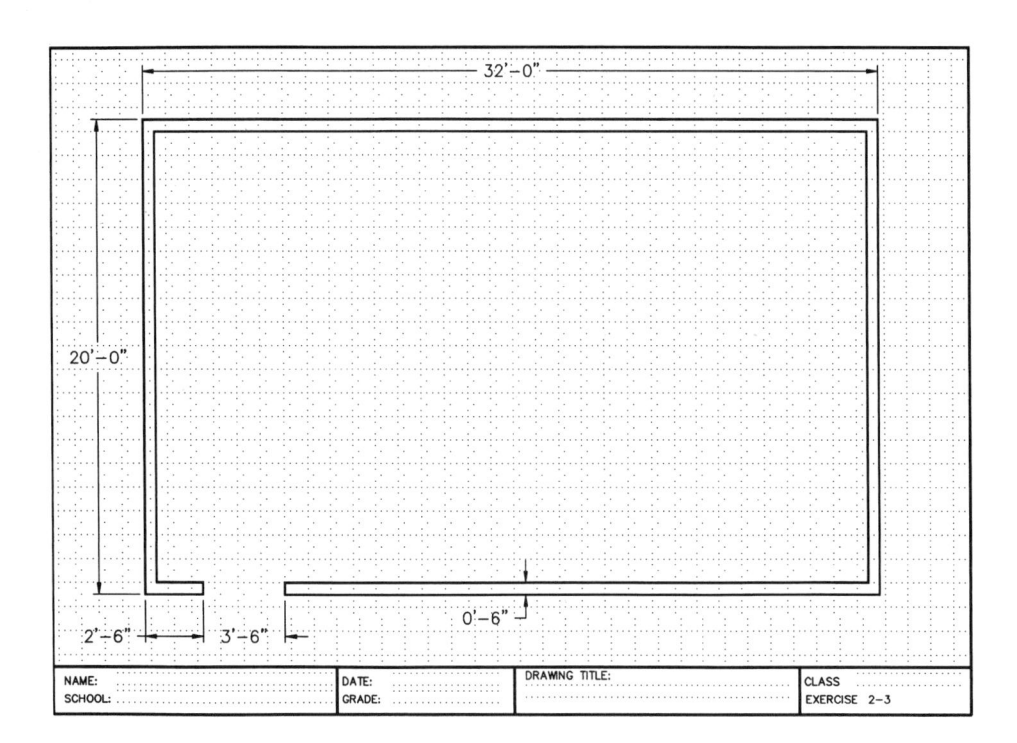

Step 2. Concentrate on making your pencil lines the same thickness and density as the printed lines.

Step 3. The lower left quarter of the drawing will require you to use a 30-60° triangle to construct a line perpendicular to the two existing ones. Draw construction lines perpendicular to the angular lines and make tick marks or dots $\frac{1}{2}''$ apart so all lines will be the same distance apart. Refer to Chapter 5 if necessary for a description of this procedure.

Step 4. Use a procedure similar to step 3 to draw the lower right quarter using a 45° triangle.

Step 5. Use your best lettering to fill in the title block with the required information using all capital letters. Title the drawing LINES. You will learn the preferred methods of lettering for technical drawing in Chapter 4.

EXERCISE 2–3   Complete the sheet labeled Exercise 2–3 by drawing the floor plan described in Figure 2–25 at a scale of $\frac{1}{4}''=1'$. Your final drawing should look like Figure 2–25 without dimensions. Use these steps:

Step 1. Draw construction lines parallel to the grid lines for the vertical lines so you will know where to stop the horizontal lines.

Step 2. Draw all horizontal lines dark, making sure they are parallel by using your 30-60° or 45° triangles to draw lines on the grid lines or parallel to them if the measurement falls between grid lines.

Step 3. Concentrate on making your pencil lines the same thickness and density as the printed lines.

Step 4. Draw the vertical lines solid to complete the drawing.

Step 5. Use your best lettering to fill in the title block with the required information, using all capital letters. Title the drawing FLOOR PLAN. You will learn the preferred methods of lettering for technical drawing in Chapter 4.

EXERCISE 2–4   Complete the sheet labeled Exercise 2–4 by drawing the site plan described in Figure 2–26 at a scale of $\frac{1}{4}''=1'$. Your final drawing should look like Figure 2–26 without dimensions. Use these steps:

Step 1. Draw construction lines by extending two of the angular lines to the right edge of the border.

Step 2. Draw a construction line parallel to the other angular line 4' from it using a scale of $\frac{1}{4}''=1'$.

Sketching Tools, Supplies, and Their Uses

FIGURE 2–26
Dimensions for Exercise 2–4

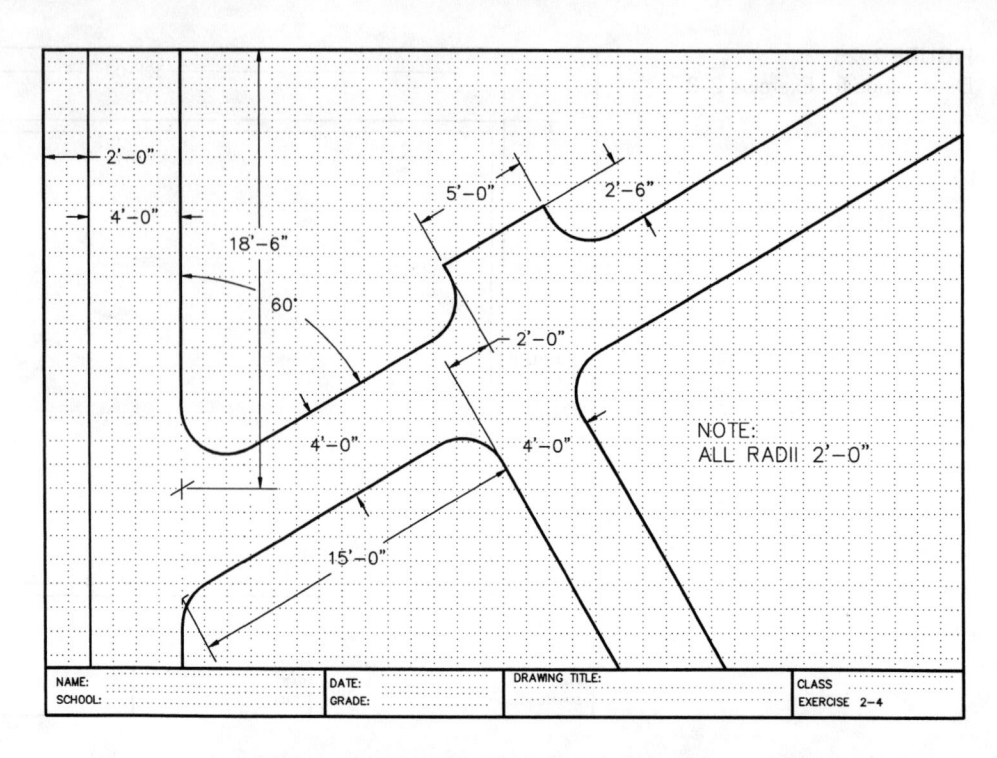

Step 3. Draw all radii in their final form. Concentrate on making your pencil lines the same thickness and density as the printed lines.

Step 4. Draw the remaining vertical line solid and darken all lines necessary to complete the drawing.

Step 5. Use your best lettering to fill in the title block with the required information, using all capital letters. Title the drawing SITE PLAN. You will learn the preferred methods of lettering for technical drawing in Chapter 4.

REVIEW QUESTIONS

Circle the best answer.

1. Which of the following is *not* a technical sketching tool?
   a. Pencil
   b. Eraser
   c. Compass
   d. Scale
   e. Triangle

2. Which of the following diameter leads should be used for sketching thin lines?
   a. 0.3 mm
   b. 0.5 mm
   c. 0.7 mm
   d. 0.9 mm
   e. Any of the above is OK for thin lines.

3. Which hardness of lead should be used for technical sketching?
   a. 7B
   b. 2B
   c. H
   d. 4H
   e. 9H

4. At what angle should a mechanical pencil be held to make a line that reflects the full diameter of the lead?
   a. 30°
   b. 45°
   c. 60°
   d. 90°
   e. Any angle will do.

5. Which of the following erasers is best for technical sketching?
   a. Pink Pearl
   b. Art gum
   c. Gray ink
   d. White
   e. Red ink

6. Which of the following diameter circles is found on common circle templates?
   a. $\frac{1}{32}''$
   b. $\frac{1}{8}''$
   c. $\frac{1}{64}''$
   d. $2\frac{1}{16}''$
   e. $3''$

7. Which of the following is found on the same edge of the architect's scale as the $\frac{1}{4}$ scale?
   a. $\frac{1}{16}$
   b. $\frac{1}{8}$
   c. $\frac{3}{8}$
   d. $\frac{1}{2}$
   e. $\frac{3}{4}$

8. When the architect's scale is used to represent $\frac{1}{8}'' = 1'$, each of the small spaces in the area to the left of 0 on the $\frac{1}{8}$ scale represents:
   a. 1 foot
   b. 1 inch
   c. 2 feet
   d. 2 inches
   e. 3 inches

9. When the civil engineer's scale is used to draw at a scale of $1'' = 2000'$, which of the following scales should be selected?
   a. 10
   b. 20
   c. 30
   d. 40
   e. Use the 60 scale and divide by 3.

10. How many millimeters are contained in 1 inch?
    a. 10
    b. 100
    c. 2.5
    d. 25.4
    e. 0.254

Fill in the blank.

11. List all the tools that you will be using for sketching in this course.

12. The smallest space on the decimal edge of the steel rule is what decimal part of an inch?

13. The smallest space on the fractional edge of the steel rule is what fractional part of an inch?

14. List six of the scales on an architect's scale.

15. List four of the scales on a civil engineer's scale.

16. List four of the common metric scales.

17. Describe how to hold a wooden pencil to draw a vertical line with a triangle.

18. Describe how to hold a mechanical pencil to draw a circle with a circle template.

19. Describe how to hold a triangle to draw a vertical line.

20. Describe what drafting powder is used for.

Sketching Tools, Supplies, and Their Uses

3 AutoCAD Fundamentals

OBJECTIVES

After completing this chapter, you will be able to

☐ Make settings for an AutoCAD or AutoCAD LT drawing to include Units, Limits, Grid, and Snap.
☐ Create layers and assign color and linetype to each layer.
☐ Use function keys F1 and F2 (flip screen), F7 (grid), and F9 (snap) to control the display screen, grid, and snap as required.
☐ Use the commands Save, SaveAs, and Quit to save work and exit AutoCAD or AutoCAD LT.

INTRODUCTION

When a project that is to be manually drafted is started, decisions are made about the number of drawings required, the appropriate sheet size, scale, and so on. Similar decisions are made when preparing to draw with AutoCAD or AutoCAD LT. This chapter describes the settings that must be made before drawing can begin and some common means of making sure you save your work. The drawings contained on the disk that came with your book already have these settings made, but knowing how to make these settings is important to your understanding of the AutoCAD program. Knowing how to save your work is not just important, it is absolutely necessary for you to continue from one step to the next.

Note: Some of the prompts you will see may vary slightly from the ones shown in the book. All your responses will be as described in the book, however.

Although you can make all the settings described in this chapter using toolbars or menus, you can also make them by typing from the keyboard. This book is designed to be used with any version of AutoCAD or AutoCAD LT, and the only common means of activating these commands is by typing the command; therefore, typing is the method used throughout this book.

The following is a hands-on, step-by-step procedure for making the setup for your first drawing. Each step is followed by an explanation of the command used. To begin, turn on the computer and start AutoCAD or AutoCAD LT.

MAKING THE SETUP FOR THE FIRST DRAWING EXERCISE

EXERCISE 3–1
Start a New Drawing

Step 1. Name the new drawing.

If you are using AutoCAD Release 14 or 2000, Click: **Start from Scratch** and **OK.** If you are using AutoCAD 2002, Click: the Create Drawings tab from the AutoCAD 2002 Today window, then Click: Start from Scratch.

Make Settings

Step 2. Select Units.

Prompt	Response
Command:	Type: **UNITS**↵

Units

Note: The number of units to the right of the decimal has no bearing on how accurately AutoCAD LT draws. It controls the display of dimensions and other values displayed on the screen such as coordinates and defaults. No matter what the decimal places setting, AutoCAD draws with extreme accuracy.

Any of the units may be used. The decimal selection can be used for any units (commonly inches or millimeters). The architectural selection allows feet and inches to be used. You will use only decimal and architectual in this book. The Precision: button allows you to set the number of digits to the right of the decimal point. The remaining settings are for measuring angles. There is no reason to change these settings. When you click OK, AutoCAD's default values, which measure angles using decimal degrees in the counterclockwise direction, are accepted.

If you get a dialog box:

Click: **OK** to accept the default values.

If you get a series of options allowing you to select values:

Press: ↵ after each option to accept the default value until you return to the Command: prompt. Then Press: F2 to return to the graphics screen.

Step 3. Set the Drawing Limits.

Prompt	Response
Command:	Type: **LIMITS**↵
Specify lower left corner or [ON/OFF] <0.000, 0.000>:	↵
Specify upper right corner <12.000, 9.000>:	Type: **8.5,11**↵

Limits

Think of drawing limits as the sheet size or sheet boundaries. Here 8.5,11 was set as the drawing limits. In AutoCAD and AutoCAD LT that value is entered as 8.5,11 using a comma with no spaces to separate the X and Y axes. AutoCAD defaults to inches (or any other basic unit of measure), so the inch symbol is not required. The X axis is first (8.5) and measures drawing limits from left to right. The Y axis is second and measures drawing limits from bottom to top. You will be drawing in a vertical 8.5",11" area similar to a standard sheet of typing paper.

The lower left corner of the drawing boundaries is 0,0. The upper right corner is 8.5,11 (Figure 3–1). These are the limits for this exercise. To turn the 8.5",11" area horizontally, the limits are entered as 11,8.5.

FIGURE 3–1
Drawing Limits

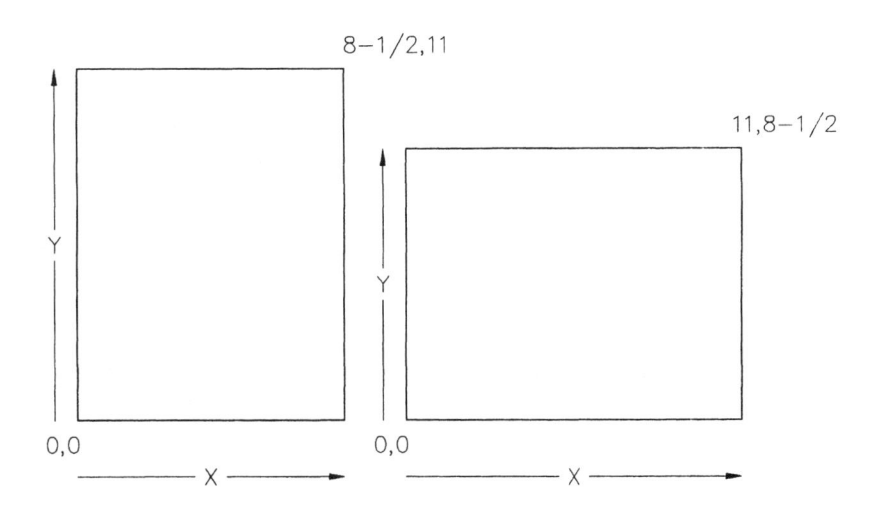

AutoCAD Fundamentals

You can also respond to the Limits: prompt "ON/OFF/<Lower left corner>:" by selecting ON or OFF on the menu. The ON mode, when activated, helps you avoid drawing outside the drawing limits by responding in the prompt line with "** Outside limits" if a point is selected outside the limits. Some drawing elements, such as a circle, may be started inside the limits and completed outside the limits without the appearance of this prompt. The OFF mode, when activated, allows you to draw outside the drawing limits.

If the drawing limits need to be changed, you may do so at any time by entering new limits to the "Upper right corner>:" prompt. Changing the drawing limits will automatically change the grid pattern to the new limits.

Step 4. Set the grid and snap spacing.

Prompt	Response
Command:	Type: **GRID**↵
Specify grid spacing(X) or ON/OFF/ Snap/Aspect <0.000>:	Type: **.5**↵
Command:	Type: **SN**↵
Specify snap spacing or ON/OFF/Aspect/ Rotate/Style <0.000>:	Type: **.125**↵

Grid

You have just set .5 as the grid spacing. The grid is the visible pattern of dots on the display screen. With a setting of .5, each grid dot is spaced .5 vertically and horizontally. The grid is not part of the drawing, but it helps in getting a sense of the size and relationship of the drawing elements. It is never plotted.

Function key F7 turns the grid ON or OFF. The grid can also be turned on or off by selecting either option in response to the prompt "Grid spacing(X) or ON/OFF/Snap/Aspect:", or by clicking GRID at the bottom of the screen.

Snap

You have set .125 as the snap spacing. Snap is an invisible pattern of dots on the display screen. As you move the mouse across the screen the crosshair will snap, or lock, to an invisible snap grid when snap is ON. With a setting of .125, each snap point is spaced .125 horizontally and vertically.

Function key F9 turns the snap ON or OFF. The snap can also be turned on or off by selecting either option in response to the prompt "Snap spacing or ON/OFF/Aspect/Rotate/Style>:" or by clicking SNAP at the bottom of the screen. Rotate and Style options are used in later chapters.

It is helpful to set the snap spacing the same as the grid spacing or as a fraction of the grid spacing so the crosshair snaps to every grid point or to every grid point and in between. The snap can be set to snap several times in between the grid points.

Some drawings or parts of drawings should never be drawn with snap off. Snap is a very important tool for quickly locating or aligning elements of your drawing. You may need to turn snap off and on while drawing, but remember that a drawing entity drawn on snap is easily moved, copied, or otherwise edited.

Drawing Aids Dialog Box and Components of All Dialog Boxes

You can also set snap and grid by using the Drawing Aids or Drafting Settings dialog box.

To locate the Drawing Aids or Drafting Settings dialog box, move the pointer across the top of the display screen and highlight each menu on the menu bar until you see the label Drawing Aids (or Drafting Settings). When you Click: Drafting Settings, the Drafting Settings dialog box appears on your screen.

There are some basic components common to all dialog boxes. The following is a description of the Snap and Grid components that appear in the Drafting Settings dialog box and also other dialog boxes if you choose to use them:

1. *Cursor:* Changes to an arrow.

2. *OK button:* Click this button to complete the command, leave the dialog box, and return to the drawing. If any changes have been made, they will remain as changes. Pressing ↵ has the same effect.

3. *Cancel button:* Click this button to cancel the command, leave the dialog box, and return to the drawing. If any changes have been made, they will be canceled and the original settings will return. Pressing the Esc key has the same effect.

4. *Input buttons:* An input button has two parts, its name and the area where changes can be made by typing new input. Click the second part of the input button, X Spacing, under Snap, and experiment with the text cursor that is attached to the point of the arrow. As you move the mouse and click a new spot, the text cursor moves also. The following editing keys can be used to edit the text in input buttons:

Backspace key: Deletes characters to the left of the text cursor one at a time as it is pressed.
Delete key: Deletes characters to the right of the text cursor one at a time as it is pressed.
Left arrow: Moves the text cursor to the left without changing the existing text.
Right arrow: Moves the text cursor to the right without changing the existing text.
Character keys: After deletion of existing settings, new settings can be typed.
Snap X Spacing input button: Enter the X spacing in this input button, and the Y spacing is automatically set to the same spacing.
Grid X Spacing input button: Enter the X spacing in this input button and the Y spacing is automatically set to the same spacing.
Snap Angle, X Base, and Y Base input buttons: These buttons relate to the Rotate option.

5. *Check buttons:* A check button has two parts: its mode name and the area that can be clicked to toggle the check mark and mode ON and OFF. A check mark in the box indicates the mode is ON.

While in this dialog box, experiment with the different editing keys to become familiar with their functions. The dialog box is a handy tool to use in setting the snap and grid spacing, but if you are a fair typist, typing these commands from the keyboard is faster. After experimenting, be sure to return the grid spacing to .5 and the snap to .125 to have the correct settings for this exercise.

Create Layers

Layers

The layer concept in AutoCAD is similar to the use of transparent overlays with a manually drafted project. Different parts of the project can be placed on separate layers. The building shell may be on one layer, the interior walls on another, the electrical on a third layer, the furniture on a fourth layer, and so on. There is no limit to the number of layers you may use in a drawing. Each is perfectly aligned with all the others. Each layer may be viewed on the display screen separately, one layer may be viewed in combination with one or more of the other layers, or all layers may be viewed together. Each layer may also be plotted separately or in combination with other layers, or all layers may be plotted at the same time. The layer name may be from 1 to 255 characters in length.

Step 5.   Create layers using the keyboard.

The LAYER prompt "[?/Make/Set/New/ON/OFF/Color/Ltype/LWeight/Plot/Freeze/Thaw/LOck/Unlock/stAte]:" allows you to select 10 options.

To use the options Make and Color:

Prompt	Response
Command:	Type: **-LA**↵

AutoCAD Fundamentals

Prompt	Response
Enter an option [?/Make/Set/New/ON/ OFF/Color/Ltype/LWeight/Plot/ Freeze/Thaw/LOck/Unlock/stAte]:	Type: **M**↵
Enter name for new layer (becomes the current layer) <0>:	Type: **R**↵
Enter an option [?/Make/Set/New/ON/ OFF/Color/Ltype/LWeight/Plot/ Freeze/Thaw/LOck/Unlock/stAte]:	Type: **C**↵
Enter color name or number (1-255):	Type: **R**↵
Enter name list of layer(s) for color 1 (red) <R>:	Type: ↵ (If the default name shown is not R, Type: **R**↵)
Enter an option [?/Make/Set/New/ON/ OFF/Color/Ltype/LWeight/Plot/ Freeze/Thaw/LOck/Unlock/stAte]:	↵ (to exit from the Layer command)

Make and Color

Note: Layer 0, color white, with a continuous linetype, is created automatically by AutoCAD. It has some special properties that will be used in later chapters. Layer 0 cannot be deleted.

With the option "Make" from the LAYER: prompt, you have just created a layer named "R." Upper- or lowercase letters may be entered for the name. AutoCAD converts all layer names to uppercase. The name may be as long as 255 characters, including numbers and letters. It may include only letters, numbers, and three special characters: $ (dollar sign), - (hyphen), and _ (underscore).

Note that the prompt is for "New current layer <0>." By using the option "Make," you may create a new layer, *and* it is automatically set as the current layer on which you will draw. The new layer name appears in the status line on the display screen as the current layer on which you will be working. If the layer name has more than eight characters, only eight characters of the name will be displayed in the status line.

Using the option "Color" from the LAYER: prompt allows you to choose a color from seven standard colors and assign it to a layer in response to the prompt "Layer name(s) for color 1 (red) <R>." "R" is the *default,* so the Enter key is used. If the option "Color" is not used, the layer will default to the color white.

The option "Ltype" was not used, so the linetype for layer "R" will default to a "continuous" linetype. Continuous is a solid line.

To use the options New, Color, and Ltype:

Prompt	Response
Command:	Type: **-LA**↵
Enter an option [?/Make/Set/New/ON/ OFF/Color/Ltype/LWeight/Plot/ Freeze/Thaw/LOck/Unlock/stAte]:	Type: **N**↵
Enter name list for new layer(s):	Type: **G,Y**↵
Enter an option [?/Make/Set/New/ON/ OFF/Color/Ltype/LWeight/Plot/ Freeze/Thaw/LOck/Unlock/stAte]:	Type: **C**↵
Enter color name or number (1-255):	Type: **G**↵
Enter name list of layer(s) for color 3 (GREEN) <R>:	Type: **G**↵
Enter an option [?/Make/Set/New/ON/ OFF/Color/Ltype/LWeight/Plot/ Freeze/Thaw/LOck/Unlock/stAte]:	Type: **C**↵
Enter color name or number (1-255):	Type: **Y**↵
Enter name list of layer(s) for color 2 (YELLOW) <G>:	Type: **Y**↵

Prompt	Response
Enter an option [?/Make/Set/New/ON/ OFF/Color/Ltype/LWeight/Plot/ Freeze/Thaw/LOck/Unlock/stAte]:	Type: **L**↵
Enter loaded linetype name (or ?) <CONTINUOUS>:	Type: **HIDDEN**↵
Enter name list of layer(s) for linetype HIDDEN <Y>:	Type: **Y**↵
Enter an option [?/Make/Set/New/ON/ OFF/Color/Ltype/LWeight/Plot/ Freeze/Thaw/LOck/Unlock/stAte]:	↵

New, Color, and Linetype

With the option "New" from the LAYER: prompt, you have just created two layers named "G" and "Y." The prompt is for "New layer name(s):". One or more layers may be created at the same time. Enter the layer names, separated by a comma. Do not use spaces.

Using the option "Color" from the LAYER: prompt, you first selected the color Green and then assigned it to layer G. You could have immediately used the option "Color" again to select and assign the color Yellow to the layer Y, but instead you went on to the option "Ltype" and returned to "Color" later.

The option "Ltype" from the LAYER: prompt allows you to choose a linetype from nine standard linetypes and assign it to a layer. In response to the prompt "Layer name(s) for linetype HIDDEN <G>:", you typed Y <enter> to assign linetype HIDDEN to layer Y. A linetype was not assigned to layer G, so it will default to a continuous linetype. Later versions of AutoCAD have three sizes for each linetype, for example, HIDDEN (the standard size), HIDDEN2 (half the standard size), and HIDDENX2 (twice the standard size).

One or more layers may be created with the option "New," but unlike with the option "Make," none of the new layers is made current. The current layer is not affected by using the option "New" to create new layers.

To use the option Set:

Prompt	Response
Command:	Type: **-LA**↵
Enter an option [?/Make/Set/New/ON/ OFF/Color/Ltype/LWeight/Plot/ Freeze/Thaw/LOck/Unlock/stAte]:	Type: **S**↵
Enter layer name to make current or <select object>:	Type: **G**↵
Enter an option [?/Make/Set/New/ON/ OFF/Color/Ltype/LWeight/Plot/ Freeze/Thaw/LOck/Unlock/stAte]:	↵

Set

The option "Set" from the LAYER: prompt allows you to set a new current layer. By responding to the prompt "New current layer <default>:" with an existing layer name, you may make any existing layer the current layer on which you will draw. The new layer name will appear in the status line on the upper left of the display screen.

Unlike "Make" and "New," the option "Set" does not *create* layers. The layer must already exist for you to use the "Set" option.

To use the option ?:

Prompt	Response
Command:	Type: **-LA**↵
Enter an option [?/Make/Set/New/ON/ OFF/Color/Ltype/LWeight/Plot/ Freeze/Thaw/LOck/Unlock/stAte]:	Type: **?**↵

AutoCAD Fundamentals

Prompt	Response
Enter layer names to list <*>:	↵

(A listing of all existing layers and the status of each appears.)

Prompt	Response
Enter an option [?/Make/Set/New/ON/ OFF/Color/Ltype/LWeight/Plot/ Freeze/Thaw/LOck/Unlock/stAte]:	↵
?	

The option "?" from the LAYER: prompt provides a checklist of the layers created, showing their names, colors, and linetypes, as well as which layer is current.

Note the 0 layer, color white, with a continuous linetype. Layer 0 is created automatically by AutoCAD. It has some special properties that will be used in later chapters. Layer 0 cannot be deleted.

ON or OFF; Freeze or Thaw

The options ON and OFF pertain to the visibility of layers. When a layer is turned OFF, it is still part of the drawing, but it is not visible on the screen, nor can it be plotted. For instance, the building shell layer, interior walls layer, and furniture layer may be turned ON and all other layers turned OFF to view, edit, or plot a furniture plan. One or more layers can be turned OFF and ON as required to get the combination of layers needed.

The options Freeze and Thaw also pertain to the visibility of layers. The difference between ON or OFF and Freeze or Thaw is a matter of how quickly the drawing *regenerates* on the display screen. If a layer is frozen, it is not visible, it cannot be plotted, and AutoCAD does not spend any time regenerating it. A layer that is turned OFF is not visible and cannot be plotted, but AutoCAD does regenerate it.

Regeneration of a layer takes time. If you are switching between layers and doing some editing on each layer, it is best to use ON and OFF. If while editing layers there are layers that you do not need to edit or see, it is best to freeze those layers to avoid their regeneration.

When a layer is created, it is automatically turned ON and is thawed.

The Layer Properties Manager Dialog Box

When you Type: **LA<enter>** or Click: **the Layer icon in the upper left of your screen**, the Layer Properties Manager dialog box appears. It is very easy to create new layers and assign colors and linetypes to them with the dialog box.

The Layer List

After you have created new layers and assigned colors and linetypes, a Layer List box appears in the upper left of your screen (on the Object Properties toolbar). You can turn layers on and off, freeze and thaw, and set a layer current using the Layer List. On and off requires you to Click: **the light bulb**, freeze and thaw requires you to Click: **the sun** or **the snowflake**, and set requires you to Click: **on the desired current layer** to close the Layer List.

Save the Drawing and Exit AutoCAD

Step 6. Save the settings and layers for Exercise 3–1 on the hard drive. The drawing name will be EX3-1(your initials).

Prompt	Response
Command:	Type: **SAVEAS**↵
The Save Drawing As dialog box appears with the File Name: highlighted:	Type: **EX3-1(your initials)** (Because the File Name: was highlighted, you were able to type in that place. If you had used any other part of the dialog box first, you would have had to click to

the left of unnamed, hold down the click button, and drag the cursor across the name to highlight it and then begin typing.) The drawing name may be from 1 to 255 characters in length.

Click: **OK**

Be sure to make note of the drive and directory where the drawing is being saved so you can retrieve it easily when you need it.

Step 7. **Save the same drawing to a floppy disk in the A drive.** (Substitute B: for A: if your computer is so labeled.)

Insert a formatted diskette into the A drive.

Prompt	**Response**
Command:	↵ (to repeat the previous command)
The Save Drawing As dialog box appears:	Click: **the down arrow in the Drives: button, highlight the a: drive, and** Click: **a:**
	or
	Click: **the Up Arrow and locate "3-1/2 Floppy Drive [A:]" and click it**
	Click: **OK**

The light should brighten on the A drive, indicating that the drawing is being saved.

Because the drawing was named when you saved it on the hard drive, you did not have to type the name again to save it with that name on the floppy disk. You could have chosen to give the drawing another name when you saved it on the floppy disk, in which case you would have had to type the new name in the File Name: input button.

Save Drawing As Dialog Box

The parts and functions of the Save Drawing As dialog box are as follows:

File Name: input button

The drawing file name that will be saved appears here.

Beneath the File Name: button is an alphabetized list of all files of the type defined by the Files of Type: button in the drive and directory specified in the Save in: list. If you want to use one of the names in this list Click: it, and that name will replace the one in the File Name: button. You will need to use the scroll bar if the list of files is longer than the area shown.

Files of Type: list

Clicking the down arrow reveals a list of file types under which the drawing may be saved. All have the extension .dwg except the drawing template type, which has the extension .dwt. Some of these file types can be used interchangeably, but others require that a specific file type be used. In this book you will use only the AutoCAD file type .dwg for drawings.

Save in: list

The drive and folder where the file will be saved is listed under the Save in: list. The area beneath this lists the drive and directories existing on that drive. The list indicates that the directory or folder is open and that it contains other subdirectories or folders. The list of files in the File Name: area shows the files with the .dwg extension that exist in the directory or folder. If you double click the folder labeled c:\ a list of the other folders on the C drive appears.

Save button

When clicked, this button executes the Save As command.

Cancel button

When clicked, this button cancels the command or closes any open button on the dialog box. The Esc key has the same effect.

Step 8.   Exit AutoCAD or AutoCAD LT.

Prompt	**Response**
Command:	Type: **QUIT**↵ to exit

Exiting AutoCAD or AutoCAD LT

Three commands, Save, SAVEAS, and Exit (or Quit), and their uses must be understood to save your work in the desired drive and directory and to exit AutoCAD after you have saved your work.

Save

When the command Save is clicked and the drawing has been named, the drawing is saved automatically to the drive and directory in which you are working, and a backup file is created with the same name but with the extension .bak. If the drawing has not been named, Save behaves in the same manner as Save As.

Save As

Save As activates the Save Drawing As dialog box whether or not the drawing has been named and allows you to save your drawing to any drive or directory you choose.

Some additional features of the Save As command are as follows:

1. A drawing file can be saved and you may continue to work, because with the Save As command, the drawing editor is not exited.
2. If the default drive is used (the drive on which you are working), and the drawing has been opened from that drive, .dwg and .bak files are created.
3. If a different drive is specified (a floppy disk in the floppy drive), only a .dwg file is created.
4. To change the name of the drawing, you may save it under a new name by typing a new name in the File Name: button.
5. If the drawing was previously saved or if a drawing file already exists with the drawing file name you typed, AutoCAD gives you the following message:

    This file already exists. Do you want to replace it?

    When a drawing file is updated, the old .dwg file is replaced with the new drawing, so the answer to click is Yes. If an error has been made and you do not want to replace the file, click No.
6. A drawing may be saved to as many floppy disks or to as many directories on the hard disk as you wish. You should save your drawing on two different floppy disks as insurance against catastrophe.

Exit

If you have not made any changes to the drawing since you last saved it, the Exit command takes you out of the AutoCAD or AutoCAD LT program. If you have made changes to the drawing and have not saved these changes, AutoCAD will give you the message "Save changes to EX3-1MK?" (or whatever the drawing name is). This is a safety feature

Warning: Remember, do not use the Exit command without first saving your work. If you answer yes to the Exit question without first saving a drawing, you will lose your work.

because the Exit command, by itself, *does not update or save a drawing*. You have three options, Yes, save the changes; No, do not save changes; or Cancel the Exit command.

If you have just entered a drawing, have made a lot of mistakes, and just want to get rid of everything, respond with No to the Save changes question. If you opened an existing drawing and use the Exit command without making any changes, the stored .dwg file and .bak files are preserved unchanged.

While you are making a new drawing, AutoCAD is creating a .dwg (drawing) file of your drawing. There is no .bak (drawing file backup) file for a new drawing.

Each time an existing drawing file is opened for editing, the original drawing file (.dwg) becomes the drawing file backup (.bak). The new edited version of the drawing becomes the .dwg file. Thus there is a copy of the drawing file as it was originally (.bak) and a copy of the new edited version (.dwg).

TIPS FOR BEGINNING AUTOCAD AND AUTOCAD LT USERS

If you make a mistake and want to back up one or more steps:

Prompt	**Response**
Command:	Type: **U**↵
	Continue pressing ↵ to repeat the UNDO command until you back up to the desired point.

If you UNDO too much:

Prompt	**Response**
Command:	Type: **REDO**↵ to redo one undo. REDO undoes only one undo.

If you have drawn a wrong line and want to back up one or more lines but still stay in the line command:	Type: **U**↵ until you are back to the desired point.
If you want to cancel a command:	Press: the **Esc** key

If you are having trouble selecting items, Zoom a window around the items to enlarge the view, then Zoom back out to complete the command.

If you Zoom-All and your drawing is very small in one corner of the display, look for a stray piece of the drawing on an opposite side of the display and erase it. You may have to pan the display to select the problem piece. After you find it and erase, Zoom-All again.

If you still have the same problem, thaw and turn on all layers and Zoom-All again to locate and move or erase the problem. To thaw all layers: Type: **-LA**↵, then **T**↵ for Thaw, then *****↵ for all layers. To turn on all layers: Type: **-LA**↵, then **ON**↵, then *****↵ for all layers.

EXERCISE

EXERCISE 3–1   Complete Exercise 3–1 using steps 1 through 8 described in this chapter.

AutoCAD Fundamentals

REVIEW QUESTIONS

Circle the best answer.

1. Which of the following is *not* on the list of units?
   a. Scientific
   b. Metric
   c. Decimal
   d. Fractional
   e. Architectural
2. The default lower left corner of the drawing limits is 8.5,11.
   a. True
   b. False
3. The function key F7, described in this chapter, does which of the following?
   a. Provides a checklist of the layers created
   b. Turns Snap ON or OFF
   c. Flips the screen from the text display to the graphics display
   d. Turns Grid ON or OFF
   e. Turns Ortho ON or OFF
4. Which of the following function keys is used to turn Snap ON or OFF?
   a. F1
   b. F2
   c. F7
   d. F8
   e. F9
5. If you want to correct a mistake made by the previous command, which of the following is the correct response?
   a. Type: U↵
   b. Press: Esc
   c. Press: Ctrl-C
   d. Type: REDO↵
   e. Press: F7
6. To make sure you have all layers thawed, which of the following selects all layers?
   a. #
   b. @
   c. *
   d. <
   e. F9
7. A frozen layer may be turned ON before it is thawed.
   a. True
   b. False
8. How many layers may be set current at the same time?
   a. 1
   b. 2
   c. 3
   d. 16
   e. Unlimited
9. When a layer is OFF, it will regenerate but is not visible.
   a. True
   b. False
10. Later versions of AutoCAD provide you with how many sizes of each standard linetype (except continuous)?
    a. 1
    b. 2
    c. 3
    d. 4
    e. As many as you want

Complete.

11. Describe the effect of using the Esc key while in a command.
    _____

12. An invisible grid to which the crosshair will lock is called _____.

13. What is the maximum number of characters that may be used in a layer name?
    _____

14. How many layers may be used in a drawing?
    _____

15. What is the maximum number of characters that may be used in a drawing name?
    _____

16. Explain what .dwg and .bak files are.

    .dwg _____

    .bak _____

17. List the three sizes for the HIDDEN linetype.
    _____

18. Describe how the number of places to the right of the decimal point affects the accuracy of a drawing.
    _____

19. What does the Quit command do when used by itself?
    _____

20. Describe how Save differs from Save As when the drawing has been named.
    _____
    _____

AutoCAD Fundamentals

4 Lettering for Pencil Sketches

OBJECTIVE

After completing this chapter, you will be able to

☐ Develop a legible, uniform lettering style that is pleasing in appearance.

INTRODUCTION

The sketches you draw in this book not only must have correct construction and good line quality and be well arranged in the field of the drawing (the field of the drawing is that area inside the border where the drawing is placed), they must also have lettering that is uniform, legible, and pleasing in appearance. A good lettering style is a result of careful planning and a controlled lettering style. The following alphabet is a good model for a legible, uniform lettering style that you can develop with a little careful practice.

THE SKETCHING ALPHABET

The letters used in the sketching alphabet are made with single strokes and are as simple as possible. This alphabet is very similar to the Simplex font, which is one of the fonts available to you in AutoCAD and AutoCAD LT. There are no serifs on the letters and all strokes are the same width (Figure 4–1). Most companies use all capital (uppercase) letters because lowercase letters are more likely to become illegible when they are copied in a reduced size.

Uppercase Letters

The sequence of strokes for each letter in the vertical alphabet is shown in Figure 4–2. Study each character until you know what it should look like. It may take a while for you to form the letters well. After you have drawn a letter, compare it with the form in Figure 4–2. If your letter does not look as good as or better than those in the figure, stop, analyze what is wrong, and correct it. *Go slowly at first and do it right.* You can pick up speed later.

Fractions

It is important that fractions be large enough to be reproduced at a reduced size and still be legible. Fractions should be twice the height of whole numbers if they are vertical (Figure 4–3A). Each number of the fraction should be almost the same size as a whole number if the fraction must be horizontal to fit a given space (Figure 4–3B). The vertical

FIGURE 4–1
The Sketching Alphabet Style

FIGURE 4–2
The Vertical Alphabet

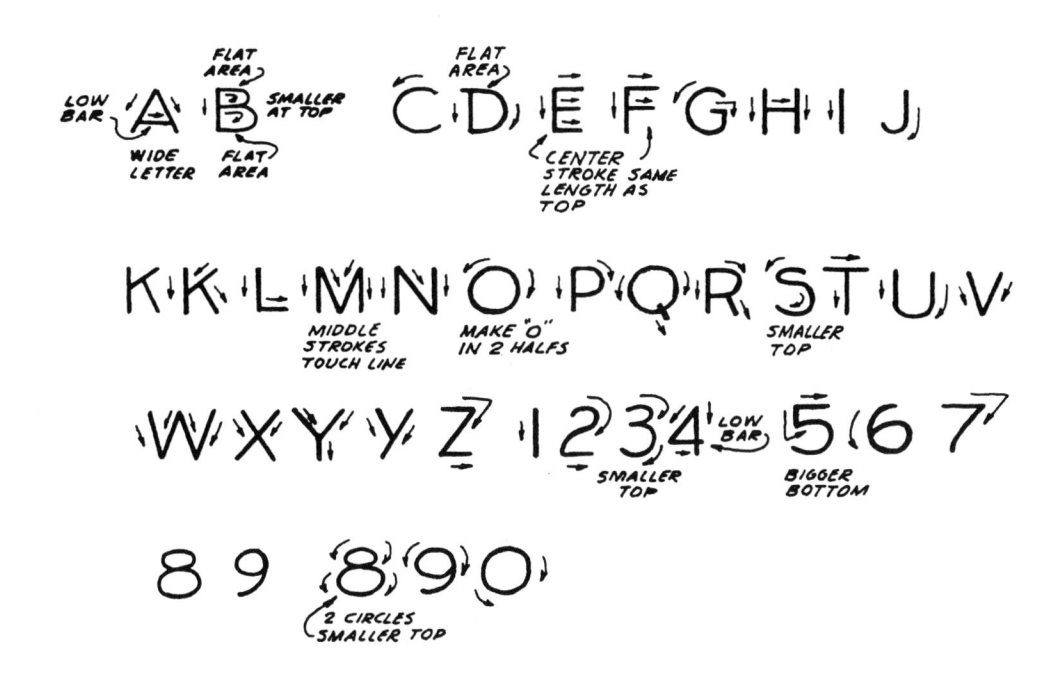

fraction is preferred. Do not let fraction numbers touch the horizontal bar, because that can make the fraction illegible (Figure 4–3C).

Slant

Letters can be vertical or inclined; either style is usually acceptable. Which one you use usually depends on your natural ability. The important characteristics of a good lettering style are that all letters are the same height and slant and that they are bold enough to reproduce well. Figure 4–4 shows the inclined alphabet.

Guidelines

Use guidelines, ruled paper, or a lettering aid like that shown in Figure 4–5 for all letters and numbers. The lettering aid shown is easy to use (Figure 4–6). Just letter inside a slot and make all letters "bounce off" the top and bottom of the slot so that they are all the same height. Guidelines must be very thin and light—so light that they are barely visible. All lettering should be very dark and thin enough to be easily read.

FIGURE 4–3
Fractions

FIGURE 4–4
The Inclined Alphabet

Lettering for Pencil Sketches

FIGURE 4–5
Lettering Guide

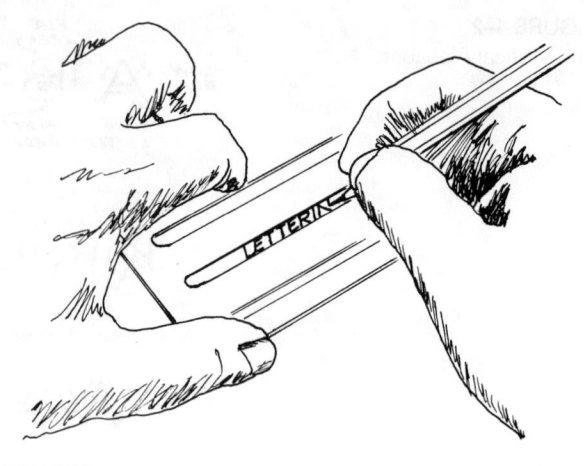

FIGURE 4–6
Using the Lettering Guide

DEVELOPING A GOOD LETTERING STYLE

The most important characteristic of lettering is legibility. Lettering must be consistent and neat but above all must be easily read so that numbers and values cannot be mistaken. The alphabets shown in Figures 4–2 and 4–4 can be done legibly with a little practice. The rules for developing a good lettering style are as follows:

1. Make sure the form of the letter is right. Do not mix uppercase and lowercase letters. Use all uppercase letters unless your instructor tells you otherwise.
2. Use guidelines, a lettering aid, or ruled paper. When drawing guidelines, make sure they are very light and thin.
3. Keep the slant of the letters the same. Use either a vertical or slanted stroke, but keep the stroke consistent.
4. Put all the letters in words as close to each other as you can while still making them look good.
5. Concentrate on keeping the characters open and easily read.
6. Do not make letters too tall for the thickness of your stroke. With a slightly blunted pencil point, letters $\frac{1}{8}''$ to $\frac{3}{16}''$ high look good.
7. Make the space between words approximately big enough for the letter I. The space between lines of lettering should be half to two-thirds the height of the letter (Figure 4–7).
8. Do not allow letters or numbers to touch any object line, border, or fraction bar. Letters and numbers should have clear space all around them (sides, top, and bottom).
9. Begin by drawing letters and numbers. *Take your time.*

FIGURE 4–7
Spacing between Words and Lines

46

Chapter 4

10. Make all letters very dark. If you must repeat a stroke to improve its density, do so. A no. 2 pencil or an H-grade lead is right for most people.
11. Work to improve your speed as soon as you have the form, density, and slant correct.

EXERCISES

EXERCISE 4–1   Using the vertical style of lettering, repeat each of the letters, numbers, fractions, words, and sentences (on the sheet in the back of this book labeled Exercise 4–1) the stated number of times.

EXERCISE 4–2   Using the inclined style of lettering, repeat each of the letters, numbers, fractions, words, and sentences (on the sheet in the back of this book labeled Exercise 4–2) the stated number of times.

REVIEW QUESTIONS

Circle the best answer.

1. The sketching alphabet described in this chapter is similar to which of the following AutoCAD fonts?
   a. Simplex
   b. Standard
   c. Txt
   d. Italic
   e. Freehand
2. Guidelines for lettering should be drawn
   a. Thin and dark
   b. Thin and light
   c. Thick and dark
   d. Thick and light
   e. Guidelines should not be used.
3. Vertical fractions should be
   a. Twice the height of whole numbers
   b. The same height as whole numbers
   c. About one and a half times the height of whole numbers
   d. A little smaller than whole numbers
   e. Height is not important.
4. The most important quality of good lettering is
   a. Consistent height
   b. Consistent slant
   c. Consistent darkness
   d. Legibility
   e. Good appearance
5. Either vertical or inclined lettering is acceptable in most companies.
   a. True
   b. False
6. The space between words should be approximately
   a. The spacing for the letter W
   b. The spacing for the letter O
   c. The spacing for the letter I
   d. Twice the height of the letters
   e. Approximately two-thirds the height of the letters
7. The space between lines of lettering should be
   a. Approximately two-thirds the height of the letters
   b. Approximately twice the height of the letters
   c. Invisible (lines of lettering should touch)
   d. Approximately one and a half times the height of the letters
   e. Approximately one-fourth the height of the letters

8. A good lead grade for lettering is
    a. H
    b. 4H
    c. F
    d. 6H
    e. B
9. It is OK for letters to touch the lines forming the title block of a drawing.
    a. True
    b. False
10. When you practice lettering, you should letter as fast as you can in the beginning.
    a. True
    b. False

5 Sketching Line Weights and Drawing Constructions

OBJECTIVES

After completing this chapter, you will be able to

- Draw dense, uniform, dark lines of the correct thickness for object, cutting plane, centerline, dimension, and extension lines.
- Draw thin, light lines for drawing construction and lettering guidelines for stated problems.
- Divide lines, angles, and arcs.
- Construct polygons.
- Draw tangents to circles, lines, and arcs.

LINES USED IN TECHNICAL DRAWINGS

Line quality is one of the most important elements of a good drawing. Many different types of lines are commonly used. All lines except construction lines for drawing, and guidelines for lettering, must be sharp and dark. Some are thicker than others, but all lines must be black and uniformly dark. Lettering guidelines and construction lines must be very thin and light—just barely dark enough to be followed for drawing the final lines. The alphabet of lines is shown in Figure 5–1. AutoCAD calls these different lines *linetypes*. These linetypes are the same as the ones that you will use to sketch the exercises in this book.

FIGURE 5–1
Alphabet of Lines

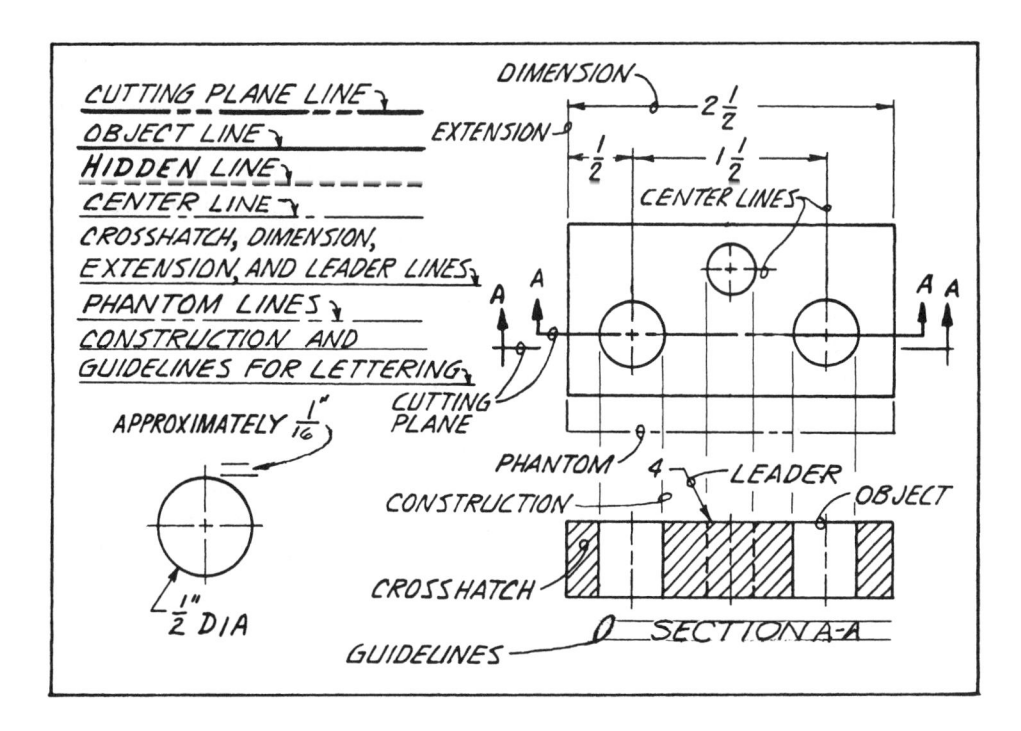

FIGURE 5–2
Preparing the Pencil Point

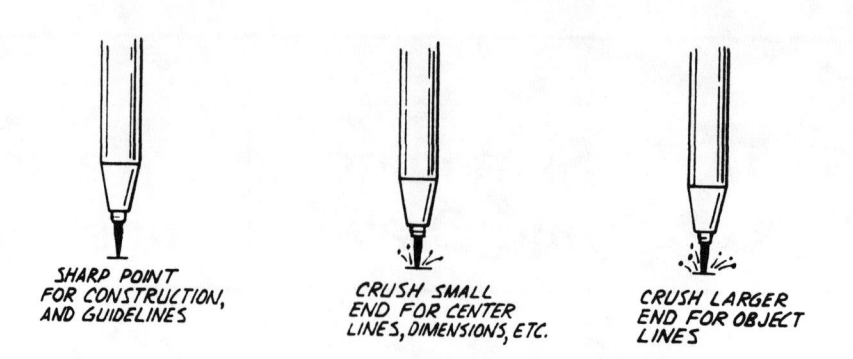

SHARP POINT FOR CONSTRUCTION, AND GUIDELINES

CRUSH SMALL END FOR CENTER LINES, DIMENSIONS, ETC.

CRUSH LARGER END FOR OBJECT LINES

The linetypes are described as follows:

Cutting-plane lines are the thickest lines. They have a long dash and two short dashes. Another form of the cutting-plane line, also shown in Figure 5–1, is used in making sectional views, which will be covered in Chapters 9 and 10.

Object lines are approximately half as thick as cutting-plane lines and are continuous (no breaks).

Hidden lines, a series of short dashes, are the same thickness or slightly thinner than object lines. Each dash is approximately $\frac{1}{8}''$ long separated by $\frac{1}{16}''$ spaces.

Hatch, dimension, extension, and *leader lines* are continuous and are approximately half as thick as object lines but are just as dark.

Centerlines are the same width as dimension lines and consist of a long dash and one short dash. The long dash is approximately $\frac{5}{8}''$ long, the short dash is approximately $\frac{1}{16}''$ long, and the space between dashes is approximately $\frac{1}{16}''$. The centerline should extend approximately $\frac{1}{16}''$ past the feature it is describing. Centerlines describe circular features and the centers of some objects.

Phantom lines are used to show some alternative position or feature that differs from the main drawing. They are similar to the centerline but with two short dashes instead of one.

Construction lines and *guidelines for lettering* are continuous and are thin and very light.

SKETCHING GOOD LINES

Using a Wooden Pencil or a Lead Holder

Drawing good lines with a wooden pencil or lead holder requires that the pencil be sharpened often. In drawing thin lines, such as centerlines and dimension lines, the lead must be sharpened to a point, as shown in Figure 5–2. In drawing object, hidden, and cutting-plane lines, use a blunter point (Figure 5–2). To draw the line, hold the pencil about 1″ from the end, at approximately a 60° angle to the paper (Figure 5–3). Roll the pencil in your hand as you draw the line from left to right. Apply enough pressure to get a solid, dark line. Go back over the lines from right to left if necessary. Use a no. 2 pencil and H or 2H lead for these lines.

Using a Mechanical Pencil

The mechanical pencil has a lead of uniform thickness and as a result must be held perpendicular to the paper. To draw object and hidden lines, use a 0.7 mm lead of H or 2H softness. To draw thin lines, such as dimension lines, centerlines, and extension lines, use a 0.5 mm lead of H or 2H softness. Use a firm, even pressure to get a solid line and to avoid breaking the lead. Go back over lines if necessary to get a solid, dark line. Construction lines and lettering guidelines can be drawn with a 0.3 mm lead of 2H or 4H softness.

FIGURE 5–3
Drawing Angle

60° ANGLE TO PAPER

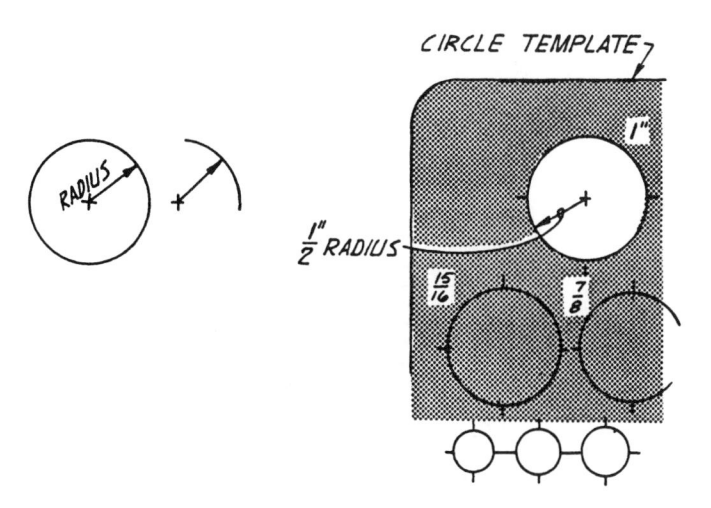

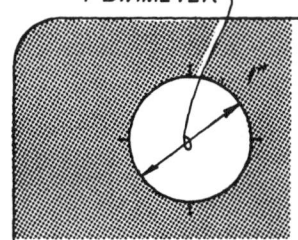

FIGURE 5–4
Radius

FIGURE 5–5
Diameter

CONSTRUCTION TERMS

Many drawing constructions are done repeatedly in technical drawings. Because many of the terms used to describe these constructions are not familiar to everyone, the most common terms are defined next.

Radius   The distance from the center of a circle or arc to the outside edge or circumference (Figure 5–4). It is the distance halfway across a circle template.
Diameter   The distance all the way across a circle, through the center (Figure 5–5). It is the size marked on circles on a circle template.
Circumference   The distance around the outside of a circle.
Intersection   Where two or more lines, circles, or arcs cross (Figure 5–6).
Parallel   Two lines or curves that are the same distance apart along all parts of the lines or curves (Figure 5–7).
Perpendicular   Lines at 90° angles to each other (Figure 5–8).
Tangent   Lines or arcs that touch at only one point. When lines or arcs are tangent, they lie exactly on top of each other at one point (Figure 5–9).
Bisected   Divided in half (Figure 5–10).

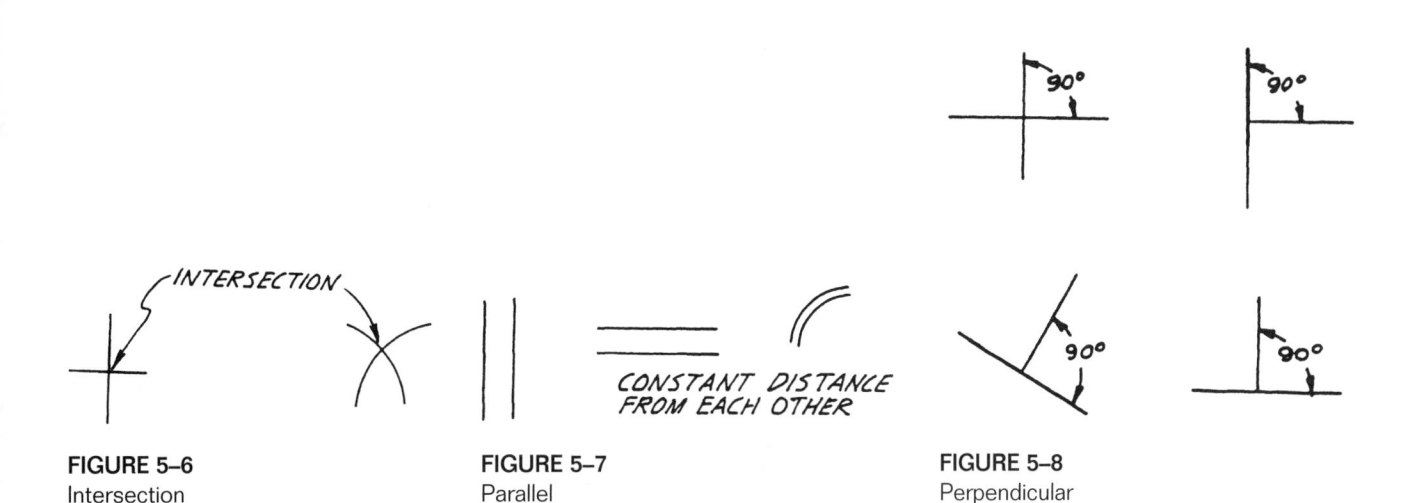

FIGURE 5–6
Intersection

FIGURE 5–7
Parallel

FIGURE 5–8
Perpendicular

Sketching Line Weights and Drawing Constructions

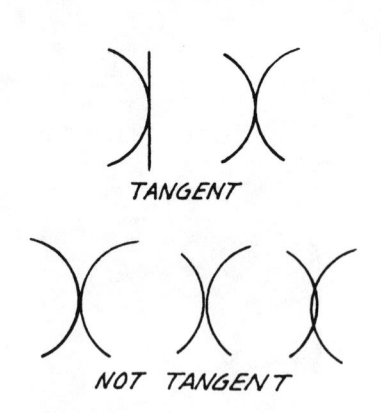

FIGURE 5–9
Tangent

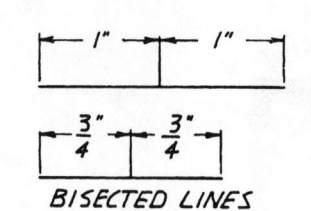

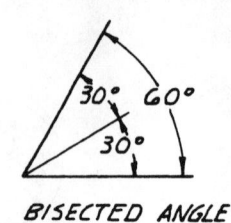

FIGURE 5–10
Bisected

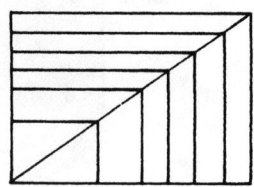

FIGURE 5–11
Proportional

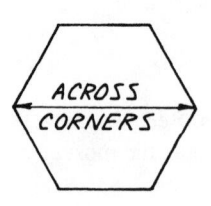

FIGURE 5–12
Across Corners

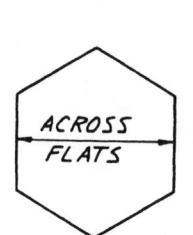

FIGURE 5–13
Across Flats

Proportional   The same ratio or proportions (Figure 5–11).
Across corners   A measurement made across corners through the center of a feature (Figure 5–12).
Across flats   A measurement made across the parallel sides of a feature (Figure 5–13). The measurement is made perpendicular to the sides.
Right angle   A 90° angle (Figure 5–14).
Acute angle   An angle less than 90° (Figure 5–15).
Obtuse angle   An angle greater than 90° (Figure 5–16).
Fillet   An inside radius (Figure 5–17).
Round   An outside radius (Figure 5–18).
Hexagon   A feature with six equal sides and angles (Figure 5–19).
Polygon   A feature enclosed with straight lines (Figure 5-20). Regular polygons have equal sides and angles. Irregular polygons have unequal sides and/or angles.

FIGURE 5–14
Right Angles

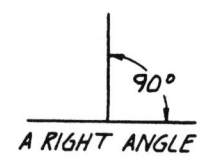

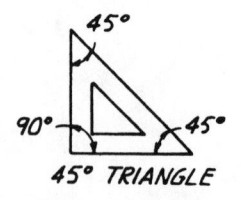

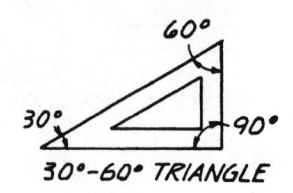

Chapter 5

FIGURE 5–15
Acute Angles

FIGURE 5–16
Obtuse Angles

FIGURE 5–17
Fillet

FIGURE 5–18
Round

FIGURE 5–19
Hexagon

FIGURE 5–20
Polygons

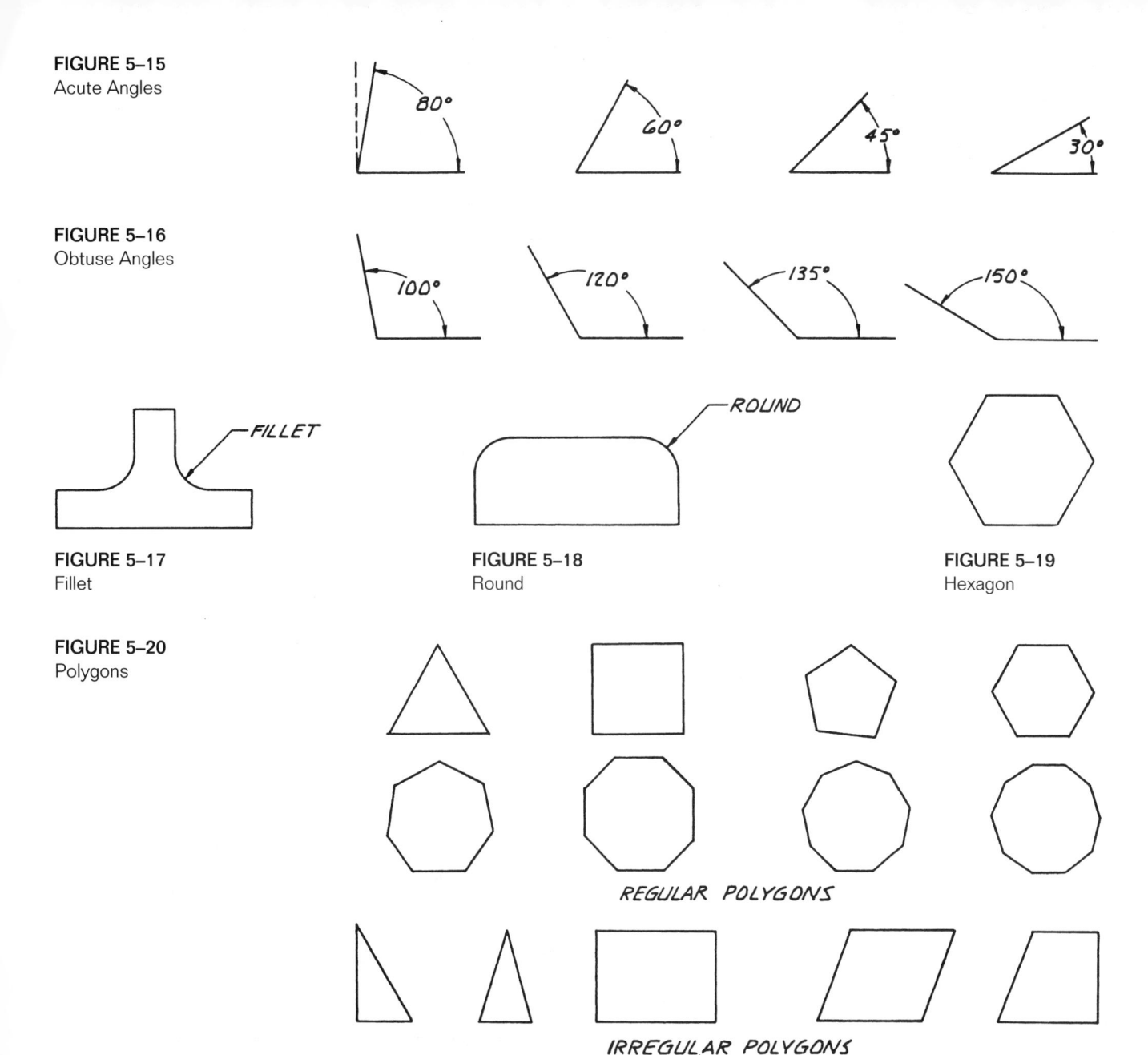

DRAWING CONSTRUCTIONS

Now that we have defined the terms, we describe the most common drawing constructions.

Sketching Parallel Lines

There are several methods for drawing parallel lines. The grid paper in your book allows you to use the grid for sketching parallel vertical and horizontal lines. Figure 5–21 shows the use of a drafting machine or T-square and triangles to sketch parallel lines. This book is designed for you to use triangles to sketch parallel lines. AutoCAD has a command called Offset that is used to draw parallel lines. To sketch parallel lines on an angle follow these steps:

Step 1. Sketch one line at the desired angle.
Step 2. Align one edge of a triangle on the newly drawn line.
Step 3. Place one edge of another triangle at the base of the first triangle as shown in Figure 5–21.
Step 4. Either tape the second triangle in place or hold it firmly as you slide the first triangle along its edge and sketch the parallel lines.

Sketching Line Weights and Drawing Constructions

FIGURE 5–21
Sketching Parallel Lines Using Drafting Machines and Triangles

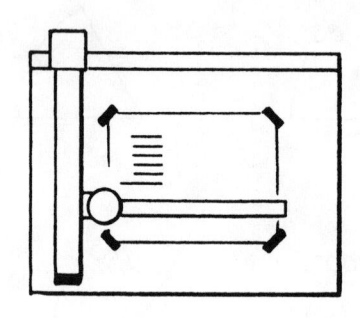

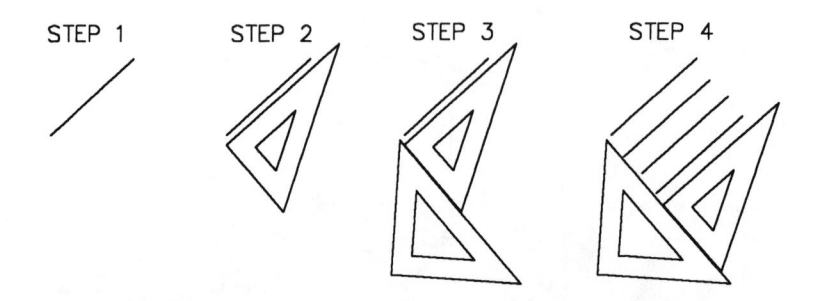

Figure 5–22 shows the steps in sketching parallel lines a given distance apart (in this case $\frac{3}{4}''$) by drawing a construction line perpendicular to the first line sketched. The perpendicular ensures that the lines are spaced exactly the distance marked. The steps are as follows:

Step 1. **Sketch one line at the desired angle.**
Step 2. **Construct a perpendicular construction line by aligning one triangle on the line and placing another triangle containing a 90° angle on it, as shown in Figure 5–22. Make sure your construction line is very thin and light.**
Step 3. **Using either a dot or a small light line, mark $\frac{3}{4}''$ spaces on the perpendicular line.**
Step 4. **Using the two-triangle method described in Figure 5–21, sketch lines through the points.**

Sketching Perpendicular Lines

In addition to the method for sketching perpendicular lines described in Figure 5–22, two methods are shown in Figure 5–23. The first method involves the use of a straight edge and a triangle. The straight edge can be a drafting machine, a T-square, or another triangle. The second method is as follows:

Step 1. **Start with a line of a specified length (in this case $2\frac{1}{2}''$).**
Step 2. **Using a radius greater than half the length of the line, draw arcs from each end of the line.**
Step 3. **Connect the intersections of the arcs to form a perpendicular.**

FIGURE 5–22
Sketching Parallel Lines Using Two Triangles

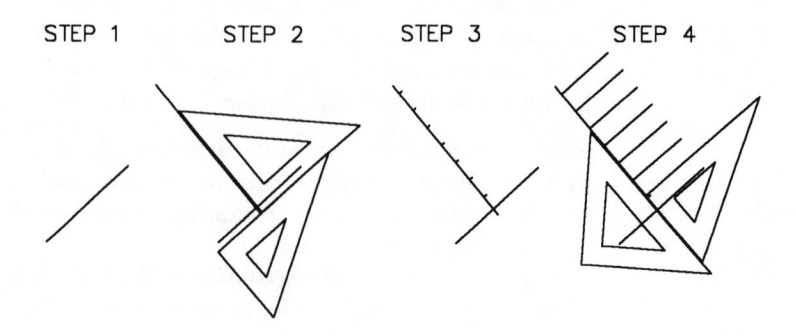

FIGURE 5–23
Sketching Perpendicular Lines

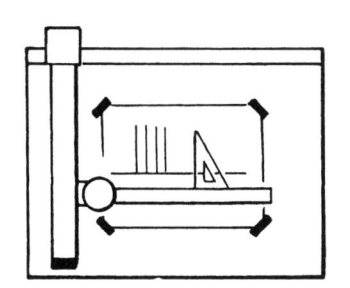

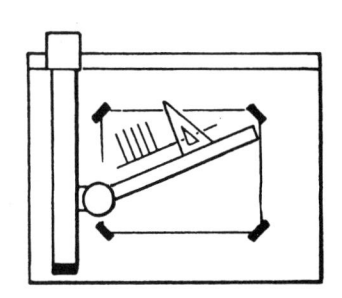

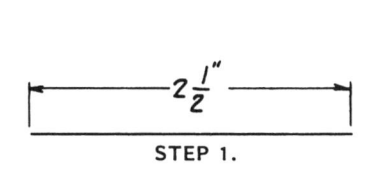

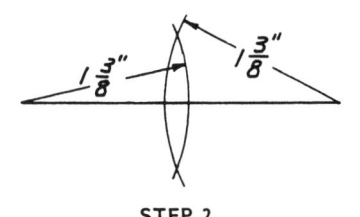

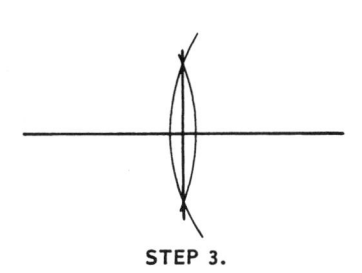

STEP 1.

STEP 2.

STEP 3.

AutoCAD uses a command called Osnap-Perpendicular to draw lines perpendicular to other lines.

Sketching Tangents

The following figures show several instances in which tangents are used and the methods for sketching them. AutoCAD uses a combination of Osnap-Tangent and arc and circle commands that use the Tangent-Tangent-Radius option to draw tangents. The Fillet command is also used to draw arcs of a specified radius tangent to two lines.

Sketching a line tangent to two circles (Figure 5–24):

Step 1. Lay a triangle so that it rests just below the outside edges of both circles. This small space allows your pencil lead to pass exactly through both circles, making the line tangent to them.
Step 2. Draw the line tangent to both circles.
Step 3. Position the triangle differently in step 1 to obtain an alternative position.

Sketching an inside arc tangent to two circles (Figure 5–25):

Although you will not use method 1 in your sketching exercises because it often requires a compass, it is important that you study this method so you will know how it is done. You will find it to be useful in understanding constructions in both sketching and AutoCAD.

Method 1

Step 1. Start with two given circles and a given radius (in the example, a larger circle with a $1\frac{1}{2}''$ radius, a smaller circle with a $1''$ radius, and a $\frac{3}{4}''$ radius

Sketching Line Weights and Drawing Constructions

FIGURE 5–24
Sketching a Line Tangent to Two Circles

GIVEN: TWO CIRCLES

STEP 1.

STEP 2.

STEP 3.

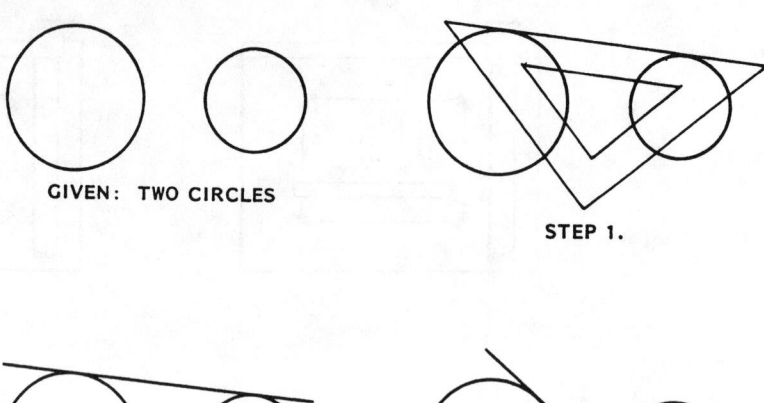

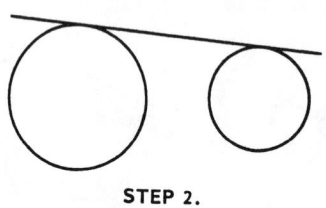

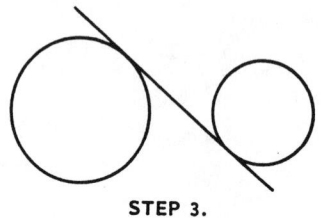

FIGURE 5–25
Sketching an Inside Arc Tangent to Two Circles

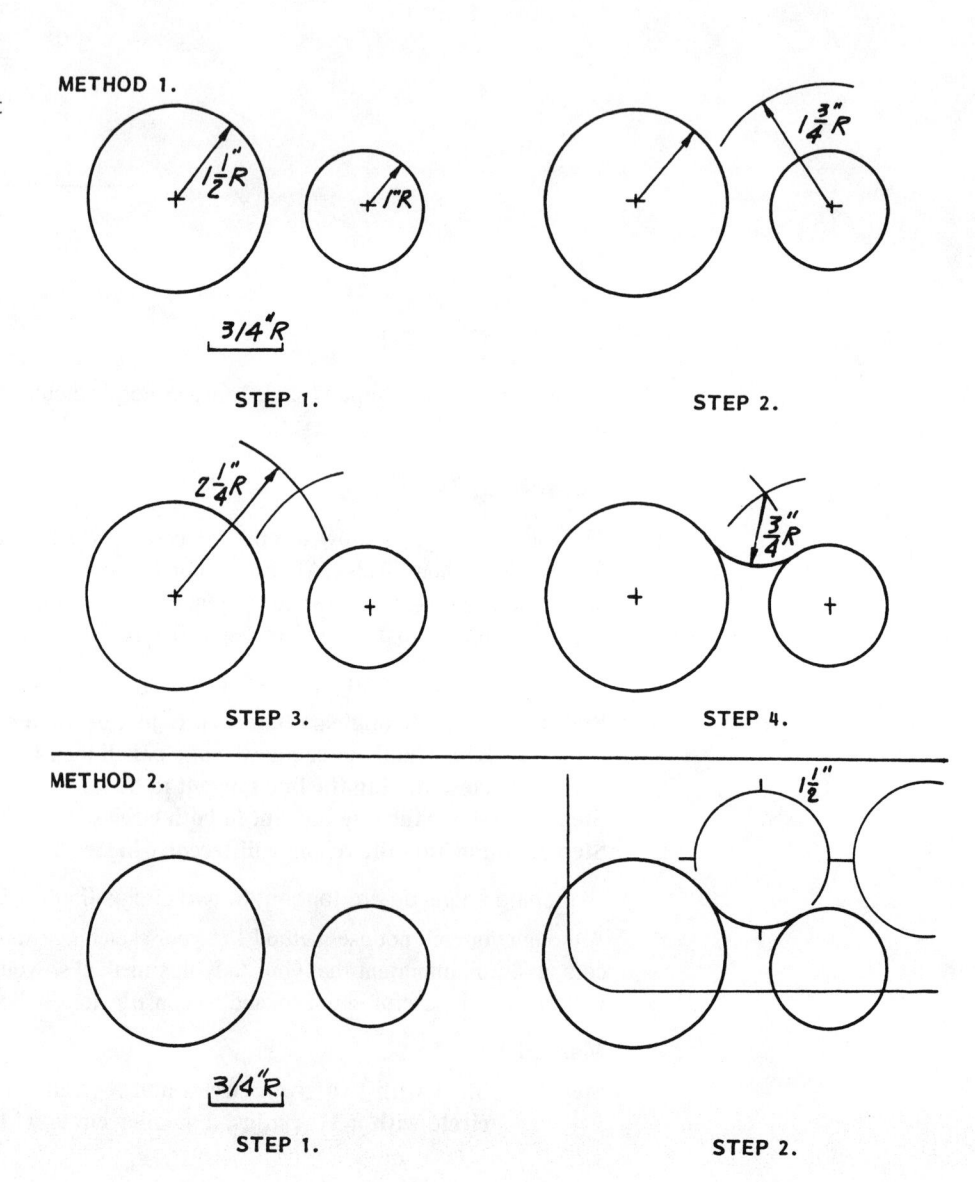

METHOD 1.

$1\frac{1}{2}"R$　$1"R$

$3/4"R$

STEP 1.

$1\frac{3}{4}"R$

STEP 2.

$2\frac{1}{4}"R$

STEP 3.

$3/4"R$

STEP 4.

METHOD 2.

STEP 1.

$3/4"R$

$1\frac{1}{2}"$

STEP 2.

56　　　　Chapter 5

for the arc to be sketched tangent). The circles must be close enough to allow the arc to touch.

Step 2. From the center of the 1″-radius circle, sketch an arc with a radius that is 1″ (the small circle radius) plus $\frac{3}{4}″$ (the radius of the arc to be sketched).

Step 3. From the center of the $1\frac{1}{2}″$-radius circle sketch an arc with a radius that is $1\frac{1}{2}″$ (large circle radius) plus $\frac{3}{4}″$ (radius of the arc to be sketched).

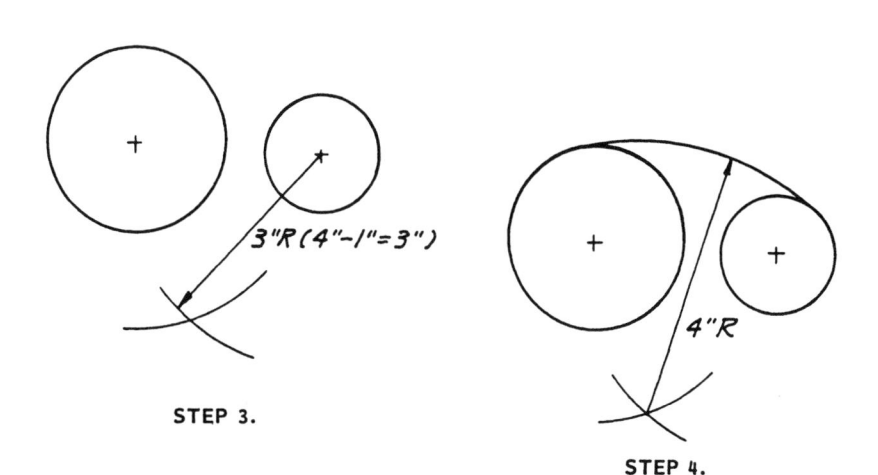

Step 4. From the intersection of the two arcs, sketch an arc with a radius of $\frac{3}{4}″$. This arc will be exactly tangent to both circles.

Method 2

Step 1. Start with the same given circles and radius used in method 1.

Step 2. With a circle template, place a circle hole with a $\frac{3}{4}″$ radius ($1\frac{1}{2}″$ diameter) tangent to both circles, and draw the radius (be sure you make allowance for the width of the pencil lead).

Sketching an outside arc tangent to two circles (Figure 5–26):

Although you will not use this technique in your sketching exercises because it requires a compass, it is important that you study this method so you will know how it is done. You will find it to be useful in understanding constructions in both sketching and AutoCAD.

Step 1. Start with two given circles and a given radius (in the example, a larger circle with a $1\frac{1}{2}″$ radius, a smaller circle with a 1″ radius, and a 4″ radius for the arc to be sketched tangent). The radius must be large enough to touch the outside edges of both circles.

Step 2. From the center of the $1\frac{1}{2}″$-radius circle, sketch an arc with a radius that is $2\frac{1}{2}″$ (the 4″ radius minus the radius of the $1\frac{1}{2}″$-radius circle).

Step 3. From the center of the 1″-radius circle sketch an arc with a radius of 3″ (the 4″-radius minus the radius of the 1″-radius circle).

FIGURE 5–26
Sketching an Outside Arc Tangent to Two Circles

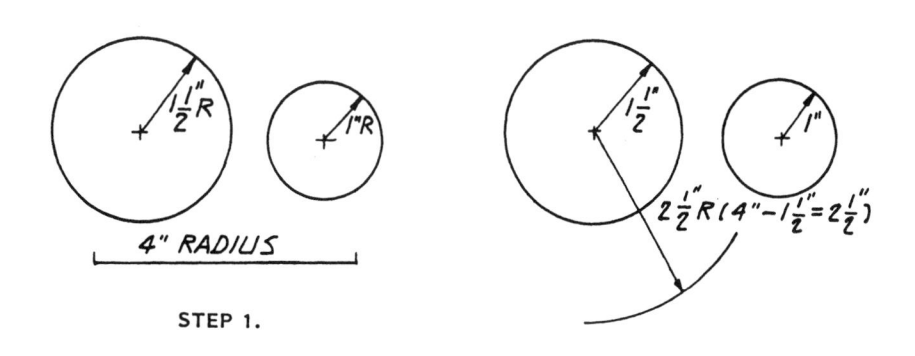

Sketching Line Weights and Drawing Constructions

Step 4. From the intersection of the two arcs, sketch an arc with a radius of 4″. This arc will be exactly tangent to both circles.

Sketching an arc (fillet) tangent to lines at right angles (Figure 5–27):

Method 1

Step 1. Start with given lines AB and CD at right angles and a given radius ($\frac{1}{2}″$ in the example).
Step 2. Draw construction lines parallel to lines AB and CD $\frac{1}{2}″$ from the lines.
Step 3. From the intersection of the construction lines, draw a $\frac{1}{2}″$ radius arc, which will form the fillet.

Method 2

Step 1. Start with the same given lines and the radius used in method 1.
Step 2. With a circle template, place a circle with a $\frac{1}{2}″$ radius (1″ diameter) in the correct position, and draw the radius tangent to lines AB and CD.

Sketching an arc (fillet) tangent to lines at any angle (Figure 5–28):

Method 1

Step 1. Start with given lines AB and CD that will intersect if extended and a given radius ($\frac{1}{2}″$ in the example).

FIGURE 5–27
Sketching a Fillet Tangent to Two Lines at Right Angles

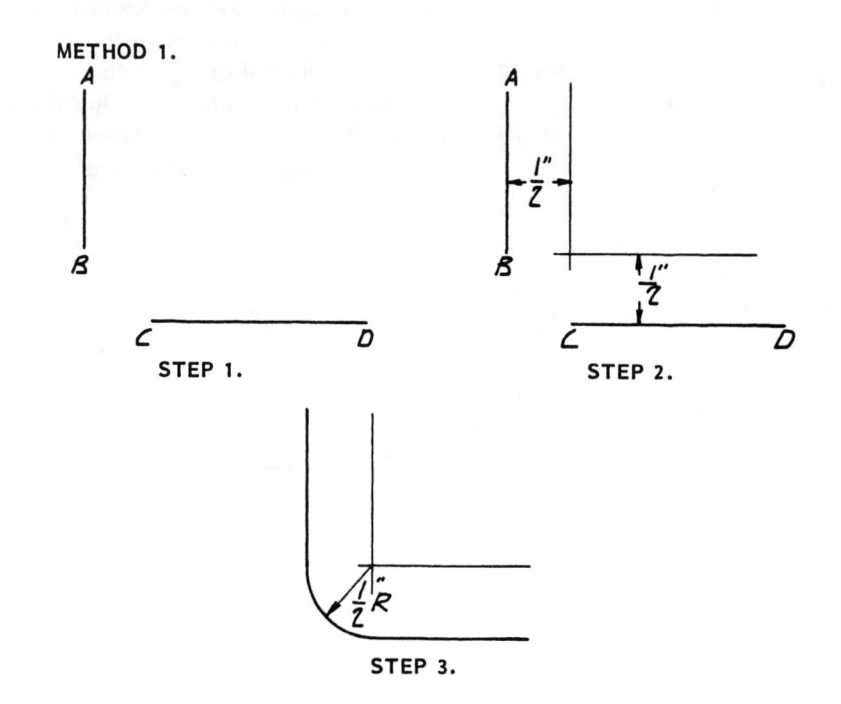

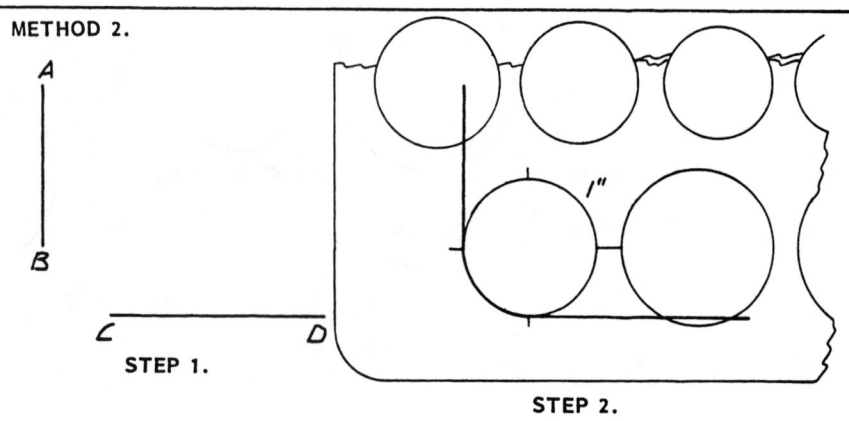

FIGURE 5-28
Sketching a Fillet Tangent to Lines at Any Angle

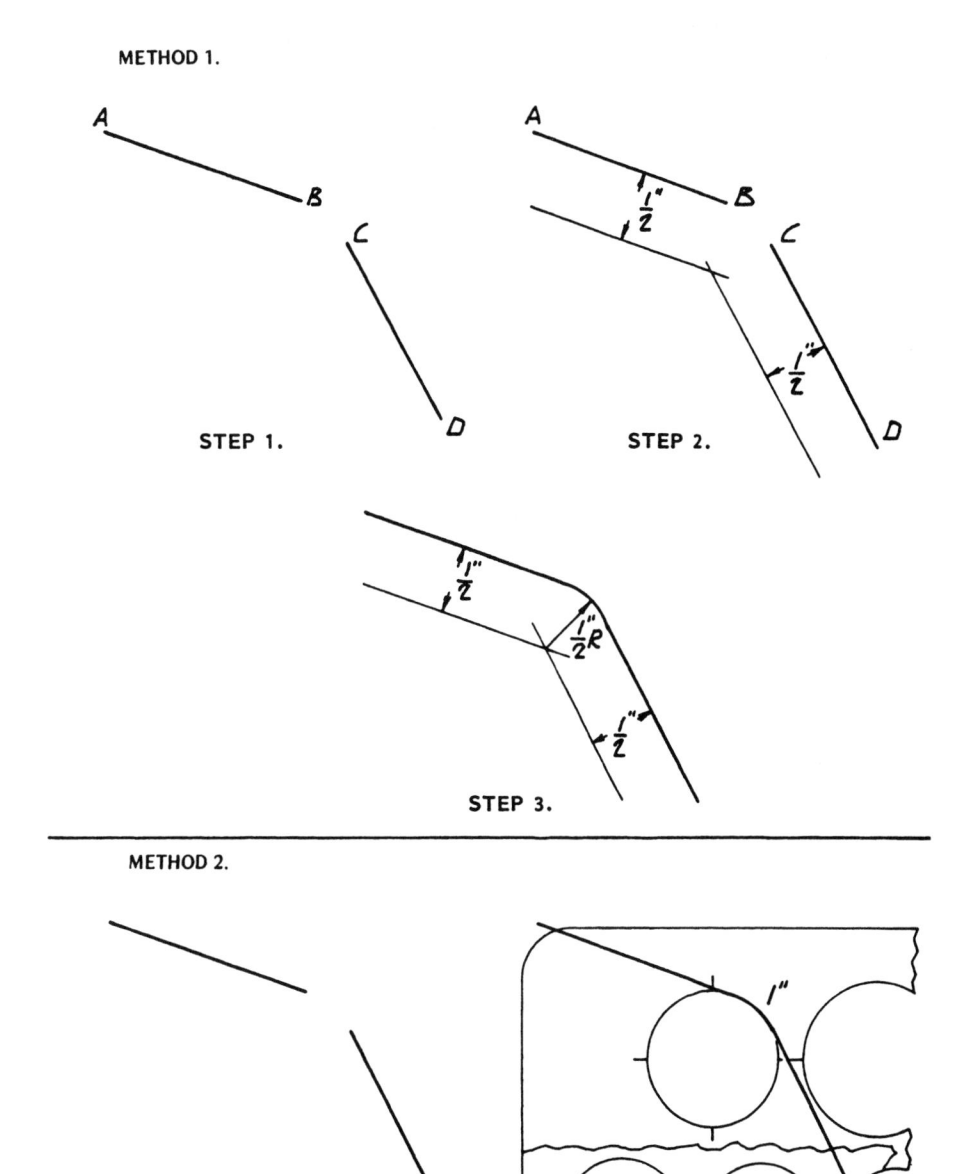

METHOD 1.

STEP 1.   STEP 2.

STEP 3.

METHOD 2.

STEP 1.   STEP 2.

Step 2. Draw construction lines parallel to lines AB and CD $\frac{1}{2}''$ from the lines.

Step 3. From the intersection of the construction lines, draw a $\frac{1}{2}''$-radius arc, which will form the fillet.

Method 2

Step 1. Start with the same given lines and the radius used in method 1.

Step 2. With a circle template, place a circle with a $\frac{1}{2}''$ radius (1" diameter) in the correct position, and draw the radius tangent to lines AB and CD.

Sketching an arc of a given radius tangent to a straight line and a circle (Figure 5–29):

Method 1

Step 1. Start with a given circle ($1\frac{1}{2}''$ radius in the example), a straight line, and a given radius ($\frac{3}{4}''$ in the example).

Step 2. From the center of the circle, draw an arc that has a radius of $1\frac{1}{2}''$ (the radius of the circle) plus $\frac{3}{4}''$ (the radius of the arc to be drawn).

Sketching Line Weights and Drawing Constructions

FIGURE 5–29
Sketching an Arc of a Given Radius Tangent to a Circle and a Line

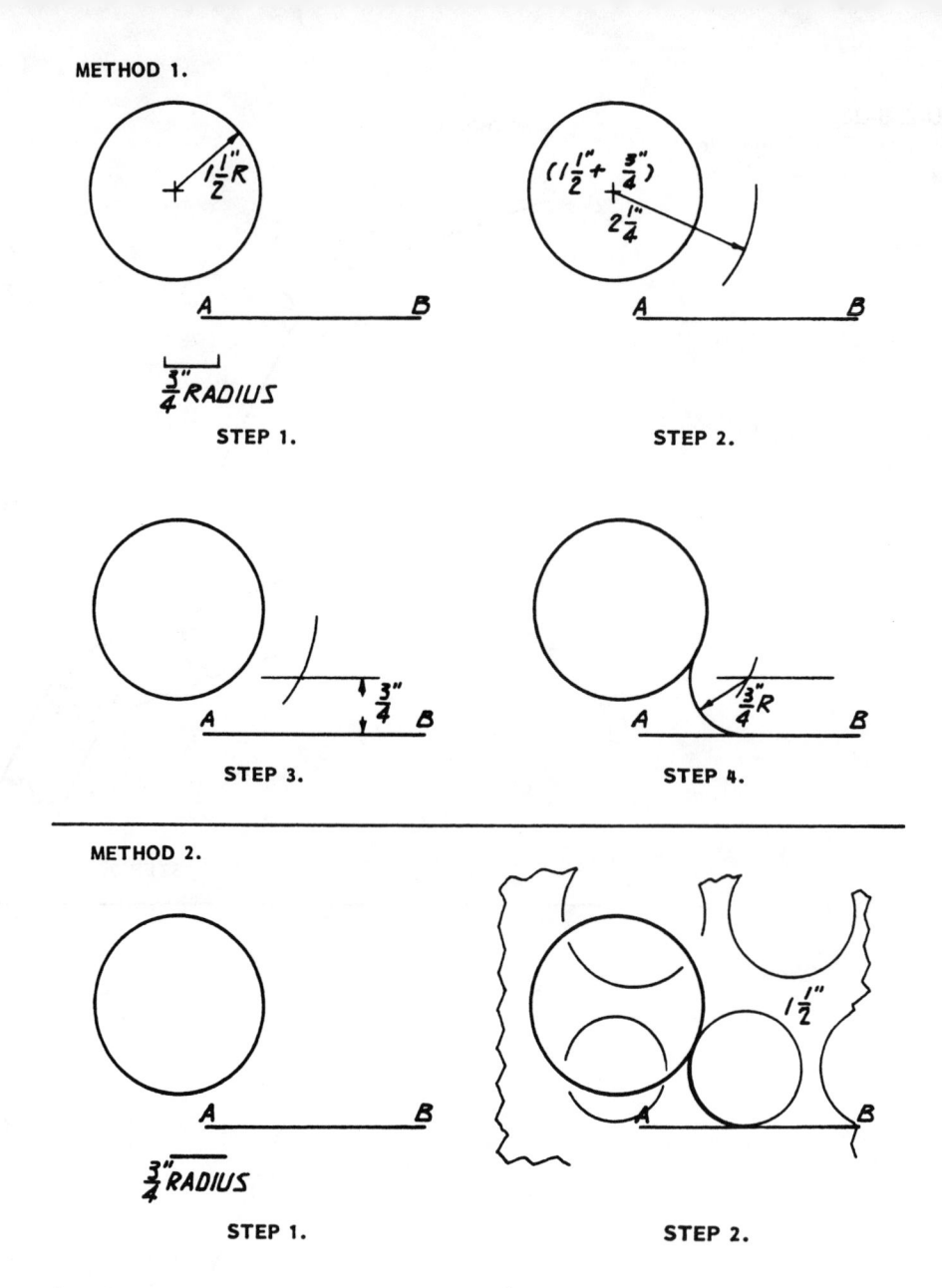

METHOD 1.

STEP 1.

STEP 2.

STEP 3.

STEP 4.

METHOD 2.

STEP 1.

STEP 2.

Step 3. Draw a line parallel to line AB ¾" away from it to intersect the arc.
Step 4. From the intersection, draw the ¾" radius tangent to the line and the circle.

Method 2

Step 1. Start with the same given circle, line, and radius used in method 1.
Step 2. With a circle template, place a circle with a ¾" radius (1½" diameter) tangent to the circle and the line and draw the ¾" radius.

Sketching an arc tangent to two parallel lines (Figure 5–30):

Method 1

Step 1. Start with two given parallel lines (2½" apart in the example).
Step 2. Draw a perpendicular that crosses both lines.
Step 3. From each intersection, draw an arc with a radius greater than half the distance between the two lines.
Step 4. Draw a line through the intersection of the two arcs. Then, using the distance AB as a radius and point A as the center, draw the arc tangent to the two parallel lines.

Method 2

Step 1. Start with the same given parallel lines used in method 1.

FIGURE 5–30
Sketching an Arc Tangent to Two Parallel Lines

METHOD 1.

STEP 1.

STEP 2.

STEP 3.

STEP 4.

METHOD 2.

STEP 1.

STEP 2.

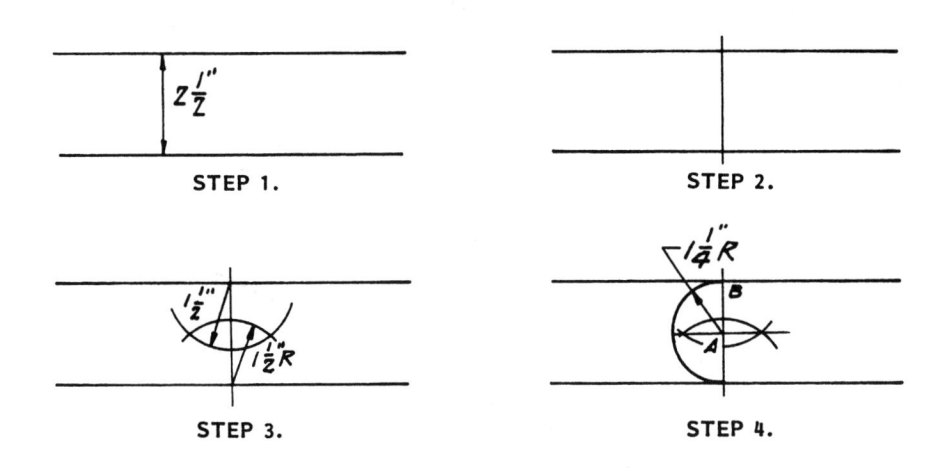

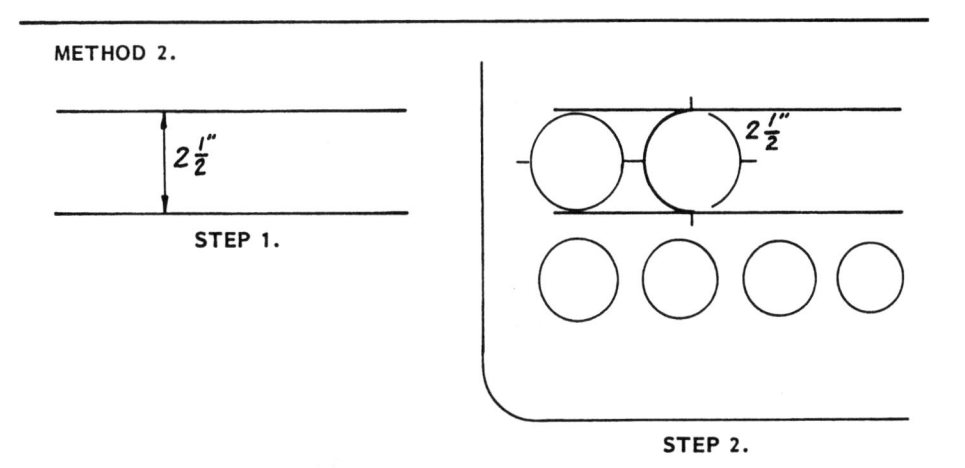

Step 2. With a circle template, place a circle with a $2\frac{1}{2}''$ diameter between the lines to draw the arc tangent to the two lines.

Dividing Lines and Angles

When you are sketching it is sometimes necessary to divide a line into a specified number of parts or to bisect an angle into two equal parts. AutoCAD makes this very easy with the Divide command, but when you are sketching, you can use the following methods:

Dividing a line into any number of equal parts (Figure 5–31):

Method 1

Step 1. Start with a line AB of any length.
Step 2. Draw a line perpendicular to one end of line AB.
Step 3. On a scale, select any convenient length that has equal units (six in the example) and place it so that one end is at point A and the other end lies on the perpendicular line. Make marks on the paper at the six unit points.
Step 4. Draw perpendicular lines through the six marks to intersect line AB and divide it into six equal parts.

Method 2

Step 1. Start with a line AB of any length.
Step 2. Place a scale with six equal units on it at a convenient angle to point A. Make marks on the paper at the six unit points.
Step 3. Connect the last unit (0 in the example) with point B, and draw lines parallel to line OB to divide line AB into six equal parts.

Sketching Line Weights and Drawing Constructions

FIGURE 5–31
Dividing a Line into Any Number of Equal Parts

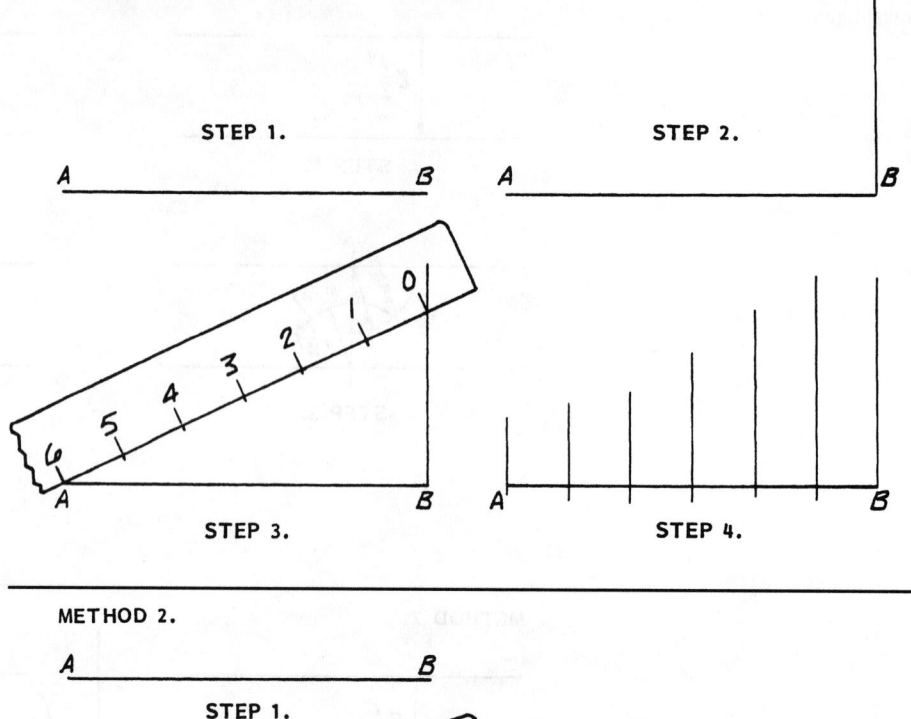

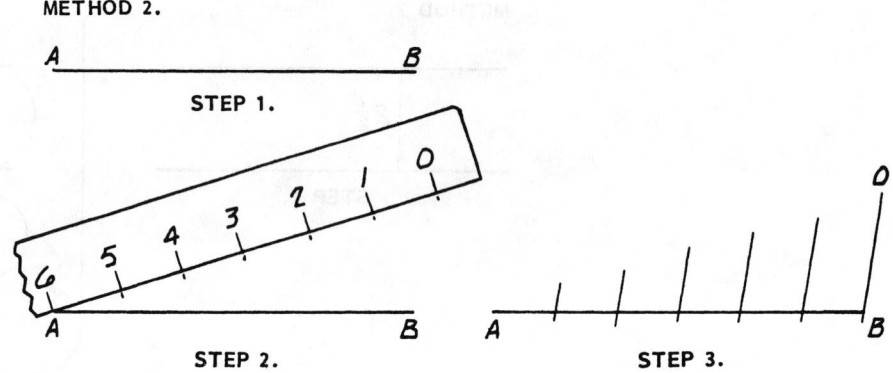

METHOD 1.

STEP 1.

STEP 2.

STEP 3.

STEP 4.

METHOD 2.

STEP 1.

STEP 2.

STEP 3.

Dividing a line into proportional parts (Figure 5–32):

Step 1.   Start with a line AB of any length.
Step 2.   Place a scale with six equal units on it at a convenient angle to point A. Make marks on the paper at the specified unit points (5 units, 3 units, 2 units, and 1 unit in the example).
Step 3.   Connect the last unit (0 in the example) with point B, and draw lines parallel to line OB to divide line AB into parts proportional to 5, 3, 2, and 1.

Bisecting an angle (Figure 5–33):

Step 1.   Start with a given angle (52° in the example).

FIGURE 5–32
Dividing a Line into Proportional Parts

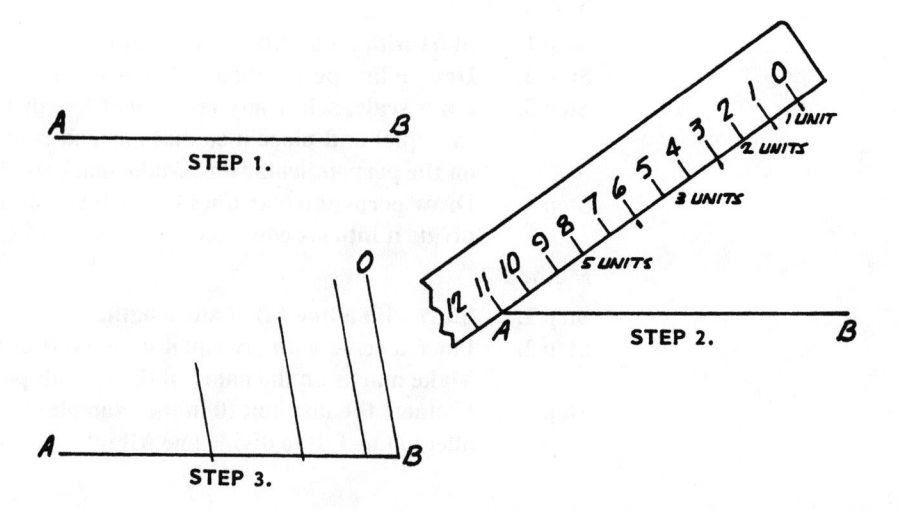

STEP 1.

STEP 2.

STEP 3.

Chapter 5

FIGURE 5–33
Bisecting an Angle

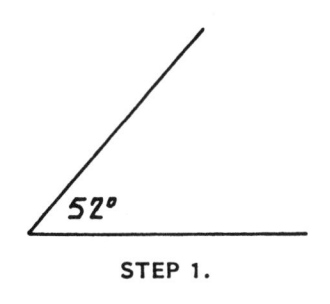

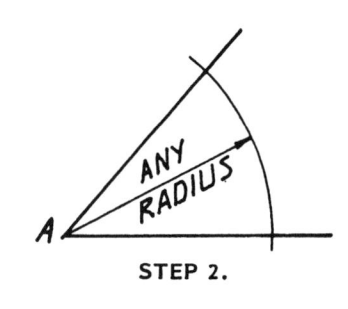

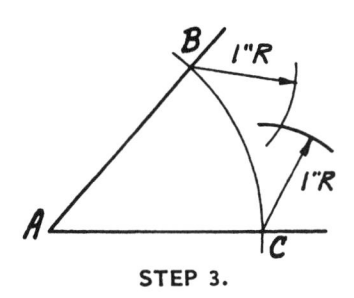

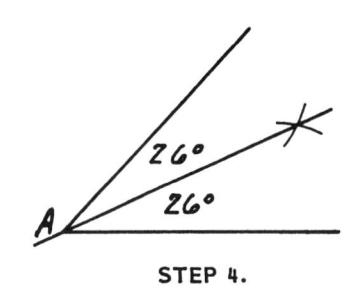

Step 2. Draw a radius from the start point of the angle (A) to intersect the sides of the angle.
Step 3. Using the intersections B and C as centers, draw arcs of equal radius.
Step 4. Draw a line from the intersection of the arcs to point A to bisect the angle.

Sketching Polygons

Sketching polygons requires considerable construction. AutoCAD has a Polygon command that makes drawing polygons very easy. The sketching technique is described next.

Inscribing a hexagon in a circle (Figure 5–34) (Inscribing means drawing the hexagon inside the circle.):

Step 1. Start with a given circle.
Step 2. Draw a horizontal line through the center of the circle to the outside edge.
Step 3. Using a 30-60° triangle as shown, draw lines from the points at which the horizontal line intersects the circle.
Step 4. Complete the hexagon by drawing horizontal lines at top and bottom.

Circumscribing a hexagon around a circle (Figure 5–35) (Circumscribing means drawing the hexagon outside the circle and tangent to it.):

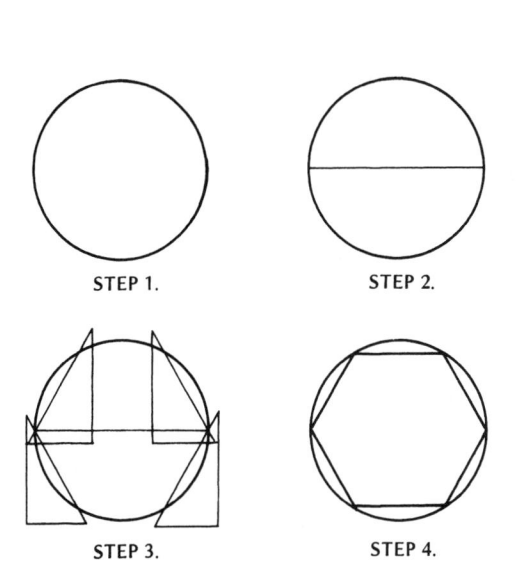

FIGURE 5–34
Inscribing a Hexagon in a Circle

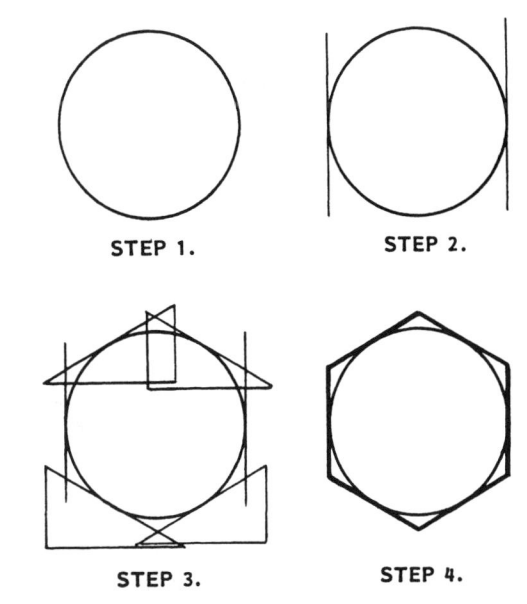

FIGURE 5–35
Circumscribing a Hexagon Around a Circle

Sketching Line Weights and Drawing Constructions

63

FIGURE 5–36
Inscribing an Octagon in a Circle

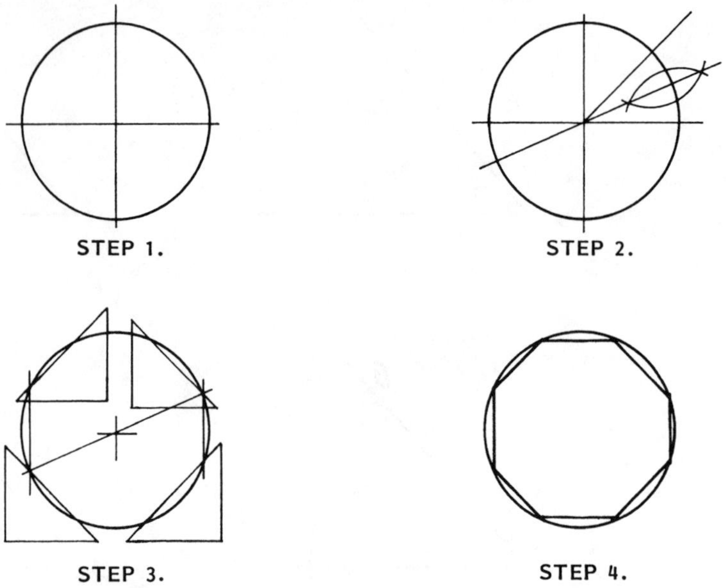

STEP 1.
STEP 2.
STEP 3.
STEP 4.

Step 1. Start with a given circle.
Step 2. Draw vertical lines tangent to the circle.
Step 3. Using a 30-60° triangle as shown, draw lines tangent to the circle.
Step 4. Darken the lines to complete the hexagon.

Inscribing an octagon in a circle (Figure 5–36):

Step 1. Start with a given circle and draw perpendicular construction lines through the center for reference.
Step 2. Draw a $22\frac{1}{2}°$ line through the center (bisect a 45° angle).
Step 3. Where the $22\frac{1}{2}°$ line intersects the circle, draw 90° and 45° angles. (Use 45° triangles as shown.)
Step 4. Draw horizontal lines at top and bottom to complete the octagon.

Circumscribing an octagon around a circle (Figure 5–37):

Step 1. Start with a given circle.
Step 2. Use the 45° triangle in the position shown to draw 45° and 90° angles tangent to the circle.
Step 3. Draw horizontal lines at top and bottom to complete the octagon.

FIGURE 5–37
Circumscribing an Octagon around a Circle

STEP 1.
STEP 2.
STEP 3.

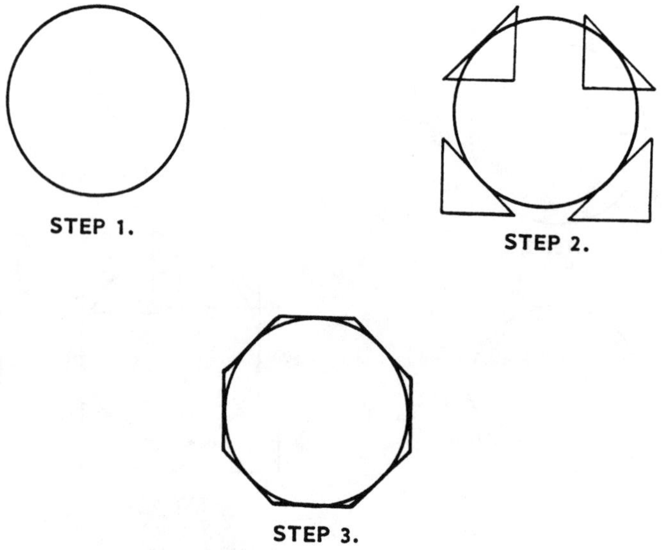

FIGURE 5-38
Dimensions for Exercise 5-1

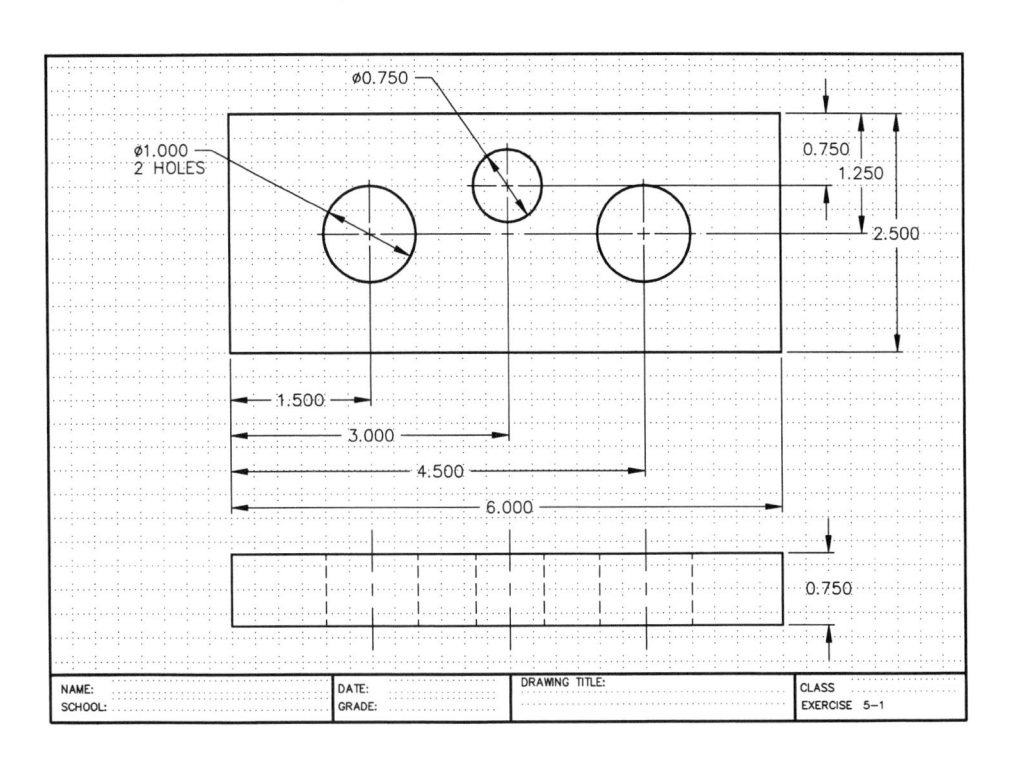

EXERCISES

EXERCISE 5-1   Use the sheet in your book labeled EXERCISE 5-1 for this exercise. Draw Exercise 5-1 on the sheet in your book using the dimensions shown in Figure 5-38. Make sure that your lines are the correct weight and are of even width and darkness. Use guidelines for lettering and make your lettering the best you can. Draw and letter everything that is shown in Figure 5-38. Refer to Figure 5-1 for correct line weights.

EXERCISE 5-2   Use the sheet in your book labeled EXERCISE 5-2 for this exercise. Divide the drawing area into four equal parts as shown in Figure 5-39 and make the required constructions. Center your construction in the areas. Darken the object lines and the division marks. Leave very light construction lines. Fill in the title block information; title the drawing LINES.

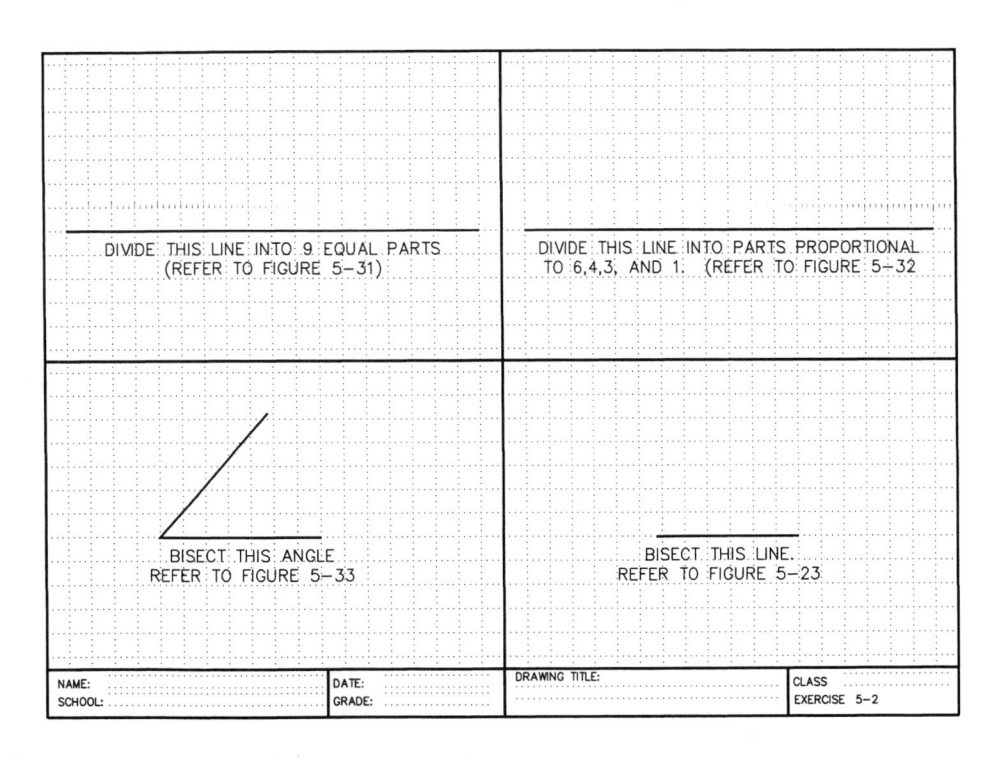

FIGURE 5-39
Instructions for Exercise 5-2

Sketching Line Weights and Drawing Constructions

FIGURE 5–40
Instructions for Exercise 5–3

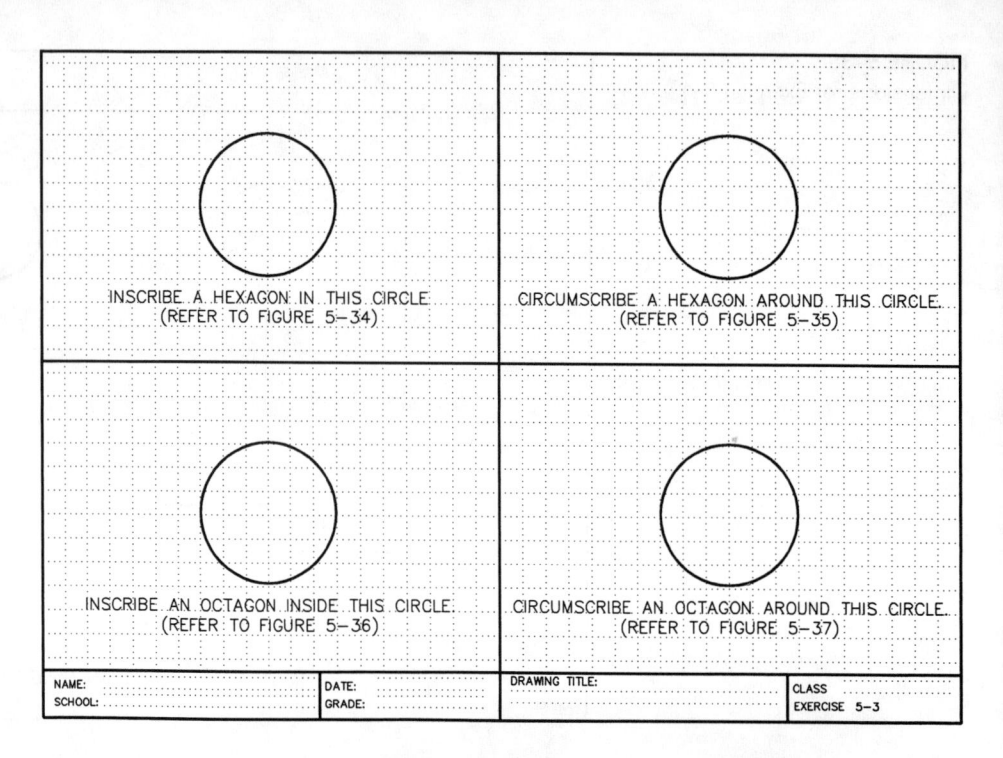

EXERCISE 5–3   Use the sheet in your book labeled EXERCISE 5–3 for this exercise. Divide the drawing area into four equal parts as shown in Figure 5–40 and draw the polygons described in that figure. Center your construction in the areas. Darken the object lines and leave very light construction lines. Fill in the title block information; title the drawing POLYGONS.

EXERCISE 5–4   Draw Exercise 5–4 on the sheet in your book using the dimensions shown in Figure 5–41. Make sure that your lines are the correct weight and are of even width and darkness. Do not show any dimensions or other lettering on the drawing itself. Fill in the title block using your best lettering; title the drawing PLATE. Use construction methods shown in Figures 5–24 through 5–30.

EXERCISE 5–5   Draw Exercise 5–5 on the sheet in your book at half scale of the dimensions shown in Figure 5–42. Make sure that your lines are the correct weight and are of even width and darkness. Do not show any dimensions or other lettering on the drawing itself. Fill in the title block using your best lettering; title the drawing GASKET. Use construction methods shown in Figures 5–24 through 5–30.

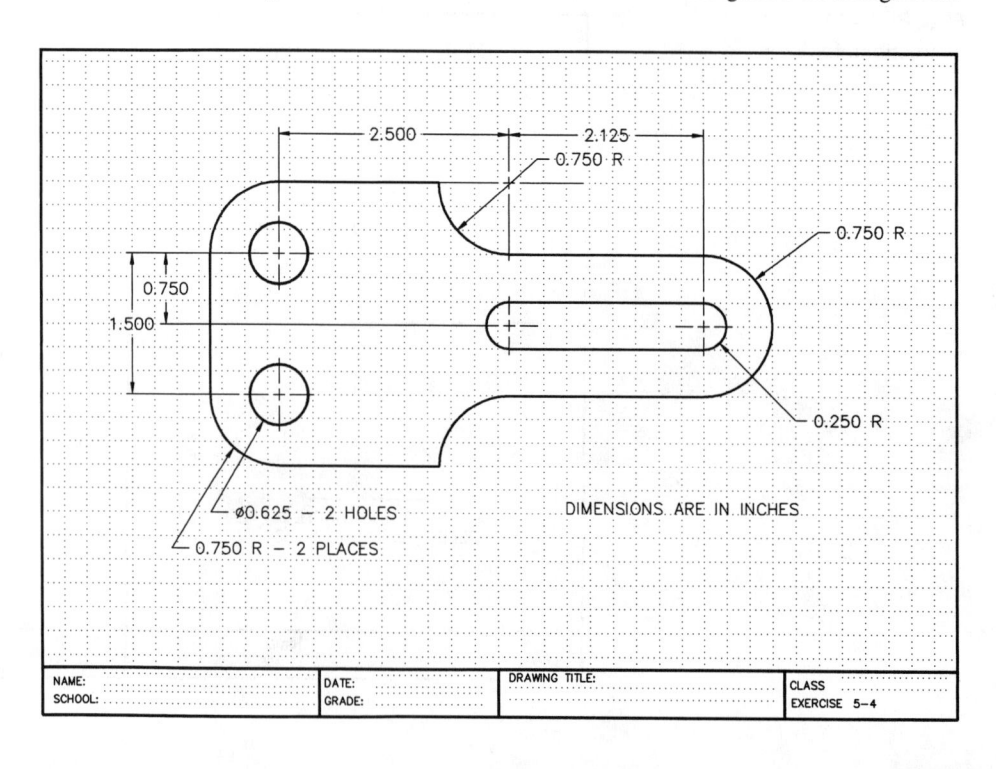

FIGURE 5–41
Dimensions for Exercise 5–4

66   Chapter 5

FIGURE 5–42
Dimensions for Exercise 5–5

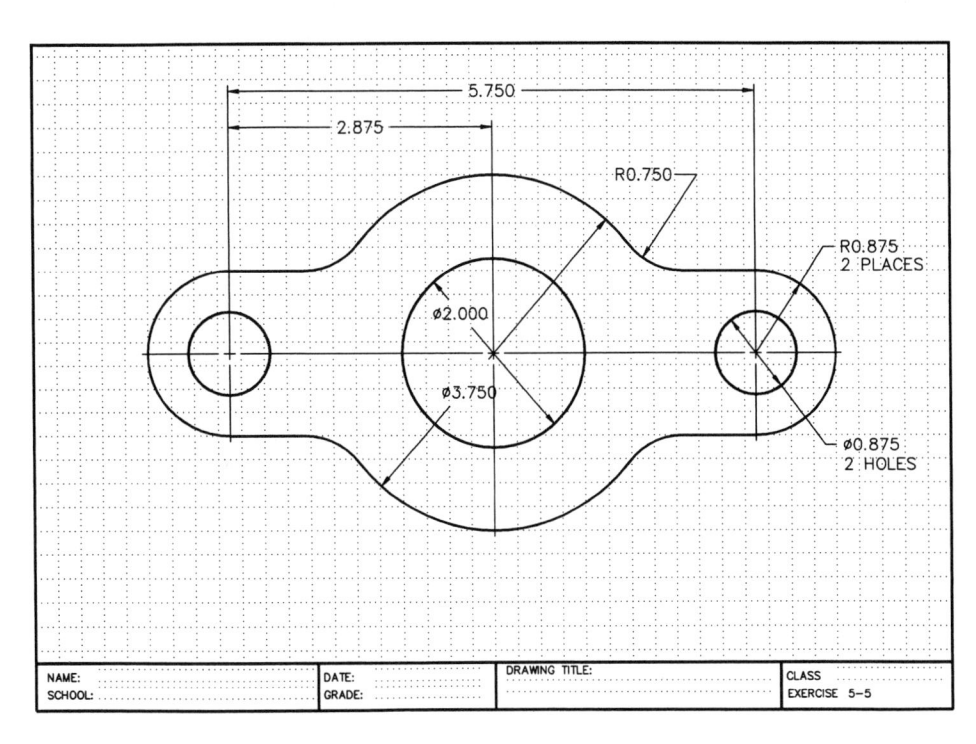

FIGURE 5–43
Dimensions for Exercise 5–6

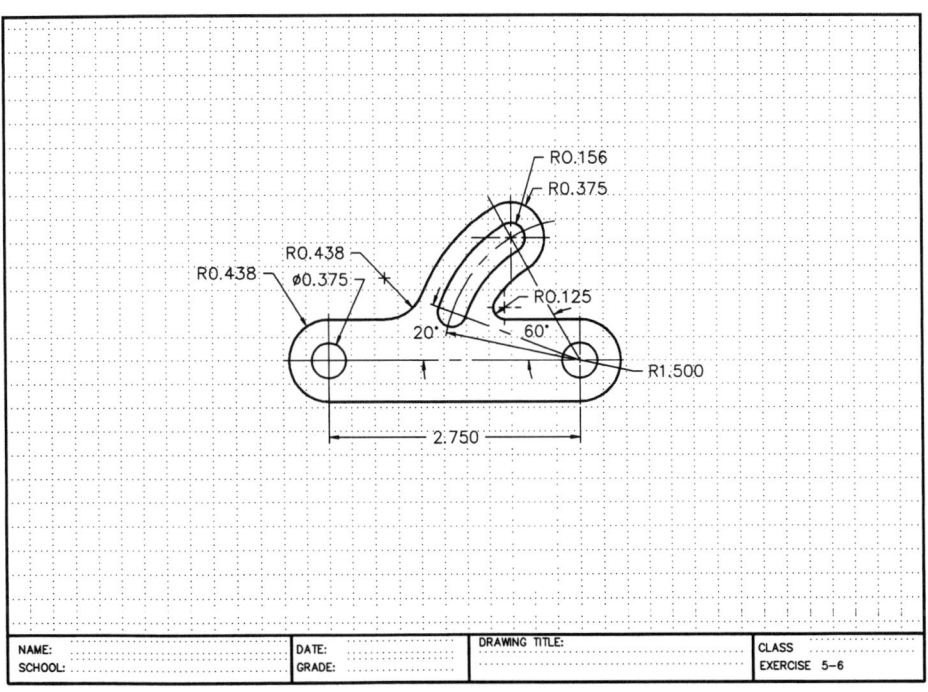

EXERCISE 5–6   Draw Exercise 5–6 on the sheet in your book using the dimensions shown in Figure 5–43. Make sure that your lines are the correct weight and are of even width and darkness. Do not show any dimensions or other lettering on the drawing itself. There are four large arcs that you will have to sketch without using a circle template. Use your scale and make several dots along the path of the arc to aid you in sketching. Fill in the title block using your best lettering; title the drawing, STOP. Use construction methods shown in Figures 5–24 through 5–30.

REVIEW QUESTIONS

Circle the best answer.

1. Construction lines and guidelines for lettering should be
   a. Thin and very dark
   b. Thick and very dark
   c. Thin and very light
   d. Thick and very light
   e. Any of the above is OK.

Sketching Line Weights and Drawing Constructions

2. Hidden lines are shown with
   a. A series of short dashes
   b. A long dash followed by a short dash
   c. Solid lines
   d. Two short dashes and a long dash
   e. Hidden lines are not shown.
3. Centerlines, hatch lines, extension lines, and dimension lines are drawn thin and very dark.
   a. True   b. False
4. A cutting-plane line is drawn
   a. Thin and very dark
   b. Very thick and very dark
   c. Thin and very light
   d. Very thick and very light
   e. Any of the above is OK.
5. A good lead softness to use for object lines is
   a. H   c. 4H   e. Any of the above is OK.
   b. B   d. 6H
6. A good lead thickness for object lines is
   a. 0.25 mm   c. 0.5 mm   e. 1.0 mm
   b. 0.35 mm   d. 0.7 mm
7. The distance from the center of a circle to the outside edge is called
   a. A diameter   c. A circumference   e. A tangent
   b. A radius   d. An arc
8. The circles on a circle template are measured in
   a. Diameters   b. Radii
9. Lines that are the same distance apart along all parts of the lines are
   a. Parallel   c. Tangent   e. Intersecting
   b. Perpendicular   d. Proportional
10. Lines that are at 90° to each other are
    a. Parallel   c. Tangent   e. Intersecting
    b. Perpendicular   d. Proportional
11. Lines that are tangent
    a. Lie very close to each other
    b. Just touch at the edges
    c. Lie exactly on top of each other at one point
    d. Do not touch
    e. Are perpendicular
12. A 90° angle that is bisected will result in
    a. Two angles greater than 90°
    b. Two 45° angles
    c. No angle at all
    d. Two 90° angles
    e. Two 30° angles
13. A right angle is
    a. 90°   c. Less than 90°   e. Either a 45° or a 30° angle
    b. Greater than 90°   d. Any angle correctly drawn
14. An obtuse angle is
    a. 90°   c. Less than 90°   e. Either a 45° or a 30° angle
    b. Greater than 90°   d. Any angle correctly drawn
15. An acute angle is
    a. 90°
    b. Greater than 90°
    c. Less than 90°
    d. Any angle correctly drawn
    e. Either a 45° or a 30° angle
16. A circumference is
    a. The distance across a circle
    b. The distance from the center of a circle to the outside edge
    c. The distance around the outside of a circle
    d. The form of an arc
    e. The area of a circle
17. A fillet is
    a. Another name for diameter
    b. An arc forming the corner of two lines
    c. An intersection
    d. A complete circle
    e. A form of polygon
18. A hexagon has
    a. four sides   c. six sides   e. Any number of sides greater than three
    b. five sides   d. eight sides
19. An octagon has
    a. four sides   c. six sides   e. Any number of sides greater than three
    b. five sides   d. eight sides
20. Which AutoCAD command will divide a line into equal parts?
    a. Divide   c. Edit   e. Offset
    b. Proportional   d. Line

6 Linetypes and Drawing Constructions Using AutoCAD

OBJECTIVES

After completing this chapter, you will be able to

- Make drawings using continuous, hidden, and center linetypes.
- Correctly answer questions regarding linetypes and drawing constructions in AutoCAD.
- Use the following commands to produce drawings:
Line	Trim
Polyline ID	Offset
Osnap	Mirror
Circle	Fillet
Zoom	Chamfer
- Answer questions regarding the commands listed above.

LINETYPES

Chapter 3 describes the concept of layers and how to assign colors and linetypes to them. This chapter shows you how to use AutoCAD commands to make drawings containing different linetypes. The drawings on the disk that came with your book already have layers created and linetypes assigned. When you need to draw lines with, for example, the HIDDEN linetype, you will change the current layer to one that has the HIDDEN linetype assigned to it. AutoCAD has many standard linetypes and a means of creating custom linetypes if needed. You will not need to create a custom linetype to complete the drawings in this book.

DRAWING CONSTRUCTIONS

There are many ways to use AutoCAD to make drawings. The commands described here are not the only ones that can be used to make the drawings in this chapter, but they are the ones commonly used. Every feature of each command is not described in detail. The purpose of these exercises is to introduce you to the AutoCAD program while you learn drafting fundamentals. The commands you will use in this chapter are described next.

Line and Pline (Polyline)

The Line and Pline commands can be used to draw lines very accurately. Lines drawn with the Line command are separate lines, so that if you click any one of them to move, erase, or otherwise modify it, only the line clicked is modified. Lines drawn with the Pline command are all connected, so that if any one of them is clicked, all of them are selected to be modified. There are four common ways to draw lines with either of these commands. There is a fifth method, called direct distance entry, which will be discussed in a later chapter.

Drawing Lines Using the Grid Marks

Lines can be drawn by snapping to the grid marks visible on the screen.

While in the Line command, if you decide you do not like the last line segment drawn, Type: **U↵** to erase it and continue on with the Specify next point: prompt. Clicking more than one undo will backtrack through the line segments in the reverse order in which they were drawn.

The Line command has a very handy feature. If you respond to the prompt Specify next point: by pressing the Enter key or the space bar, the line will start at the end of the most recently drawn line. The "continue" feature of the Line command will do the same.

Drawing Lines Using Absolute Coordinates

Remember, 0,0 is the lower left corner of the page. When using absolute coordinates to draw, you enter the X axis coordinate first. It identifies a location on the horizontal axis. You enter the Y axis coordinate second. It identifies a location on the vertical axis. An absolute coordinate of 1,2 commands the line to move to a location that is 1″ in the X direction and 2″ in the Y direction from the lower left corner of the page.

Drawing Lines Using Relative Coordinates

Relative coordinates are taken from a point entered. (Relative to what? Relative to the point just entered.) After clicking a point on the drawing, enter relative coordinates by typing "@", followed by the X,Y coordinates. For example, after entering a point to start a line, type and enter **@1,0** to draw the line 1″ in the X direction, 0″ in the Y direction.

A minus sign (−) is used for negative line location with relative coordinates. Negative is to the left for the X axis or down for the Y axis.

Drawing Lines Using Polar Coordinates

Absolute and relative coordinates are extremely useful in some situations; however, polar coordinates are used for many design applications. Be sure you understand how to use all three types of coordinates.

Polar coordinates are relative to the last point entered also. They are typed starting with an @, followed by a distance and angle of direction. Figure 6–1 shows the polar coordinate angle directions. The angle direction is preceded by a < sign. A polar coordinate of @4<0 commands the line to move 4″ to the right of the previous point. A polar coordinate of @4<270 commands the line to move 4″ downward from the previous point.

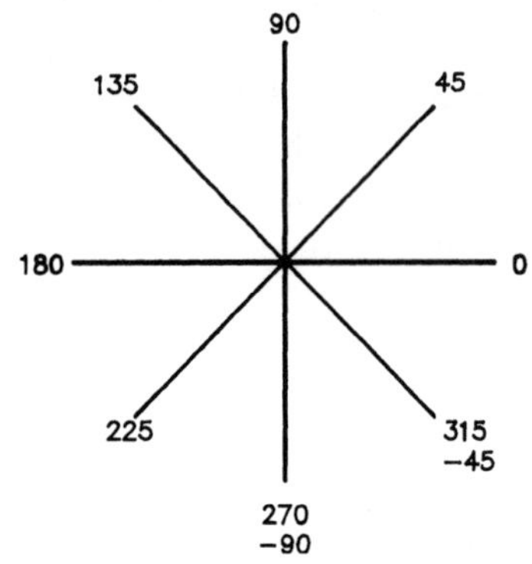

ID

ID identifies a location from which relative or polar coordinates may be given. If, for example, you need to draw a circle that is 1.05″ in the X direction and 2.12″ in the Y direction from the intersection of two lines, you can specify that location by first identifying to AutoCAD the intersection by using the ID command and then typing the coordinates @1.05,2.12 when AutoCAD asks for the location of the circle center.

FIGURE 6–1
AutoCAD Polar Directions

OSNAP

It is very important to become familiar with and use Object Snap modes while drawing. If an existing drawing entity is not drawn on a snap point, it is nearly impossible to snap a line or other drawing entity exactly to it. You may try, and think that the two points are connected, but a close examination will reveal they are not. Object snap modes such as Endpoint, Midpoint, and Center can be used to snap exactly to specific points of existing objects in a drawing. Object snap modes need to be used constantly for complete accuracy while drawing. The exercises in this book direct you to use Osnap modes whenever they are needed. You can type the first three letters of any osnap mode to activate it. Osnap modes can also be selected from one of the menus in AutoCAD or AutoCAD LT.

Running Osnap Modes

You may also set a "running" OSNAP mode to be constantly in effect while drawing, until it is disabled. A running OSNAP mode may include one or more modes. When you right-click **OSNAP** from the STATUS bar at the bottom of the screen and Click: **Settings**, the Drafting Settings dialog box appears, with the Object Snap tab active. Any number of OSNAP modes may be checked to be active. For instance if INTersection, MIDpoint, and ENDpoint are all checked, all of them will be in effect. When one or more modes are entered this way, they become the running object snap modes. The running mode can be disabled by clicking OSNAP at the bottom of the screen so all running OSNAP modes are disabled. You can also override the running mode for a single point by selecting another mode or typing the first three letters (followed by ↵) of another mode not included in the running mode. The running mode returns after the single override.

CIRCLE

The Circle command allows you to draw circles with extreme accuracy. There are several options to the Circle command that you will use in the exercises in this book. AutoCAD asks you first to locate the center of the circle, then to specify the radius. If you do not want to divide the diameter of the circle to be specified by 2 to obtain the radius, you can Type: **D**↵ to tell AutoCAD you want to specify the diameter of the circle. Other options of the Circle command will be explained as they are used in the exercises.

TRIM

The Trim command is one that is very useful. It allows you to trim many objects to one or more cutting edges. AutoCAD asks you first, to select cutting edges; second, to press Enter to move to the next phase of the command; third, to select objects to trim. Watch the TRIM prompts carefully. Not until all cutting edges have been selected and the prompt "<Select object to trim>/Undo: appears can you click the objects to trim. When you have many lines to trim, use the TRIM Crossing or All options to select the cutting edges. If you are unable to trim an entity because it does not intersect a cutting edge, use the Erase command.

OFFSET

The Offset command allows you to copy and enlarge or shrink a shape through a point or at a specified distance. AutoCAD asks you first, to specify a distance or select the through default; second, to click the object to be offset; third, to pick the side of the selected object on which you want the shape to appear. You will use the Offset command often in the exercises.

MIRROR

The Mirror command is used to produce a mirror image of one or more selected objects. AutoCAD asks you first, to select the object(s) to be mirrored; second, to define a line that acts as the edge about which the object will be flopped; third, whether you want to delete old objects (the ones you originally selected). You will find the Mirror command to be very useful.

FILLET

The Fillet command is used to connect two lines, arcs, or circles with an arc of a specified radius. Lines cannot be parallel in versions of AutoCAD prior to Release 13, and the specified radius must be smaller than the selected lines. The Polyline option of the Fillet command allows you to fillet all the intersections of a polyline at the same time. The Polyline command will be used in later chapters.

ERASE

You will have occasion to erase something you have done. You can activate the Erase command by typing **E↵**. AutoCAD then asks you to select objects to be erased. You may select each individual object, or you may use a window or crossing window to select objects to be erased. If you select one or more objects that you do not want to erase, you can type **R↵** (or hold down the Shift key) and remove objects from those selected to be erased by clicking them individually or using a window.

ZOOM

The different Zoom commands (All, Center, Extents, Previous, Window, and Scale) control how you view the drawing area on the display screen. Only three of these Zoom commands—Window, All, and Previous—are described because they are the most commonly used.

Zoom-Window

The Zoom-Window command allows you to pick two opposite corners of a rectangular window on the screen. The crosshair of the pointer changes to form a rubber band that shows the size of the window on the screen. The size of the window is controlled by the movement of the pointer. The part of the drawing inside the windowed area is magnified to fill the screen when the second corner of the window is clicked.

Zoom-All

Zoom-All provides a view of the entire drawing area.

Zoom-Previous

Zoom-Previous is a very convenient feature. AutoCAD and AutoCAD LT remember up to 10 previous views. This is especially helpful and timesaving if you are working on a complicated drawing.

Now that you have read some descriptions of several commands, you will complete three exercises that use these commands.

EXERCISE 6–1
Making a Drawing Using Continuous, Hidden, and Center Linetypes

Your final drawing will look like the drawing in Figure 6–2 without dimensions.

Step 1. To begin Exercise 6–1, turn on the computer and start AutoCAD or AutoCAD LT.
Step 2. Open drawing EX6-1 supplied on the disk that came with your book.
Step 3. Use Zoom-All to view the entire drawing area.
Step 4. Use the Line command to draw the object lines in the top and front views.

Prompt	Response
Command:	Type: **L↵**
Specify first point:	Type: **END↵**
of	Click: **the top end of the long vertical line**
Specify next point or [Undo]:	Type: **@6<180↵**
Specify next point or [Undo]:	Type: **@2.5<270↵**
Specify next point or [Close/Undo]:	Type: **END↵**
of	Click: **the bottom end of the longer vertical line**
Specify next point or [Close/Undo]:	↵

FIGURE 6–2
Dimensions for Exercise 6–1

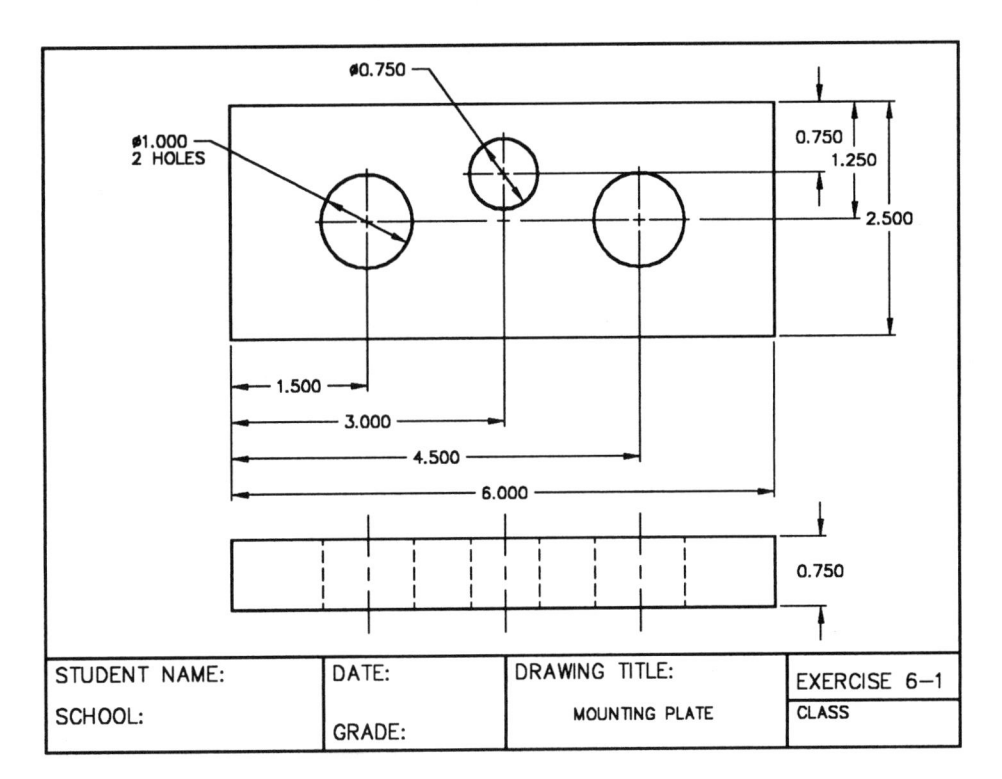

Prompt	Response
Command:	↵
Specify first point:	Type: **END**↵
of	Click: **the top end of the short vertical line**
Specify next point or [Undo]:	Type: **@6<180**↵
Specify next point or [Undo]:	Type: **@.75<270**↵
Specify next point or [Close/Undo]:	Type: **END**↵
of	Click: **the bottom end of the short vertical line**
Specify next point or [Close/Undo]:	↵

Step 5. Use the ID and Circle commands to draw holes in the top view.

Prompt	Response
Command:	Type: **ID**↵
Specify point:	Type: **END**↵
of	Click: **D1** (Figure 6–3)
Command:	Type: **C**↵
Specify center point for circle or [3P/2P/Ttr (tan tan radius)]:	Type: **@1.5,-1.25**↵
Specify radius of circle: [Diameter]:	Type: **.5**↵
Command:	Type: **ID**↵
Specify point:	Type: **END**↵
of	Click: **D1** (again)
Command:	Type: **C**↵
Specify center point for circle or [3P/2P/Ttr (tan tan radius)]:	Type: **@3,-.75**↵
Specify radius of circle or [Diameter]:<0.5000>:	Type: **D**↵

Linetypes and Drawing Constructions Using AutoCAD

FIGURE 6-3

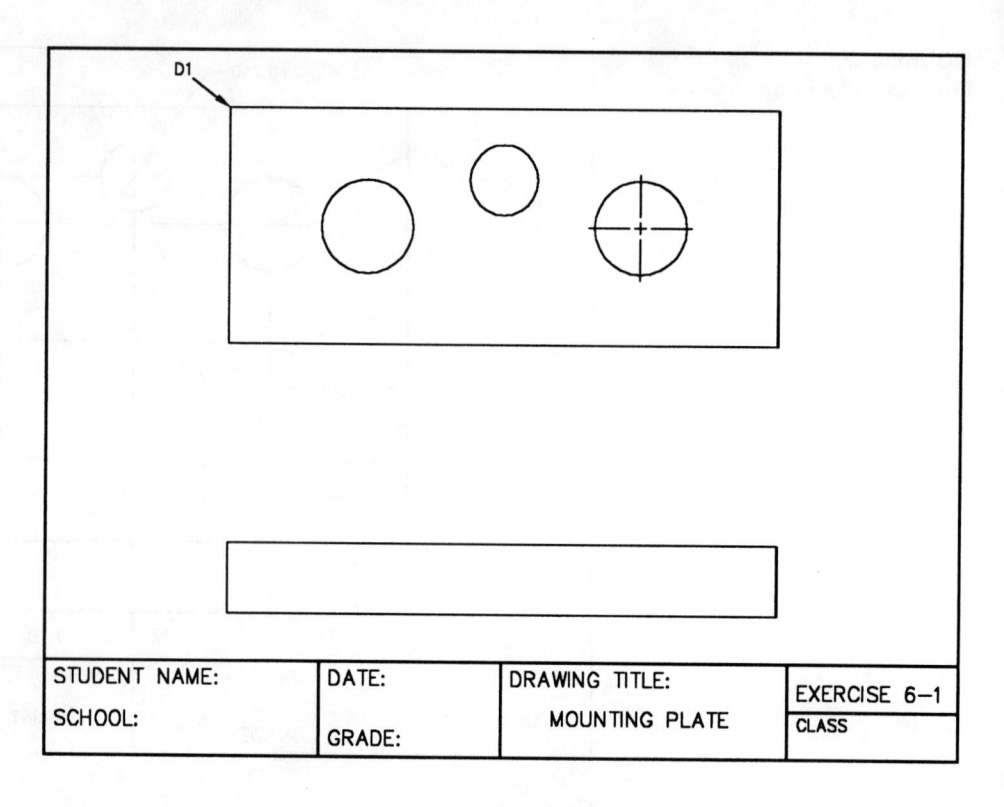

[Figure 6-3: Mounting Plate drawing with title block — STUDENT NAME, SCHOOL, DATE, GRADE, DRAWING TITLE: MOUNTING PLATE, EXERCISE 6-1, CLASS. D1 indicated at upper-left corner of plate.]

Important: Make sure Snap and Ortho are ON (press the function keys F8 and F9 once or twice and test it) so you can accurately line up hidden lines with the circle quadrants.

| Specify diameter of circle <1.000>: | Type: **.75**⏎ |

Step 6. Set the HID layer current and draw hidden lines in the front view.

Prompt	Response
Command:	Type: **-LA**⏎
Enter an option [?/Make/Set/New/ON/OFF/Color/Ltype/LWeight/Plot/Freeze/Thaw/LOck/Unlock/stAte]:	Type: **S**⏎
Enter layer name to make current or <select object>:	Type: **HID**⏎
Enter an option [?/Make/Set/New/ON/OFF/Color/Ltype/LWeight/Plot/Freeze/Thaw/LOck/Unlock/stAte]:	⏎
Command:	Type: **L**⏎
Specify first point:	Click: **D1** (Figure 6–4)
Specify next point or [Undo]:	Click: **D2**
Specify next point or [Undo]:	⏎
Command:	⏎
Specify first point:	Click: **D3**
Specify next point or [Undo]:	Click: **D4**
Specify next point or [Undo]:	⏎
Command:	⏎
Specify first point:	Click: **D5**
Specify next point or [Undo]:	Click: **D6**
Specify next point or [Undo]:	⏎
Command:	⏎

FIGURE 6-4

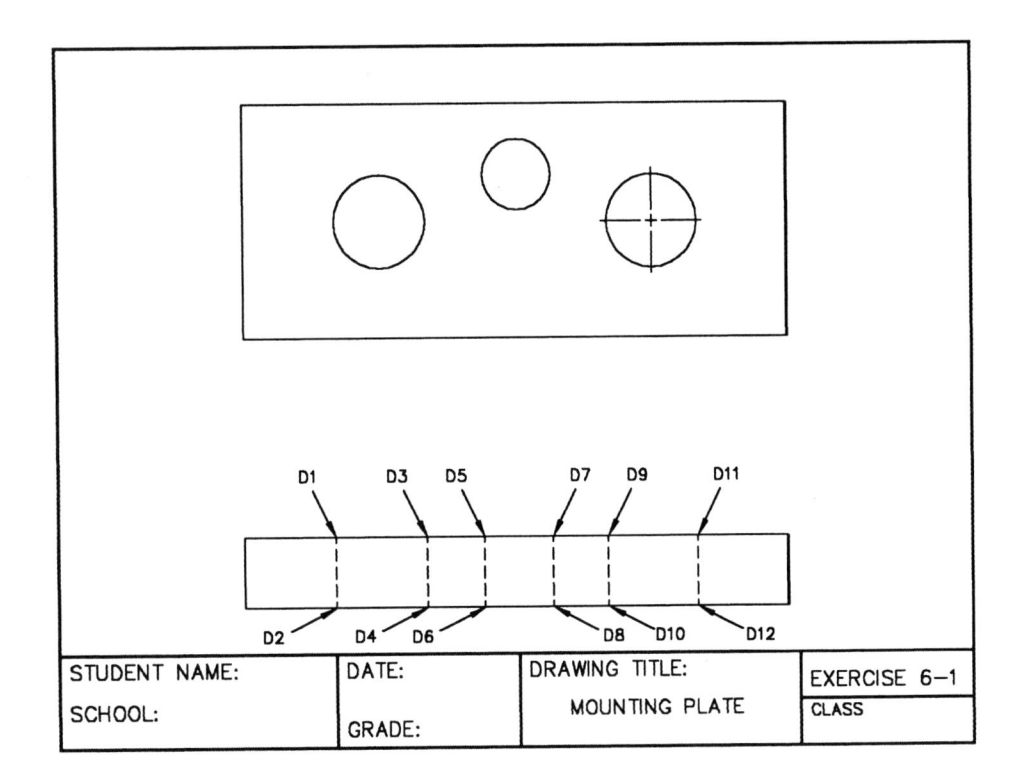

Prompt	Response
Specify first point:	Click: **D7**
Specify next point or [Undo]:	Click: **D8**
Specify next point or [Undo]:	↵
Command:	↵
Specify first point:	Click: **D9**
Specify next point or [Undo]:	Click: **D10**
Specify next point or [Undo]:	↵
Command:	↵
Specify first point:	Click: **D11**
Specify next point or [Undo]:	Click: **D12**
Specify next point or [Undo]:	↵

Step 7. On your own: Set the CEN layer current.
Step 8. Draw centerlines to complete the drawing.

Prompt	Response
Command:	Type: **DIM**↵
Dim:	Type: **CEN**↵
Select arc or circle:	Click: **D1** (Figure 6-5)
Dim:	↵
Select arc or circle:	Click: **D2**
Dim:	Type: **E**↵
Command:	Type: **L**↵
Specify first point:	Click: **D3** (Figure 6-5)

Important: Be sure Snap is ON before you draw the following center lines.

Linetypes and Drawing Constructions Using AutoCAD

FIGURE 6–5

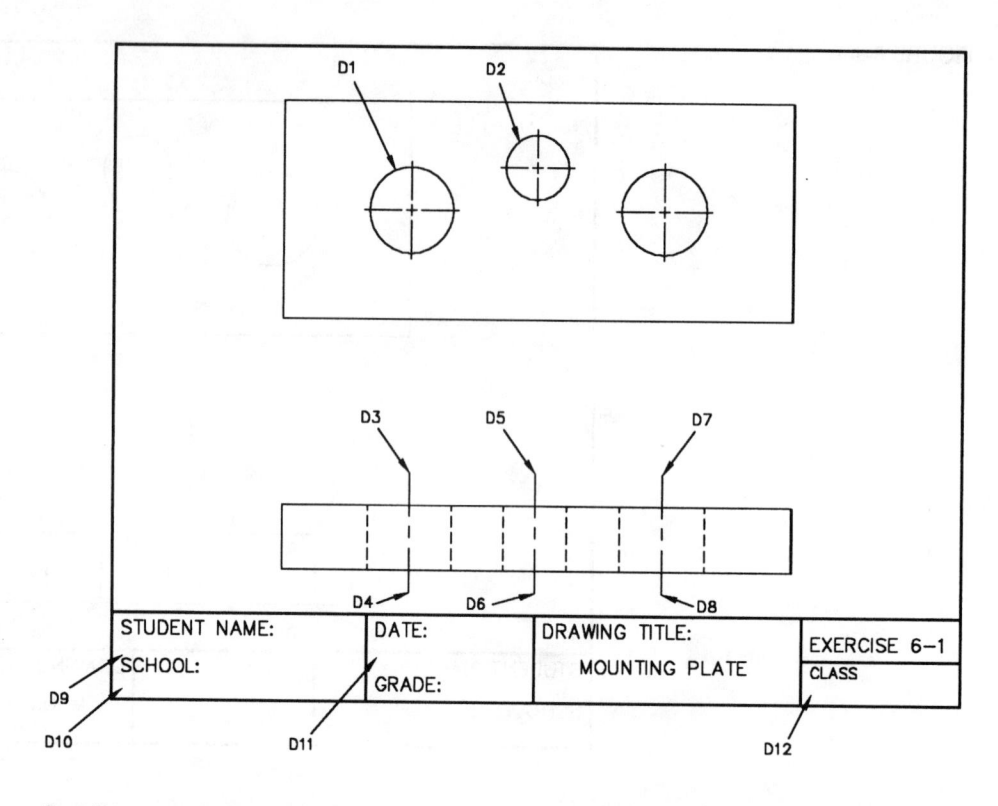

Prompt	Response
Specify next point or [Undo]:	Click: **D4**
Specify next point or [Undo]:	↵
Command:	↵
Specify first point:	Click: **D5**
Specify next point or [Undo]:	Click: **D6**
Specify next point or [Undo]:	↵
Command:	↵
Specify first point:	Click: **D7**
Specify next point or [Undo]:	Click: **D8**
Specify next point or [Undo]:	↵

Step 9. Use the Dtext command to complete the title block. On your own: Set the object layer current.

Prompt	Response
Command:	Type: **DT**↵
Specify start point of text or [Justify/Style]:	Click: **D9** (Figure 6–5)
Specify height <0.120>:	Type: **.12**↵ (or just ↵ if the default is .12)
Specify rotation angle of text <0>:	↵
Enter text:	Type: **YOUR FIRST INITIAL AND YOUR LAST NAME**↵ (Make sure all lettering in the title block are capital letters.)
Enter text:	Click: **D10** (Figure 6–4) and Type: **YOUR SCHOOL NAME**↵
Enter text:	Click: **D11** and Type: **TODAY'S DATE**↵
Enter text:	Click: **D12** and Type: **YOUR CLASS NAME OR NUMBER**↵
Enter text:	↵

Step 10. Use the SAVEAS command to save your drawing as EX6-1(your initials) on a floppy disk and again on the hard drive of your computer.
Step 11. Plot or print your drawing full size on an 11" × 8½" sheet.

EXERCISE 6–2
Making a Drawing Containing Several Circles, Fillets, and Parallel Lines

Your final drawing will look like the drawing in Figure 6–6 without dimensions.

Step 1. To begin Exercise 6–2, turn on the computer and start AutoCAD or AutoCAD LT.
Step 2. Open drawing EX6-2 supplied on the disk that came with your book.
Step 3. Use Zoom-All to view the entire drawing area.
Step 4. Use the Circle and ID commands to draw circles on the right and in the center of the part.

Prompt	Response
Command:	Type: **C**↵
Specify center point for circle or [3P/2P/Ttr (tan tan radius)]:	Type: **CEN**↵
of	Click: **D1** (Figure 6–7)
Specify radius of circle or [Diameter]:	Type: **.875**↵
Command:	Type: **ID**↵
Specify point:	Type: **CEN**↵
of	Click: **D1** (again)
Command:	Type: **C**↵
Specify center point for circle or [3P/2P/Ttr (tan tan radius)]:	Type: **@-2.875,0**↵

FIGURE 6–6
Dimensions for Exercise 6–2

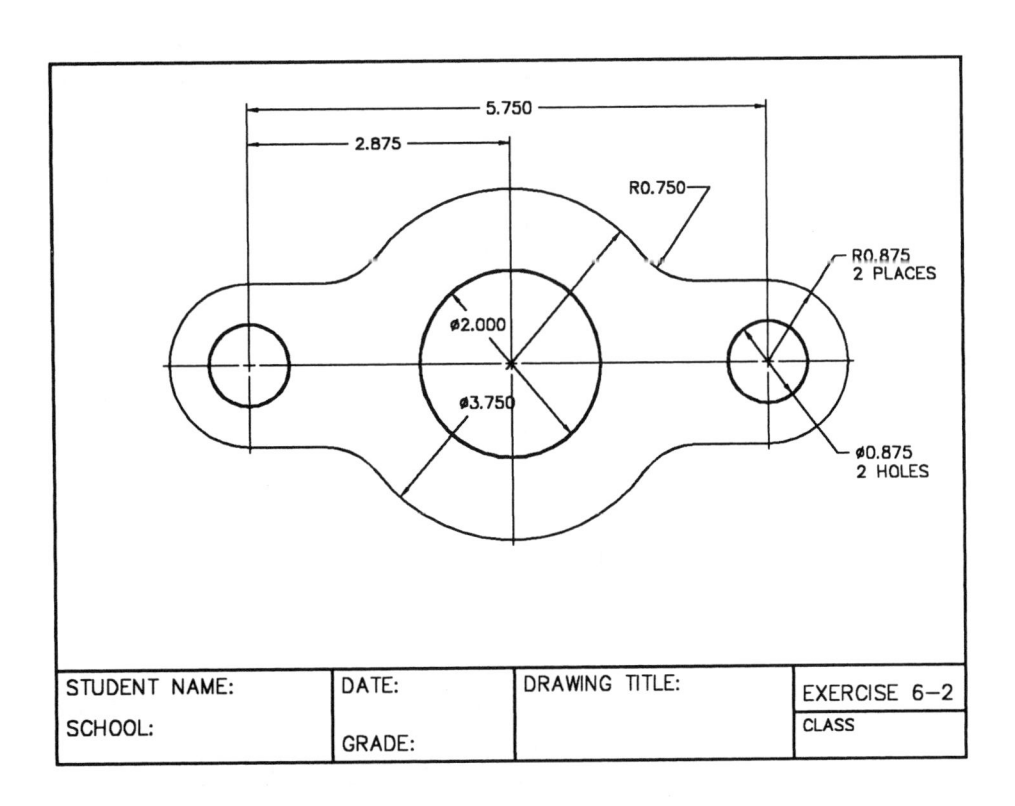

Linetypes and Drawing Constructions Using AutoCAD

Prompt	Response
Specify radius of circle or [Diameter]: <0.875>:	Type: **D**↵
Specify diameter of circle <1.750>:	Type: **3.75**↵
Command:	↵
Specify center point for circle or [3P/2P/Ttr (tan tan radius)]:	Type: **CEN**↵
of	Click: **D2** (Figure 6–7)
Specify radius of circle or [Diameter]:	Type: **1**↵

Step 5. Use the Line and Offset commands to draw parallel lines.

Prompt	Response
Command:	Type: **L**↵
Specify first point:	Type: **QUA**↵
of	Click: **D1** (Figure 6–8)
Specify next point or [Undo]:	Type: **@1.25<180**↵
Specify next point or [Undo]:	↵
Command:	Type: **O**↵
Specify offset distance or [Through]:	Type: **T**↵
Select object to offset or <exit>:	Click: **D2** (Figure 6–8)
Specify through point:	Type: **QUA**↵
of	Click: **D3**
Select object to offset or <exit>:	↵

Step 6. Use the Trim command to remove circle excesses.

Prompt	Response
Command:	Type: **TR**↵
Select cutting edges:	

FIGURE 6–7

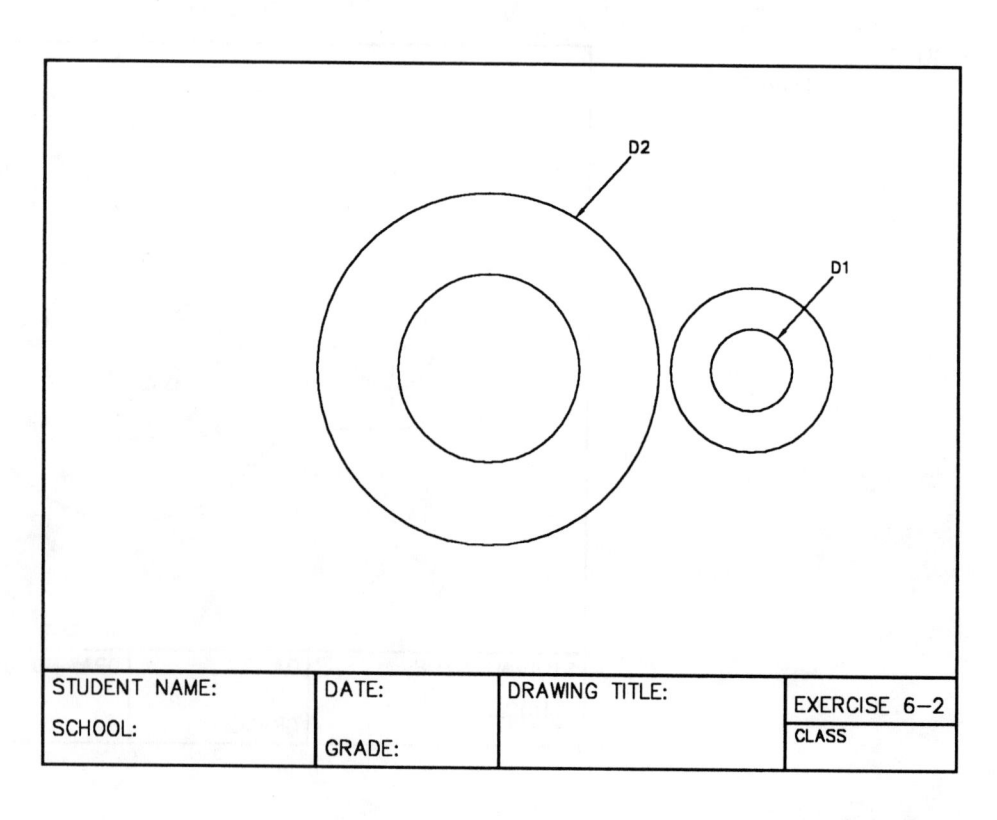

FIGURE 6–8

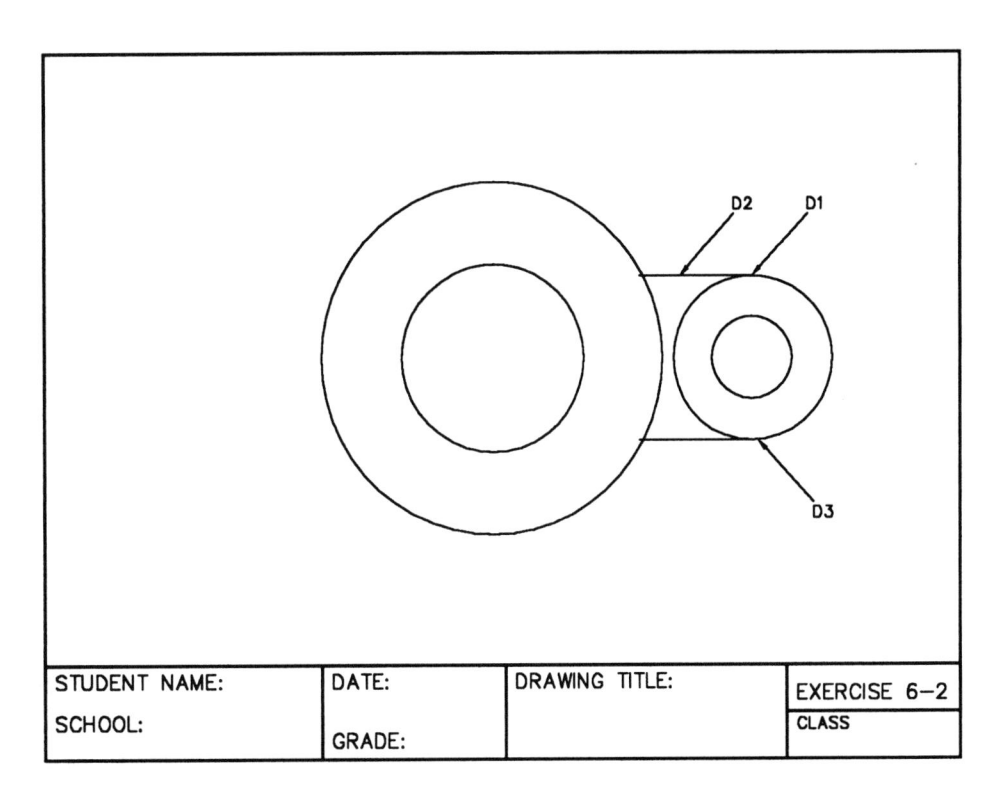

Select objects:	Click: **D1 and D2** (Figure 6–9)
Select objects:	↵
Select object to trim or shift-select to extend or [Project/Edge/Undo]:	Click: **D3 and D4**
Select object to trim or shift-select to extend or [Project/Edge/Undo]:	↵

Step 7. Use the Mirror command to draw the left side of the figure.

FIGURE 6–9

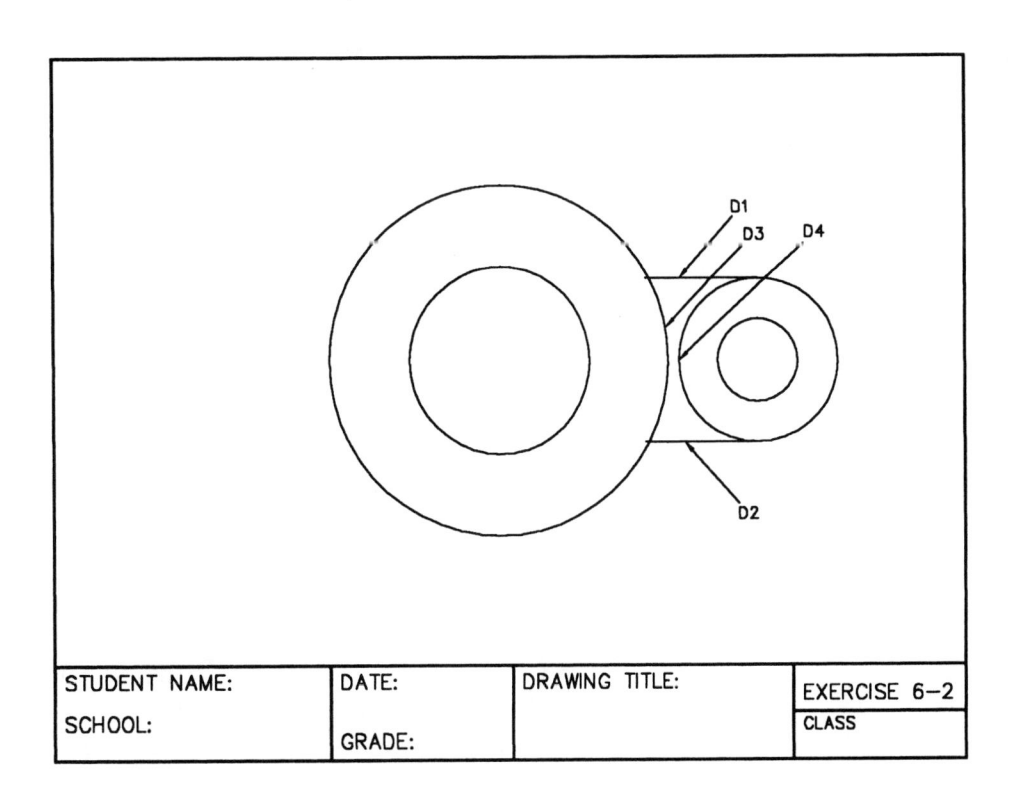

Linetypes and Drawing Constructions Using AutoCAD

79

FIGURE 6-10

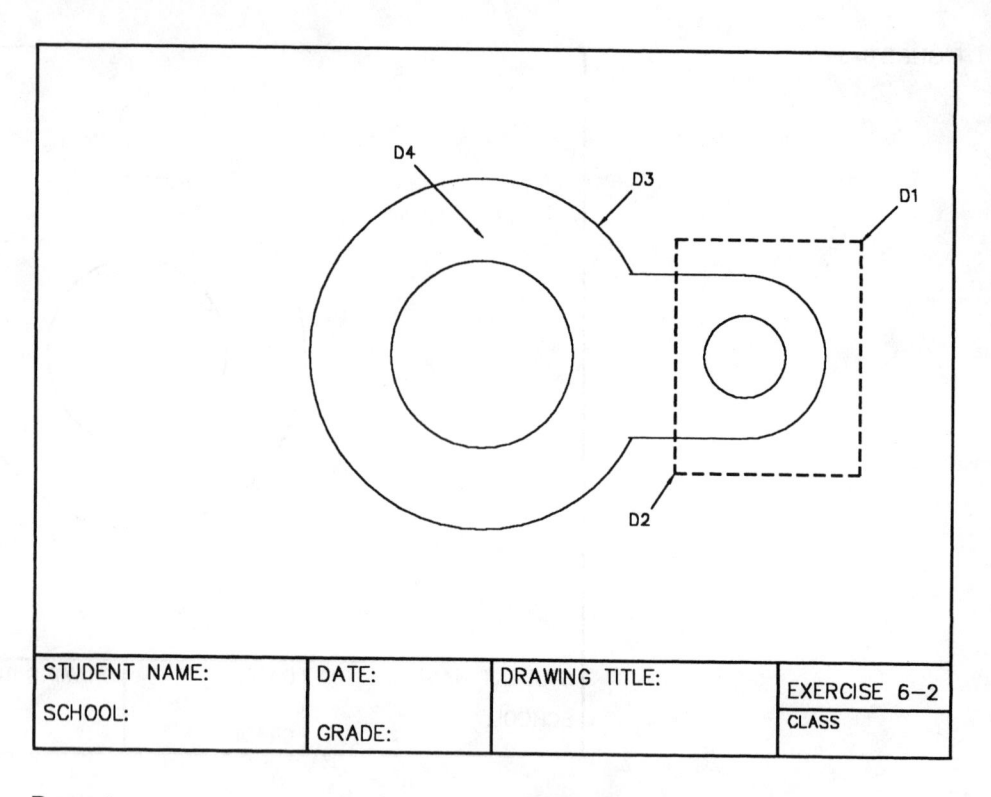

[Figure 6-10: Drawing showing a circular part with D3, D4 labels on a circle, and D1, D2 marking a dashed selection window around a smaller feature to the right. Title block: STUDENT NAME, SCHOOL, DATE, GRADE, DRAWING TITLE, EXERCISE 6-2, CLASS]

Prompt	Response
Command:	Type: **MI**↵
Select objects:	Click: **D1** (Figure 6-10)
Specify opposite corner:	Click: **D2**
Select objects:	↵
Specify first point of mirror line:	Type: **CEN**↵
of	Click: **D3**

FIGURE 6-11

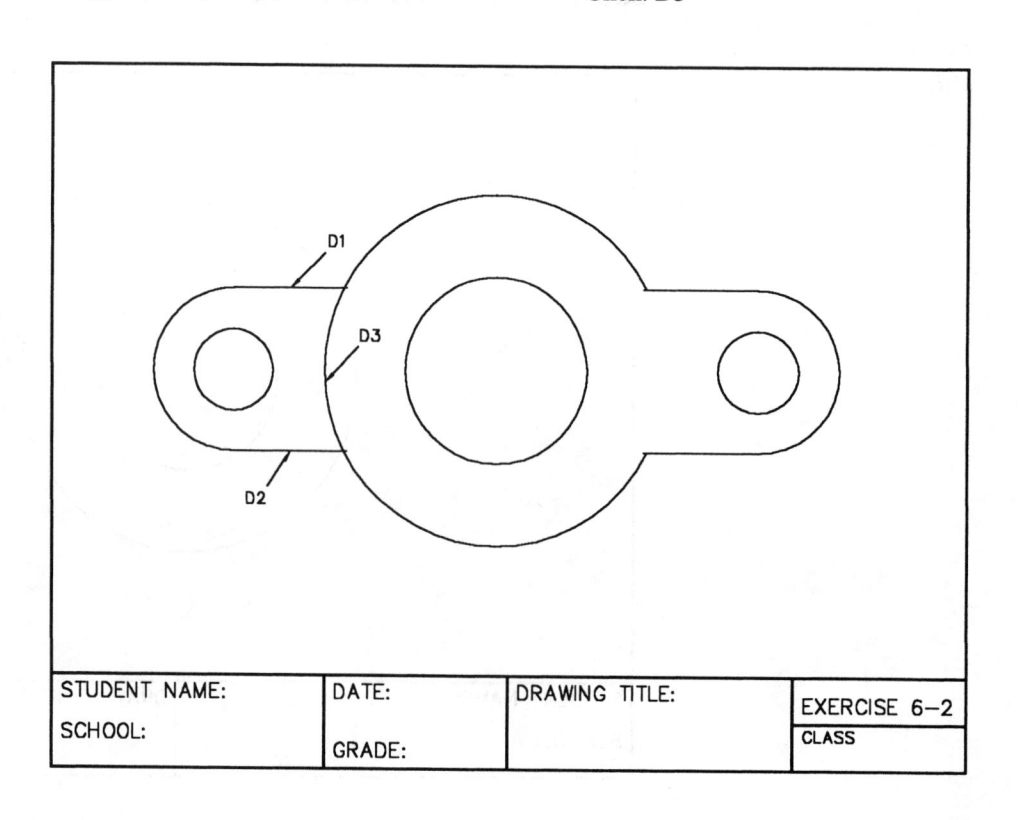

[Figure 6-11: Resulting drawing after mirror operation, showing symmetric part with two small circles on either side of central large circle, labeled D1, D2, D3. Title block: STUDENT NAME, SCHOOL, DATE, GRADE, DRAWING TITLE, EXERCISE 6-2, CLASS]

80   Chapter 6

FIGURE 6–12

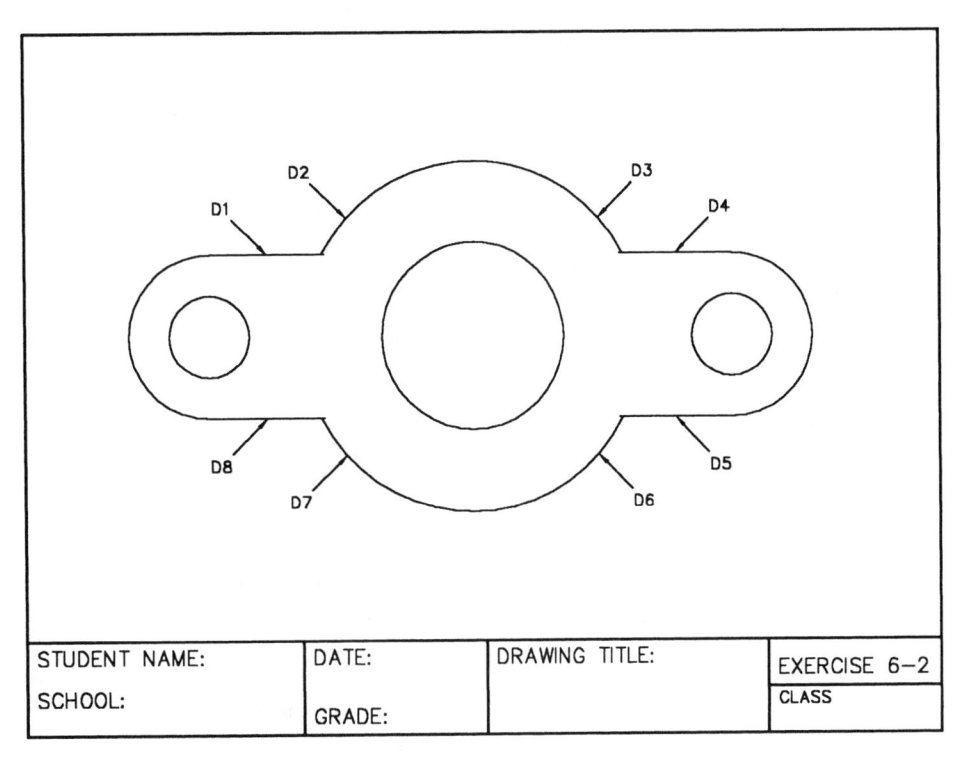

| STUDENT NAME: | DATE: | DRAWING TITLE: | EXERCISE 6–2 |
| SCHOOL: | GRADE: | | CLASS |

Specify second point of mirror line: **Make sure Ortho is ON** and Click: **D4** (any point directly above or below the center)

Delete source objects? [Yes/No] <No>: ↵

Step 8. Use the Trim command to remove circle excess on the left side of the figure.

Prompt	Response
Command:	Type: **TR**↵
Select cutting edges:	
Select objects:	Click: **D1 and D2** (Figure 6–11)
Select objects:	↵
Select object to trim or shift-select to extend or [Project/Edge/Undo]:	Click: **D3**
Select object to trim or shift-select to extend or [Project/Edge/Undo]:	↵

Step 9. Use the Fillet command to complete the figure.

Prompt	Response
Command:	Type: **F**↵
Select first object or [Polyline/Radius/Trim]:	Type: **R**↵
Specify fillet radius <0.0000>:	Type: **.75**↵
Select first object or [Polyline/Radius/Trim]:	Click: **D1** (Figure 6–12)
Select second object:	Click: **D2**
Command:	↵
Select first object or [Polyline/Radius/Trim]:	Click: **D3** (Figure 6–12)
Select second object:	Click: **D4**
Command:	↵
Select first object or [Polyline/Radius/Trim]:	Click: **D5**
Select second object:	Click: **D6**

Linetypes and Drawing Constructions Using AutoCAD

Prompt	Response
Command:	↵
Select first object or [Polyline/Radius/Trim]:	Click: **D7**
Select second object:	Click: **D8**

Step 10. Use the Dtext command to complete the title block as you did for EX6-1. Title the drawing GASKET.

Step 11. Use the SAVEAS command to save your drawing as EX6-2(your initials) on a floppy disk and again on the hard drive of your computer.

Step 12. Plot or print your drawing full size on an 11″ × 8½″ sheet.

EXERCISE 6–3
Drawing a Complex Shape Containing a Curved Slot

Your final drawing will look similar to the drawing in Figure 6–13 without dimensions.

Step 1. To begin Exercise 6–3, turn on the computer and start AutoCAD or AutoCAD LT.

Step 2. Open drawing EX6-3 supplied on the disk that came with your book.

Step 3. Use Zoom-All to view the entire drawing area.

Step 4. Use the Line command to draw the 2.75 line forming the bottom of the shape.

Command:	Type: **L**↵
Specify first point:	Type: **END**↵
of	Click: **D1** (Figure 6–14)
Specify next point or [Undo]:	Type: **@2.75<180**↵
Specify next point or [Undo]:	↵

Step 5. Use the Mirror command to draw circles on the left side of the part.

Prompt	Response
Command:	Type: **MI**↵

FIGURE 6–13
Dimensions for Exercise 6–3

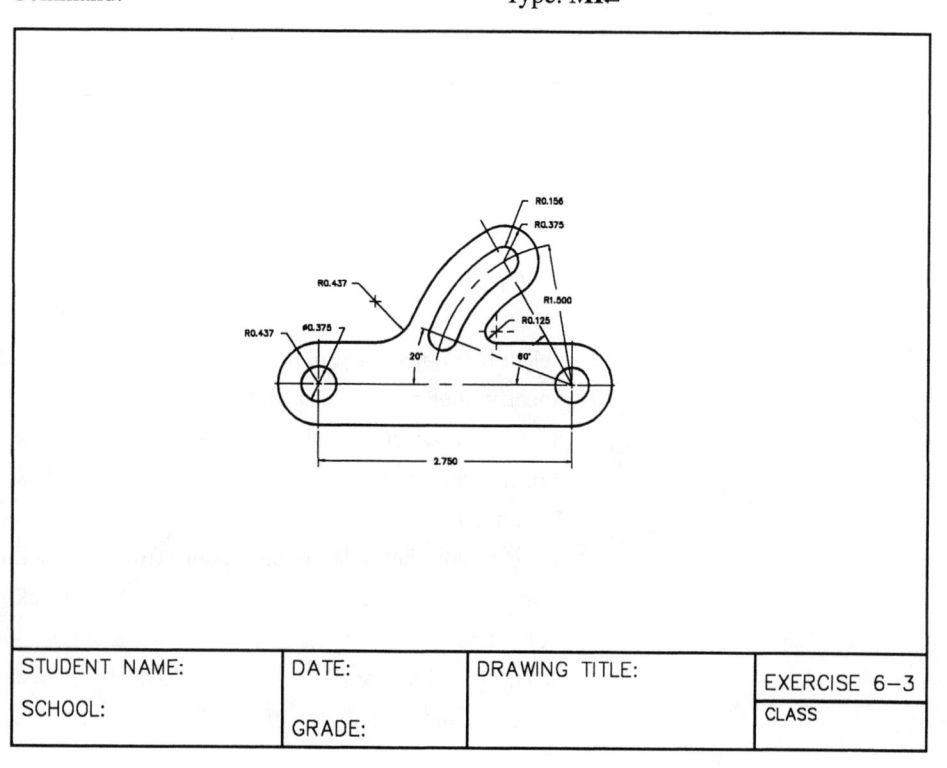

Select objects:	Click: **D2** (Figure 6–14)
Specify opposite corner:	Click: **D3**
Select objects:	↵
Specify first point of mirror line:	Type: **MID**↵
of	Click: **D4**
Specify second point of mirror line:	**Make sure Ortho is ON** and Click: **D5** (any point directly above or below D4)
Delete source objects? [Yes/No] <No>:	↵

Step 6. Use the ID and Circle commands to draw circles at the top and bottom of the curved slot.

Prompt	Response
Command:	Type: **ID**↵
Specify point:	Type: **CEN**↵
of	Click: **D1** (Figure 6–15)
Command:	Type: **C**↵
Specify center point for circle or [3P/2P/Ttr (tan tan radius)]:	Type: **@1.5<120**↵
Specify radius of circle or [Diameter]:	Type: **.375**↵
Command:	↵
Specify center point for circle or [3P/2P/Ttr (tan tan radius)]:	Type: **CEN**↵
of	Click: **D2** (Figure 6–15)
Specify radius of circle or [Diameter]<0.3750>:	Type: **.156**↵
Command:	Type: **ID**↵

FIGURE 6–14

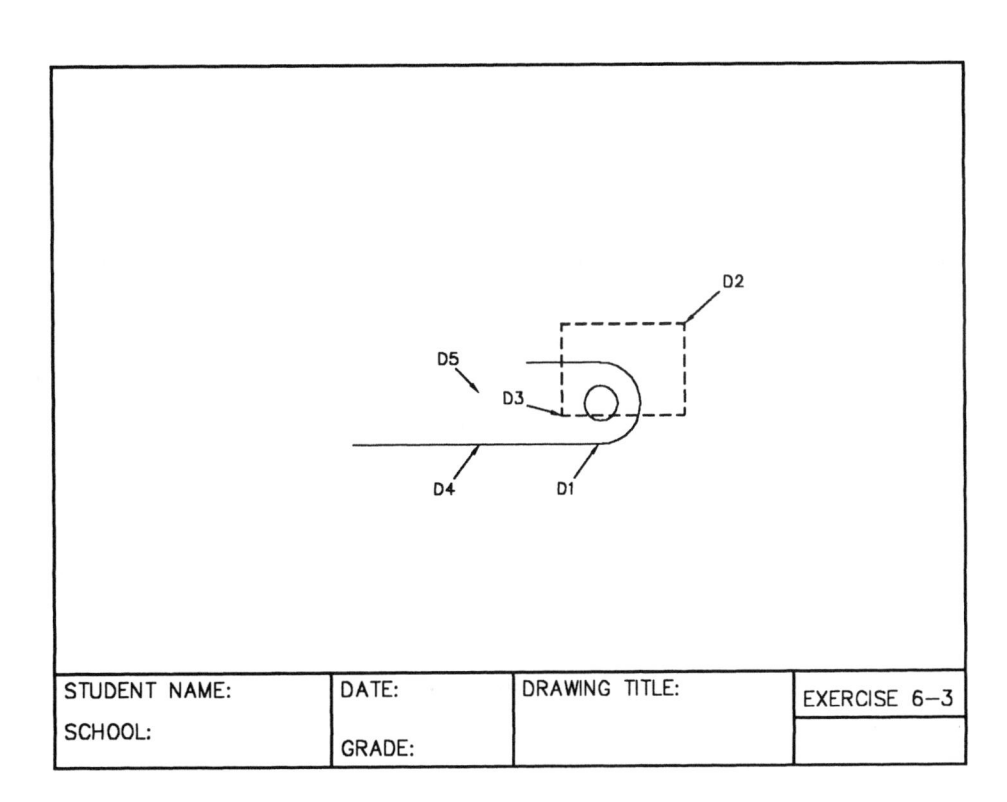

Linetypes and Drawing Constructions Using AutoCAD

FIGURE 6–15

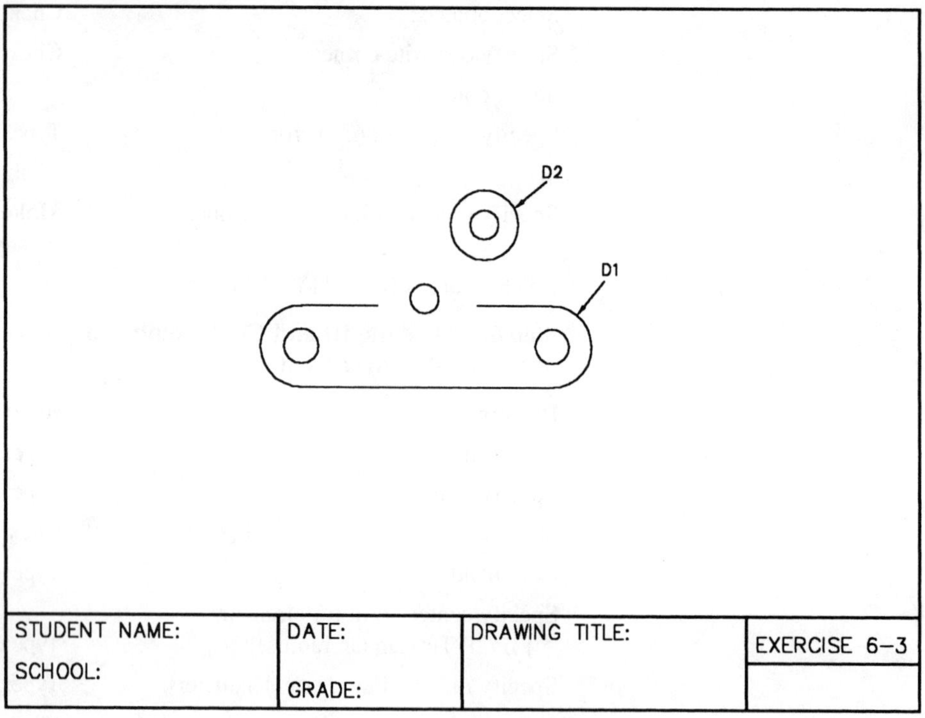

STUDENT NAME:	DATE:	DRAWING TITLE:	EXERCISE 6–3
SCHOOL:	GRADE:		

Specify point:	Type: **CEN**↵
of	Click: **D1** (again)
Command:	Type: **C**↵
Specify center point for circle or [3P/2P/Ttr (tan tan radius)]:	Type: **@1.5<160**↵
Specify radius of circle or [Diameter]<0.1560>:	Type: **.156**↵

Because the drawing you are using to make your CAD drawing has several dimensions on it, the easiest way to make the CAD drawing is to use those dimensions without adding and subtracting. Therefore, if you make a circle through the center of the slot, you can use the Offset command and the radii shown to make the other circles.

Step 7. Use the Circle and Offset commands to draw circles for the slot.

Prompt	**Response**
Command:	Type: **C**↵
Specify center point for circle or [3P/2P/Ttr (tan tan radius)]:	Type: **CEN**↵
of	Click: **D1** (Figure 6–16)
Specify radius of circle or [Diameter]<0.1560>:	Type: **1.5**↵
Command:	Type: **O**↵
Specify offset distance or [Through]:	Type: **.156**↵
Select object to offset or <exit>:	Click: **D2** (Figure 6–16)
Specify point on side to offset:	Click: **D3**
Select object to offset or <exit>:	Click: **D2 (again)**
Specify point on side to offset:	Click: **D4**

84

Chapter 6

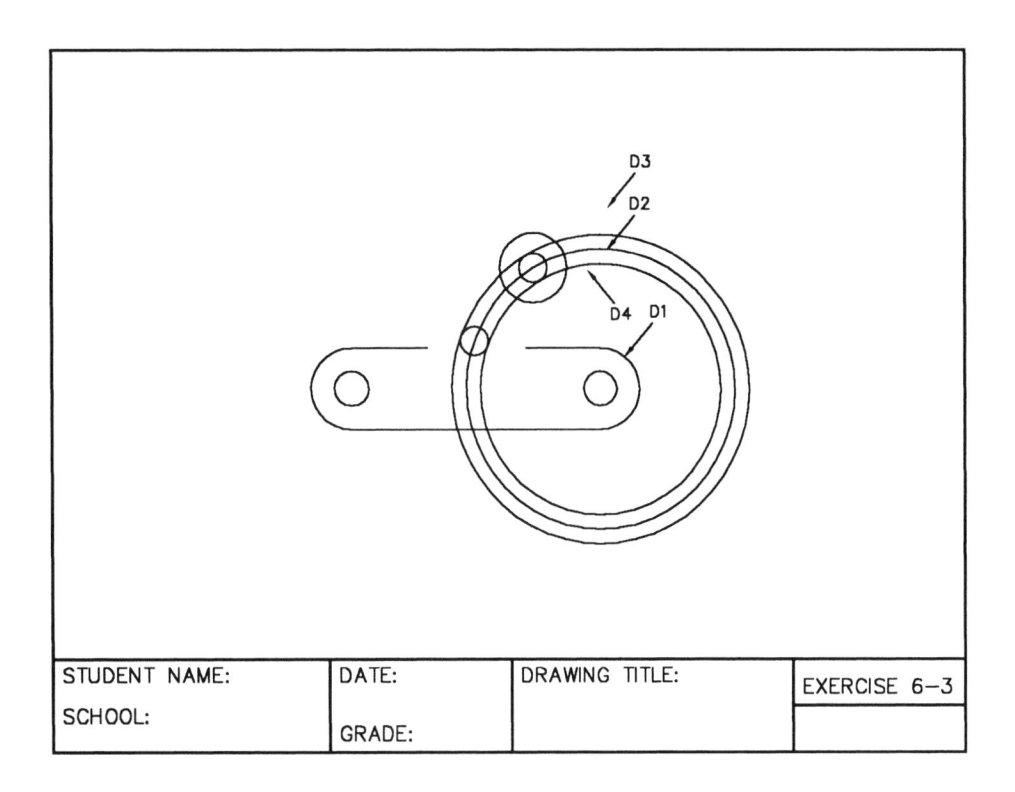

FIGURE 6-16

[Figure showing Exercise 6-3 drawing with circles labeled D1, D2, D3, D4 and title block with STUDENT NAME, SCHOOL, DATE, GRADE, DRAWING TITLE, EXERCISE 6-3]

Note: You may need to use Zoom-Window to zoom in closer so the smaller circles can be clicked. Turn Snap off also if you need to.

Prompt	Response
Select object to offset or <exit>:	↵

Step 8. Use the Trim command to remove circle excess.

Prompt	Response
Command:	Type: **TR**↵
Select cutting edges:	
Select objects:	Click: **D1** (Figure 6–17)
Select objects:	Click: **D2**
Select objects:	Click: **D3**
Select objects:	Click: **D4**
Select objects:	↵
Select object to trim or shift-select to extend or [Project/Edge/Undo]:	Click: **D2** (again)
Select object to trim or shift-select to extend or [Project/Edge/Undo]:	Click: **D3** (again)
Select object to trim or shift-select to extend or [Project/Edge/Undo]:	Click: **D5**
Select object to trim or shift-select to extend or [Project/Edge/Undo]:	Click: **D6**
Select object to trim or shift-select to extend or [Project/Edge/Undo]:	↵

Step 9. Use the Offset command to draw the remaining circles.

Prompt	Response
Command:	Type: **O**↵
Specify offset distance or [Through]:	Type: **.375**↵
Select object to offset or <exit>:	Click: **D1** (Figure 6–18)

Linetypes and Drawing Constructions Using AutoCAD

FIGURE 6–17

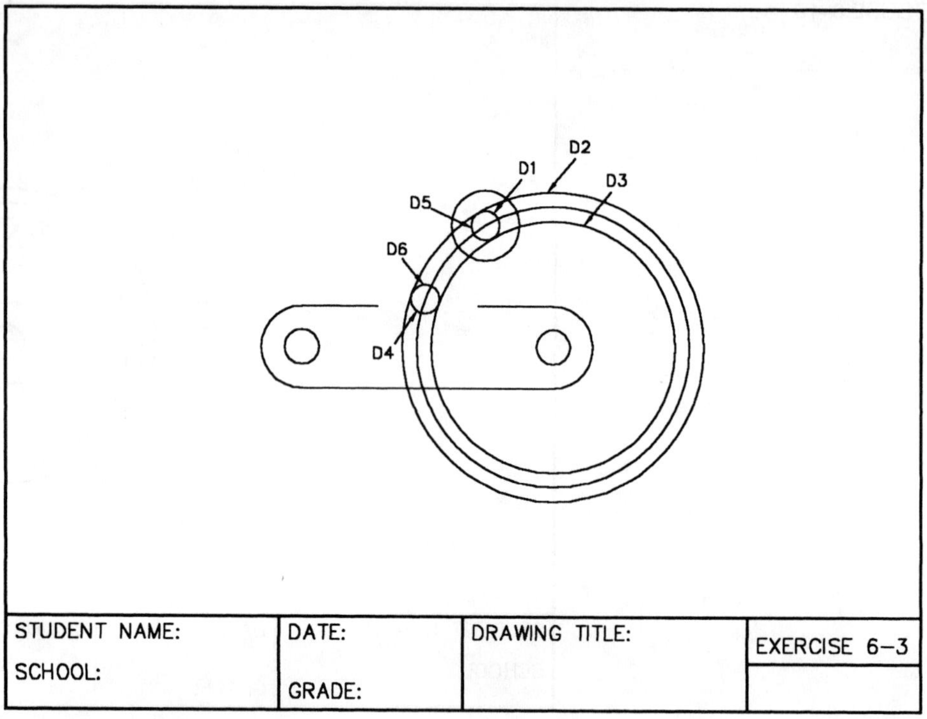
D2
D1
D3
D5
D6
D4
STUDENT NAME:
SCHOOL:
DATE:
GRADE:
DRAWING TITLE:
EXERCISE 6–3

Specify point on side to offset:	Click: **D2**
Select object to offset or <exit>:	Click: **D1** (again)
Specify point on side to offset:	Click: **D3**
Select object to offset or <exit>:	↵

Step 10. Use the Fillet command to draw fillets on each side of the curve.

Prompt	Response
Command:	Type: **F**↵

FIGURE 6–18

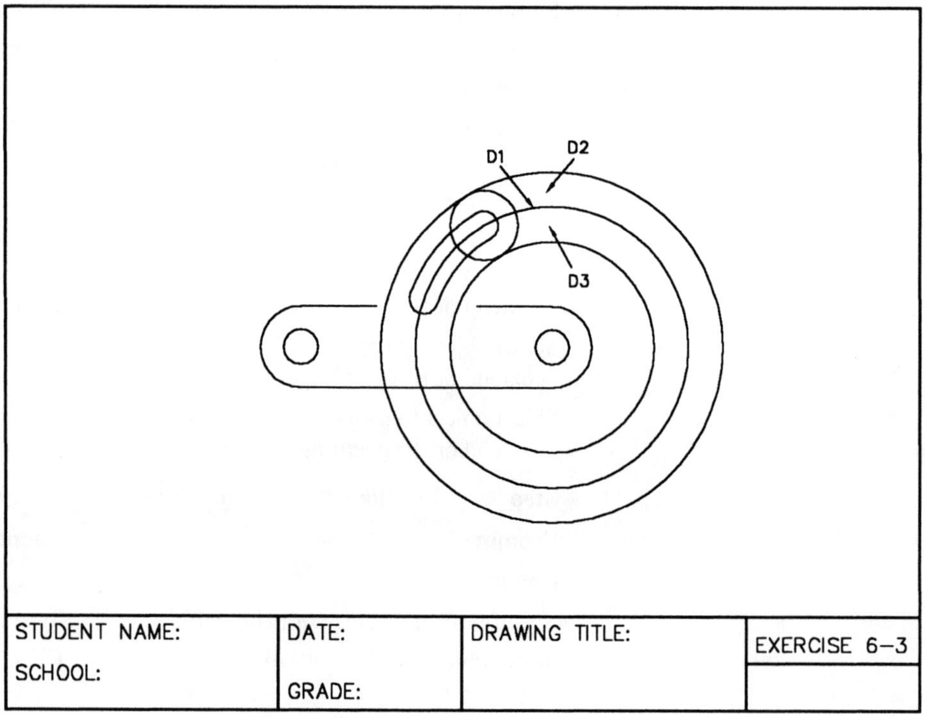
D1
D2
D3
STUDENT NAME:
SCHOOL:
DATE:
GRADE:
DRAWING TITLE:
EXERCISE 6–3

Select first object or [Polyline/Radius/Trim]:	Type: **R**↵
Specify fillet radius <0.0000>:	Type: **.437**↵
Select first object or [Polyline/Radius/Trim]:	Click: **D1** (Figure 6–19)
Select second object:	Click: **D2**
Command:	↵
Select first object or [Polyline/Radius/Trim]:	Type: **R**↵
Enter fillet radius <0.437>:	Type: **.125**↵
Select first object or [Polyline/Radius/Trim]:	Click: **D3** (Figure 6–19)
Select second object:	Click: **D4**

Step 11. Use the Trim and Erase commands to remove circle excesses and complete the drawing.

Prompt	Response
Command:	Type: **TR**↵
Select cutting edges:	
Select objects:	Click: **D1** (Figure 6–20)
Select objects:	Click: **D2**
Select objects:	Click: **D3**
Select objects:	Click: **D4**
Select objects:	Click: **D5**
Select objects:	↵
Select object to trim or shift-select to extend or [Project/Edge/Undo]:	Click: **D2** (again)
Select object to trim or shift-select to extend or [Project/Edge/Undo]:	Click: **D4** (again)
Select object to trim or shift-select to extend or [Project/Edge/Undo]:	Click: **D6**

FIGURE 6–19

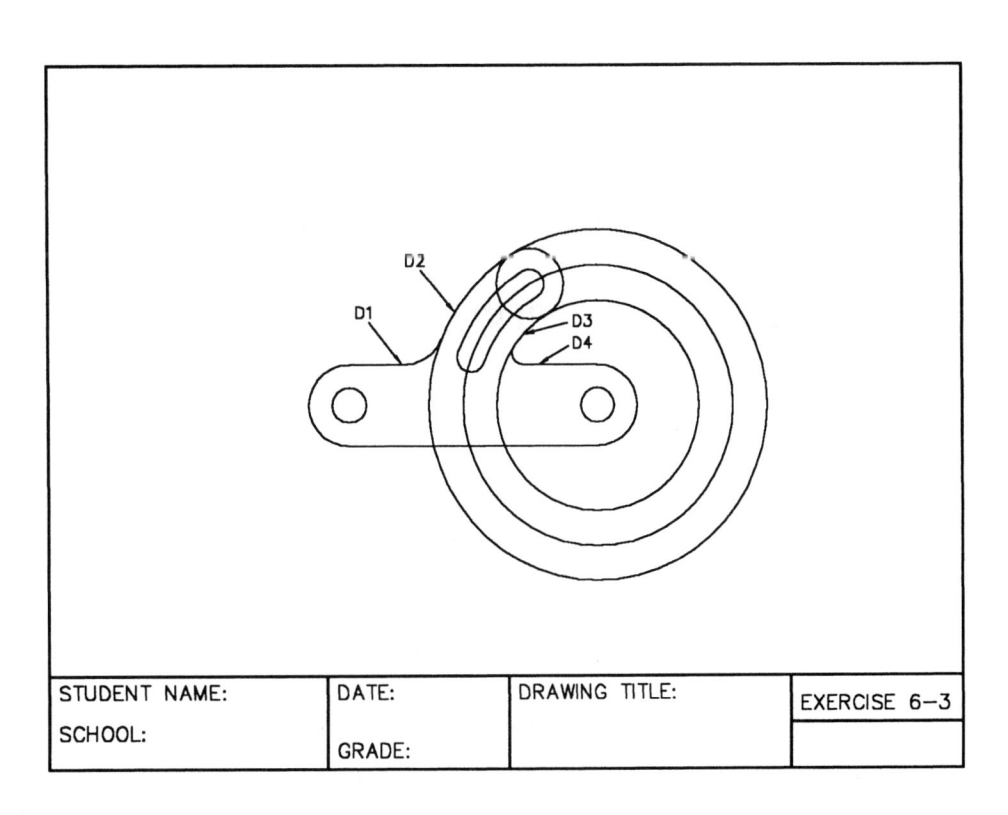

Linetypes and Drawing Constructions Using AutoCAD

FIGURE 6–20

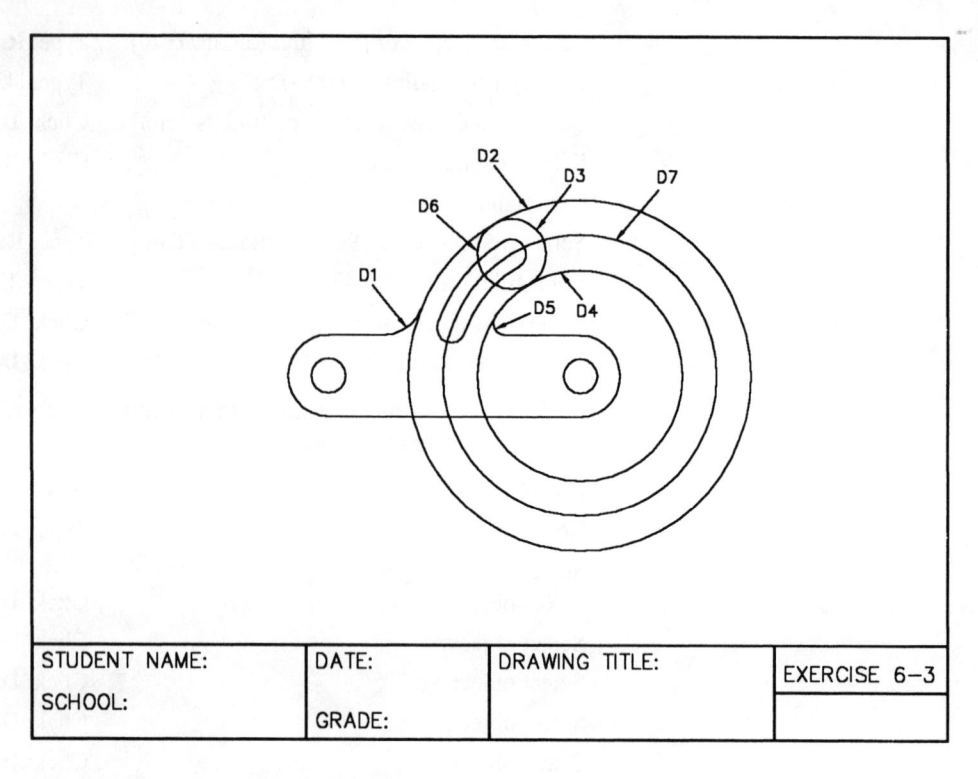

STUDENT NAME:	DATE:	DRAWING TITLE:	EXERCISE 6–3
SCHOOL:	GRADE:		

Select object to trim or shift-select to extend or [Project/Edge/Undo]:	↵
Command:	Type: **E**↵
Select objects:	Click: **D7** (Figure 6–20)
Select objects:	↵

Step 12. Use the Dtext command to complete the title block as you did for EX6-1. Title the drawing SWIVEL PLATE.

Step 13. Use the SAVEAS command to save your drawing as EX6-3(your initials) on a floppy disk and again on the hard drive of your computer.

Step 14. Plot or print your drawing full size on an $11'' \times 8\frac{1}{2}''$ sheet.

EXERCISES

EXERCISE 6–1   Complete Exercise 6–1 using steps 1 through 10 described in this chapter.
EXERCISE 6–2   Complete Exercise 6–2 using steps 1 through 12 described in this chapter.
EXERCISE 6–3   Complete Exercise 6–3 using steps 1 through 14 described in this chapter.

REVIEW QUESTIONS

Circle the best answer.

1. Which of the following will produce a line $3\frac{1}{2}''$ long downward from a point?
   a. 3.5 × 90
   b. @3.5<90
   c. @0<3.5
   d. @3.5<-90
   e. 90<3.5
2. Which of the following is used to identify a point that may be used for one command?
   a. Status
   b. Point
   c. ID
   d. Line
   e. Dist

3. A rounded corner may be obtained most easily with which of the following commands?
   a. Chamfer
   b. Fillet
   c. Draw
   d. Offset
   e. Break
4. Which of the following circles is produced if .5 is entered in response to the circle prompt "Specify radius of circle or [Diameter]:"?
   a. .50 diameter
   b. .25 radius
   c. 1.00 radius
   d. 1.00 diameter
   e. .25 diameter
5. Which command is used to trim lines between cutting edges?
   a. Edit
   b. Trim
   c. Erase
   d. Copy
   e. Extend
6. From the Line prompt "Specify next point," which is the correct response to draw a horizontal line 4.501" to the right of the starting point?
   a. 4.501<180
   b. 4.501<0
   c. <180<4.501
   d. @4.501<0
   e. <0<4.501
7. To draw a line parallel to another line, which of the following commands should be used?
   a. Line parallel
   b. Parallel
   c. Offset
   d. Offset parallel
   e. LP
8. To draw a line perpendicular to another line, which of the following options should be used?
   a. Square
   b. Rt Angle
   c. 90 Angle
   d. Osnap-Perpendicular
   e. @90
9. If you have just drawn a line at the wrong angle and you want to return to the starting point of the line, which command should you enter?
   a. Redo
   b. R
   c. U
   d. Ctrl-C
   e. Erase
10. To produce a circle with a diameter of .500, what is the correct response to the prompt "Specify radius of circle or [Diameter]:"?
    a. D, then .5
    b. R, then .500
    c. .50
    d. Circle-diameter
    e. .500
11. Describe the differences among absolute, relative, and polar coordinates.

_____

_____

_____

Linetypes and Drawing Constructions Using AutoCAD

12. Write a sequence of "Line responses" to the "Specify next point" prompt that will produce a 4.5" × 3" rectangle using polar coordinates.

13. Describe the function of the ID command.

14. Describe what is meant by a running Osnap mode.

15. The following set of coordinates is an example of _____ coordinates: 2,1.

16. The following set of coordinates is an example of _____ coordinates: @2,1.

17. The following set of coordinates is an example of _____ coordinates: @2<0.

18. List the command that asks you to select objects and then specify a line about which the selected objects are to be copied as a mirror image.

19. What letter must you type to remove an object from a set of objects selected to be erased?

20. AutoCAD stores up to _____ views for the Zoom-Previous command.

7 Reading and Sketching Orthographic Views

OBJECTIVES

After completing this chapter, you will be able to

- Correctly identify surfaces in two-dimensional views from given three-dimensional views.
- Correctly sketch two-dimensional views from given three-dimensional views.
- Correctly answer questions regarding the orthographic projection method of drawing.

ORTHOGRAPHIC PROJECTION

Orthographic projection is a system of drawing that is used throughout the world. Orthographic views are two-dimensional or flat views of objects. One of the reasons orthographic drawing is used instead of pictorial drawing is that it is easy to place dimensions on these drawings, and so confusion about measurements is avoided (Figure 7–1).

Two forms of orthographic projection are commonly used: first-angle and third-angle (Figure 7–2). Third-angle projection is used in the United States and many other countries. First-angle projection is used in several countries in Europe. This chapter presents third-angle orthographic projection in detail but first describes how first-angle projection is different.

Differences Between First-Angle and Third-Angle Orthographic Projections

In the third-angle projection shown in Figure 7–2A, think of the pyramid shape as being attached in the center and swinging to all four sides to give right-side, left-side, top, and bottom views. First-angle projection is viewed as if the object were tipped to the sides to give the four views. Third-angle orthographic projection is used in the rest of this book, and no further reference will be made to first-angle projection.

Third-Angle Projection Theory

One of the best ways to explain third-angle projection is to use the transparent box theory (Figure 7–3), in which an imaginary cube is placed around the object. The surfaces you would see if you looked into each side of the cube are registered on the cube. When the cube is unfolded, the views are arranged as shown in Figure 7–4. This arrangement of views is understood throughout the world. Your drawings must follow the same arrangement, with two exceptions, as shown in Figure 7–5.

Views

The transparent box theory shows that every object has six possible views: front, back, top, bottom, left side, and right side. Often an object can be fully described using three of these views, or sometimes fewer. Many objects such as screws and other fasteners need only one view to completely describe them. Other complex objects require all six views and several auxiliary, section, and detail views for complete description. These other types of views will be presented in later chapters.

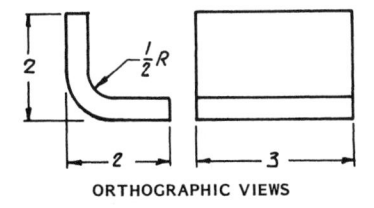

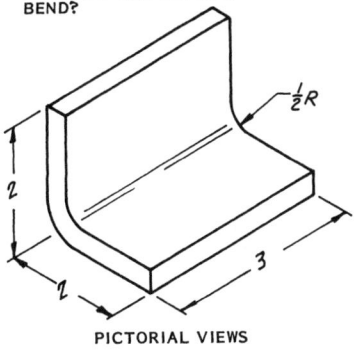

FIGURE 7–1
Orthographic and Pictorial Views

FIGURE 7–2
Third- and First-Angle Projection

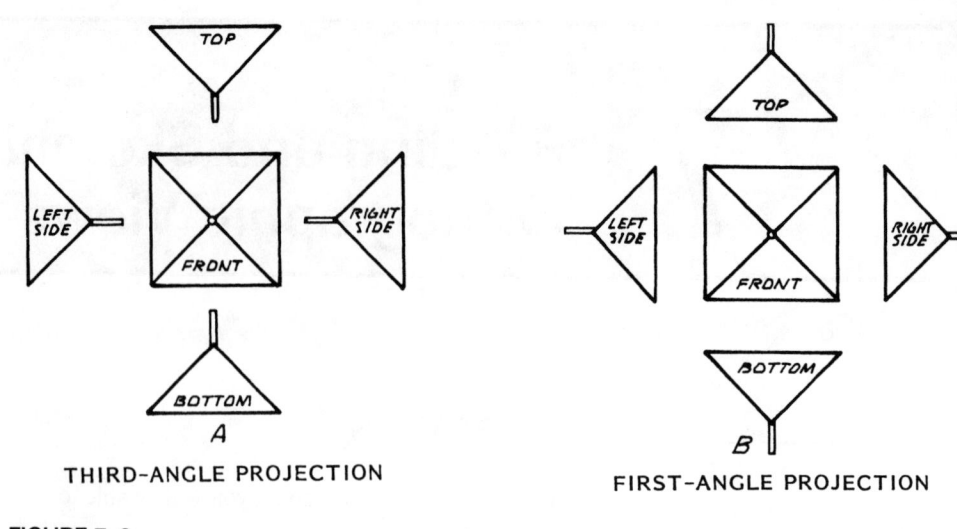

THIRD-ANGLE PROJECTION

FIRST-ANGLE PROJECTION

FIGURE 7–3
Transparent Box Theory

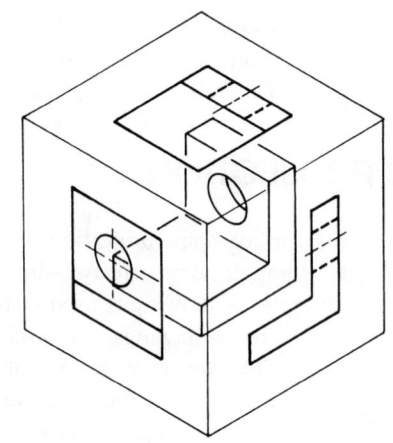

Height, Width, and Depth Dimensions

Each view of the object contains only two of its three dimensions: top and bottom views show width and depth; right and left side views show height and depth; and front and back views show width and height (Figure 7–6). The most commonly used combination of views is the top, front, and right-side views, and the next most commonly used combi-

FIGURE 7–4
Arrangement of Views

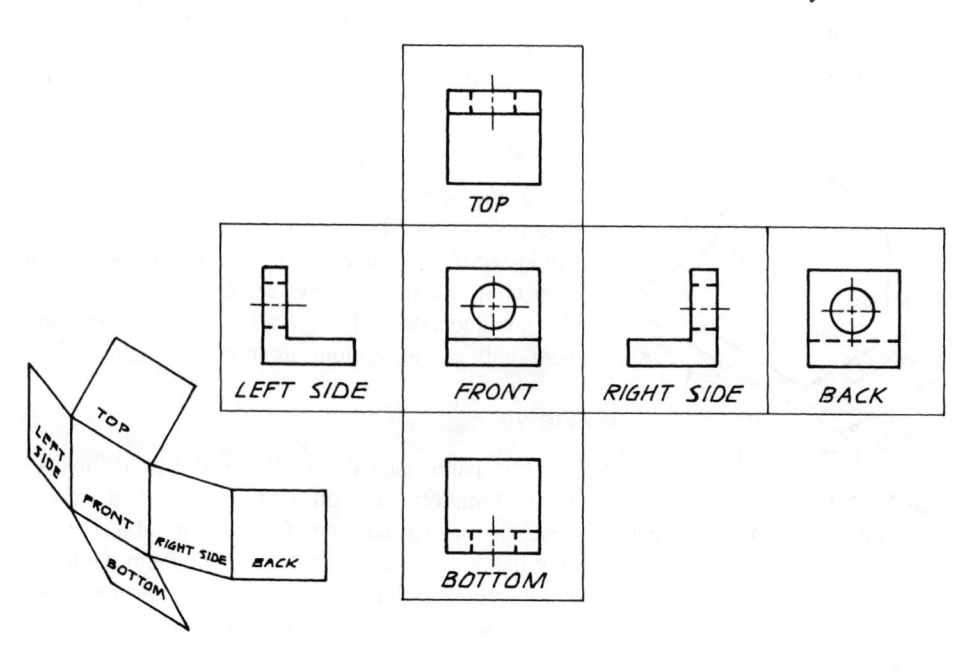

92 Chapter 7

FIGURE 7–5
Exceptions to Standard View Arrangement

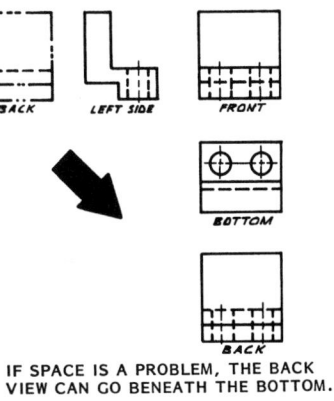

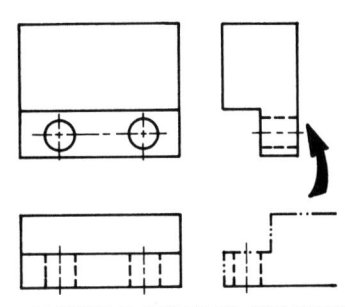

IF SPACE IS A PROBLEM, THE BACK VIEW CAN GO BENEATH THE BOTTOM.

IF SPACE IS A PROBLEM, THE RIGHT SIDE CAN GO TO THE RIGHT OF THE TOP VIEW (BUT AVOID ANY EXCEPTIONAL ARRANGEMENT).

nation is top, front, and left side (Figure 7–7). Usually the front view is drawn first, and then either the top or side view is drawn. Height and width dimensions are easily projected into the adjacent view, but depth dimensions either must be transferred using a scale or a piece of paper or they must be projected through a 45° angle (Figure 7–8).

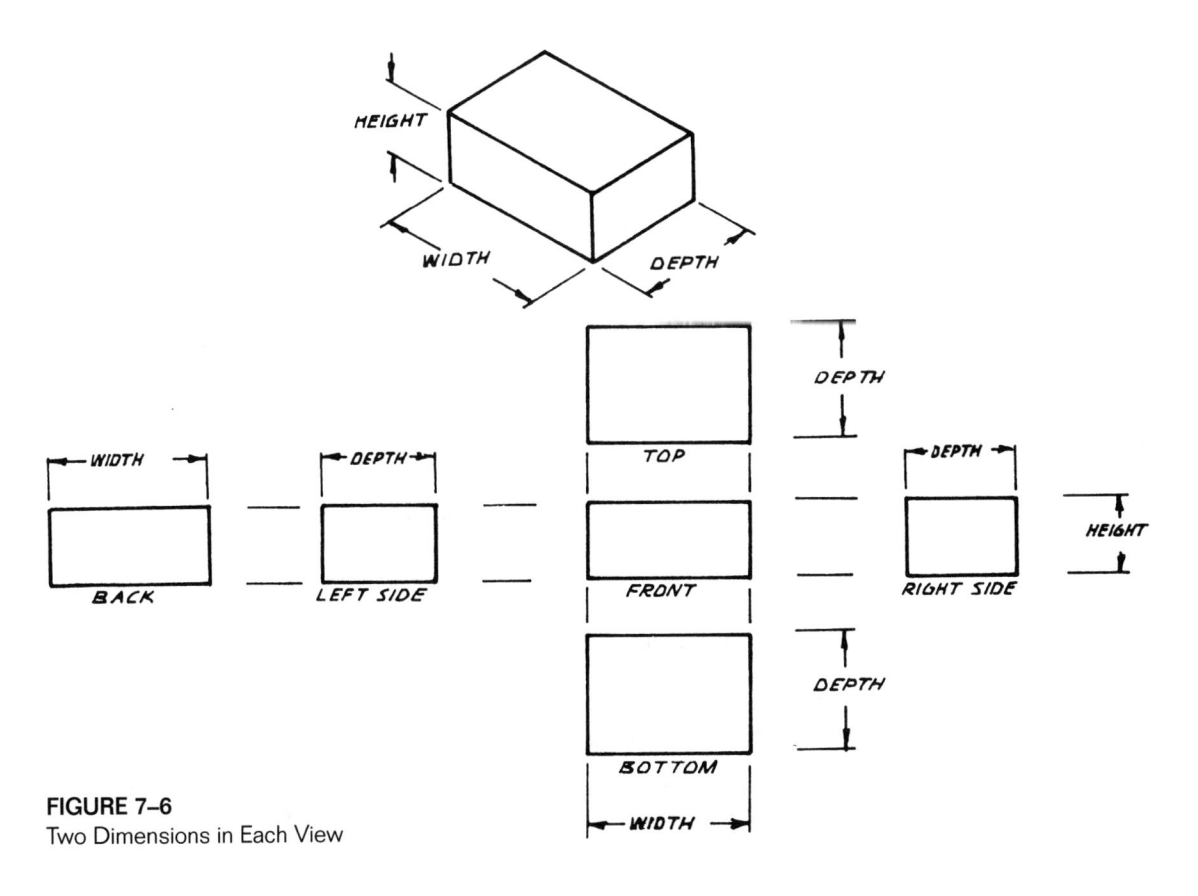

FIGURE 7–6
Two Dimensions in Each View

Reading and Sketching Orthographic Views

93

FIGURE 7–7
The Most Common Combination of Views

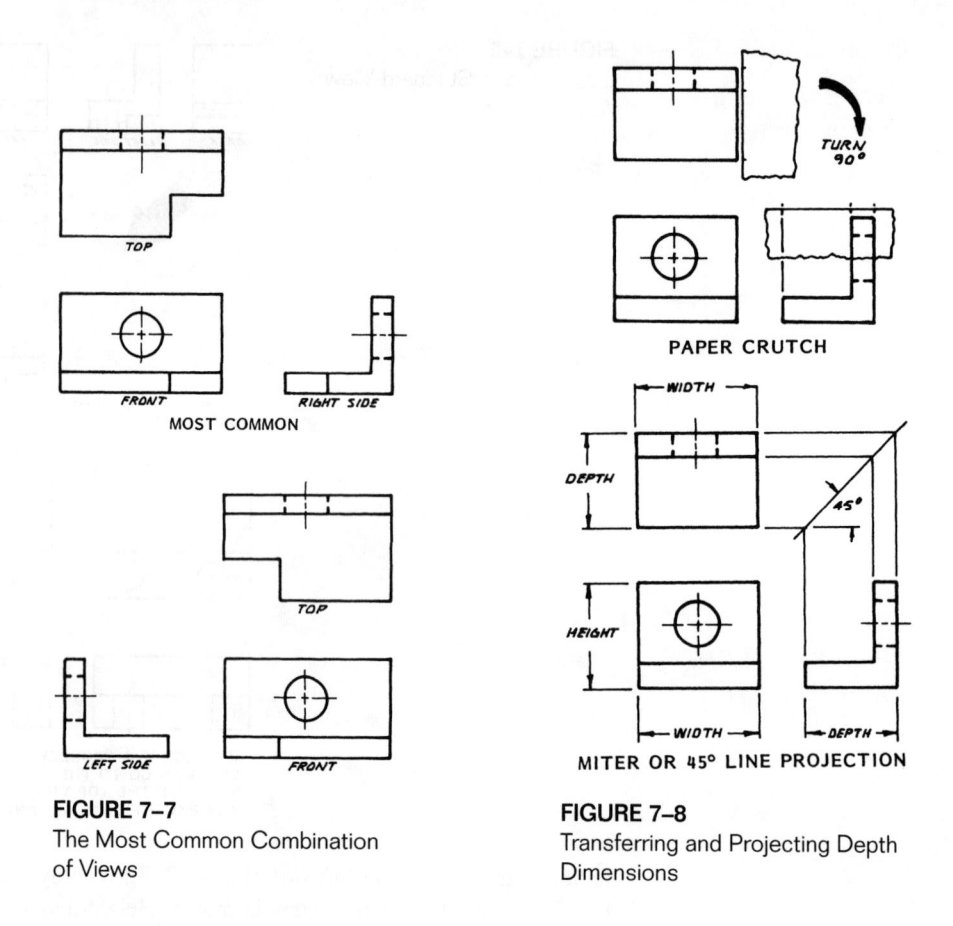

FIGURE 7–8
Transferring and Projecting Depth Dimensions

IDENTIFYING SURFACES AND FEATURES

To make two-dimensional drawings from three-dimensional drawings or an actual part, you must learn to identify surfaces, edges, and other features. The following paragraphs and exercises will give you practice in doing that.

Normal Surfaces

The easiest surfaces to read are the flat, unslanted surfaces. These are called *normal* (normal means perpendicular) surfaces because they are perpendicular to your line of sight when you look at the object. Figure 7–9 shows an object that has all normal surfaces. Exercise 7–1 will test your skill in identifying normal surfaces from a three-dimensional drawing.

EXERCISE 7–1
Identifying Normal Surfaces

Step 1. Remove the sheet labeled Exercise 7–1 from your book.
Step 2. Identify the numbered surfaces in the normal views in the spaces provided. Use the numbers from the pictorial view. Use your best lettering with guidelines top and bottom.
Step 3. Fill in the title block with your best lettering. Title the drawing NORMAL SURFACES.

Inclined or Slanted Surfaces

An inclined or slanted surface is perpendicular to two of the normal surfaces, but at an angle other than 90° to the other four normal surfaces. The slanted surface in Figure 7–10

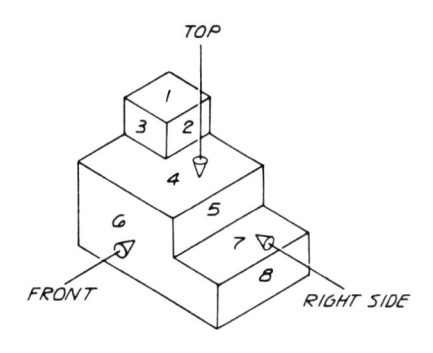

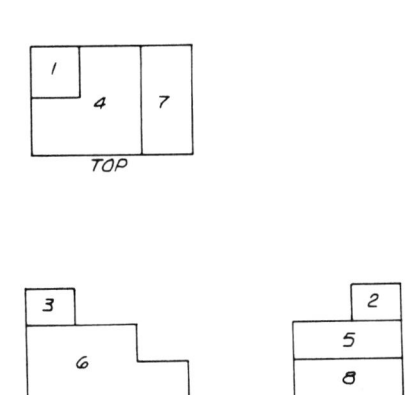

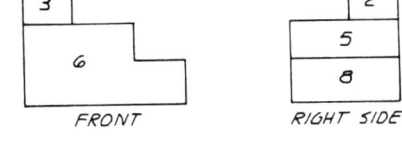

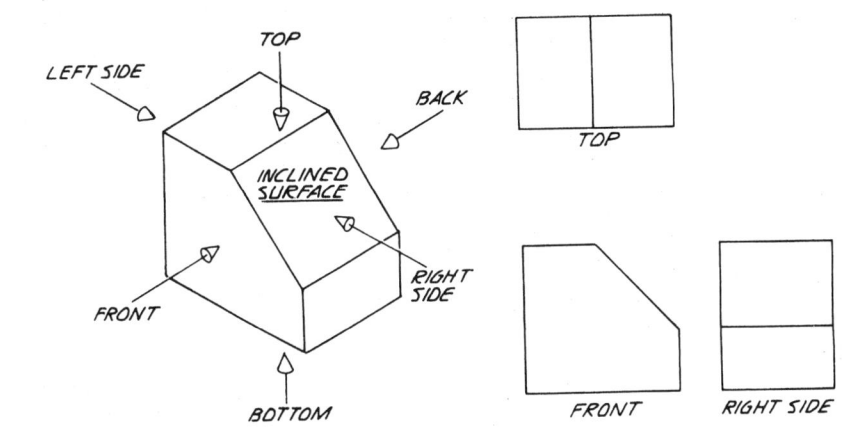

FIGURE 7–9
Normal Surfaces

FIGURE 7–10
Inclined Surfaces

is perpendicular to the front and back surfaces but inclined to the top, bottom, right side, and left side. The surface is said to be *foreshortened* when its true length is not shown. When you view this inclined surface, you see it slanted in only the front view. (You could see it slanted in the back view, but since it would appear exactly as it does in the front view, only the front is shown.)

Complete Exercise 7–2 now to see how well you can identify inclined and normal surfaces.

EXERCISE 7–2
Identifying Inclined Surfaces

Step 1. Remove the sheet labeled Exercise 7–2 from your book.
Step 2. Identify the numbered surfaces in the normal views in the spaces provided. Use the numbers from the pictorial view. Use your best lettering with guidelines top and bottom.
Step 3. Fill in the title block with your best lettering. Title the drawing **INCLINED SURFACES**.

Oblique Surfaces

Slanted surfaces that are inclined to all the normal surfaces are called *oblique surfaces* (Figure 7–11). Notice that the oblique surface appears in all three views and that only one line of the surface is true length in each view. You will read more about oblique surfaces in the chapters on auxiliary views. Try Exercise 7–3 to identify oblique and normal surfaces.

Reading and Sketching Orthographic Views

FIGURE 7–11
Oblique Surfaces

FIGURE 7–12
Edges on a Thin Sheet

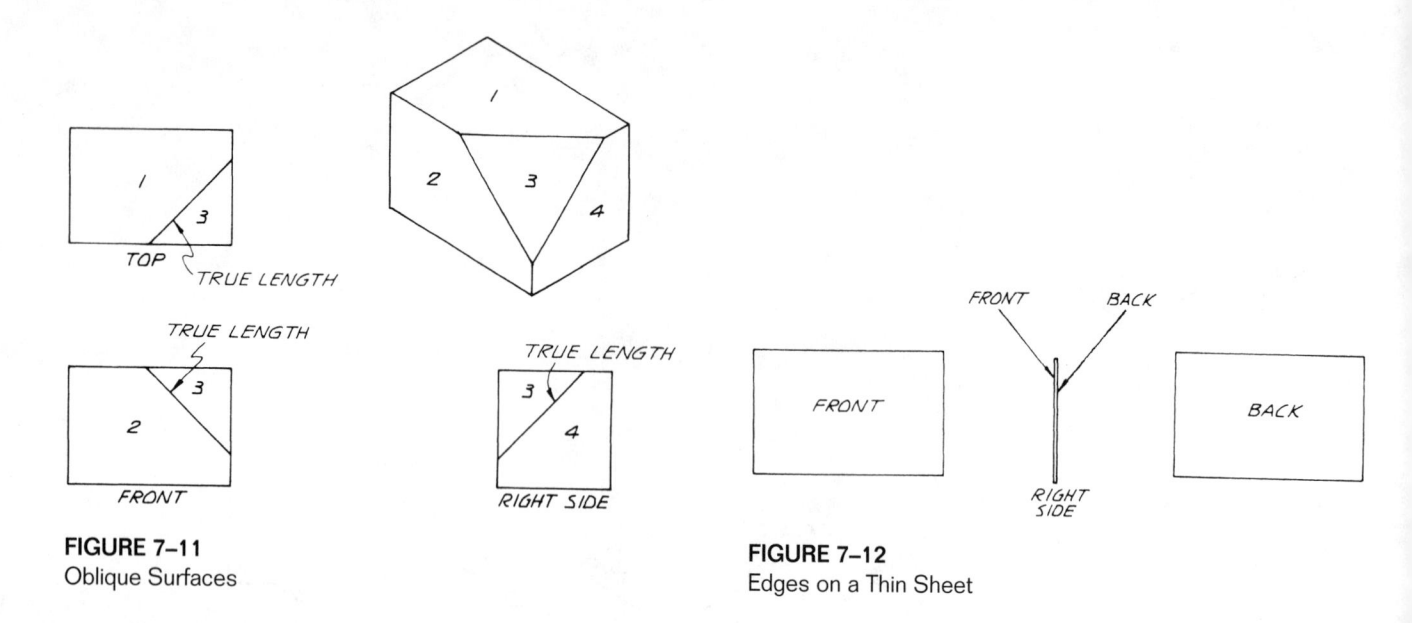

EXERCISE 7–3
Identifying Oblique Surfaces

Step 1. Remove the sheet labeled Exercise 7–3 from your book.

Step 2. Identify the numbered surfaces in the normal views in the spaces provided. Use the numbers from the pictorial view. Use your best lettering with guidelines top and bottom.

Step 3. Fill in the title block with your best lettering. Title the drawing **OBLIQUE SURFACES**.

Identifying Edges

In reading two-dimensional drawings, it is often helpful to identify surfaces where they appear as an edge. As shown in Figure 7–12, when you look at a very thin sheet of metal from the front, you see it in the true shape of its front surface. When you view it from its right side, you see it as an edge. Then, when you flop it over, you see the true shape of its back surface.

If the metal is thicker, as in Figure 7–13, the edges are farther apart, and the side view takes on a shape. In Figure 7–14 notice that edges in one view line up with edges in the view adjacent to it. Surfaces 2, 5, 7, and 10 in the top view, for example, line up with the

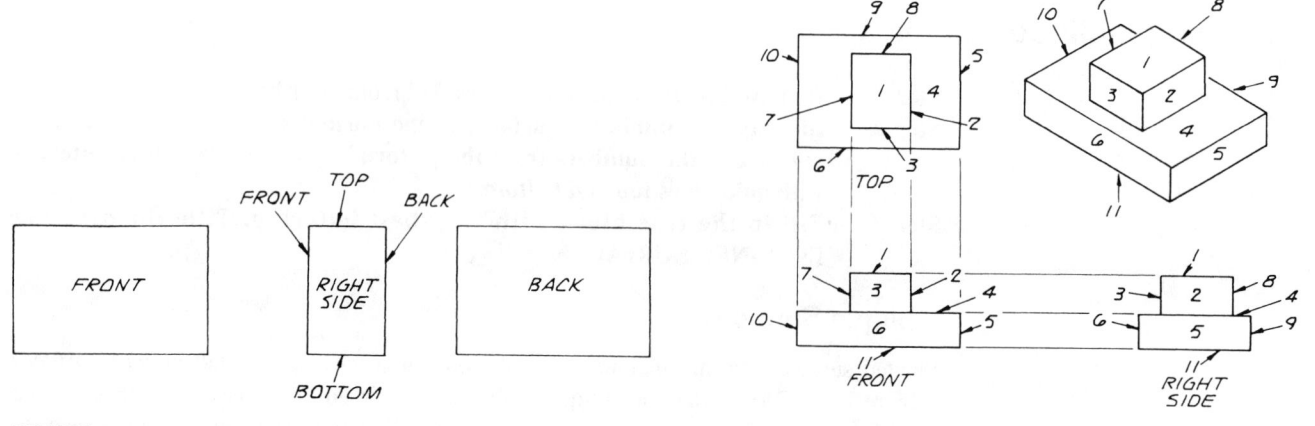

FIGURE 7–13
Edges on a Thicker Part

FIGURE 7–14
Identifying Edges

Chapter 7

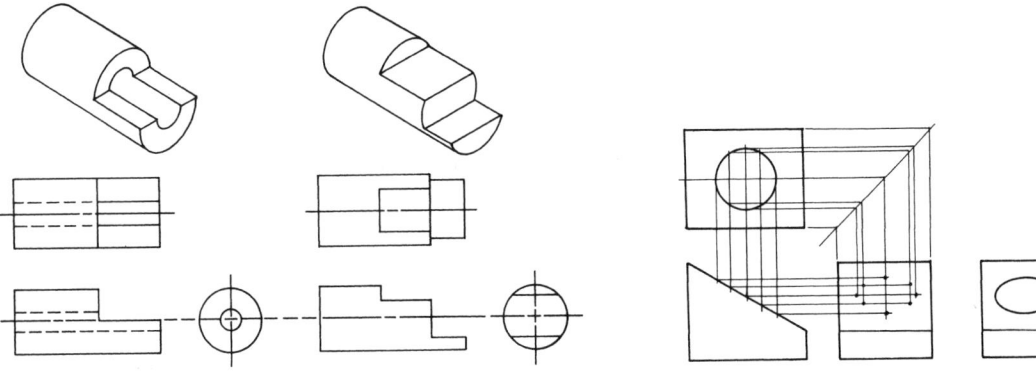

FIGURE 7–15
Cylinders Cut at Right Angles

FIGURE 7–16
Cylinders Cut at an Angle

same surfaces in the front view. Surfaces 1, 4, and 11 line up with the same surfaces in the right-side view.

Study Figure 7–14 for a few minutes until you feel certain that you understand these views and can number edges correctly. After you feel confident about Figure 7–14, complete Exercise 7–4.

EXERCISE 7–4
Identifying Edges

Step 1.   Remove the sheet labeled Exercise 7–4 from your book.
Step 2.   Identify the numbered surfaces as edges in the normal views in the spaces provided. Use the numbers from the pictorial view. Use your best lettering with guidelines top and bottom.
Step 3.   Fill in the title block with your best lettering. Title the drawing EDGES.

Cut Cylinders

When cylinders are cut at right angles, their shapes in adjacent views do not change (Figure 7–15). When they are cut at an angle, their shapes appear as ellipses in one of the adjacent views (Figure 7–16).

Runouts

A *runout* is a surface that blends into another surface without forming an edge at the point where the runout ends. Figure 7–17 shows a runout on a casting. Castings and other molded parts often have runouts. Castings also often have rounded edges because of the way they are manufactured. The only sharp edges on most castings are surfaces that have been machined to allow a part to fit or function better. Some rounds of the cast parts are

FIGURE 7–17
A Runout

Reading and Sketching Orthographic Views

FIGURE 7-18
Two Shapes with Runouts

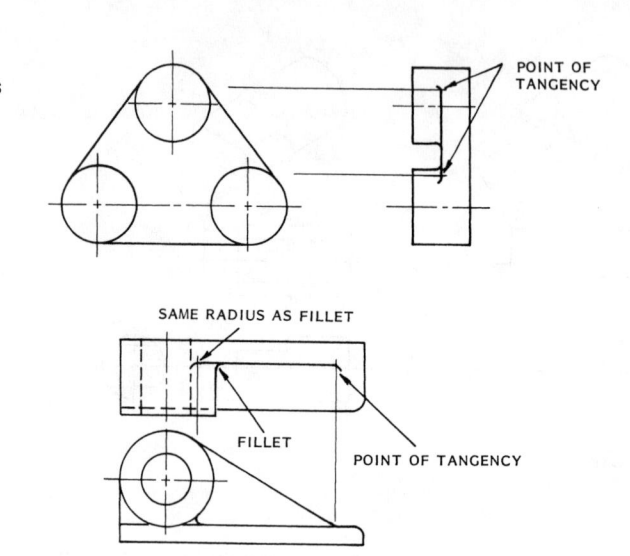

usually found connected to the runout. Figure 7–18 shows two shapes with runouts. Runouts are drawn just as you see them in this figure. The edge view line ends with the radius of the fillet drawn at the end. The point of tangency, where the flat surface meets the cylindrical surface, is where the runout ends.

Different Shapes That Look the Same in One or More Views

In reading two-dimensional views, it is easy to make a mistake about the shape of an object if you look at only two views. Figure 7–19 shows several objects that have the same appearance in two views. Only the right-side view shows the true shape of each object.

Hidden Features

To describe many objects fully, hidden surfaces often must be shown. These surfaces are shown with hidden lines (Figure 7–20). Notice that the right-side view would not be complete if the hidden surface were not shown with a hidden line.

When hidden lines are drawn at a corner, they should form an L shape (Figure 7–21). When they intersect, they should form a T or a cross if one line crosses over the other. If they are a continuation of a solid line, there should be a break between the solid line and the first dash of a hidden line. The dashes should be about $\frac{1}{8}''$ long, with $\frac{1}{16}''$ between dashes. They can be longer on larger drawings.

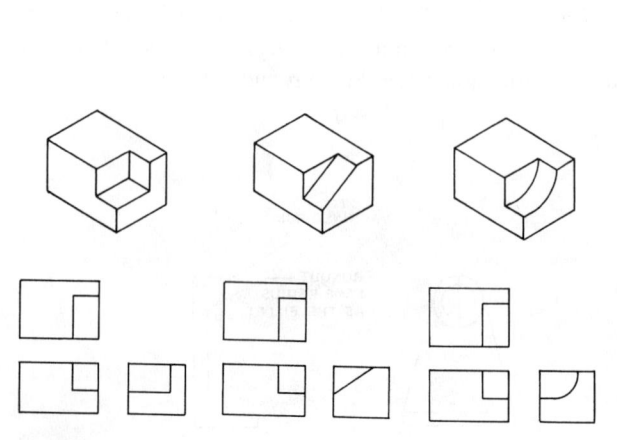

FIGURE 7-19
Different Shapes with the Same Front and Top Views

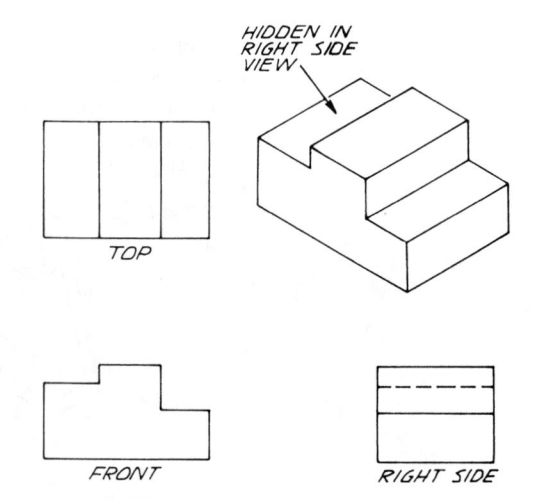

FIGURE 7-20
Hidden Surface Shown with a Hidden Line

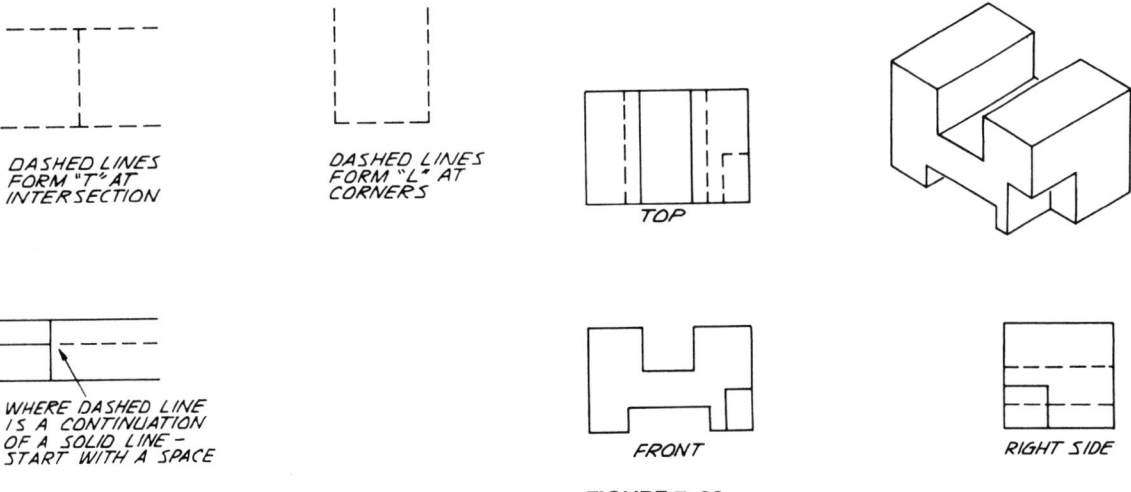

FIGURE 7–21
Rules for Drawing Hidden Lines

FIGURE 7–22
Views with Hidden Lines

Figures 7–22 and 7–23 show several hidden lines. In Figure 7–23 notice that the solid line in the top view takes precedence over the hidden line that shows the hidden slot. On complex objects, the number of hidden lines often can be very confusing, so it is generally understood that only one layer of hidden lines is shown (Figure 7–24).

Normal Cylinders

Cylinders that are perpendicular to the normal surfaces of an object are called *normal cylinders*. These cylinders may be round holes or round rods. Figure 7–25 shows how normal cylinders look in two-dimensional views. Notice that the edges of the holes are shown with hidden lines in the view where they are hidden. The hidden lines in the front view that show the holes in front exactly cover the hidden lines that show the holes in the back.

When holes do not go all the way through the object, the bottom of the hole is shown with a hidden line as in Figure 7–26. Notice that the hole in Figure 7–26B is slanted so that the circle at the bottom of the hole is smaller than the circle at the top of the hole.

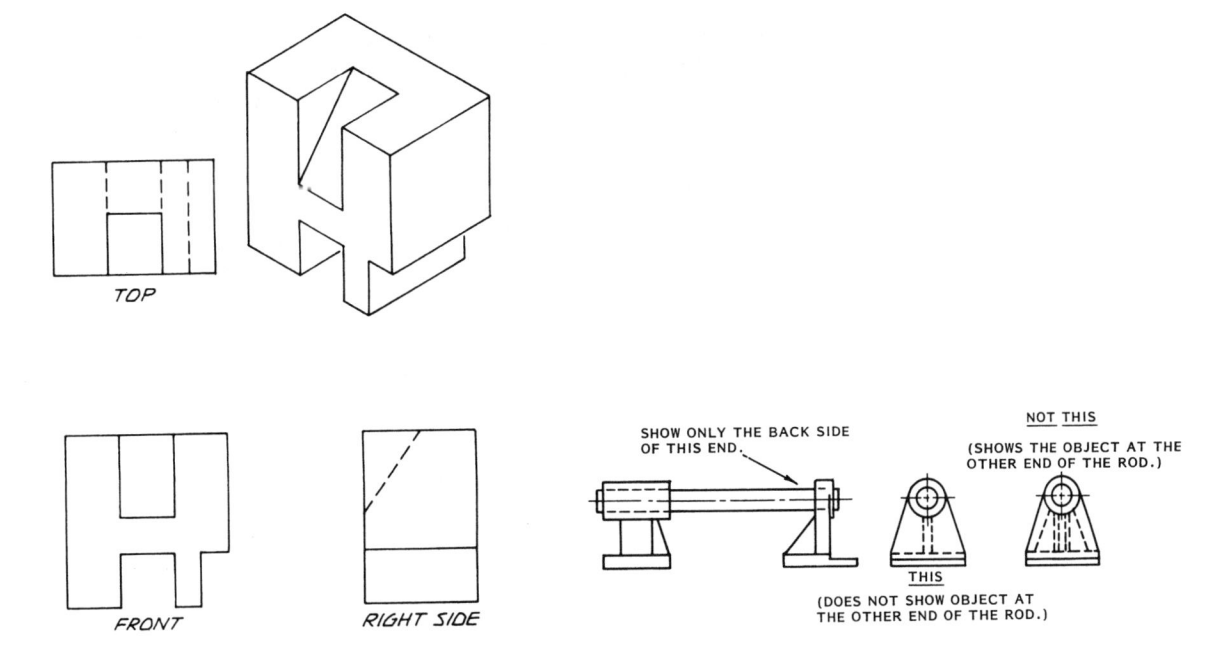

FIGURE 7–23
More Views with Hidden Lines

FIGURE 7–24
Showing Only One Layer of Hidden Features

Reading and Sketching Orthographic Views

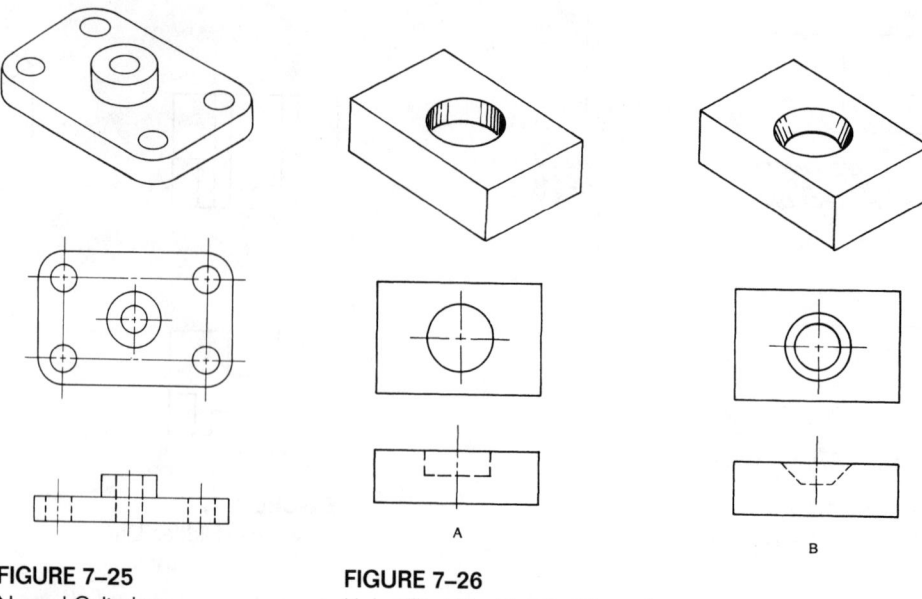

FIGURE 7–25
Normal Cylinders

FIGURE 7–26
Holes That Do Not Go Through

Countersinks and Counterbores

Figure 7–27 shows the features countersink and counterbore. The countersink feature is used to allow a flathead screw that has the same shape as the countersunk hole to fit below the surface of the part. A hole is drilled in the part first and then the countersink is added. The counterbore comprises two holes, one larger than the other. The smaller hole is drilled first, and then the counterbore is added. Notice that the bottom of the larger hole, which does not go through the part, is shown with a hidden line. A counterbore is used to allow parts to fit deeper into the material. A variation of the counterbore, called a *spotface*, is a shallow counterbore that makes the surface smoother so that the head of a screw or bolt fits better. Now use your hidden line skills to fill in the hidden lines on the views in Exercise 7–5.

EXERCISE 7–5
Sketching Hidden Lines

Step 1. **Remove the sheets labeled Exercise 7–5, Sheets 1 and 2, from your book.**
Step 2. **Add missing object lines and the missing hidden lines in the top and front views of sheet 1, and in the front and right-side views of sheet 2. Make the**

FIGURE 7–27
Counterbore and Countersink

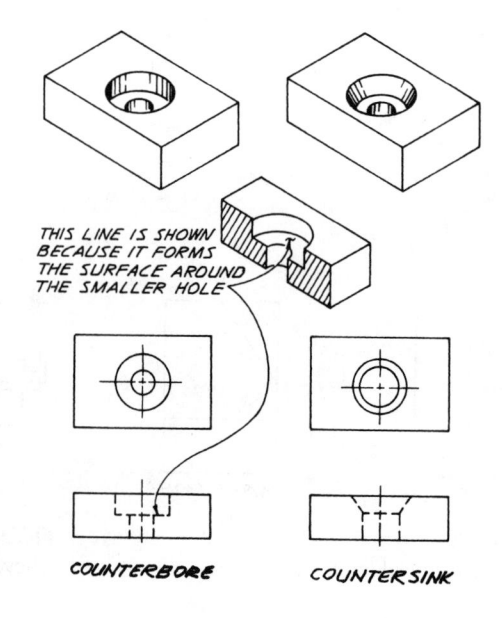

dashes about $\frac{1}{8}''$ long with a $\frac{1}{16}''$ space between dashes. Make sure the hidden lines are aligned with features in the adjacent views and that your hidden lines follow the rules described in Figure 7–21. Use the pictorial drawing as a guide.

Step 3. Fill in the title blocks with your best lettering. Title the drawings HIDDEN SURFACES.

SKETCHING ORTHOGRAPHIC VIEWS

Although you have already begun sketching in the previous chapters, we need to review some of the common practices in technical sketching and add some other information that will make your sketches professional quality.

Materials

The materials you will need have already been listed in a previous chapter, namely, paper (the exercises in this book), pencils, triangle, eraser, and circle template. If you are making sketches on your own, you may prefer gridded paper for some objects, but you will find that plain paper is just as easy to use for some drawings. Be sure the eraser you use makes a clean erasure because you will make some mistakes, and erasing is not only OK, it is encouraged.

Lines

You have already begun using lines in your sketches. If you are not certain about how they should appear, review Chapter 5 before you complete the remaining exercises in this chapter.

Circles

Some of the circles you will sketch will be bigger than the largest circle on your circle template. These circles will not look as good as the ones you draw with the circle template, but they should be as accurate as you can make them. Figure 7–28 shows an easy method for drawing circles.

Step 1. Determine the diameter of the circle you want to sketch.
Step 2. Using light lines, draw a square the size of the diameter of the circle.
Step 3. Find the center of the circle by drawing diagonals across the corners of the square.
Step 4. Using the center as a guide, estimate the midpoint of each side and mark it.
Step 5. Along the diagonals, mark the radius of the circle from its center.
Step 6. Connect the construction points using a dark sketch line.

Figure 7–29 shows another method for drawing circles.

Step 1. Determine the diameter of the circle you want to sketch.
Step 2. Draw horizontal, vertical, and 45° construction lines through a center point.
Step 3. Mark the radius of the circle on the construction lines.
Step 4. Connect the construction points using a dark sketch line.

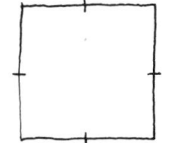

FIGURE 7–28
Sketching a Circle

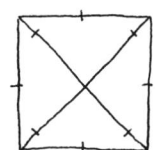

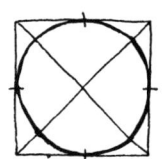

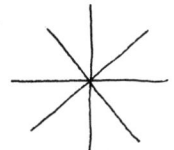

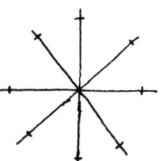

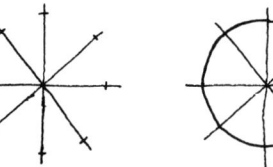

FIGURE 7–29
Another Method for Sketching a Circle

Reading and Sketching Orthographic Views

FIGURE 7–30
Sketching an Arc

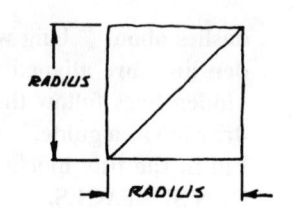

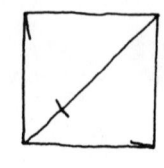

Arcs

Arcs larger than your circle template are sketched in a manner similar to the one used for circles. To draw an arc using the method shown in Figure 7–30, follow these steps:

Step 1. Sketch a square the size of the radius of the arc.
Step 2. Draw a construction line across a diagonal of the square.
Step 3. Mark the radius on the diagonal.
Step 4. Sketch the radius using a dark sketch line from one corner of the square through the mark on the diagonal to the other corner.

Aligning Views

As noted earlier, all features of an object in one view must be lined up with those same features in another view. When you make your sketches, drawing will be much easier if you are sure to keep all the views lined up with one another. Often, one view will have a feature that must be completed before it can be projected into the adjacent view, as shown in Figure 7–31. The 45° miter line method shown in Figure 7–32 can be used to project depth dimensions from the top view to the right-side view.

FIGURE 7–31
Aligning Features in Adjacent Views

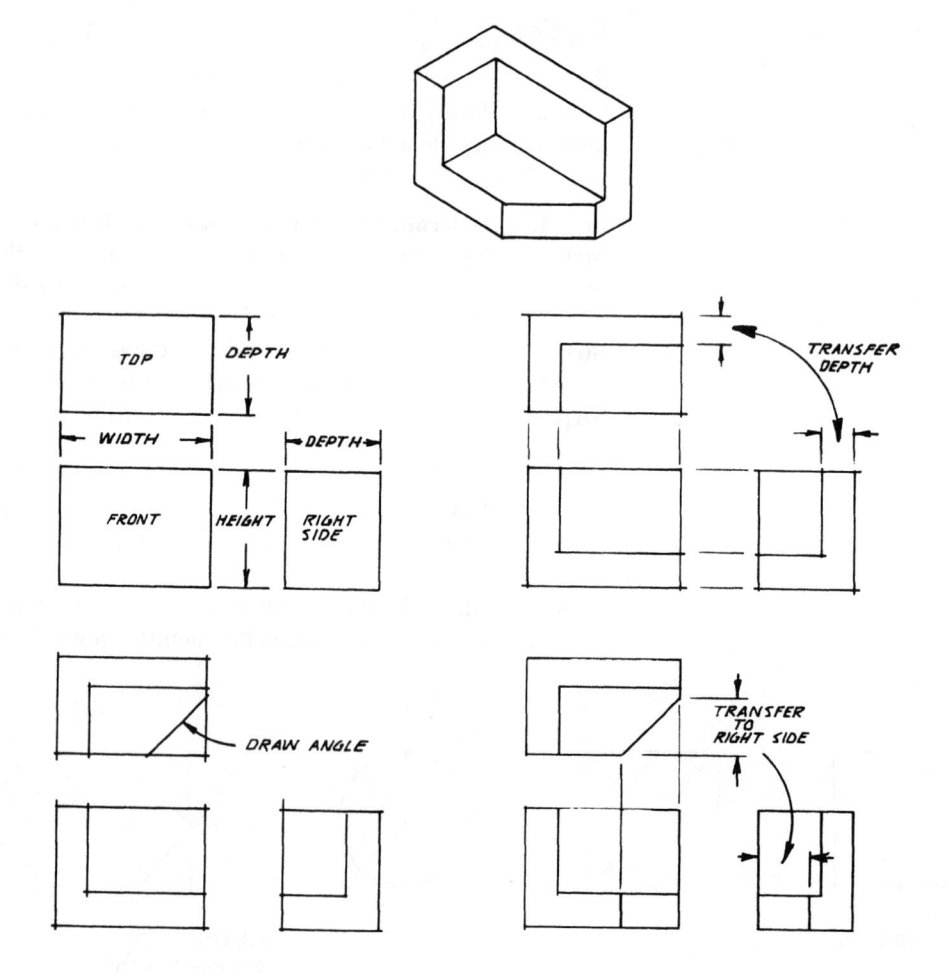

102 Chapter 7

FIGURE 7–32
Miter-Line Method for Projecting Depth

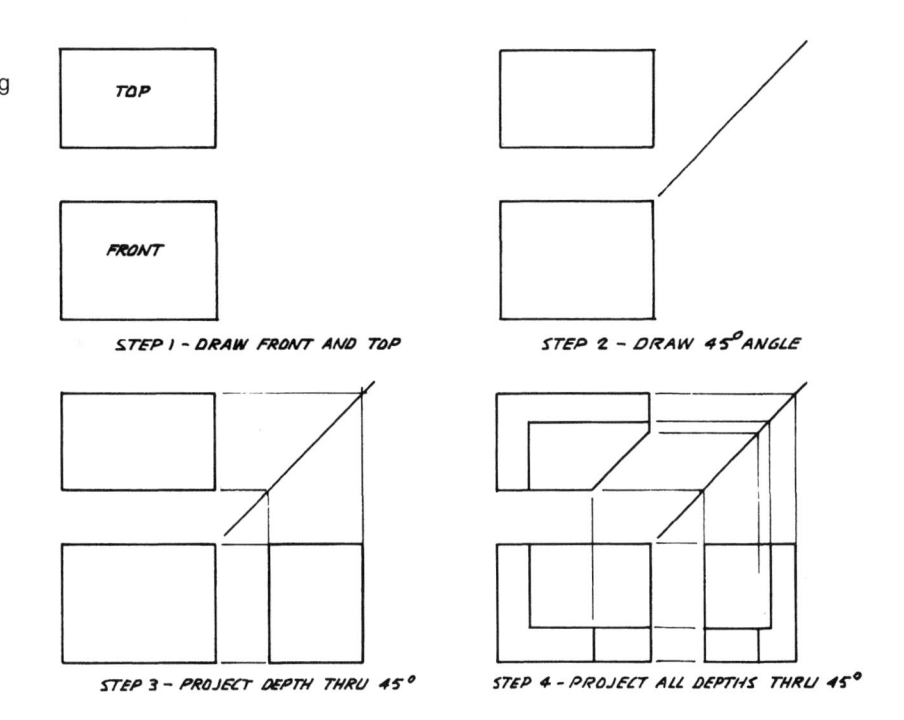

STEP 1 – DRAW FRONT AND TOP

STEP 2 – DRAW 45° ANGLE

STEP 3 – PROJECT DEPTH THRU 45°

STEP 4 – PROJECT ALL DEPTHS THRU 45°

Selecting Views

Selecting the views that best describe the object completely using the least number of views is very important. The views that show the least number of hidden lines and fully describe all contours of all surfaces should be selected. Figure 7–33 shows several objects and the correct view selection for them.

In Figure 7–33A, top, front, and right-side views show no hidden surfaces. Any other view selection would have shown hidden lines and would not have been as clear. The top view is necessary to show the shape of the surface, which could have had rounded corners, for example. The same description applies to the objects in Figure 7–33B, and C.

The objects in Figure 7–33D, E, and F are more complex, so additional views are necessary to describe fully the features that appear only as hidden lines in all other views.

The object in Figure 7–33G is an example of a flat object that can be described with only one view by placing a note on the drawing giving the thickness of the object.

The object in Figure 7–33H is an example of a round object that can be described with either two views or one view by showing the diameter on that view. The notation DIA (or ∅) means that the only shape the object can have is round.

Selecting the Front View

The front view is usually the starting point for all orthographic drawings. There are three factors to consider when you are deciding which surfaces to use as the front view.

Place the longest side horizontally on the front (Figure 7–34).
Place the most complex feature on the front (Figure 7–35).
Sketch the object in the position in which it is used or most often seen (Figure 7–36).

Drawing Circles and Curves on a Slanted Surface

Circular holes that have been cut at other than a 90° angle to their center appear as ellipses. An ellipse has the same major diameter as the circle but a smaller minor diameter. In other words, an ellipse is a circle that has been flattened. Figure 7–37 shows two methods for sketching an ellipse:

In Figure 7–37A, the circle has been divided into four parts so that points on the circle can be identified. These points are then projected onto the slanted surface in the front view and then into the right-side view. Depth measurements are taken from the center of the circle on the top view and transferred to the right-side view.

Reading and Sketching Orthographic Views

FIGURE 7–33
View Selection

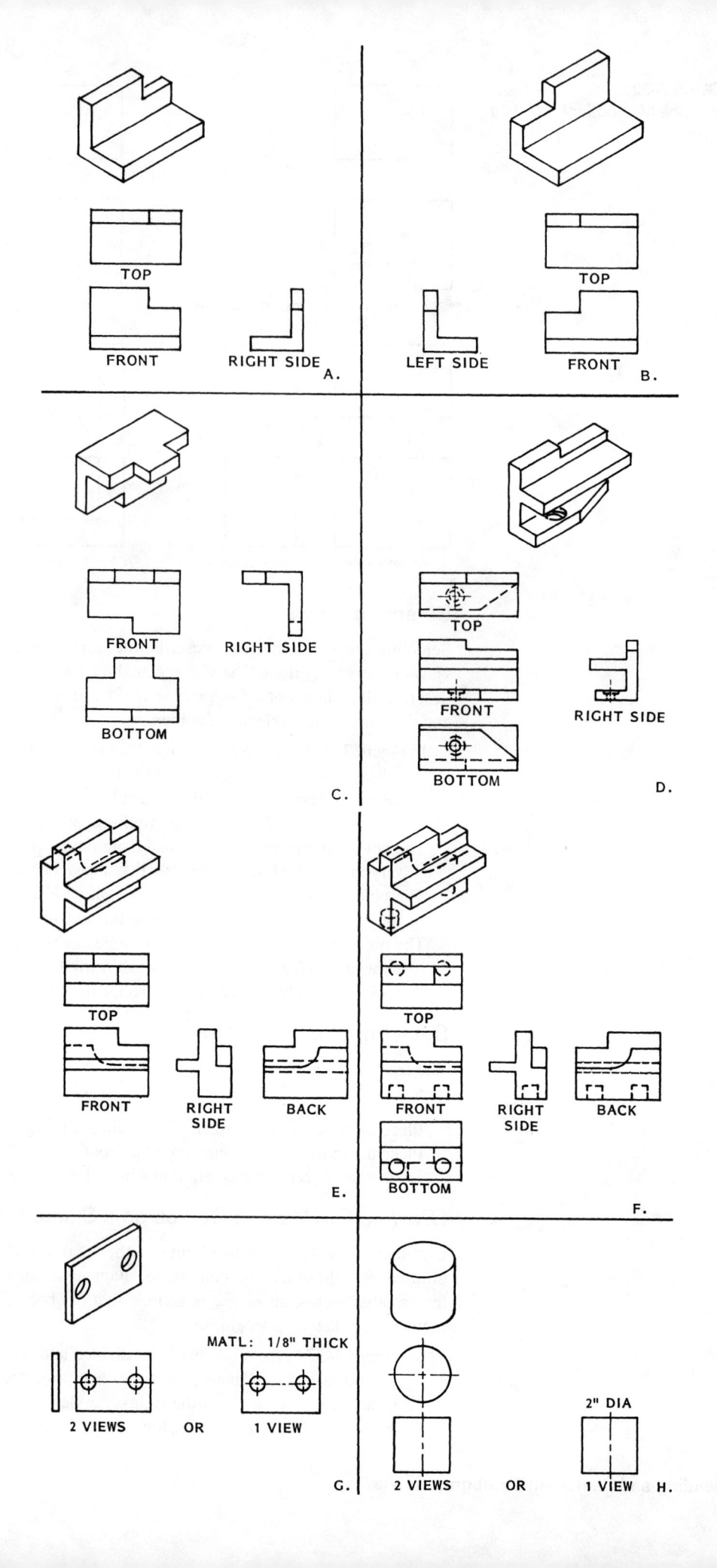

104

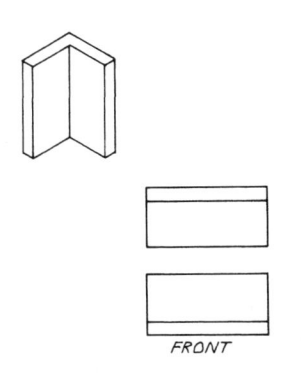

FIGURE 7–34
Placing the Longest Side Horizontally on the Front

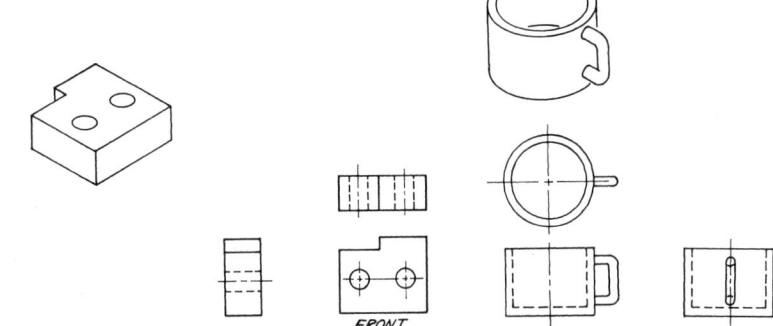

FIGURE 7–35
Placing the Most Complex Feature on the Front

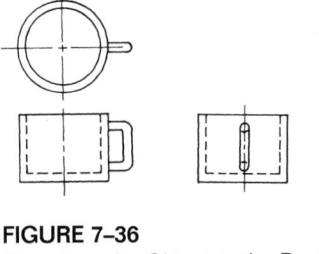

FIGURE 7–36
Sketching the Object in the Position in Which It Is Used or Most Often Seen

In Figure 7–37B, the miter-line method was used to transfer depth dimensions. Ellipses and other curved shapes can be sketched more accurately by dividing the circle or other shape into a greater number of parts.

Figure 7–38 shows a curved surface that has been drawn using the same methods as those shown in Figure 7–37. The only difference is that depth dimensions in Figure 7–38A were taken from the back surface.

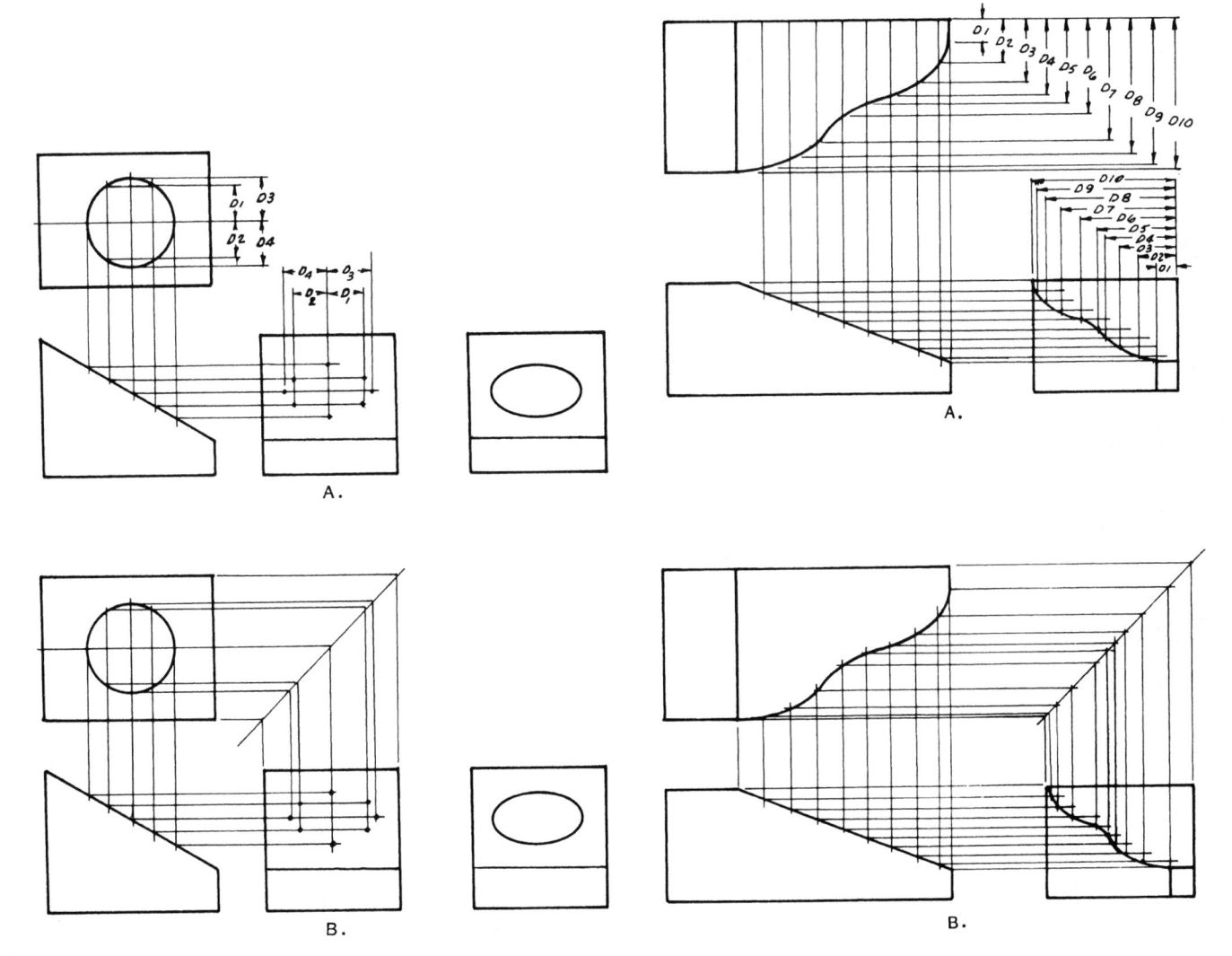

FIGURE 7–37
Sketching Ellipses

FIGURE 7–38
Sketching Curved Surfaces

Reading and Sketching Orthographic Views

Placing Views on the Drawing

Placing of the views on the page is important to the appearance of the drawing. At first, you should center the views in the drawing. Later, notes, dimensions, revisions, and parts lists will require a different arrangement. First you will have to know how to center the drawing before you can use the other arrangements.

As an example of centering, assume that the drawing will contain top, front, and right-side views of the object shown in Figure 7–39.

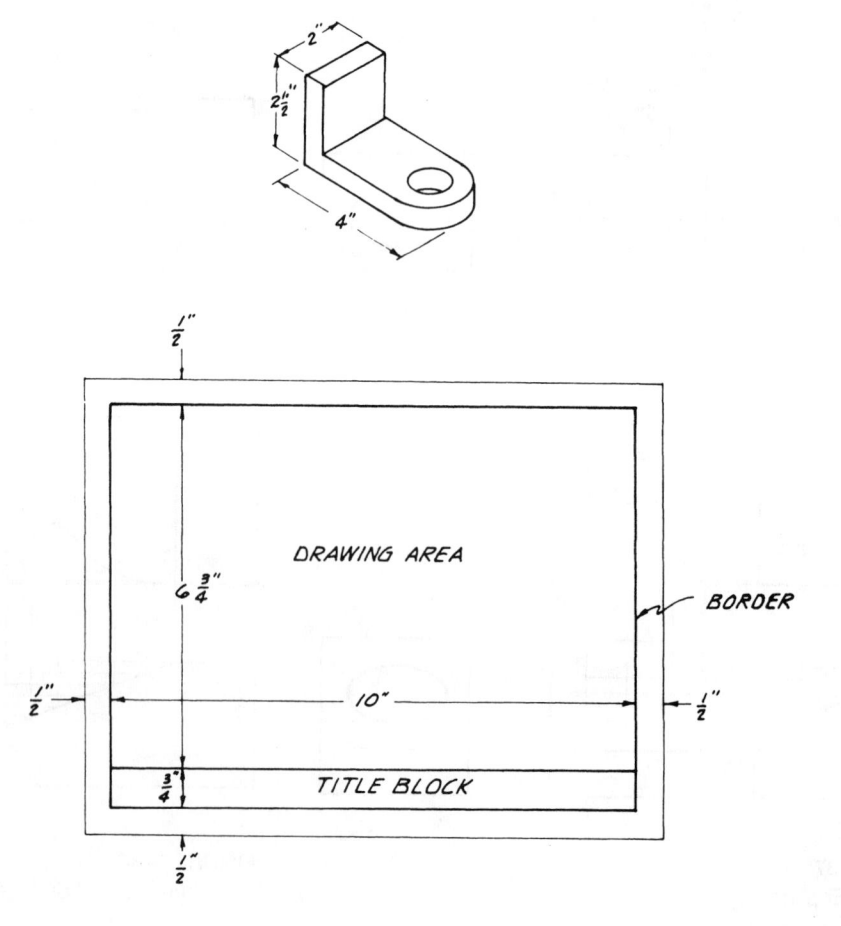

Step 1. Calculate the area called the field of the drawing that will contain the complete drawing. The drawing will be placed on an 11″ × 8½″ sheet. The border is ½″ on all sides, and the title block measures ¾″ deep. The drawing area is 10″ × 6¾″.

Step 2. Calculate where to start the left side of the front and top views (Figure 7–40). Add the width and depth dimensions (4″ + 2″ = 6″). Subtract this total from 10″, the width of the drawing field (10″ − 6″ = 4″). Divide this number by 4 (4″/4 = 1″). Place one-quarter of the space (1″) on the left, one-quarter on the right (1″), and half the available space between the views (2″). The 1″ dimension is not critical. You can make it a little more or less if you choose.

Step 3. Calculate where to start the bottom of the front and right-side views. Add the height and depth dimensions (2½″ + 2″ = 4½″). Subtract this total from 6¾″, the height of the drawing field (6¾″ − 4½″ = 2¼″). Divide this number by 4 (2¼″/4 = 9/16″). Place one-quarter of the space above the top view, one-quarter below the front view, and half the available space between views. The 9/16″ dimension is not critical. You can make the top and bottom spaces ½″, 5/8″, or ¾″ if you choose.

Please be aware that views must be moved farther apart when dimensions and notes are added. The preceding example is a guide only. Many other methods will work just as well.

FIGURE 7–39
The Object and the Drawing Sheet

106 Chapter 7

FIGURE 7–40
Centering Views in the Field of the Drawing

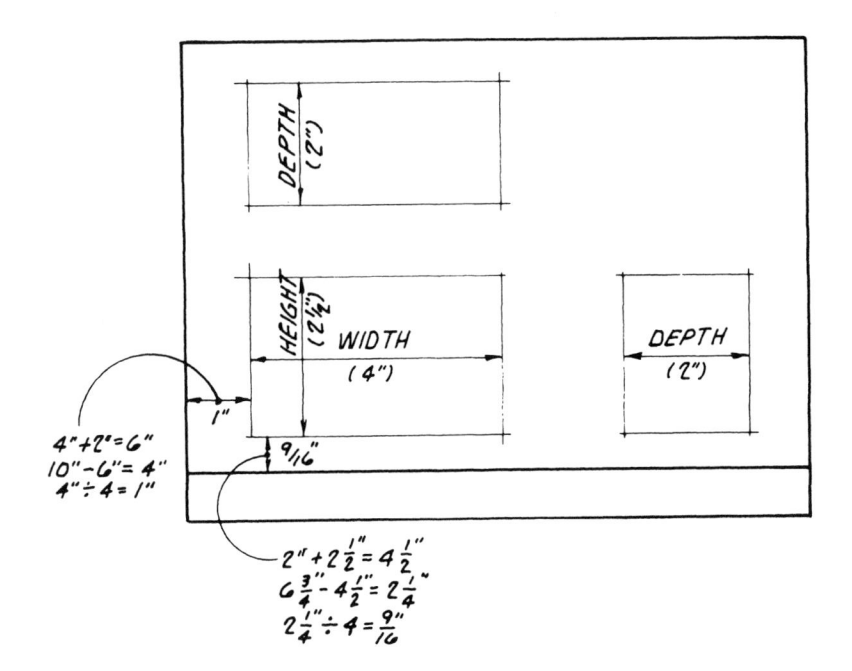

Order of Sketching

The order of sketching is important to maintain accurate sizes and to improve speed (Figure 7–41).

Step 1. Block in the overall dimensions of all views, using very light construction lines.
Step 2. Locate and mark the centerlines of all circles and arcs using light construction lines.
Step 3. Darken all circles and arcs.
Step 4. Darken all other object lines.
Step 5. Add hidden lines by projecting the surfaces from views where the features are visible.
Step 6. Darken the border and title block if the form is not preprinted, and letter the title block using guidelines and your best lettering.

Now use what you have learned about sketching to complete Exercises 7–6 through 7–10.

FIGURE 7–41
Order of Drawing

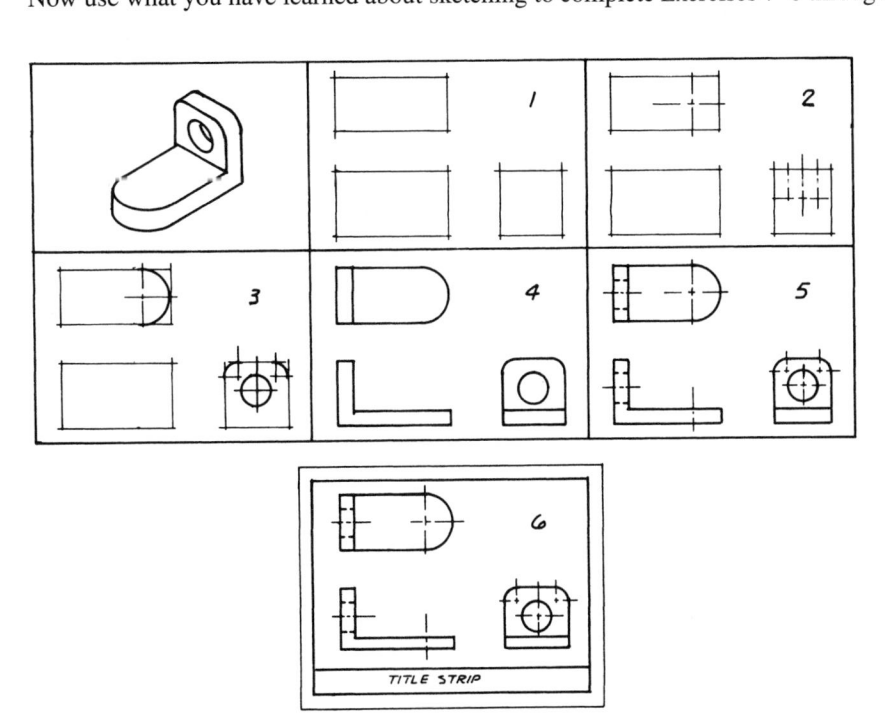

Reading and Sketching Orthographic Views

107

EXERCISES

EXERCISE 7–1	Complete Exercise 7–1 on the sheet in your book using steps 1 through 3 described in this chapter. Fill in the title block with your best lettering. Title the drawing NORMAL SURFACES.
EXERCISE 7–2	Complete Exercise 7–2 on the sheet in your book using steps 1 through 3 described in this chapter. Fill in the title block with your best lettering. Title the drawing INCLINED SURFACES.
EXERCISE 7–3	Complete Exercise 7–3 on the sheet in your book using steps 1 through 3 described in this chapter. Fill in the title block with your best lettering. Title the drawing OBLIQUE SURFACES.
EXERCISE 7–4	Complete Exercise 7–4 on the sheet in your book using steps 1 through 3 described in this chapter. Fill in the title block with your best lettering. Title the drawing EDGES.
EXERCISE 7–5	Complete Exercise 7–5 on the sheets in your book using steps 1 through 3 described in this chapter. Fill in the title block with your best lettering. Title the drawing HIDDEN SURFACES.
EXERCISE 7–6	Complete Exercise 7–6 on the sheet in your book using the steps described.

Step 1. Remove the sheet labeled Exercise 7–6 from your book.

Step 2. Complete the object lines in the front and right-side views of Exercise 7–6.

(A) Use one unit on the three-dimensional drawing to equal one unit on the two-dimensional sketch. (B) Be sure to line up all features of the object with the adjacent views. (C) Make sure that your lines are the correct weight and are of even width and darkness. Try to match the thickness and darkness of the existing lines.

Step 3. Draw hidden lines in the top and right-side views.

(A) Be sure to line up all features of the object with the adjacent views. (B) Make hidden lines the same width and darkness as the object lines.

Step 4. Fill in the title block with your best lettering. Title the drawing SKETCH 1.

EXERCISE 7–7 Complete Exercise 7–7 using the steps described.

Step 1. Remove the sheet labeled Exercise 7–7 from your book.

Step 2. Complete the object lines in the top, front, and right-side views of Exercise 7–7.

(A) Use one unit on the three-dimensional drawing to equal one unit on the two-dimensional sketch. (B) Be sure to line up all features of the object with the adjacent views. (C) Make sure that your lines are the correct weight and are of even width and darkness. Try to match the thickness and darkness of the existing lines.

Step 3. Draw hidden lines in the right-side view.

(A) Be sure to line up all features of the object with the adjacent views. (B) Make hidden lines the same width and darkness as the object lines.

Step 4. Fill in the title block with your best lettering. Title the drawing SKETCH 2.

EXERCISE 7–8 Complete Exercise 7–8 using the steps described.

Step 1. Remove the sheet labeled Exercise 7–8 from your book.

Step 2. Complete the object lines in the top, front, and right-side views of Exercise 7–8.

(A) Use one unit on the three-dimensional drawing to equal one unit on the two-dimensional sketch. (B) Be sure to line up all features of the object with the adjacent views. (C) Make sure that your lines are the correct weight and are of even width and darkness. Try to match the thickness and darkness of the existing lines.

Step 3. Draw hidden lines in the top view.

(A) Be sure to line up all features of the object with the adjacent views. (B) Make hidden lines the same width and darkness as the object lines.

Step 4. Fill in the title block with your best lettering. Title the drawing SKETCH 3.

EXERCISE 7–9 Complete Exercise 7–9 using the steps described.

Step 1. Remove the sheet labeled Exercise 7–9 from your book. Use light construction lines to sketch the top, front, and right-side views of step 1 shown in Figure 7–42.

FIGURE 7–42
Dimensions for Exercise 7–9

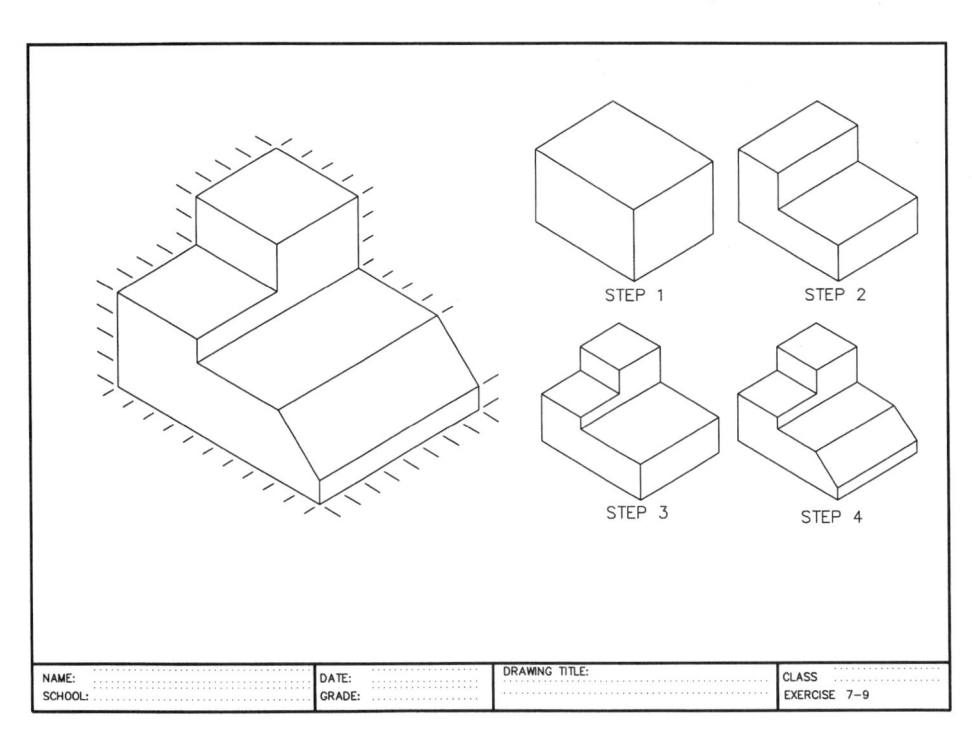

(A) Use one unit on the three-dimensional drawing to equal one unit on the two-dimensional sketch. (B) Be sure to line up all features of the object with the adjacent views.

Step 2. Use light construction lines to sketch the L-shape in the front view of step 2 in Figure 7–42. Project the edges of the L-shape into the top and right-side views with light construction lines.

Step 3. Use light construction lines to draw the notch in the right-side view as shown in step 3. Project the edges of the notch into the top and front views with light construction lines.

Step 4. Use light construction lines to draw the slanted surface in the front view of step 4, and project the edges defining the slanted surface into the top and right-side views. (You can draw these lines with dark object lines if you are sure of their locations.) Darken all lines as follows:

(A) Sprinkle drafting powder over your drawing before you begin to darken lines. (B) Darken all other object lines in all views.

Step 5. Fill in the title block with your best lettering. Title the drawing **SKETCH 4**.

EXERCISE 7–10   Complete Exercise 7–10 using the steps described.

Step 1. Remove the sheet labeled Exercise 7–10 from your book. Use light construction lines to sketch the top, front, and right-side views of step 1 shown in Figure 7–43.

(A) Use one unit on the three-dimensional drawing to equal one unit on the two-dimensional sketch. (B) Be sure to line up all features of the object with the adjacent views.

Step 2. Use light construction lines to sketch the L-shape in the front view of step 2 of Figure 7–43. Project the edges of the L-shape into the top and right-side views with light construction lines.

Step 3. Use light construction lines to draw the visible lines of the notch in the upper part of the object in the right-side and top views as shown in step 3. Use light construction lines to draw the notch in the base in the top view. Project the edges of the notch into the front and right-side views.

Step 4. Use light construction lines to draw the hidden lines showing the shape of the slanted surface in the front view of step 4, and project the edges defining the slanted surface into the top and right-side views. (You can draw these lines with dark object lines if you are sure of their locations.) Draw the small notch in the bottom of the base in the right-side view using light construction lines. Project the edges of the notch into the front and top views as hidden lines. Darken all lines as follows:

Reading and Sketching Orthographic Views

FIGURE 7–43
Dimensions for Exercise 7–10

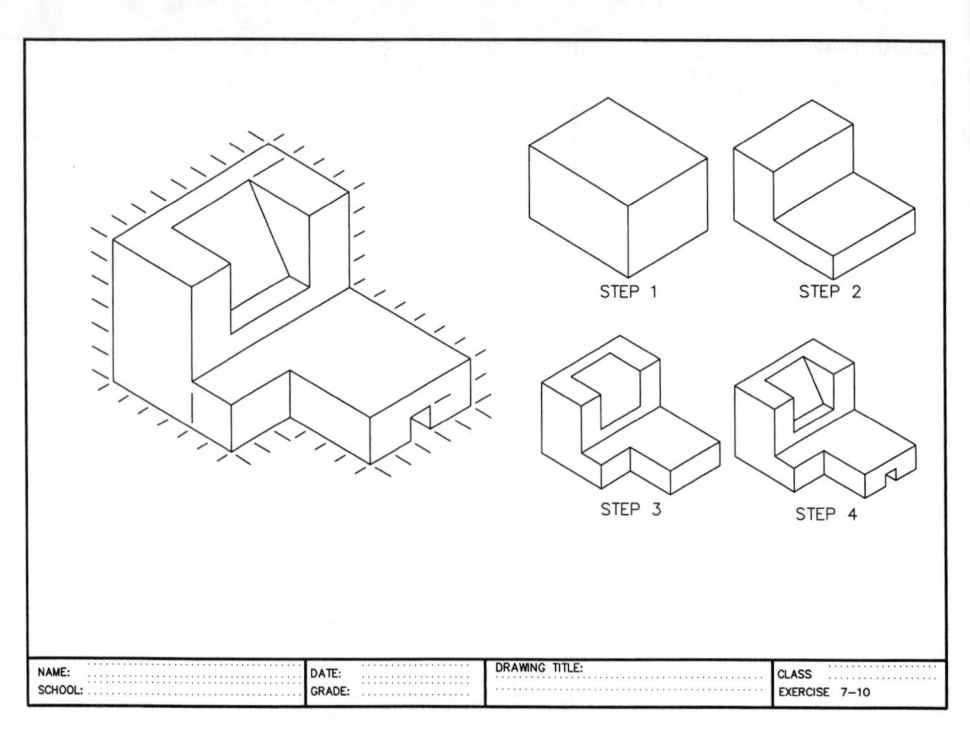

(A) Sprinkle drafting powder over your drawing before you begin to darken lines. (B) Darken all lines in all views.

Step 5. Fill in the title block with your best lettering. Title the drawing SKETCH 5.

EXERCISE 7–11   Complete Exercise 7–11 using the steps described.

Step 1. Remove the sheet labeled Exercise 7–11 from your book. Use light construction lines to sketch the top, front, and right-side views of step 1 shown in Figure 7–44.

(A) Use one unit on the three-dimensional drawing to equal one unit on the two-dimensional sketch. (B) Be sure to line up all features of the object with the adjacent views.

Step 2. Use light construction lines to sketch the L-shape in the front view of step 2 of Figure 7–44. Project the edges of the L-shape into the top and right-side views with light construction lines.

Step 3. Use light construction lines to locate centers for the circles in the top view as shown in step 3. Draw the chamfers in the right-side view using light construction lines, and project these features into the top and right-side views. (You can draw the chamfers with dark object lines if you are sure of their locations.)

Step 4. Draw the circles and the radii with dark lines, and darken all other lines as follows:

(A) Sprinkle drafting powder over your drawing before you begin to darken lines. (B) Darken all other object lines in all views.

Step 5. Draw hidden lines in the top and right-side views.

(A) Be sure to line up all features of the object with the adjacent views. (B) Make hidden lines the same width and darkness as the object lines.

Step 6. Fill in the title block with your best lettering. Title the drawing SKETCH 6.

EXERCISE 7–12   Complete Exercise 7–12 using the steps described.

Step 1. Remove the sheet labeled Exercise 7–12 from your book. Use light construction lines to sketch the top, front, and right-side views of step 1 shown in Figure 7–45.

(A) Use one unit on the three-dimensional drawing to equal one unit on the two-dimensional sketch. (B) Be sure to line up all features of the object with the adjacent views.

Step 2. Use light construction lines to sketch the L-shape in the right-side view of step 2 of Figure 7–45. Project the edges of the L-shape into the front and top views with light construction lines.

Step 3. Use light construction lines to locate centers for all circles in the top and front views as shown in step 3. Draw the chamfers in the top view using light construction lines, and project these features

FIGURE 7–44
Dimensions for Exercise 7–11

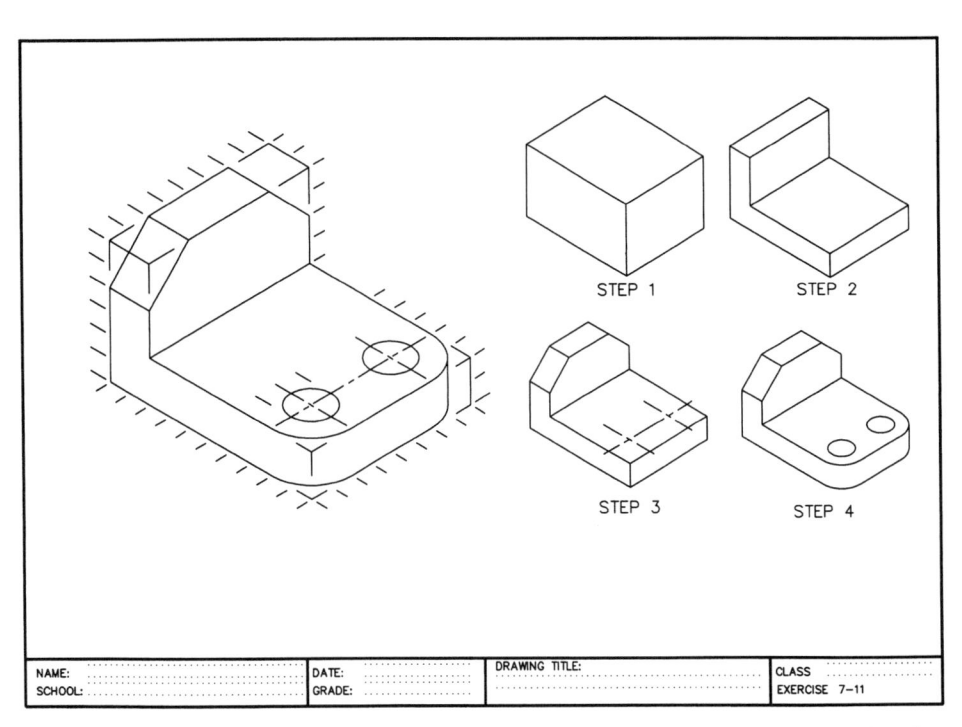

into the front and right-side views. (You can draw the chamfers with dark object lines if you are sure of their locations.)

Step 4. **Use light construction lines to draw the top half of the $1\frac{1}{2}''$-diameter circle in the front view. Draw all remaining lines and darken all lines as follows:**

(A) Sprinkle drafting powder over your drawing before you begin to darken lines. (B) Draw tangents to the $1\frac{1}{2}''$ circle from the left and right corners of the base using dark object lines. (C) Darken all remaining circles and other object lines in all views.

Step 5. **Draw hidden lines in the top and right-side views.**

(A) Be sure to line up all features of the object with the adjacent views. (B) Make hidden lines the same width and darkness of the object lines.

Step 6. **Fill in the title block with your best lettering. Title the drawing SKETCH 8.**

EXERCISE 7–13 Complete Exercise 7–13 using the steps described.

Step 1. **Remove the sheet labeled Exercise 7–13 from your book. Use light construction lines to sketch the top, front, and right-side views of step 1 shown in Figure 7–46.**

FIGURE 7–45
Dimensions for Exercise 7–12

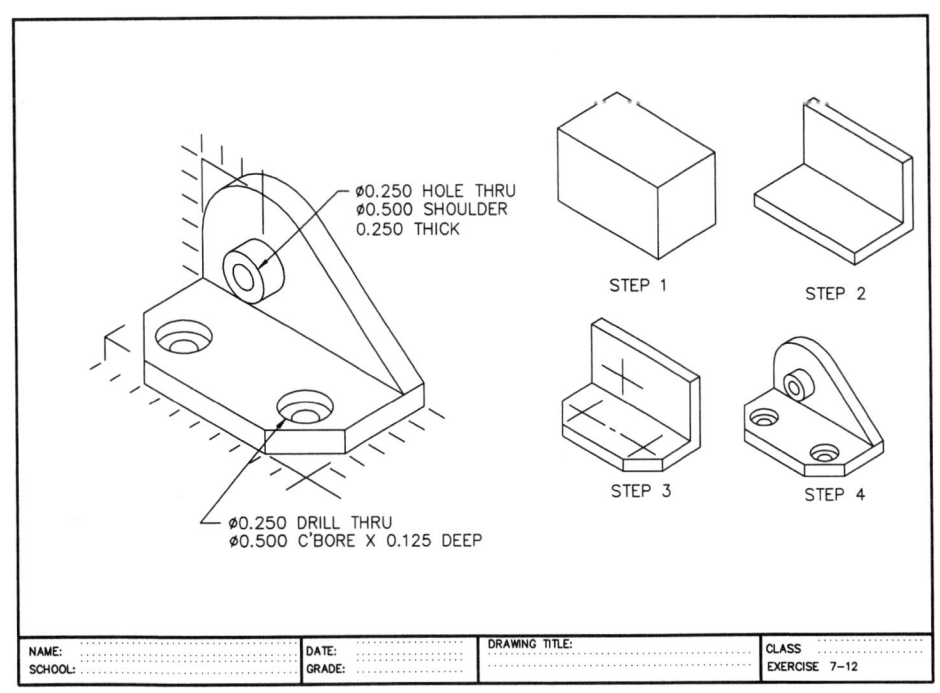

Reading and Sketching Orthographic Views

111

FIGURE 7–46
Dimensions for Exercise 7–13

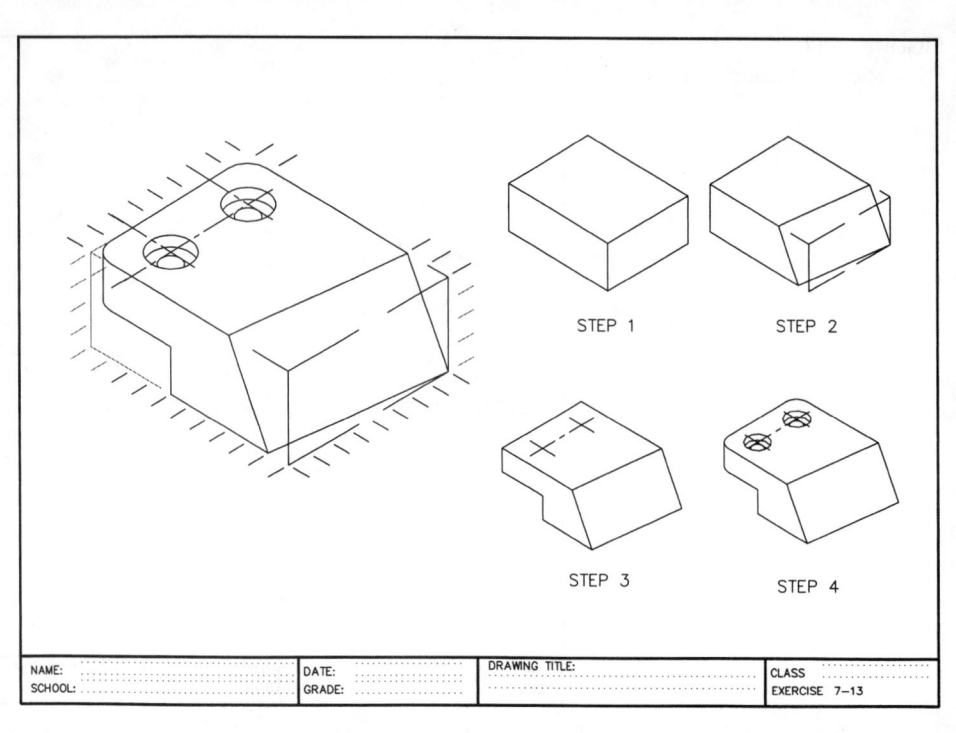

(A) Use one unit on the three-dimensional drawing to equal one unit on the two-dimensional sketch. (B) Be sure to line up all features of the object with the adjacent views.

Step 2. Use light construction lines to sketch the oblique surface in the top and front views of step 2 of Figure 7–46. There will be two angular lines in both views. Locate each end of each line using the measurements shown in the pictorial view, then sketch the line.

Step 3. Use light construction lines to locate centers for the circles in the top view as shown in step 3.

Step 4. Draw the circles and radii as dark object lines and darken all other lines as follows:
(A) Sprinkle drafting powder over your drawing before you begin to darken lines. (B) Darken all other object lines in all views.

Step 5. Draw hidden lines showing the drilled holes and the counterbores in the front and right-side views.
(A) Be sure to line up all features of the object with the adjacent views. (B) Make hidden lines the same width and darkness as the object lines.

Step 6. Fill in the title block with your best lettering. Title the drawing **SKETCH 9**.

EXERCISE 7–14 Complete Exercise 7–14 using the steps described.

Step 1. Remove the sheet labeled Exercise 7–14 from your book.

Step 2. Fill in the missing information in the views indicated. Use the miter-line method to project depth dimensions. Add missing hidden lines as well as missing object lines. Be sure to line up all features of the object with the adjacent views.

Step 3. Fill in the title block with your best lettering. Title the drawing **MISSING LINES**.

REVIEW QUESTIONS

Circle the best answer.

1. The method of technical drawing used in the United States is
   a. First-angle orthographic projection
   b. Second-angle orthographic projection
   c. Third-angle orthographic projection
   d. Fourth-angle orthographic projection
   e. All four are used about the same amount.

2. The top view of an object should be drawn
   a. To the right of the front view
   b. Directly above the front view
   c. To the left of the front view
   d. Anywhere on the same sheet with a label
   e. On a separate sheet

3. Projection theory is known as
    a. The black box theory
    b. The transparent box theory
    c. The sandbox theory
    d. The object box theory
    e. Isometric drawing
4. Lines of sight are at what angle to the sides of the projection box?
    a. 30°
    b. 45°
    c. 90° (perpendicular)
    d. 100°
    e. 180°
5. The adjacent sides of the transparent box are at what angles to each other?
    a. 30°
    b. 45°
    c. 90° (perpendicular)
    d. 100°
    e. 180°
6. When the box unfolds, where is the right-side view in relation to the front view?
    a. To the right of the front view
    b. To the left of the front view
    c. Above the front view
    d. Below the front view
    e. Behind the front view
7. What three dimensions are used in referring to the measurements of an object?
    a. Height, width, and length
    b. Height, width, and thickness
    c. Height, width, and depth
    d. Depth, thickness, and length
    e. Width, thickness, and length
8. The total number of possible normal views in orthographic projection is
    a. one
    b. two
    c. four
    d. six
    e. nine
9. As few as _____ view(s) may be drawn if adequate information is given.
    a. one
    b. two
    c. four
    d. six
    e. nine
10. A surface seen true size in the top view will appear as a(n) _____ in the front view.
    a. Surface
    b. Edge
    c. Point
    d. Inclined plane
    e. Oblique plane
11. Hidden surfaces are shown with
    a. Shaded areas
    b. Light lines
    c. Colored areas
    d. Short dashed lines
    e. A short dash and a longer dash
12. The first view chosen for most drawings should show
    a. Contour or shape
    b. Length
    c. Height
    d. Width
    e. Depth
13. The first view chosen for most drawings should be used as
    a. The right-side view
    b. The top view
    c. The left-side view
    d. The back view
    e. The front view
14. A surface that is seen foreshortened in two views and appears as a line in the third view is called
    a. A normal surface
    b. An inclined surface
    c. An oblique surface
    d. A plane surface
    e. Either an inclined or an oblique surface
15. An object line (visible line) should be
    a. Very thick and dark
    b. Thin and dark
    c. Of medium thickness and dark
    d. Thin and light
    e. Of medium thickness and light
16. Dimension lines, extension lines, and centerlines should be
    a. Very thick and dark
    b. Thin and dark
    c. Of medium thickness and dark
    d. Thin and light
    e. Of medium thickness and light
17. All lines except construction lines and projection lines should be
    a. Dark
    b. Light
18. Centerlines for holes should extend outside the feature
    a. About $\frac{1}{16}''$
    b. About $\frac{1}{4}''$
    c. About $\frac{1}{2}''$
    d. About $1''$
    e. They should be varied to provide interest.
19. If two views of an object give the same information, should both views be drawn?
    a. Yes
    b. No
20. Which of the following is a good rule to follow in deciding how many views to draw of an object?
    a. Draw only the views that seem natural to you.
    b. Draw only the front and top views of any object.
    c. Draw only the front and right-side views of any object.
    d. Draw front, top, and right-side views always.
    e. Draw as many views as are needed to fully describe the object.

8 Making Orthographic Views with AutoCAD

OBJECTIVES

After completing this chapter, you will be able to:

☐ Make orthographic drawings containing one to three views using continuous, hidden, and center linetypes.
☐ Correctly answer questions regarding how to make orthographic drawings in Auto-CAD.
☐ Use the following commands to produce drawings:
ID	Break
Mirror	Circle
Dtext	Osnap
Offset	Zoom
Chamfer	Trim
Fillet	Extend
Copy	Pline
☐ Answer questions regarding the commands listed above.

TWO-DIMENSIONAL DRAWINGS IN AUTOCAD

As you learned in the previous chapter, two-dimensional drawings are those showing only two of the three dimensions of an object in any one view. Figure 8–1 illustrates the three most commonly used two-dimensional views. The top view shows width and depth. The front view shows width and height. The right-side view shows height and depth.

AutoCAD and AutoCAD LT have excellent capabilities for drawing in two dimensions. The drawings can be extremely accurate and can be dimensioned in a manner to ensure correct results, as you will learn in a later chapter. In this chapter you will begin with three drawings that have a single view and then make several drawings that have three views.

The first exercise you will draw in this chapter using the AutoCAD program is the floor plan you sketched in Chapter 2.

FIGURE 8–1
The Three Most Commonly Used Views

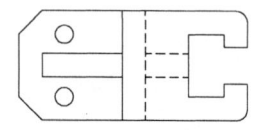

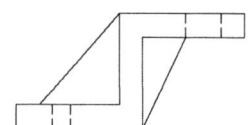

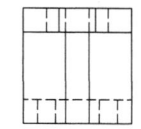

FIGURE 8-2
Dimensions for Exercise 8-1

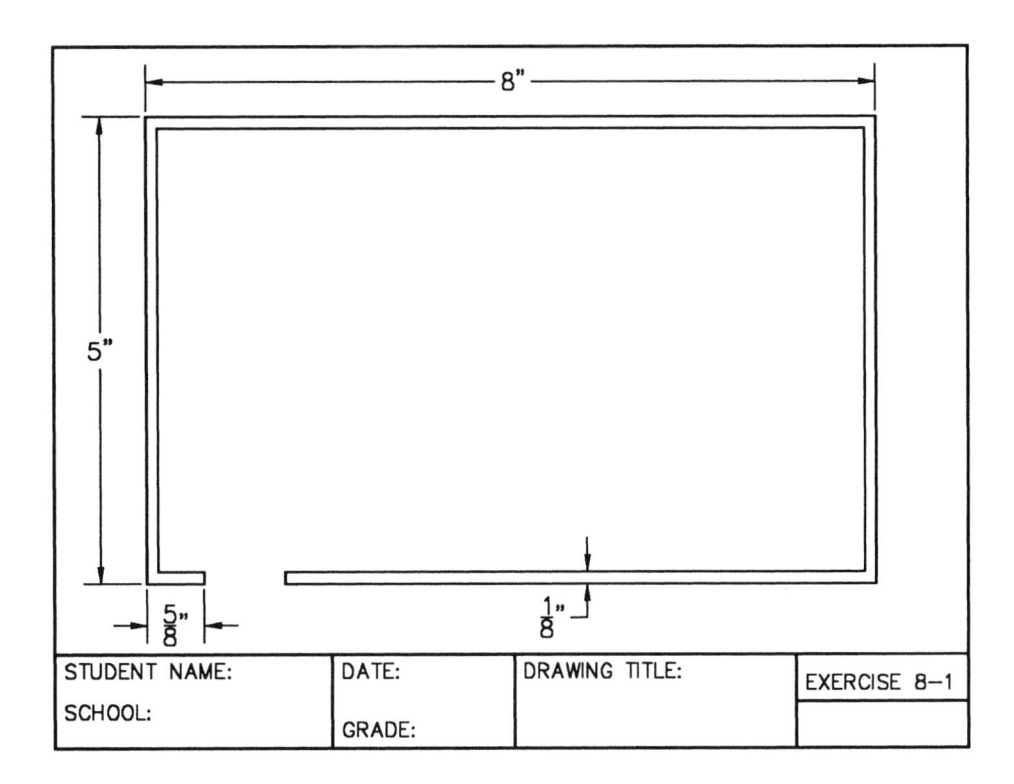

EXERCISE 8-1
Drawing a Floor Plan Using the Polyline, Offset, Line, and Trim Commands

Your final drawing will look similar to the drawing in Figure 8-2 without dimensions.

Step 1. To begin Exercise 8-1, turn on the computer and start AutoCAD or AutoCAD LT.
Step 2. Open drawing EX8-1 supplied on the disk that came with your book.
Step 3. Use Zoom-All to view the entire drawing area.
Step 4. Use the Polyline command to draw the outside edge of the floor plan.

Prompt	Response
Command:	Type: **PL**↵
Specify first point:	Type: **END**↵
of	Click: **the right end of the lower horizontal line**
Specify next point or [Arc/Halfwidth/Length/Undo/Width]:	Type: **@5<90**↵
Specify next point or [Arc/Halfwidth/Length/Undo/Width]:	Type: **@8<180**↵
Specify next point or [Arc/Close/Halfwidth/Length/Undo/Width]:	Type: **@5<270**↵
Specify next point or [Arc/Close/Halfwidth/Length/Undo/Width]:	Type: **@5/8<0**↵
Specify next point or [Arc/Close/Halfwidth/Length/Undo/Width]	Press: ↵

Step 5. Use the Offset command to draw the inside edge of the floor plan.

Making Orthographic Views with AutoCAD

Prompt	Response
Command:	Type: **O**↵
Specify offset distance or [Through]:	Type: **.125**↵
Select object to offset or <exit>:	Click: **the polyline you just drew**
Specify point on side to offset:	Click: **any point on the inside of the polyline**
Select object to offset or <exit>:	↵

Step 6. Use the Pline command to close the left side of the door opening (Figure 8–3).

Prompt	Response
Command:	Type: **PL**↵
Specify start point:	Type: **END**↵
of	Click: **D1** (Figure 8–3)
Specify next point or [Arc/Halfwidth/Length/Undo/Width]:	Type: **END**
of	Click: **D2**
Specify next point or [Arc/Close/Halfwidth/Length/Undo/Width]:	↵

Step 7. Use the Trim command to clean up the bottom right corner of the floor plan (Figure 8–4).

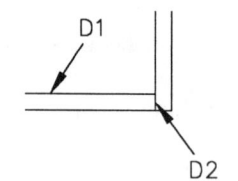

Prompt	Response
Command:	Type: **TR**↵
Select cutting edges:	
Select objects:	Click: **D1** (Figure 8–4)
Select objects:	↵
Select object to trim or shift-select to extend or [Project/Edge/Undo]:	Click: **D2**
Select object to trim or shift-select to extend or [Project/Edge/Undo]:	↵

Step 8. Use the Dtext command to complete the title block.

Prompt	Response
Command:	Type: **DT**↵
Specify start point of text or [Justify/Style]:	Click: **D1** (Figure 8–5)
Specify height <0.120>:	Type: **.12**↵ (or just ↵ if the default is .12)
Specify rotation angle of text <0>:	↵
Enter text:	Type: **YOUR FIRST INITIAL AND YOUR LAST NAME**↵ (Make sure all lettering in the title block are all capital letters.)

FIGURE 8–3
Drawing the End Cap

FIGURE 8–4
Trimming the Wall

FIGURE 8–5
Filling in the Title Block

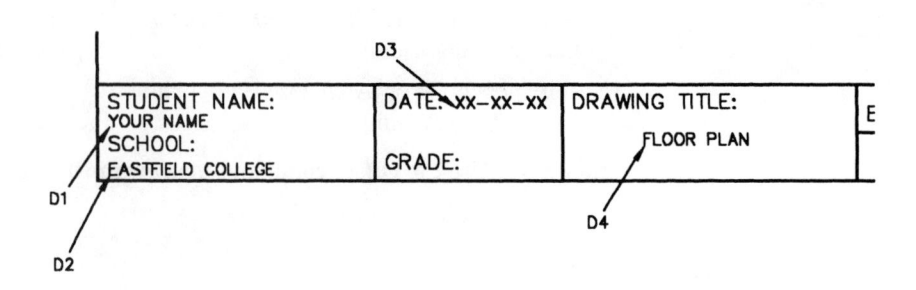

Enter text: Click: **D2** (Figure 8–5) and
                                           Type: **YOUR SCHOOL NAME**↵
Enter text: Click: **D3** and Type: **TODAY'S DATE**↵
Enter text: Click: **D4** and Type: **FLOOR PLAN**↵
Enter text: ↵

Step 9. Use the SAVEAS command to save your drawing as EX8-1(your initials) on a floppy disk and again on the hard drive of your computer.

Step 10. Plot or print your drawing full size on an 11″ × 8½″ sheet.

The second exercise in this chapter will give you practice in using the Fillet, Offset, Extend, Break, and Copy (also called Duplicate in later versions) commands.

EXERCISE 8–2
Drawing a Site Plan Using the Fillet, Offset, Extend, Break, and Copy Commands

Your final drawing will look similar to the drawing in Figure 8–6 without dimensions.

Step 1. To begin Exercise 8–2, turn on the computer and start AutoCAD or AutoCAD LT.

Step 2. Open drawing EX8-2 supplied on the disk that came with your book.

Step 3. Use Zoom-All to view the entire drawing area.

Step 4. Use the Fillet command to draw radii on existing lines.

Prompt	Response
Command:	Type: **F**↵
Select first object or [Polyline/Radius/Trim]	Type: **R**↵
Specify fillet radius <0.000>:	Type: **.5**↵
Select first object or [Polyline/Radius/Trim]	Click: **D1** (Figure 8–7)
Select second object:	Click: **D2**

FIGURE 8–6
Dimensions for Exercise 8–2

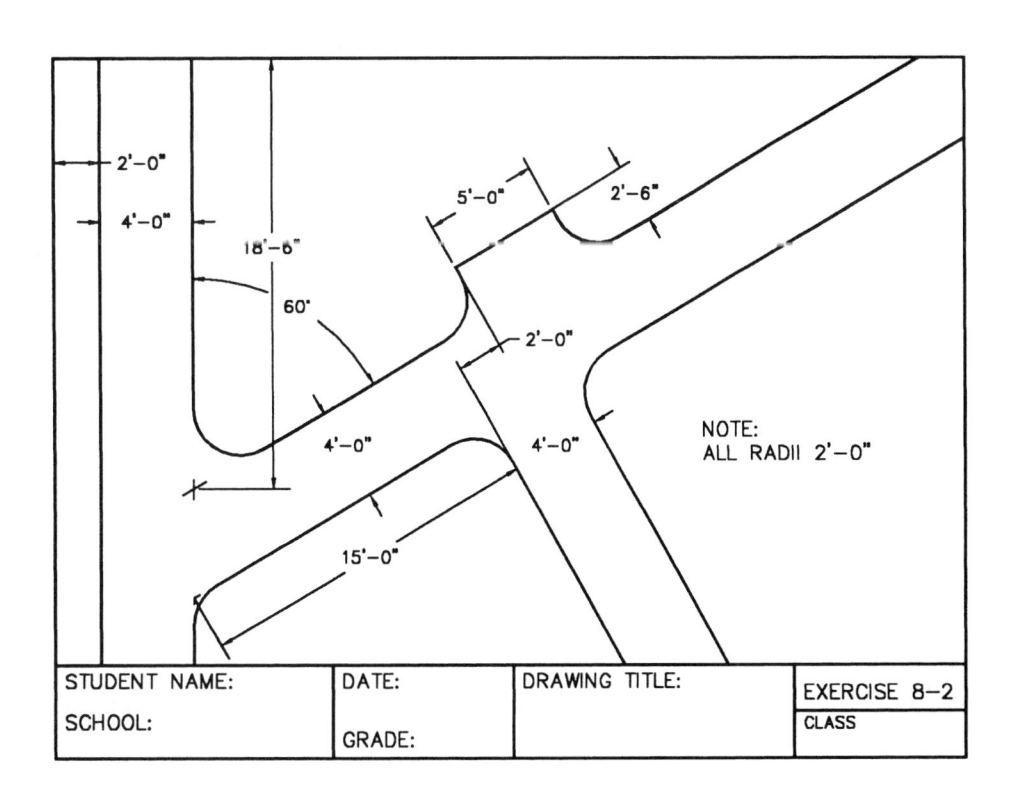

Making Orthographic Views with AutoCAD

FIGURE 8–7
Using the Fillet Command to Draw Radii on Existing Lines

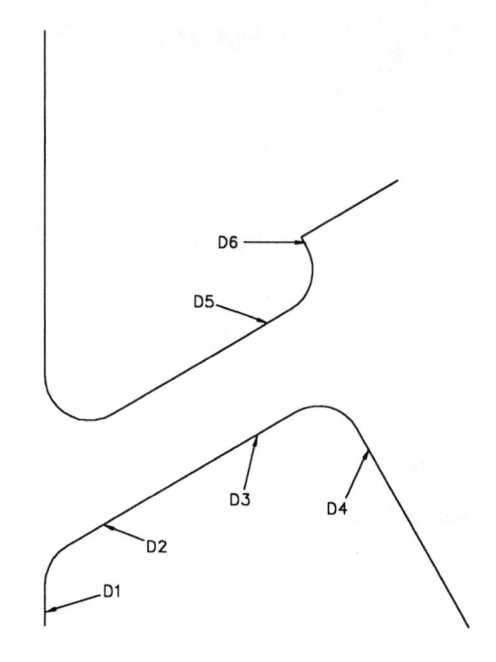

Prompt	Response
Command:	↵
Select first object or [Polyline/Radius/Trim]:	Click: **D3**
Select second object:	Click: **D4**
Command:	↵
Select first object or [Polyline/Radius/Trim]:	Click: **D5**
Select second object:	Click: **D6**

Step 5. Use the Offset command to draw parallel lines.

Prompt	Response
Command:	Type: **O**↵
Specify offset distance or [Through]:	Type: **1**↵
Select object to offset or <exit>:	Click: **D1** (Figure 8–8)
Specify point on side to offset:	Click: **D2 (any point on the left side of the selected object)**
Select object to offset or <exit>:	Click: **D3**
Specify point on side to offset:	Click: **D4 (any point on the right side of the selected object)**
Select object to offset or <exit>:	↵

Step 6. Use the Extend command to extend lines to the borders.

Prompt	Response
Command:	Type: **EX**↵
Select boundary edges...	
Select objects:	Click: **D1, D2** (Figure 8–9)
Select objects:	↵
Select object to extend or shift-select to trim or [Project/Edge/Undo]:	Click: **D3, D4, D5, D6** ↵

Step 7. Use the Break command to break diagonal lines into two segments so they can be filleted.

FIGURE 8–8
Using the Offset Command to Draw Parallel Lines

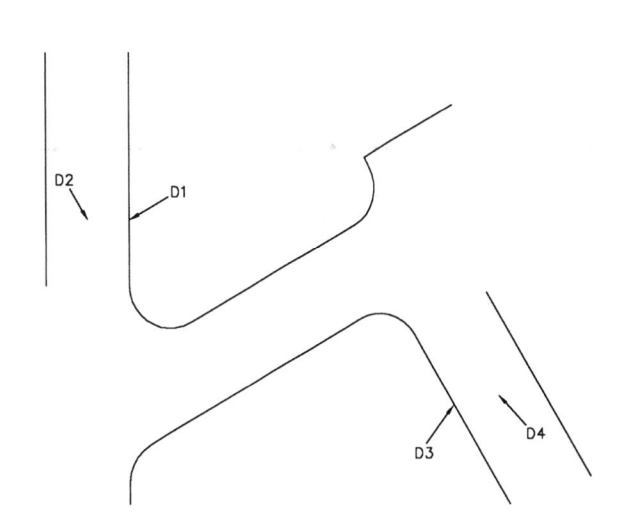

Prompt	Response
Command:	Type: **BR**↵
Select object:	Click: **D1** (Figure 8–10)
Specify second break point or [first point]:	Type: **F**↵
Specify first break point:	Type: **INT**↵
of	Click: **D2**
Specify second break point:	Click: **D3** (in the approximate location shown. The exact location is not important.)
Command:	↵
Select object:	Click: **D4**
Specify second break point [for first point]:	Type: **F**↵
Specify first break point:	Type: **INT**↵
of	Click: **D5**

FIGURE 8–9
Using the Extend Command to Extend Lines to the Border

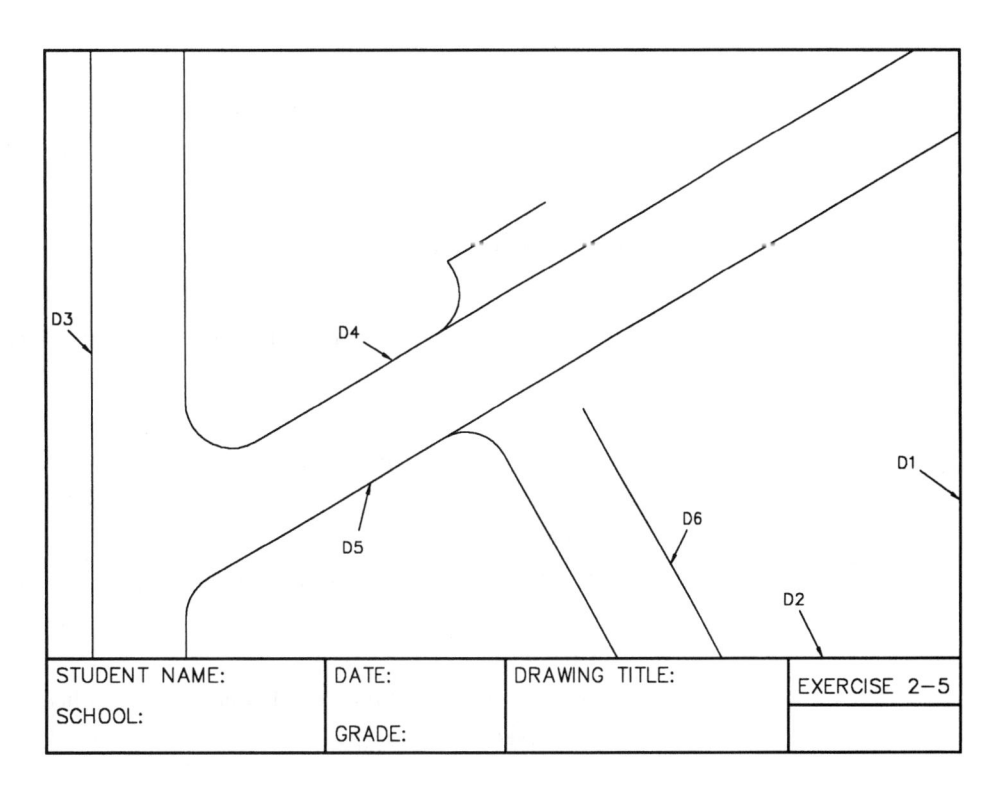

Making Orthographic Views with AutoCAD

FIGURE 8–10
Using the Break Command to
Break Diagonal Lines So Fillets
Can Be Drawn

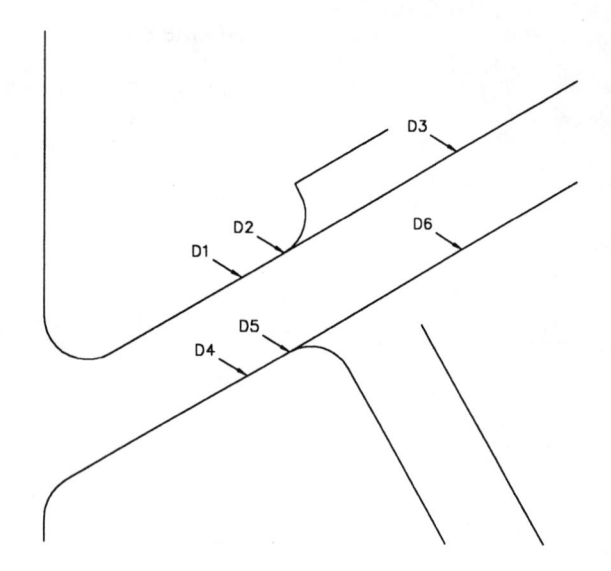

Prompt	Response
Specify second break point:	Click: **D6** (in the approximate location shown. The exact location is not important.)

Step 8. Zoom in and use the Copy command to copy the short line on the top of the drawing.

Prompt	Response
Command:	Type: **Z↵**
[All/Center/Extents/Previous/Scale/Window]<real time>:	Click: **D1** (Figure 8–11)
Specify opposite corner:	Click: **D2**
Command:	Type: **CP↵** (for COPY)
Select objects:	Click: **D1** (Figure 8–12)
Select objects:	↵
Specify base point or displacement or [Multiple]:	Type: **END↵**
of	**D2**
Specify second point of displacement or <use first point as displacement>:	Type: **END↵**
of	**D3**

Step 9. Zoom-All and use the Fillet command to complete the drawing.

Prompt	Response
Command:	Type: **Z↵**
[All/Center/Extents/Previous/Scale Window]<real time>:	Type: **A↵**
Command:	Type: **F↵**
Select first object or [Polyline/Radius/Trim]	Click: **D1** (Figure 8–13)
Select second object:	Click: **D2**
Command:	↵
Select first object or [Polyline/Radius/Trim]	Click: **D3**
Select second object:	Click: **D4**

Chapter 8

FIGURE 8–11
Zooming in on the Intersection

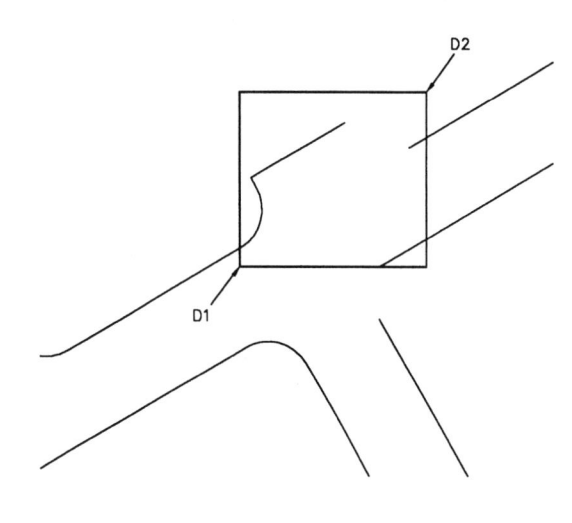

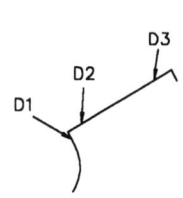

FIGURE 8–12
Using the Copy Command to Copy the Short Line

FIGURE 8–13
Using the Fillet Command to Draw the Remaining Radii

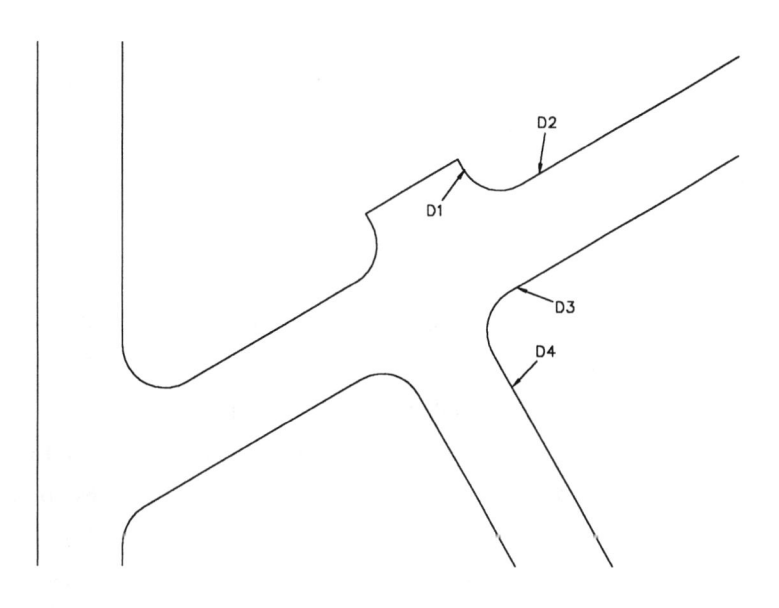

Step 10. Use the Dtext command to complete the title block as you did for Exercise 8–1. Title the drawing SITE PLAN.
Step 11. Use the SAVEAS command to save your drawing as EX8-2(your initials) on a floppy disk and again on the hard drive of your computer.
Step 12. Plot or print your drawing full size on an 11" × 8½" sheet.

EXERCISE 8–3
Drawing a Plate Using the Line, Circle, and Mirror Commands.

Your final drawing will look similar to the drawing in Figure 8–14 without dimensions.

Step 1. To begin Exercise 8–3, turn on the computer and start AutoCAD or AutoCAD LT.
Step 2. Open drawing EX8-3 supplied on the disk that came with your book.
Step 3. Use Zoom-All to view the entire drawing area.
Step 4. Use the Line command to draw lines forming the outside edges of the plate. Notice that you are drawing the plate at half scale.

Making Orthographic Views with AutoCAD

FIGURE 8–14
Dimensions for Exercise 8–3

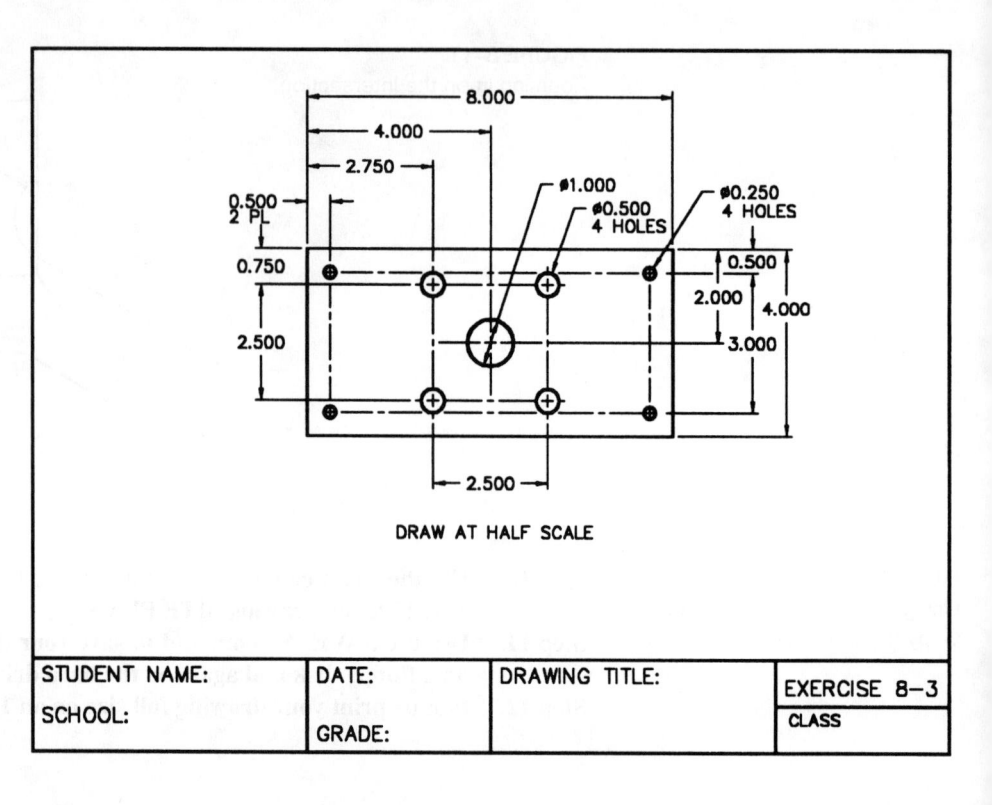

Prompt	Response
Command:	Type: **L**↲
Specify first point:	Type: **END**↲
of	**D1** (Figure 8–15)
Specify next point or [Undo]:	Type: **@4<0**↲
Specify next point or [Undo]:	Type: **@2<90**↲
Specify next point or [Close/Undo]:	Type: **END**↲
of	**D2**
Specify next point or [Close/Undo]:	↲

Step 5. Use the ID command to identify a point from which you may draw circles.

Prompt	Response
Command:	Type: **ID**↲
Specify point:	Type: **END**↲
of	Click: **D1** (Figure 8–16)
Command:	Type: **C**↲
Specify center point for circle or [3P/2P/Ttr (tan tan radius)]:	Type: **@2,-1**↲
Specify radius of circle or [Diameter]:	Type: **.25**↲
Command:	Type: **ID**↲
Specify point:	Type: **END**↲
of	Click: **D1** (Figure 8–16)
Command:	Type: **C**↲
Specify center point for circle or [3P/2P/Ttr (tan tan radius)]:	Type: **@1.375,-1.625**↲
Specify radius of circle or [Diameter]:	Type: **.125**↲

FIGURE 8–15
Drawing the Outside Edges

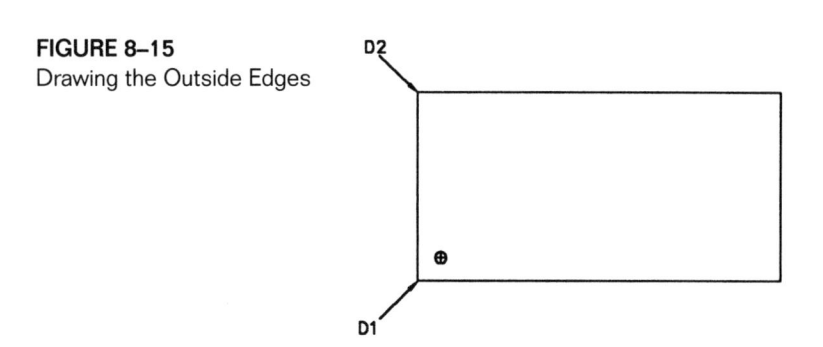

Step 6. **Use the Mirror command to complete the drawing.**

Prompt	Response
Command:	Type: **MI**↵
Select objects:	Click: **the two smaller circles, D1 and D2** (Figure 8–17)
Select objects:	↵
Specify first point of mirror line:	Type: **MID**↵
of	Click: **D3**
Specify second point of mirror line:	Type: **MID**↵
of	Click: **D4**
Delete source objects? [Yes/No]<N>:	↵
Command:	↵
Select objects:	Click: **the four smaller circles, D1, D2, D3, D4** (Figure 8–18)
Select objects:	↵
Specify first point of mirror line:	Type: **MID**↵
of	Click: **D5**
Specify second point of mirror line:	Type: **MID**↵
of	Click: **D6**
Delete source objects? [Yes/No]<N>:	↵

Step 7. Use the Dtext command to complete the title block as you did for Exercise 8–1. Title the drawing MOUNTING PLATE.
Step 8. Use the SAVEAS command to save your drawing as EX8-3(your initials) on a floppy disk and again on the hard drive of your computer.
Step 9. Plot or print your drawing full size on an 11″ × 8½″ sheet.

FIGURE 8–16
Using the ID Command to Identify a Point for a Single Command

Making Orthographic Views with AutoCAD

FIGURE 8–17
Using the Mirror Command to Copy the Two Smaller Holes

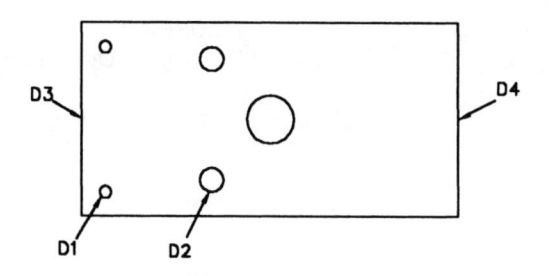

FIGURE 8–18
Using the Mirror Command to Copy the Four Smaller Holes

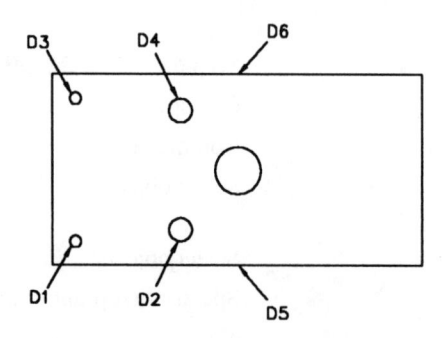

Because there are several different ways to plot drawings using different versions of Auto-CAD, consult with the laboratory instructor or your AutoCAD Manual on how to do that.

In Exercise 8–4 you will use the same commands you used in the earlier exercises in this chapter to make a drawing that has countersunk and counterbored holes that must be shown with hidden lines.

EXERCISE 8–4
Making a Drawing Using Continuous, Hidden, and Center Linetypes

Your final drawing will look like the drawing in Figure 8–19 without the pictorial view.

Step 1. To begin Exercise 8–4, turn on the computer and start AutoCAD or AutoCAD LT.
Step 2. Open drawing EX8-4 supplied on the disk that came with your book.
Step 3. Use Zoom-All to view the entire drawing area.
Step 4. Use the Fillet command to draw fillets in the top views.

Prompt	Response
Command:	Type: **F**⏎
Select first object or [Polyline/Radius/Trim]:	Type: **R**⏎
Specify fillet radius <0.000>:	Type: **.25**⏎
Select first object or [Polyline/Radius/Trim]:	Click: **D1** (Figure 8–20)
Select second object:	Click: **D2**
Command:	⏎
Select first object or [Polyline/Radius/Trim]:	Click: **D3**
Select second object:	Click: **D4**

Step 5. Set a running osnap mode of Intersection and use the Circle command to draw holes in the top view.

FIGURE 8-19
Circular Dimensions for Exercise 8-4

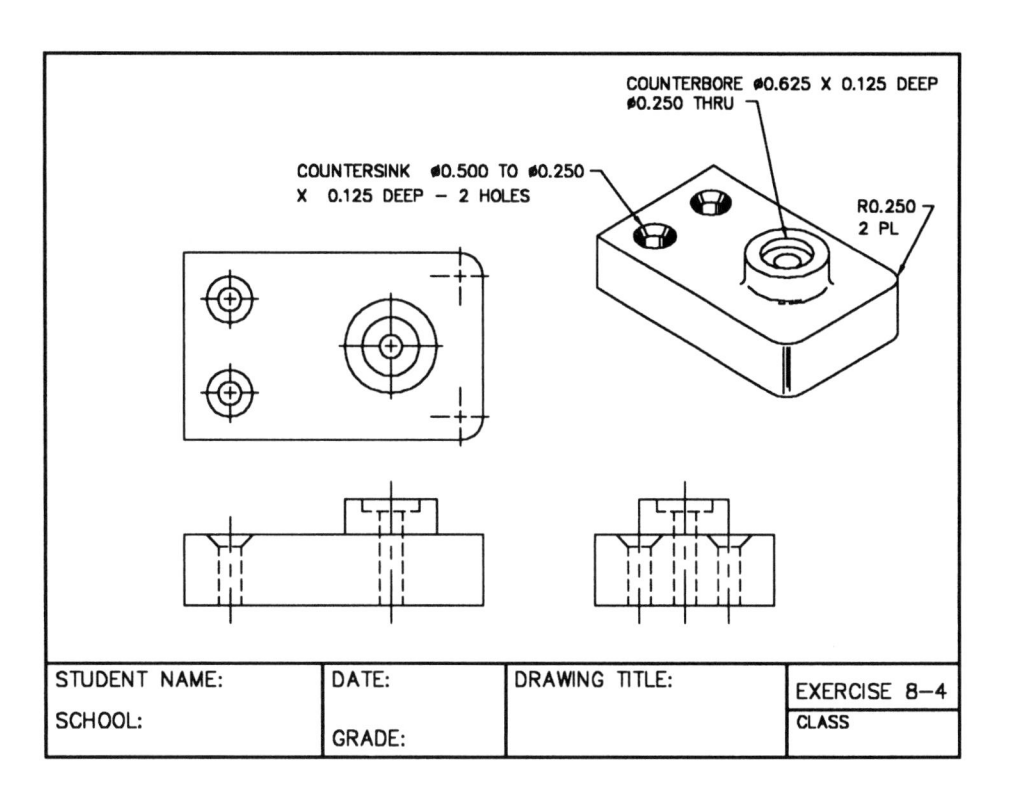

Prompt	Response
Command:	Type: **-OSNAP**↵)
Enter list of object snap modes:	Type: **INT**↵
Command:	Type: **C**↵
Specify center point for circle or [3P/2P/Ttr (tan tan radius)]:	Click: **D1** (Figure 8-21)
Specify radius of circle or [Diameter]:	Type: **D**↵
Specify diameter of circle:	Type: **.5**↵
Command:	↵
Specify center point for circle or [3P/2P/Ttr (tan tan radius)]:	Click: **D2**
Specify radius of circle or [Diameter] <0.2500>:	↵
Command:	↵
Specify center point for circle or [3P/2P/Ttr (tan tan radius)]:	Click: **D1** (again)

FIGURE 8-20
Using the Fillet Command to Draw Radii in the Top View

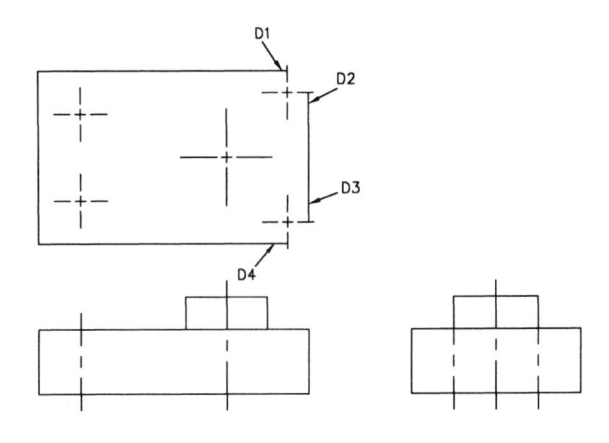

Making Orthographic Views with AutoCAD

FIGURE 8–21
Using Osnap-Intersection and the Circle Command to Draw Circles in the Top View

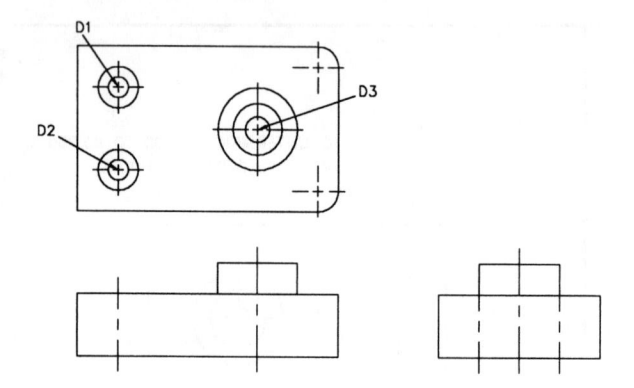

Prompt	Response
Specify radius of circle or [Diameter]:	Type: **D**↵
Specify diameter of circle:	Type: **.25**↵
Command:	↵
Specify center point for circle or [3P/2P/Ttr (tan tan radius)]:	Click: **D2** (again)
Specify radius of circle or [Diameter] <0.1250>:	↵
Command:	↵
Specify center point for circle or [3P/2P/Ttr (tan tan radius)]:	Click: **D3**
Specify radius of circle or [Diameter] <0.1250>:	↵
Command:	↵
Specify center point for circle or [3P/2P/Ttr (tan tan radius)]:	Click: **D3** (again)
Specify radius of circle or [Diameter] <0.1250>:	Type: **D**↵
Specify diameter of circle <0.2500>:	Type: **.625**↵
Command:	↵
Specify center point for circle or [3P/2P/Ttr (tan tan radius)]:	Click: **D3** (again)
Specify radius of circle or [Diameter] <0.3125>:	Type: **.5**↵

Important: Make sure Snap and Ortho are ON (press the function keys F8 and F9 once or twice and test it or read the status line) so you can accurately line up hidden lines with the circle quadrants. You must also remove the running osnap mode of Intersection as described below.

Step 6. Turn OFF the running osnap mode, set the HIDDEN layer current, and draw hidden lines in the front view.

Prompt	Response
Command:	Type: **-OSNAP**↵
Object snap modes:	Type: **NONE**↵
Command:	Type: **-LA**↵
Enter an option [?/Make/Set/New/ON/OFF/Color/Ltype/LWeight/Plot/Freeze/Thaw/LOck/Unlock/stAte]:	Type: **S**↵
Enter layer name to make current or <select object>:	Type: **HIDDEN**↵
Enter an option [?/Make/Set/New/ON/OFF/Color/Ltype/LWeight/Plot/Freeze/Thaw/LOck/Unlock/stAte]:	↵

Command:	Type: **L**↵
Specify first point:	Click: **D1** (Figure 8–22)
Specify next point or [Undo]:	Type: **@.125<-90**↵
Specify next point or [Undo]:	Type: **@.625<0**↵
Specify next point or [Close/Undo]:	Type: **@.125<90**↵
Specify next point or [Close/Undo]:	↵
Command:	↵
Specify first point:	Click: **D2**
Specify next point or [Undo]:	Click: **D3**
Specify next point or [Undo]:	↵
Command:	↵
Specify first point:	Click: **D4**
Specify next point or [Undo]:	Click: **D5**
Specify next point or [Undo]:	↵
Command:	↵
Specify first point:	Click: **D6**
Specify next point or [Undo]:	Type: **@.125,-.125**↵
Specify next point or [Undo]:	Type: **@.25<0**↵
Specify next point or [Close/Undo]:	Type: **@.125,.125**↵
Specify next point or [Close/Undo]:	↵
Command:	↵
Specify first point:	Click: **D7**
Specify next point or [Undo]:	Click: **D8**
Specify next point or [Undo]:	↵
Command:	↵
Specify first point:	Click: **D9**
Specify next point or [Undo]:	Click: **D10**
Specify next point or [Undo]:	↵

Step 7. **Copy hidden lines from the front view to the right-side view using the Copy and Mirror commands to complete the drawing.**

Copy the counterbored hole first.

Prompt	**Response**
Command:	Type: **CP**↵

FIGURE 8–22
Drawing Hidden Lines in the Front View

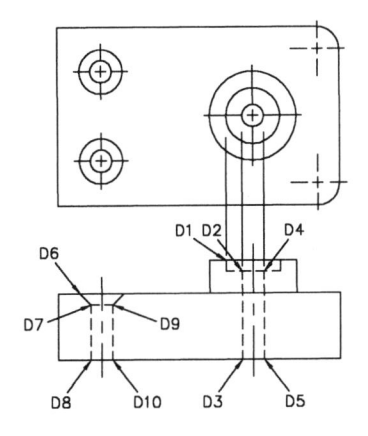

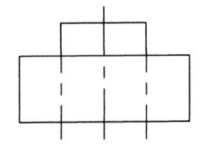

Making Orthographic Views with AutoCAD

FIGURE 8–23
Copying Hidden Lines of the Counterbore into the Right-Side View

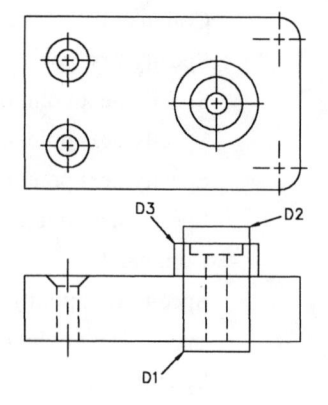

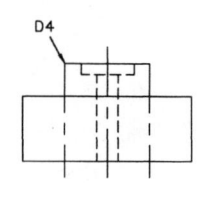

Prompt	Response
Select objects:	Click: **D1** (Figure 8–23)
Specify opposite corner:	Click: **D2**
Select objects:	↵
Specify base point or displacement or [Multiple]:	Type: **END**↵
of	Click: **D3**
Specify second point of displacement or <use first point as displacement>:	Type: **END**↵
of	Click: **D4**

Because the countersunk holes are the same distance from the front edge and the left edge in the top view, you can use the upper left corner of the front view as a base point for copying one countersunk hole. Then use the Mirror command to copy the other countersunk hole.

Prompt	Response
Command:	Type: **CP**↵

Notice that your selection window does not include the center line.

Select objects:	Click: **D1** (Figure 8–24)
Specify opposite corner:	Click: **D2**
Select objects:	↵
Specify base point or displacement or [Multiple]:	Type: **END**↵
of	Click: **D3**
Specify second point of displacement or <use first point as displacement>:	Type: **END**↵
of	Click: **D4**
Command:	Type: **MI**↵
Select objects:	Click: **D5** (Figure 8–24)
Specify opposite corner:	Click: **D6**
Select objects:	↵
Specify first point of mirror line:	Type: **END**↵
of	Click: **D7**
Specify second point of mirror line:	**Make sure Ortho is ON** and Click: **D8** (any point directly above or below D7)
Delete source objects? [Yes/No] <No>:	↵

Chapter 8

FIGURE 8-24
Copying Hidden Lines of the Countersink into the Right-Side View

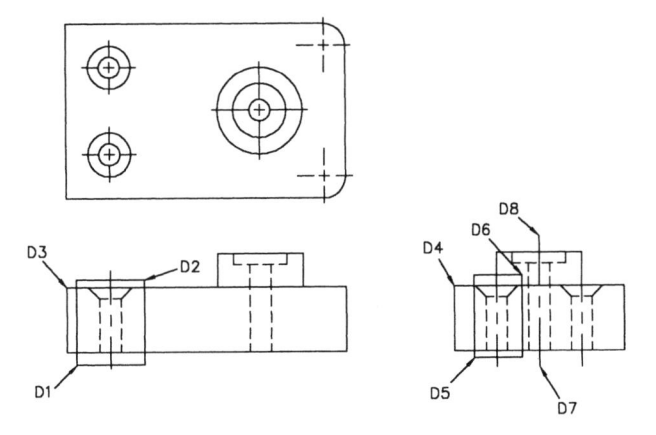

Step 8. **Use the Dtext command to complete the title block as you did for previous exercises. Name the drawing BRACKET.**

Step 9. **Use the SAVEAS command to save your drawing as EX8-4(your initials) on a floppy disk and again on the hard drive of your computer.**

Step 10. **Plot or print your drawing full size on an 11″ × 8½″ sheet.**

EXERCISE 8-5
Making a Drawing Containing Top, Front, and Right-Side Views of an Object with Uniform Thickness

Your final drawing will look like Figure 8–25 without the pictorial sketch. Each space on the sketch is $\frac{1}{4}''$.

Step 1. **To begin Exercise 8–5, turn on the computer and start AutoCAD or AutoCAD LT.**

Step 2. **Open drawing EX8-5 supplied on the disk that came with your book.**

Step 3. **Use Zoom-All to view the entire drawing area.**

FIGURE 8-25
Quarter-Inch Measurements for Exercise 8–5

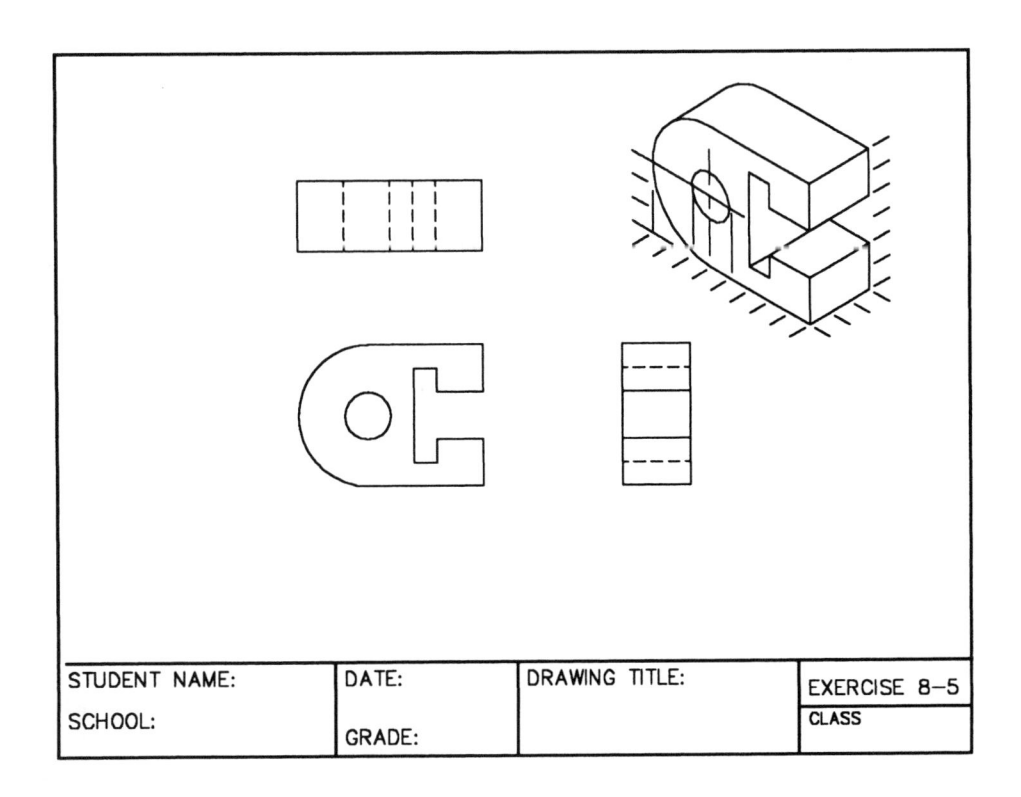

Making Orthographic Views with AutoCAD

FIGURE 8–26
Using the Pline and Arc Commands to Complete the Front View

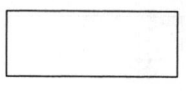

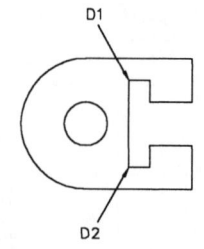

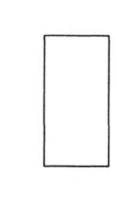

Step 4. Use the Pline and Arc commands to draw the complete front view.

Prompt	Response
Command:	Type: **PL**↵
Specify start point:	Type: **END**↵
of	Click: **D1** (Figure 8–26)
Specify next point or [Arc/Halfwidth/Length/Undo/Width]:	Type: **@.25<0**↵
Specify next point or [Arc/Halfwidth/Length/Undo/Width]:	Type: **@.25<-90**↵
Specify next point or [Arc/Close/Halfwidth/Length/Undo/Width]:	Type: **@.5<0**↵
Specify next point or [Arc/Close/Halfwidth/Length/Undo/Width]:	Type: **@.5<90**↵
Specify next point or [Arc/Close/Halfwidth/Length/Undo/Width]:	Type: **@1.25<180**↵
Specify next point or [Arc/Close/Halfwidth/Length/Undo/Width]:	Type: **A**↵
Specify Endpoint of arc:	Type: **@1.5<270**↵
Specify Endpoint of arc:	Type: **L**↵
Specify next point or [Arc/Close/Halfwidth/Length/Undo/Width]:	Type: **@1.25<0**↵
Specify next point or [Arc/Close/Halfwidth/Length/Undo/Width]:	Type: **@.5<90**↵
Specify next point or [Arc/Close/Halfwidth/Length/Undo/Width]:	Type: **@.5<180**↵
Specify next point or [Arc/Close/Halfwidth/Length/Undo/Width]:	Type: **@.25<-90**↵
Specify next point or [Arc/Close/Halfwidth/Length/Undo/Width]:	Type: **END**↵
of	Click: **D2**
Specify next point or [Arc/Close/Halfwidth/Length/Undo/Width]:	↵

Step 5. Use the Offset command to draw object lines in the right-side view.

Prompt	Response
Command:	Type: **O**↵
Specify offset distance or [Through]:	Type: **.5**↵
Select object to offset or <exit>:	Click: **D1** (Figure 8–27)
Specify point on side to offset:	Click: **D2** (any point below D1)

FIGURE 8–27
Using the Offset Command to Draw Object Lines in the Right-Side View

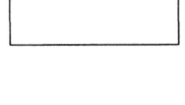

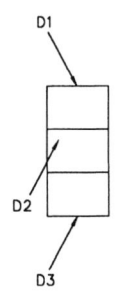

Prompt	Response
Select object to offset or <exit>:	Click: **D3**
Specify point on side to offset:	Click: **D2** (any point above D3)
Select object to offset or <exit>:	↵

Step 6. **Set the HIDDEN layer current and use the Line command to draw hidden lines in the top and right-side views.** (Refer to step 6 of Exercise 8–4 to set layer HIDDEN current.)

Be sure Snap and Ortho are ON so you can line up the hidden lines with the surfaces in the adjacent views.

Prompt	Response
Command:	Type: **L**↵
Specify first point:	Click: **D1** (Figure 8–28)
Specify next point or [Undo]:	Click: **D2**
Specify next point or [Undo]:	↵
Command:	↵
Specify first point:	Click: **D3**
Specify next point or [Undo]:	Click: **D4**
Specify next point or [Undo]:	↵
Command:	↵
Specify first point:	Click: **D5**
Specify next point or [Undo]:	Click: **D6**
Specify next point or [Undo]:	↵
Command:	↵
Specify first point:	Click: **D7**
Specify next point or [Undo]:	Click: **D8**

FIGURE 8–28
Drawing Hidden Lines in the Top and Right-Side Views

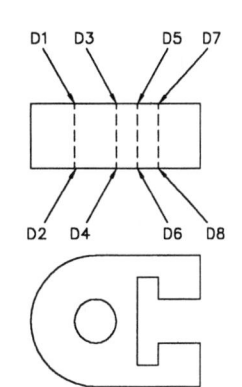

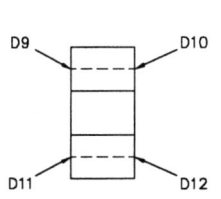

Making Orthographic Views with AutoCAD

Prompt	Response
Specify next point or [Undo]:	↵
Command:	↵
Specify first point:	Click: **D9**
Specify next point or [Undo]:	Click: **D10**
Specify next point or [Undo]:	↵
Command:	↵
Specify first point:	Click: **D11**
Specify next point or [Undo]:	Click: **D12**
Specify next point or [Undo]:	↵

Step 7. Use the Dtext command to complete the title block as you have done previously. Title the drawing HANGER.

Step 8. Use the SAVEAS command to save your drawing as EX8-5(your initials) on a floppy disk and again on the hard drive of your computer.

Step 9. Plot or print your drawing full size on an 11″ × 8½″ sheet.

EXERCISE 8–6
Making a Drawing Containing Top, Front, and Right-Side Views of a Curved Object with Uniform Thickness

Your final drawing will look like Figure 8–29 without the pictorial sketch. Each space on the sketch is $\frac{1}{4}''$.

Step 1. To begin Exercise 8–6, turn on the computer and start AutoCAD or AutoCAD LT.

Step 2. Open drawing EX8-6 supplied on the disk that came with your book.

Step 3. Use Zoom-All to view the entire drawing area.

Step 4. Use the Pline command to draw the outside lines of the front view.

Prompt	Response
Command:	Type: **PL**↵
Specify first point:	Type: **END**↵
of	Click: **D1** (Figure 8–30)
Specify next point or [Arc/Halfwidth/Length/Undo/Width]:	Type: **@.5<90**↵
Specify next point or [Arc/Halfwidth/Length/Undo/Width]:	Type: **A**↵
Specify Endpoint of Arc:	Type: **@-.5,.5**↵
Specify Endpoint of Arc:	Type: **L**↵
Specify next point or [Arc/Close/Halfwidth/Length/Undo/Width]:	Type: **END**↵
of	Click: **D2**

Step 5. Use the ID and Line commands to draw object lines in the front and top views.

Prompt	Response
Command:	Type: **ID**↵
Specify point:	Type: **END**↵
of	Click: **D1** (Figure 8–31)
Command:	Type: **L**↵

FIGURE 8–29
Quarter-Inch Measurements for Exercise 8–6

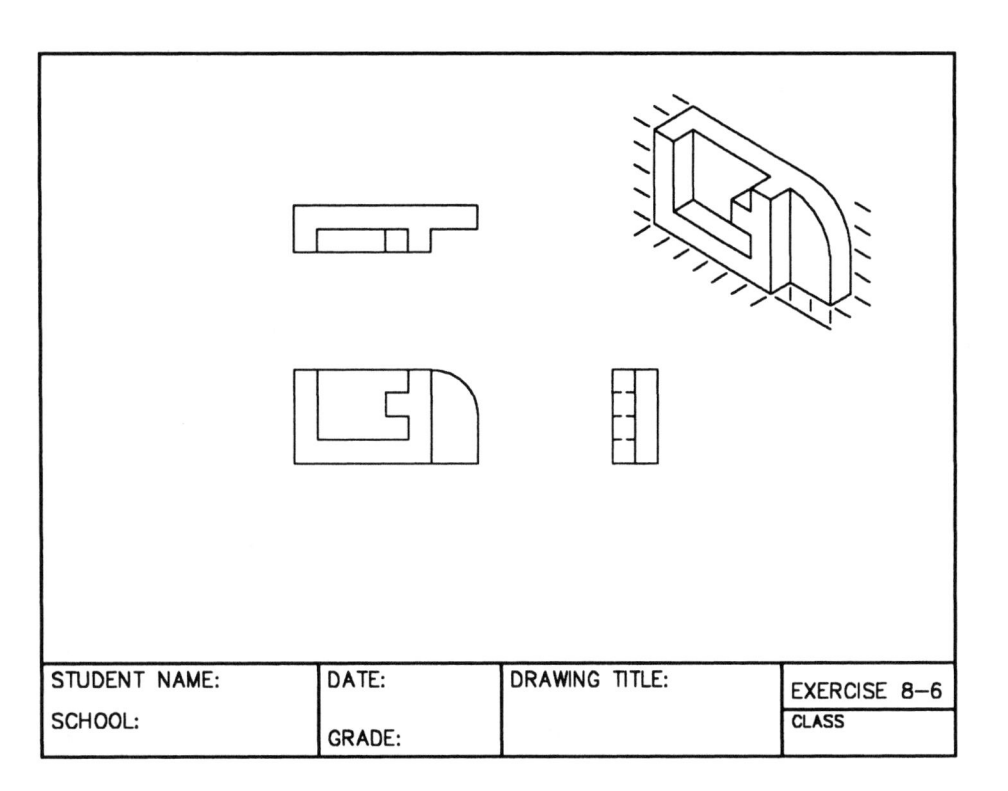

STUDENT NAME:	DATE:	DRAWING TITLE:	EXERCISE 8–6
SCHOOL:	GRADE:		CLASS

FIGURE 8–30
Using the Pline Command to Draw Outside Lines in the Front View

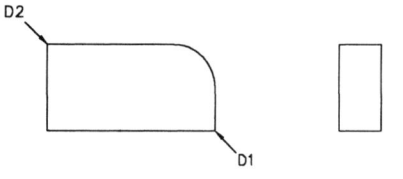

Prompt	Response
Specify first point:	Type: **@.25<0**↵
Specify next point or [Undo]:	Type: **@.75<-90**↵
Specify next point or [Undo]:	Type: **@1<0**↵
Specify next point or [Close/Undo]:	Type: **@.25<90**↵
Specify next point or [Close/Undo]:	Type: **@.25<180**↵
Specify next point or [Close/Undo]:	Type: **@.25<90**↵
Specify next point or [Close/Undo]:	Type: **@.25<0**↵
Specify next point or [Close/Undo]:	Type: **PER**↵
to	Click: **D2**
Specify next point or [Close/Undo]:	↵
Command:	↵
Specify first point:	Type: **END**↵
of	Click: **D3**
Specify next point or [Undo]:	Type: **@1.5<0**↵
Specify next point or [Undo]:	Type: **@.25<90**↵
Specify next point or [Close/Undo]:	Type: **@.5<0**↵
Specify next point or [Close/Undo]:	Type: **END**↵
to	Click: **D4**
Specify next point or [Close/Undo]:	↵
Command:	Type: **ID**↵

Making Orthographic Views with AutoCAD

FIGURE 8–31
Using the ID and Line Commands to Draw Object Lines in the Front and Top Views

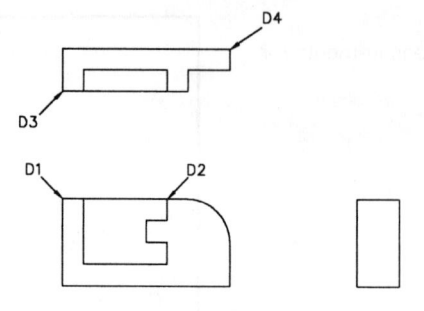

Prompt	Response
Specify point:	Type: **END**↵
of	Click: **D3** again (Figure 8–31)
Command:	Type: **L**↵
Specify first point:	Type: **@.25<0**↵
Specify next point or [Undo]:	Type: **@.25<90**↵
Specify next point or [Undo]:	Type: **@1<0**↵
Specify next point or [Close/Undo]:	Type: **@.25<-90**↵
Specify next point or [Close/Undo]:	↵

Step 6. Use the Offset command to draw object lines in the front, top, and right-side views.

Prompt	Response
Command:	Type: **O**↵
Specify offset distance or [Through]:	Type: **1.5**↵
Select object to offset or <exit>:	Click: **D1** (Figure 8–32)
Specify point on side to offset:	Click: **D2**
Select object to offset or <exit>:	↵
Command:	↵
Specify offset distance or [Through]:	Type: **.25**↵
Select object to offset or <exit>:	Click: **D3**
Specify point on side to offset:	Click: **D4**
Select object to offset or <exit>:	Click: **D5**
Specify point on side to offset:	Click: **D6**
Select object to offset or <exit>:	Click: **D7**
Specify point on side to offset:	Click: **D8**
Select object to offset or <exit>:	Click: **D9**
Specify point on side to offset:	Click: **D10**
Select object to offset or <exit>:	Click: **D11**
Specify point on side to offset:	Click: **D12**
Select object to offset or <exit>:	↵

FIGURE 8–32
Using the Offset Command to Draw Object Lines in the Front, Top, and Right-Side Views

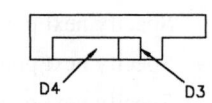

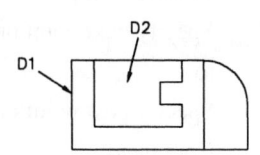

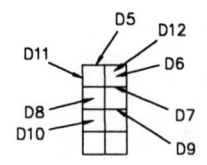

FIGURE 8–33
Using the Chprop and Trim Commands to Change Lines to the HIDDEN Layer and Trim Them

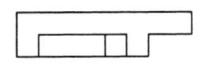

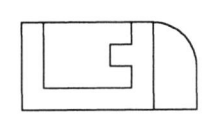

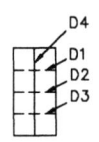

Step 7. Use the Chprop (Change properties) command to change the object lines in the right-side view to hidden lines and trim them to the correct length using the Trim command.

Prompt	Response
Command:	Type: **CHPROP**↵
Select objects:	Click: **D1** (Figure 8–33)
Select objects:	Click: **D2**
Select objects:	Click: **D3**
Select objects:	↵
Enter property to change [Color/LAyer/LType/ltScale/LWeight/Thickness]:	Type: **LA**↵
Enter new layer name <OBJECT>:	Type: **HIDDEN**↵
Enter property to change [Color/LAyer/LType/ltScale/LWeight/Thickness]:	↵
Command:	Type: **TRIM**↵
Select cutting edges: Select objects:	Click: **D4** (Figure 8–33)
Select objects:	↵
Select object to trim or shift-select to extend or [Project/Edge/Undo]:	Click: **D1**
Select object to trim or shift-select to extend or [Project/Edge/Undo]:	Click: **D2**
Select object to trim or shift-select to extend or [Project/Edge/Undo]:	Click: **D3**
Select object to trim or shift-select to extend or [Project/Edge/Undo]:	↵

Step 8. Use the Dtext command to complete the title block. Title the drawing STOP.
Step 9. Use the SAVEAS command to save your drawing as EX8-6(your initials) on a floppy disk and again on the hard drive of your computer.
Step 10. Plot or print your drawing full size on an 11″ × 8½″ sheet.

EXERCISE 8–7
Making a Drawing Containing Top, Front, and Right-Side Views from an Isometric Sketch

Your final drawing will look like Figure 8–34 without the pictorial sketch. Each space on the sketch is ¼″.

Step 1. To begin Exercise 8–7, turn on the computer and start AutoCAD or AutoCAD LT.

FIGURE 8–34
Quarter-Inch Measurements for Exercise 8–7

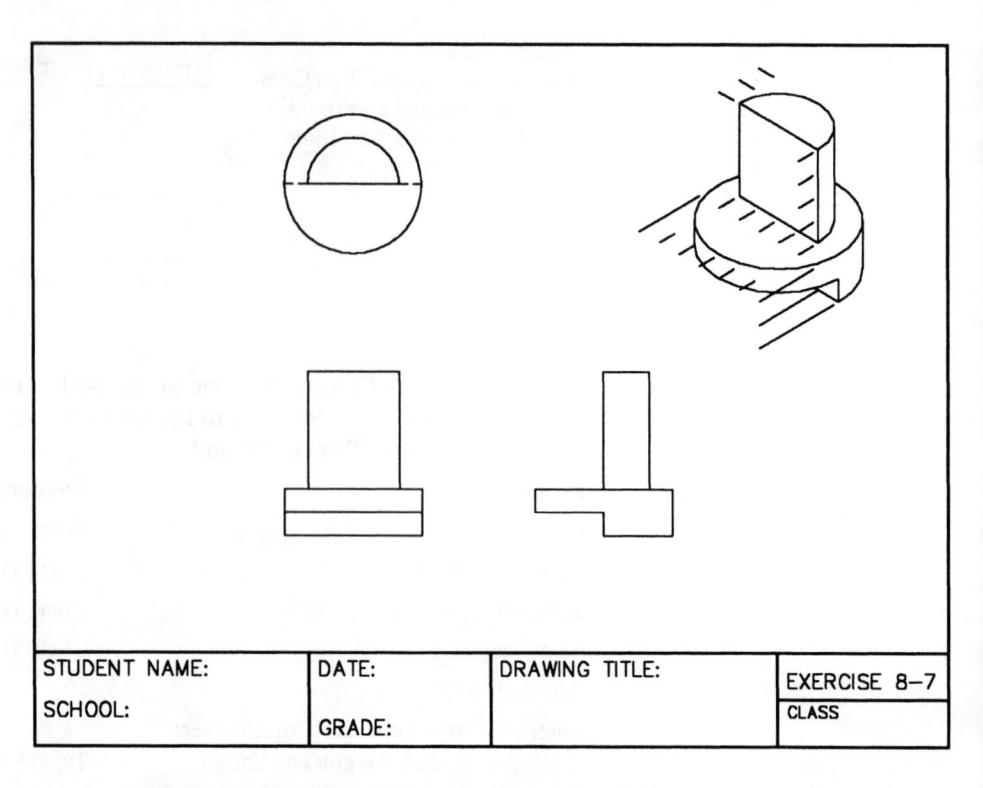

Important: Make sure Snap and Grid are ON so you can line up features in adjacent views.

Step 2. Open drawing EX8-7 supplied on the disk that came with your book.
Step 3. Use Zoom-All to view the entire drawing area.
Step 4. Use the Line command to draw the outside lines of the base in the front view.

Prompt	Response
Command:	Type: **L**↵
Specify first point:	Click: **D1** (Figure 8–35)
Specify next point or [Undo]:	Type: **@.5<-90**↵
Specify next point or [Undo]:	Type: **@1.5<0**↵
Specify next point or [Close/Undo]:	Type: **@.5<90**↵
Specify next point or [Close/Undo]:	↵

Step 5. Extend the existing line to both vertical lines.

Prompt	Response
Command:	Type: **EX**↵
Select boundary edges...	
Select objects:	Click: **D1** (Figure 8–36)
Select objects:	Click: **D2**
Select objects:	↵
Select object to extend or shift-select to trim or [Project/Edge/Undo]:	Click: **D3**
Select object to extend or shift-select to trim or [Project/Edge/Undo]:	Click: **D4**
Select object to extend or shift-select to trim or [Project/Edge/Undo]:	↵

Step 6. Use the Offset command to complete the front view.

Prompt	Response
Command:	Type: **O**↵

Chapter 8

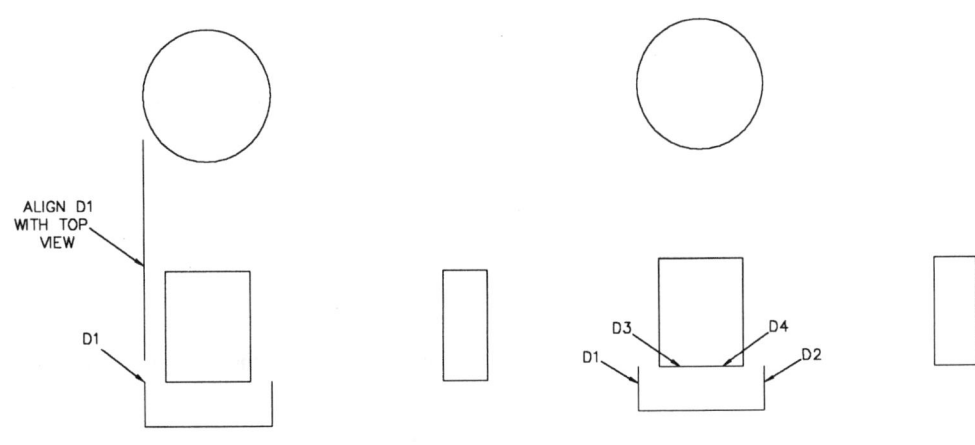

FIGURE 8–35
Using the Line Command to Draw the Outside Lines of the Base in the Front View

FIGURE 8–36
Extending an Existing Line to Both Vertical Lines

Prompt	Response
Specify offset distance or [Through]:	Type: **.25**↵
Select object to offset or <exit>:	Click: **D1** (Figure 8–37)
Specify point on side to offset:	Click: **D2**
Select object to offset or <exit>:	↵

Step 7. Use the ID and Line commands to draw the base in the right-side view.

Prompt	Response
Command:	Type: **ID**↵
Specify point:	Click: **D1** (Figure 8–38)
Command:	Type: **L**↵
Specify first point:	Type: **@.75<180**↵
Specify next point or [Undo]:	Type: **@.25<-90**↵
Specify next point or [Undo]:	Type: **@.75<0**↵
Specify next point or [Close/Undo]:	Type: **@.25<-90**↵
Specify next point or [Close/Undo]:	Type: **@.75<0**↵
Specify next point or [Close/Undo]:	Type: **@.5<90**↵
Specify next point or [Close/Undo]:	↵

Step 8. Extend the existing line in the right-side view to both vertical lines.

Prompt	Response
Command:	Type: **EX**↵
Select objects:	Click: **D1** (Figure 8–39)

FIGURE 8–37
Using the Offset Command to Complete the Front View

FIGURE 8–38
Using the ID and Line Commands to Draw the Base in the Right-Side View

Making Orthographic Views with AutoCAD

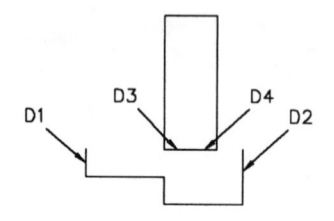

FIGURE 8–39
Extending an Existing Line in the Right-Side View to Both Vertical Lines

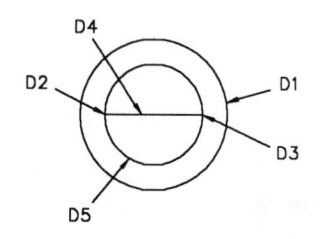

FIGURE 8–40
Using the Circle, Line, and Trim Commands to Draw the Semicircular Shape in the Top View

Prompt	Response
Select objects:	Click: **D2**
Select objects:	↵
Select object to extend or shift-select to trim or [Project/Edge/Undo]:	Click: **D3**
Select object to extend or shift-select to trim or [Project/Edge/Undo]:	Click: **D4**
Select object to extend or shift-select to trim or [Project/Edge/Undo]:	↵

Step 9. Use the Circle, Line, and Trim commands to draw the semicircular shape in the top view.

Prompt	Response
Command:	Type: **C**↵
Specify center point for circle or [3P/2P/Ttr (tan tan radius)]:	Type: **CEN**↵
of	Click: **D1** (Figure 8–40)
Specify radius of circle or [Diameter]:	Type: **.5**↵
Command:	Type: **L**↵
Specify first point:	Type: **QUA**↵
of	Click: **D2**
Specify next point or [Undo]:	Type: **QUA**↵
of	Click: **D3**
Specify next point or [Undo]:	↵
Command:	Type: **TR**↵
Select cutting edges:	
Select objects:	Click: **D4** (Figure 8–40)
Select objects:	↵
Select object to trim or shift-select to extend or [Project/Edge/Undo]:	Click: **D5**
Select object to trim or shift-select to extend or [Project/Edge/Undo]:	↵

Step 10. Set the HIDDEN layer current and draw hidden lines in the top view to complete the drawing. (Be sure Snap is ON before you start drawing.)

Prompt	Response
Command:	Type: **-LA**↵
Enter an option [?/Make/Set/New/ON /OFF/Color/Ltype/LWeight/Plots/ Freeze/Thaw/LOck/Unlock/stAte]:	Type: **S**↵
Enter layer name to make current or <selectobject>:	Type: **HIDDEN**↵
[?/Make/Set/New/ON/OFF/Color/ Ltype/LWeight/Plots/Freeze/Thaw/ LOck/Unlock/stAte]:	↵
Command:	Type: **L**↵
Specify first point:	Click: **D1** (Figure 8–41)
Specify next point or [Undo]:	Click: **D2**
Specify next point or [Undo]:	↵
Command:	↵
Specify first point:	Click: **D3**

138   Chapter 8

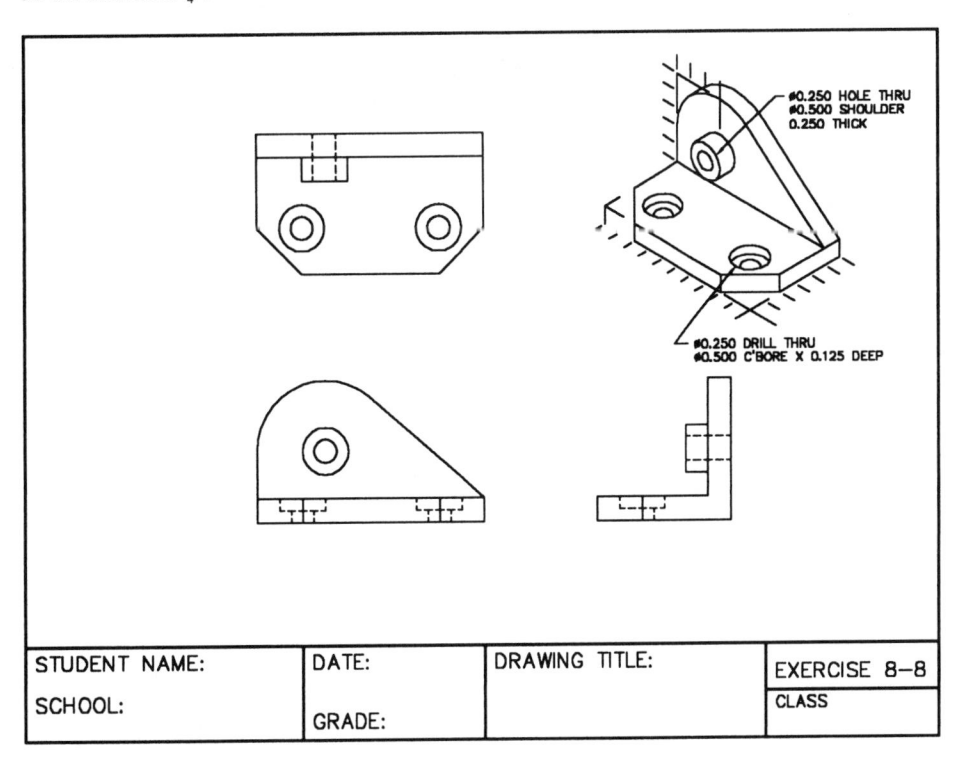

FIGURE 8–41
Drawing Hidden Lines in the Top View

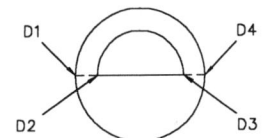

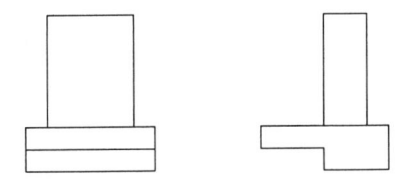

Prompt	Response
Specify next point or [Undo]:	Click: **D4**
Specify next point or [Undo]:	↵

Step 11. Set the OBJECT layer current and use the Dtext command to complete the title block. Title the drawing STOP.

Step 12. Use the SAVEAS command to save your drawing as EX8-7(your initials) on a floppy disk and again on the hard drive of your computer.

Step 13. Plot or print your drawing full size on an $11'' \times 8\frac{1}{2}''$ sheet.

EXERCISE 8–8
Making a Drawing Containing Top, Front, and Right-Side Views from an Isometric Sketch

Now you will begin to make drawings on your own with suggested commands. The Prompt–Response format will not be used in this exercise. Refer to previous exercises if you need to remember how to use a suggested command.

Your final drawing will look like Figure 8–42 without the pictorial sketch. Each space on the sketch is $\frac{1}{4}''$.

FIGURE 8–42
Quarter-Inch Measurements for Exercise 8–8

Making Orthographic Views with AutoCAD

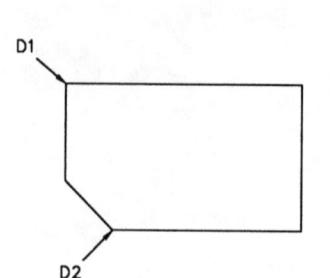

FIGURE 8–43
Drawing Outside Lines in the Top View

Step 1. To begin Exercise 8–8, turn on the computer and start AutoCAD or AutoCAD LT.

Step 2. Open drawing EX8-8 supplied on the disk that came with your book.

Step 3. Use Zoom-All to view the entire drawing area.

Step 4. Use the Line command starting with D1 (use Osnap-Endpoint) (Figure 8–43) as the "First point" to draw the outside lines in the top view. You will have one line 2.5" to the right, another 1.5" down, and another ending on the existing chamfer with Osnap-Endpoint, D2. Use polar coordinates to draw these lines:

@2.5<0↵

@1.5<270↵

Type: **END**↵, Click: **D2**, and Press: ↵

Step 5. Use the Line command starting with D1 (use Osnap-Endpoint) (Figure 8–44) as the "First point" to draw the outside lines of the base in the front view. You will have one line .25" up, another 2.5" to the right, and another ending on the existing line with Osnap-Endpoint, D2. Use polar coordinates to draw these lines:

@.25<90↵

@2.5<0↵

Type: **END**↵, Click: **D2**, and Press: ↵

Step 6. Use the Line command starting with D1 (use Osnap-Endpoint) (Figure 8–45) as the "First point" to draw the outside lines in the right-side view, ending on the existing line with Osnap-Endpoint, D2. Use polar coordinates to draw these lines:

@.25<90↵

@1.25<0↵

@1.25<90↵

@.25<0↵

Type: **END**↵, Click: **D2**, and Press: ↵

FIGURE 8–44
Drawing Outside Lines of the Base in the Front View

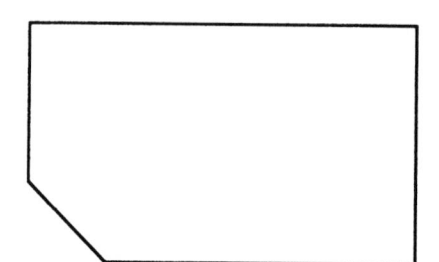

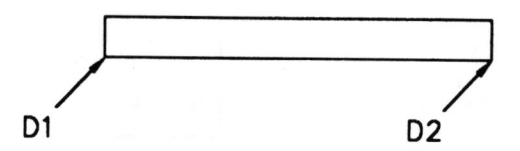

FIGURE 8–45
Drawing the Outside Lines in the Right-Side View

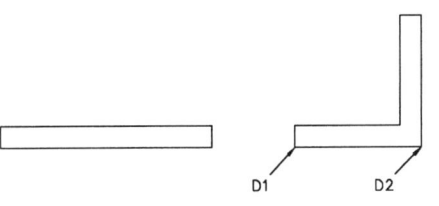

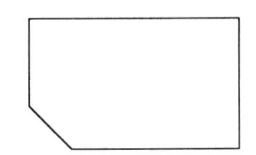

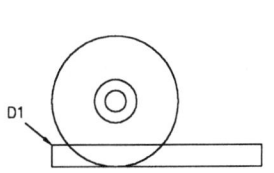

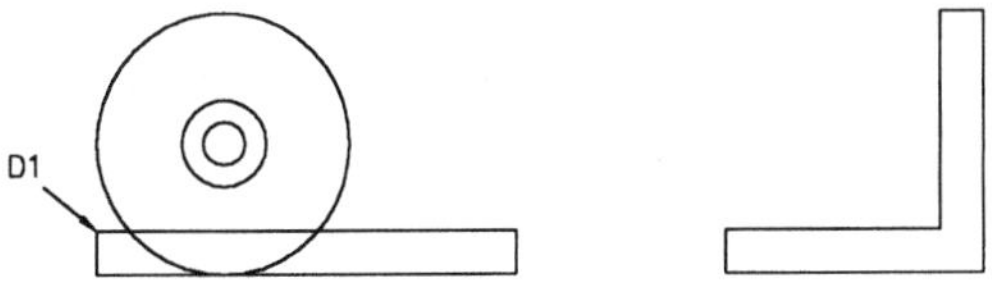

FIGURE 8–46
Using the ID and Circle Commands to Draw Three Circles in the Front View

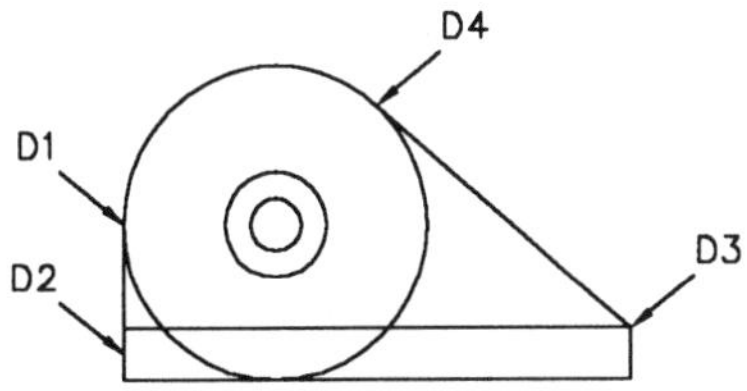

FIGURE 8–47
Using the Extend and Line Commands to Draw Tangents to the Largest Circle

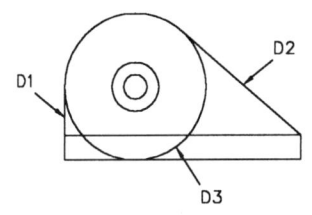

FIGURE 8–48
Using the Trim Command to Remove the Circle Excess

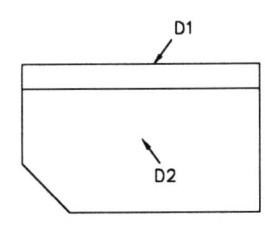

FIGURE 8–49
Using the Offset Command to Draw the Horizontal Line in the Top View

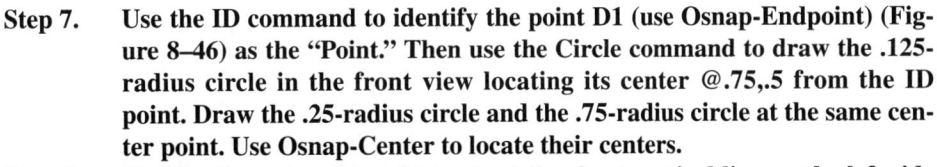

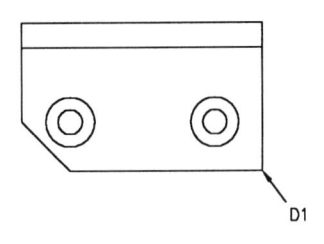

FIGURE 8–50
Using the ID and Circle Commands to Draw the Two Circles on the Right Side and Then Copying Them 1.5″ to the Left

Step 7. Use the ID command to identify the point D1 (use Osnap-Endpoint) (Figure 8–46) as the "Point." Then use the Circle command to draw the .125-radius circle in the front view locating its center @.75,.5 from the ID point. Draw the .25-radius circle and the .75-radius circle at the same center point. Use Osnap-Center to locate their centers.

Step 8. Use the Extend command to extend the short vertical line on the left side of the front view to the .75-radius circle. Click the circle, D1 (Figure 8–47), as the boundary edge, then click the line D2 as the object to extend.

Use the Line command to draw the line from the endpoint (Osnap-Endpoint) D3 tangent to the circle, D4 (use Osnap-Tangent by typing TAN↵ at the "next point" prompt).

Step 9. Use the Trim command to remove the circle excess. Click D1 and D2 (Figure 8–48) as the cutting edges, Press: ↵, then click D3 as the object to trim.

Step 10. Use the Offset command to draw the horizontal line in the top view. Set the offset distance as .25, click D1 (Figure 8–49) as the object to offset, then click D2 as the side to offset.

Step 11. Use the ID command to identify the point D1 (use Osnap-Endpoint) (Figure 8–50) as the "Point." Then draw the .125-radius circle in the top view locating its center @-.5,.5 from the ID point. Draw the .25-radius circle at the same center point. Use Osnap-Center to locate its center.

Copy the two circles 1.5″ to the left. Pick any point as the base point, then use polar coordinates to specify the second point of displacement:

@1.5<180↵

Step 12. Use the Chamfer command to chamfer the lower right corner of the top view. Set the first and second chamfer distances to .5, then click D1 and

Making Orthographic Views with AutoCAD 141

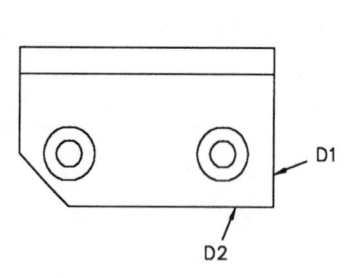

FIGURE 8–51
Using the Chamfer Command to Chamfer the Lower Right Corner

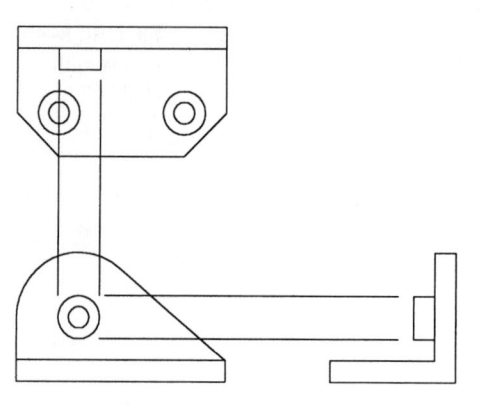

FIGURE 8–52
Using the Line Command with Snap and Ortho on to Draw the Boss in the Top and Right-Side Views

D2 (Figure 8–51) as the lines to be selected. You will probably have to Press: ↲ after you have set the distance.

Step 13. With Snap and Ortho on, use the Line command to draw the lines showing the boss in the top and right-side views. Use the .25-radius circle in the front view as a guide (Figure 8–52).

Step 14. Use the Offset command to draw lines showing the chamfer in the front and right-side views. Set the offset distance to .5, then click D1 (Figure 8–53) as the object to offset, D2 as the side to offset, D3 as the object to offset, D4 as the side to offset, D5 as the object to offset, D6 as the side to offset.

Use the Trim command to trim the offset vertical line on the left side of the front view. Click D7 as the cutting edge and D8 as the object to trim.

Step 15. Set layer HIDDEN current [Type: -LA↲ S↲ HIDDEN↲ and Press: ↲ again], activate the Line command and draw hidden lines in the top and right-side views to represent the hole in the boss that is hidden in those views. Be sure Snap and Ortho are on so your hidden lines are aligned with the quadrants of the circle in the front view (Figure 8–54).

Step 16. Continue to draw hidden lines in the front view to represent the counterbored hole in the left side of the base. With Snap on, activate the Line command, align the cursor crosshairs with the left quadrant of the counterbore (Figure 8–55), and click D1. After you click D1, use the following polar coordinates and osnap mode.

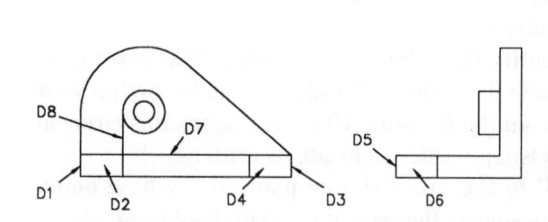

FIGURE 8–53
Using the Offset and Trim Commands to Draw the Chamfer Edges in the Front and Right-Side Views

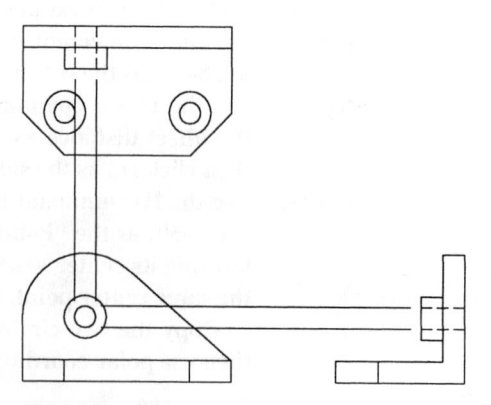

FIGURE 8–54
Using the Line Command with Layer HIDDEN Current to Draw Hidden Lines for the Hole

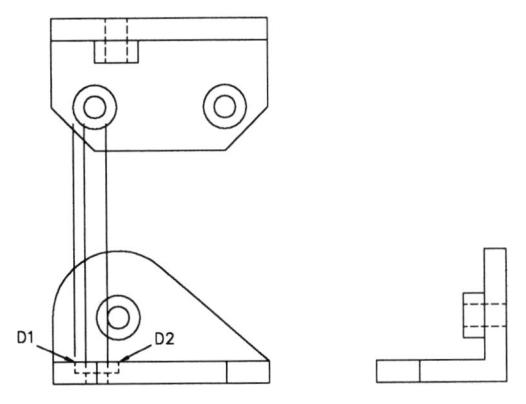

FIGURE 8–55
Drawing Hidden Lines for the Counterbore

@.125<270

@.5<0

Type: **PER↵** and Click: **D2**

Important: Make sure Snap and Ortho are on so you can line up features in adjacent views.

Activate the Line command, align the cursor crosshair with the left and right quadrants of the .125-radius hole in the top view, and draw the short vertical hidden lines showing the drilled hole in the counterbore.

Step 17. **Use the Copy command to copy the hidden lines representing the counterbored hole and drilled hole using the following as the second point of displacement. Click any point as the base point:**

@1.5<0↵

Use the Line command to draw a horizontal construction line from the front edge of the top view and a vertical line from the front edge of the right-side view so they meet as shown in Figure 8–56. From that intersection draw another construction line 2″ long at a 45° angle (the length is not critical).

Draw a line from the bottom quadrant of the counterbored hole in the top view so it intersects the 45° line. From that intersection draw a vertical line toward the right-side view. Copy the hidden lines representing the counterbored hole on the left side of the front view to the right-side view using D1 as the base point and D2 as the second point of displacement. Align D2 with the vertical construction line as shown.

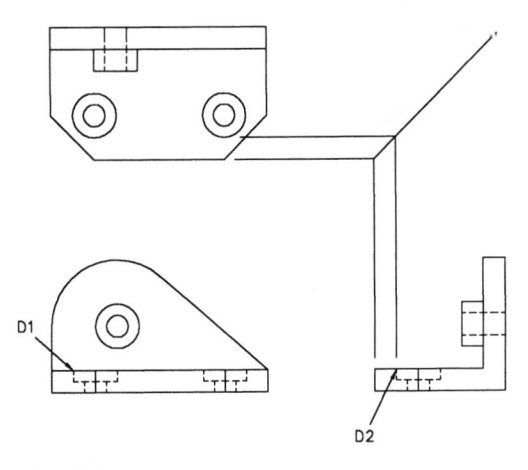

FIGURE 8–56
Using the Copy Command to Copy Hidden Lines for Counterbored Holes

Making Orthographic Views with AutoCAD

Step 18. Set the OBJECT layer current and use the Dtext command to complete the title block. Title the drawing SHAFT SUPPORT.

Step 19. Use the SAVEAS command to save your drawing as EX8-8(your initials) on a floppy disk and again on the hard drive of your computer.

Step 20. Plot or print your drawing full size on an 11" × 8½" sheet.

EXERCISE 8–9
Making a Drawing Containing Top, Front, and Right-Side Views from an Isometric Sketch

Now you will begin to make drawings on your own with suggested commands. The Prompt–Response format will not be used in this exercise. Refer to previous exercises if you need to remember how to use a suggested command.

Your final drawing will look like Figure 8–57 without the pictorial sketch. Each space on the sketch is ¼".

Step 1. To begin Exercise EX8-9, turn on the computer and start AutoCAD or AutoCAD LT.

Step 2. Open drawing EX8-9 supplied on the disk that came with your book.

Step 3. Use Zoom-All to view the entire drawing area.

Step 4. Use the Line command starting with D1 (use Osnap-Endpoint) (Figure 8–58) as the "First point" to draw the outside lines in the top view. You will have one line 2.25" to the right, another 2" down and .25" to the left, and another ending on the existing fillet with Osnap-Endpoint, D2. Use relative coordinates to draw these lines:

@2.25,0↵

@-.25,-2↵

Type: **END**↵, Click: **D2**, and Press: ↵

Step 5. Use the Line command starting with D1 (use Osnap-Endpoint) (Figure 8–59) as the "First point" to draw the outside lines of the base in the front

FIGURE 8–57
Quarter-Inch Measurements for Exercise 8–9

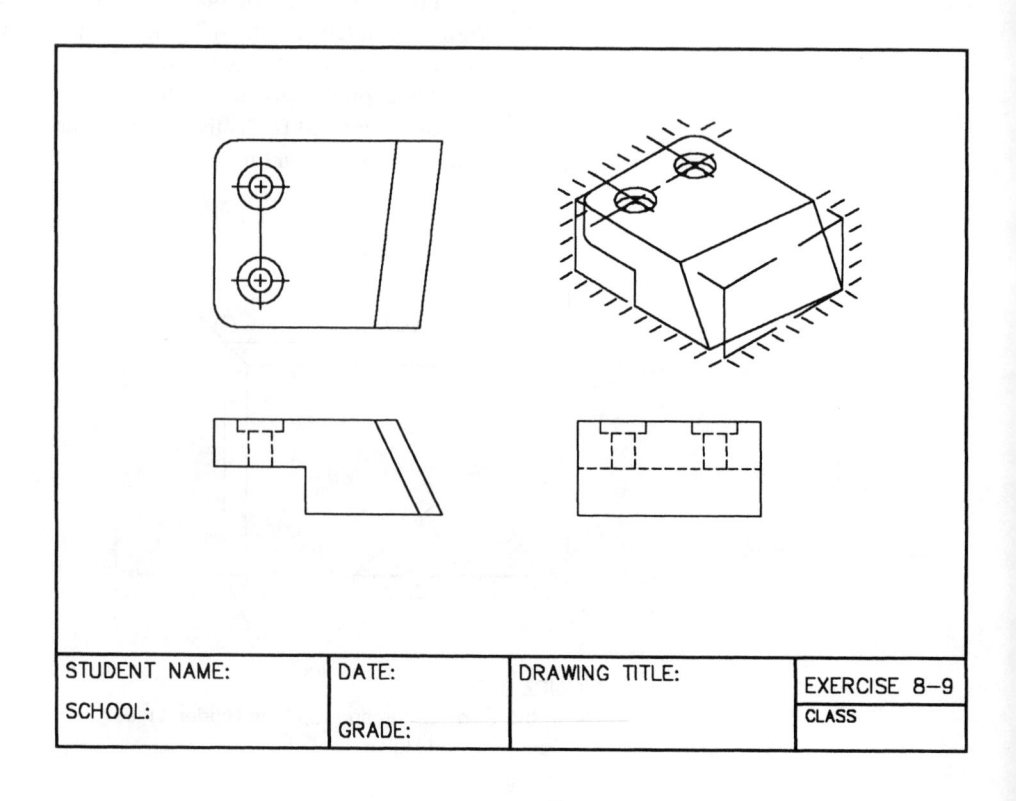

144 Chapter 8

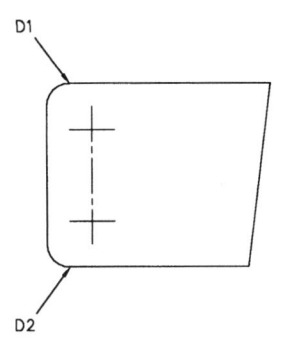

FIGURE 8–58
Using Relative Coordinates to Draw the Outside Lines of the Top View

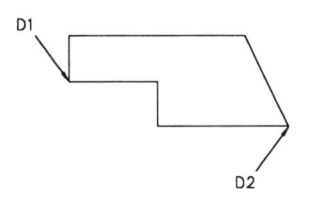

FIGURE 8–59
Using Relative Coordinates to Draw the Outside Lines of the Front View

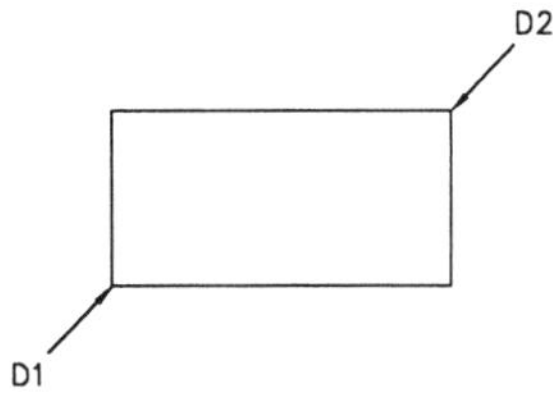

view. You will have one line .5″ up, another 2″ to the right, and another ending on the existing line with Osnap-Endpoint, D2. Use relative coordinates to draw these lines:

@0,.5↵

@2,0↵

Type: **END**↵ and Click: **D2** and Press: ↵

Step 6. Use the Line command starting with D1 (use Osnap-Endpoint) (Figure 8–60) as the "First point" to draw the outside lines in the right-side view, ending on the existing line with Osnap-Endpoint D2. Use relative coordinates to draw these lines:

@0,1↵

Type: **END**↵, Click: **D2**, and Press: ↵

Step 7. With Snap ON and Ortho OFF use the Line command to draw inclined lines in the top and front views starting with D1 (Figure 8–61).

Count 2 grids to the left of the top right corner of the top view and Click: **D1**.
Count 2 grids to the left of the bottom right corner of the top view and Click: **D2**.
Count 1 grid to the left of the top right corner of the front view and Click: **D3**.
Count 1 grid to the left of the bottom right corner of the front view and Click: **D4**.

Step 8. Use the Circle command to draw the .125-radius circle in the top view locating its center using Osnap-Intersection and clicking D1 (Figure 8–62). Draw the .25-radius circle at the same center point.

Use the Copy command to copy the two circles 1″ down. Zoom a window around the top view so you can click the two circles easily when you copy them.

Click any point as the base point, then use polar coordinates to specify the second point of displacement:

@1<-90↵

Step 9. Set layer HIDDEN current [Type: **-LA**↵ **S**↵ **HIDDEN**↵ and Press: ↵ again], activate the Line command and draw hidden lines in the front view to represent the counterbored holes. With Snap ON, activate the Line command, align the cursor crosshair with the left quadrant of the counterbore (Figure 8–63), and click D1. After you click D1, use the following polar coordinates and osnap mode.

@.125<270

FIGURE 8–60
Using Relative Coordinates to Draw the Outside Lines of the Right-Side View

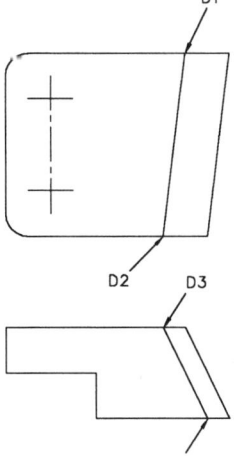

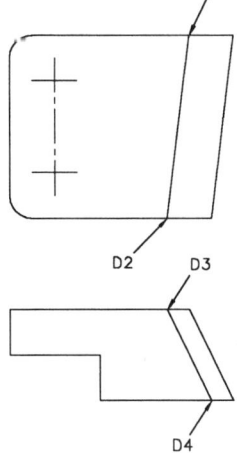

FIGURE 8–61
Using the Grid with Snap On and Ortho Off to Draw Inclined Lines

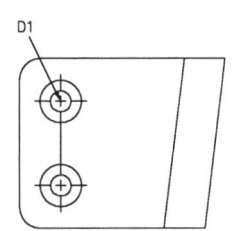

FIGURE 8–62
Drawing Circles in the Top View

Making Orthographic Views with AutoCAD

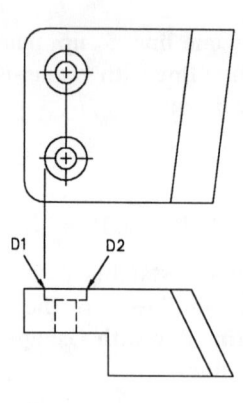

FIGURE 8–63
Drawing Hidden Lines of the Counterbore in the Front View with Layer HIDDEN Current

FIGURE 8–64
Drawing the Hidden Line in the Right-Side View

@.5<0

Type: **PER↵** and Click: **D2**

Activate the Line command, align the cursor crosshair with the left and right quadrants of the .125-radius hole in the top view and draw the short vertical hidden lines showing the drilled hole in the counterbore.

Step 10. With Snap and Ortho ON, use the Line command to draw the hidden line in the right-side view (Figure 8–64).

Step 11. Set layer OBJECT current [Type: **-LA↵ S↵ OBJECT↵** and Press: ↵ again], then use the Line command to draw a horizontal construction line from the front edge of the top view and a vertical line from the front edge of the right-side view so they meet as shown in Figure 8–65. From that intersection draw another construction line 2″ long at a 45° angle (the length is not critical).

Draw a line from the bottom quadrant of the counterbored hole in the top view so it intersects the 45° line. From that intersection draw a vertical line toward the right-side view. Copy the hidden lines representing the counterbored hole on the left side of the front view to the right-side view using D1 as the base point and D2 as the second point of displacement. Align D2 with the vertical construction line as shown.

Step 12. Use the Copy command to copy the counterbored hole in the right-side view 1″ to the right to complete the drawing (Figure 8–66). Click any

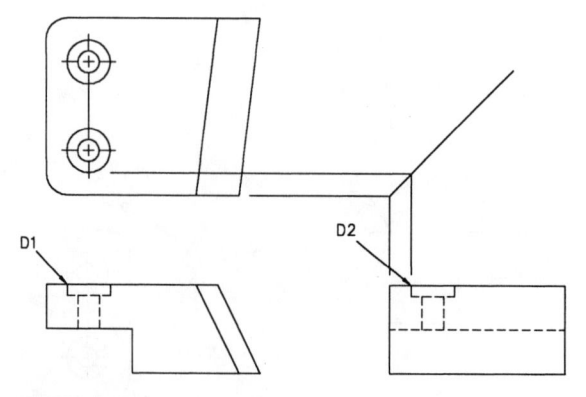

FIGURE 8–65
Using the Miter-Line Method to Locate the Counterbored Hole in the Right-Side View

146 Chapter 8

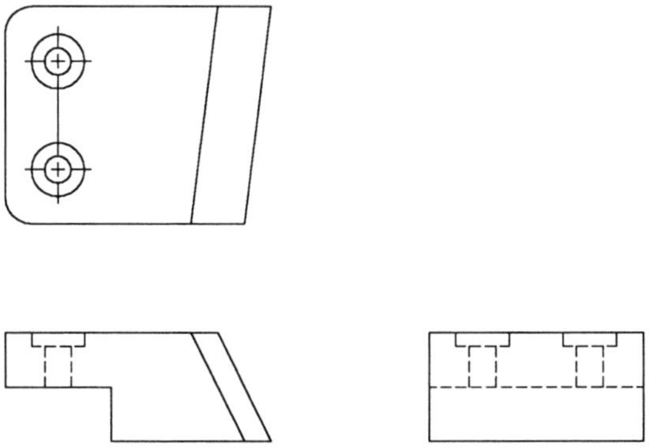

FIGURE 8–66
Copying the Counterbored Hole 1" to the Right

          point as the base point and use the following relative coordinates as the second point of displacement:

          @1,0↵

Step 13.   Use the Dtext command to complete the title block. Title the drawing ANGLE JAW.

Step 14.   Use the SAVEAS command to save your drawing as EX8-9(your initials) on a floppy disk and again on the hard drive of your computer.

Step 15.   Plot or print your drawing full size on an $11'' \times 8\frac{1}{2}''$ sheet.

EXERCISES

EXERCISE 8–1	Complete Exercise 8–1 using steps 1 through 10 described in this chapter.
EXERCISE 8–2	Complete Exercise 8–2 using steps 1 through 12 described in this chapter.
EXERCISE 8–3	Complete Exercise 8–3 using steps 1 through 9 described in this chapter.
EXERCISE 8–4	Complete Exercise 8–4 using steps 1 through 10 described in this chapter.
EXERCISE 8–5	Complete Exercise 8–5 using steps 1 through 9 described in this chapter.
EXERCISE 8–6	Complete Exercise 8–6 using steps 1 through 10 described in this chapter.
EXERCISE 8–7	Complete Exercise 8–7 using steps 1 through 13 described in this chapter.
EXERCISE 8–8	Complete Exercise 8–8 using steps 1 through 20 described in this chapter.
EXERCISE 8–9	Complete Exercise 8–9 using steps 1 through 15 described in this chapter.

REVIEW QUESTIONS

Circle the best answer.

1. The top view shows which of the following dimensions?
    a. Height and width
    b. Height and depth
    c. Width and depth
    d. Length and width
    e. Height, Width, and depth
2. Which of the following commands draws several lines that are a single entity?
    a. Line
    b. Pline
    c. Offset
    d. Chamfer
    e. Break

3. The polar coordinate @5<180 draws a line
   a. 5" to the right
   b. 5" up
   c. 5" down
   d. 5" to the left
   e. 5" at a 120° angle to the right
4. Which of the following commands can be used to produce parallel plines?
   a. Parallel
   b. Draw
   c. Offset
   d. Change
   e. Osnap
5. A rounded corner can be obtained most easily with which of the following commands?
   a. Chamfer
   b. Fillet
   c. Draw
   d. Ellipse
   e. Offset
6. A 45° angle at a corner may be obtained most easily with the use of which of the following commands?
   a. Chamfer
   b. Fillet
   c. Draw
   d. Ellipse
   e. Offset
7. Which of the following will produce a vertical line 3.5" downward from a point?
   a. 3.5 × 90
   b. @3.5<0
   c. @3.5<270
   d. @3.5<90
   e. −90<3.5
8. Which of the following is used to identify a point from which you can specify a point for a single command?
   a. Status
   b. Point
   c. ID
   d. Line
   e. Dist
9. Which of the following circles is produced if ".5" is entered in response to the circle prompt "Specify circle radius or [Diameter]:"?
   a. .5 diameter
   b. .25 radius
   c. 1.00 radius
   d. 1.00 diameter
   e. .25 diameter
10. Which command is used to trim lines between cutting edges?
    a. Edit
    b. Trim
    c. Break
    d. Erase
    e. Copy (Duplicate)
11. Describe the use of the Dtext command.

12. Describe how the Extend command is used.

13. If you type F↵ after you have selected objects to be broken with the Break command, what are you asked to specify next?

14. After you have selected objects to be copied with the Copy command and pressed Enter, what are you asked to specify next?

15. After you have selected objects to be copied with the Mirror command and pressed Enter, what are you asked to specify next?

16. Describe what a running osnap mode is.

17. Which of the layer options sets a new layer current?

18. Why do you change layers to draw different features?

19. What purpose does the ID command serve?

20. What happens when you type PER↵ and click a point on a line in response to the Line prompt "To point"?

9 Sketching Sectional Views

OBJECTIVES

After completing this chapter, you will be able to:

- Correctly sketch to scale full-section and half-section orthographic drawings from unsectioned two-dimensional or three-dimensional drawings.
- Correctly sketch to scale half-section and full-section orthographic drawings from manufactured parts.
- Correctly sketch to scale a sectioned assembly of three to eight parts from unsectioned two-dimensional or three-dimensional drawings.
- Correctly answer questions regarding sectional drawings.

USES OF SECTIONAL DRAWINGS

It is often necessary to use a drawing technique known as *sectioning*. Sectional drawings are used to show the internal construction of parts (Figure 9–1) or external features that cannot be understood easily with external views (Figure 9–2). In many cases, sectional drawings are used to show not only how to make a part but also how several parts function or how to assemble parts (Figures 9–3 and 9–4). The use of sectional views requires many conventional practices and symbols, which are described in this chapter.

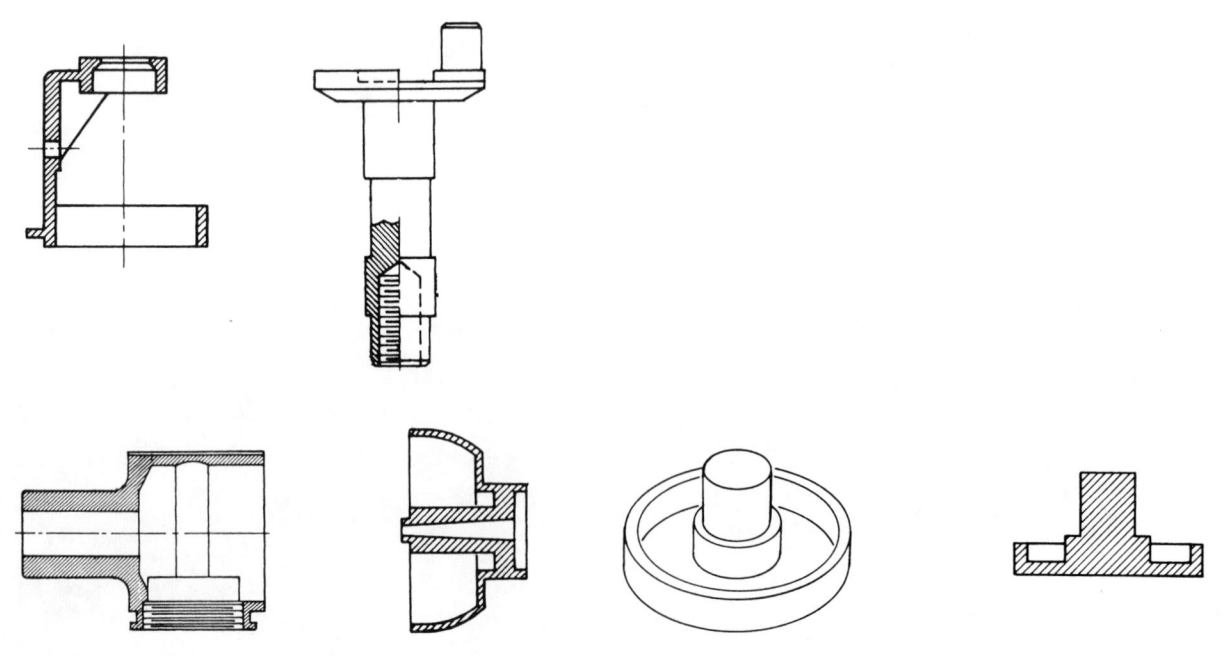

FIGURE 9–1
Sectional Views Showing Internal Construction

FIGURE 9–2
Sectional View Showing External Construction

FIGURE 9–3
Sectional Views Showing How Parts Are Assembled or Function

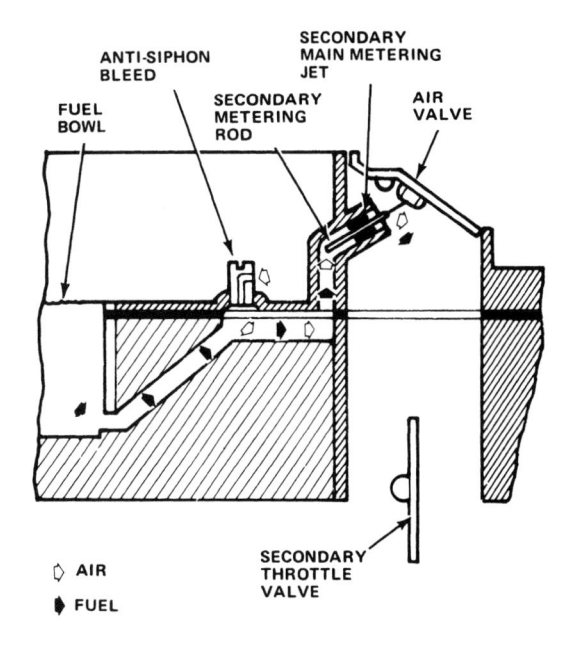

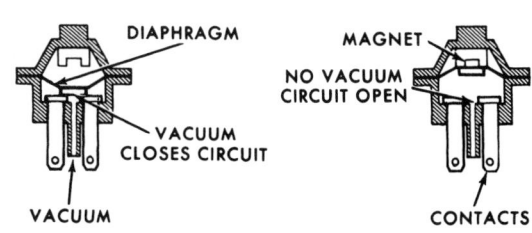

FIGURE 9–4
Sectional View Showing How a Wall Is Constructed

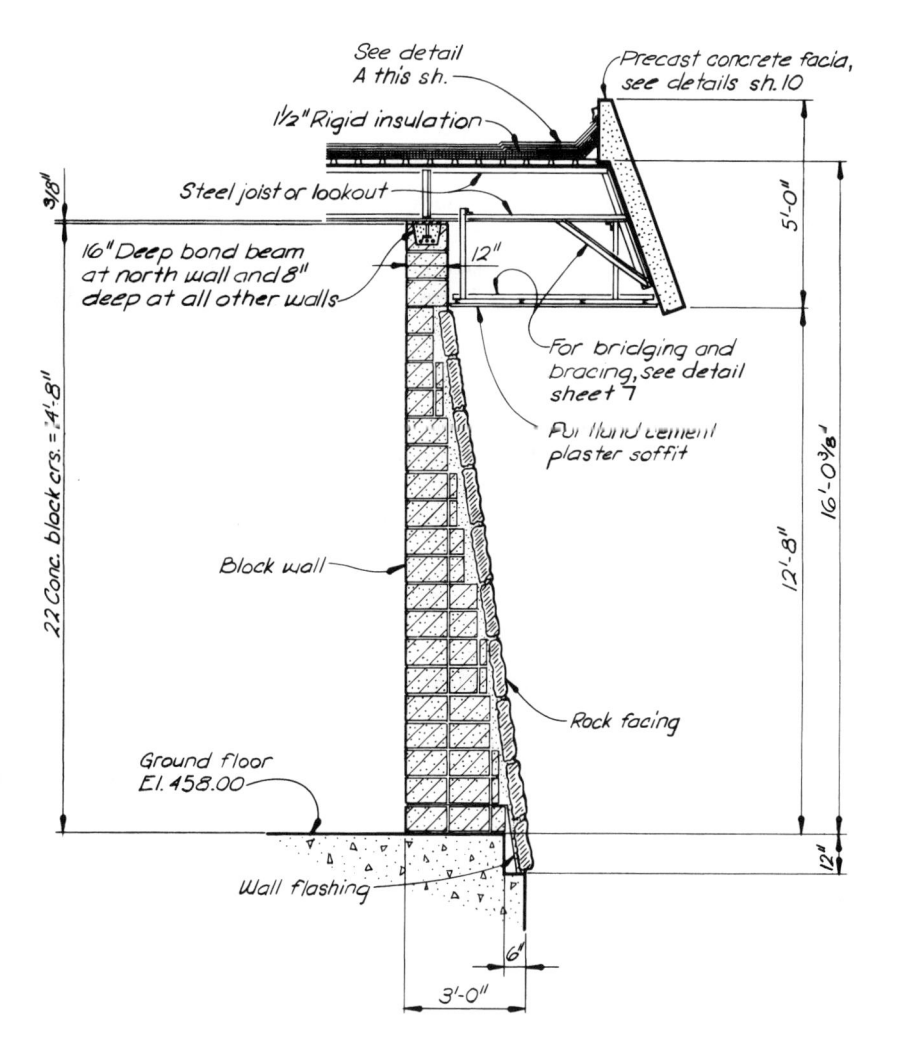

Sketching Sectional Views

151

FIGURE 9–5
Steps in Constructing a Sectional View

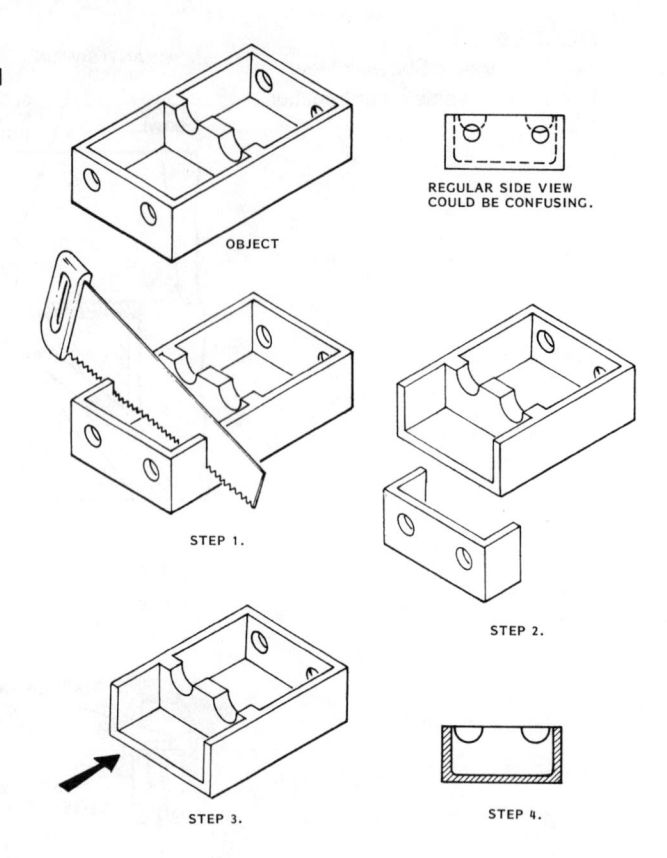

CONSTRUCTING A SECTIONAL VIEW

The object shown in Figure 9–5 is a complex shape. Its features could be misunderstood if only external views were used. Therefore, to avoid any misunderstanding, a sectional view is constructed. Sectional views are easy to construct if you follow these steps:

Step 1. Decide which view will best show the hidden feature. In your mind, cut off the part that is hiding the feature. The cut, in this case, should be straight and should extend completely across the object.

Step 2. Throw away the part you cut off, and do not think of it again. It is easy to be confused about which part to draw. Throwing away the cut part eliminates that confusion.

Step 3. Look into the part that is left. Your line of sight should be perpendicular to or straight into the remaining piece.

Step 4. Draw the shape of what you see. Draw section or hatch lines on the part that was cut, as if you are drawing saw marks on the part of the object that the saw touched when the cut was made. The parts untouched by the saw are sketched without section lining.

ELEMENTS OF SECTIONAL DRAWINGS

Now that you have read the steps in constructing a sectional view, descriptions of the elements of sectional views are needed. Let's start with cutting-plane lines.

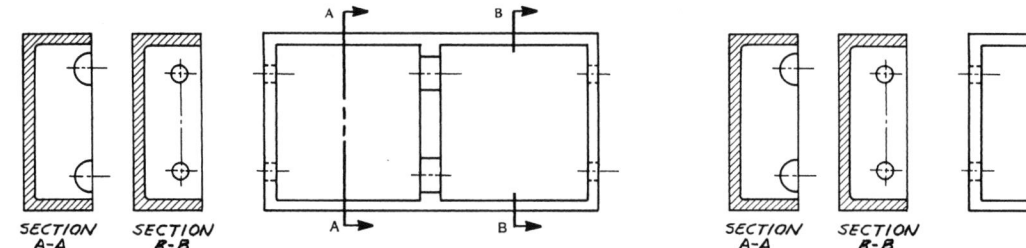

FIGURE 9–6
Cutting-Plane Lines

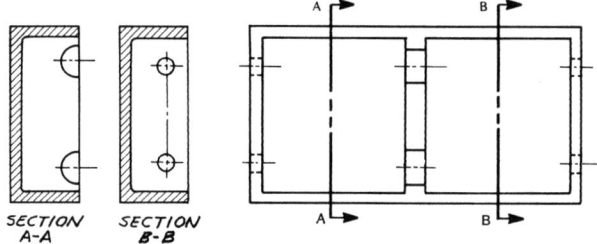

FIGURE 9–7
Labeling Sectional Views

Cutting-Plane Lines

To show someone exactly where a cut was made, place an extra-heavy line with two dashes in it and arrows on the ends, showing the line of sight (Figure 9–6). This is a cutting-plane line. Another version of the cutting-plane line is also shown in this figure. This cutting-plane line, which does not extend across the object, is preferred by many companies because it does not hide other lines as often as the complete cutting-plane line does. Notice that the left-side view has been sectioned to show the shape of the object at the point where the cutting plane passes through the object.

Using More Than One Sectional View on a Single Part

If more than one section is used on the same drawing, the sections are identified by letters placed at the base of the arrow (Figure 9–7). Double letters (A-A and B-B) are used to label the sectional views.

Exceptions to Conventional Drawing Practices

To avoid confusion, it is often necessary to treat some features of parts differently than according to standard practices. The following are common exceptions to standard practices.

Eliminating Hidden Lines in Sectional Views

Sectional views are usually much clearer if hidden lines are not shown. The hidden lines are rarely necessary (Figure 9–8). You should eliminate hidden lines on sectional views unless they are absolutely necessary. In Figure 9–8, the hidden lines not only are unnecessary but would be confusing if they were drawn.

Not Sectioning Thin Features of a Part

To eliminate confusion, it is best not to section thin features, such as ribs and spokes (Figure 9–9). Although the cutting plane does cut through the ribs, the object looks like a shortened cone if it is drawn with hatch lines on the ribs.

FIGURE 9–8
Eliminating Hidden Lines on Sectional Views

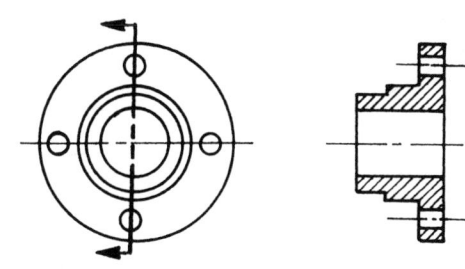

CORRECT SECTIONAL VIEW

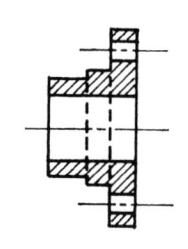

AS IT ACTUALLY APPEARS

Sketching Sectional Views

FIGURE 9–9
Do Not Section Ribs and Spokes

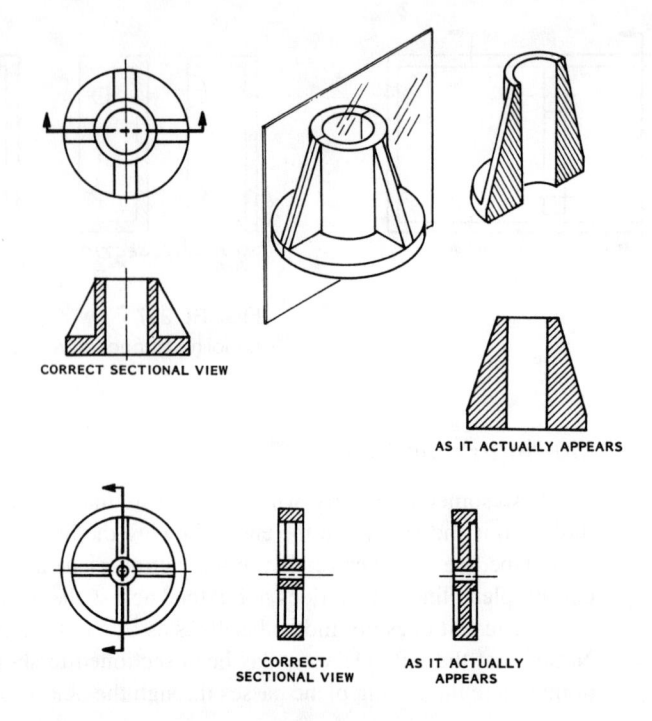

Offsetting Cutting-Plane Lines

To show the best sectional view, it is sometimes necessary to show the cutting plane offset (Figure 9–10). The imaginary line made by the offset in the cut is not shown on the sectional view. The offset cutting-plane line was needed to allow both holes to be shown in the sectional view.

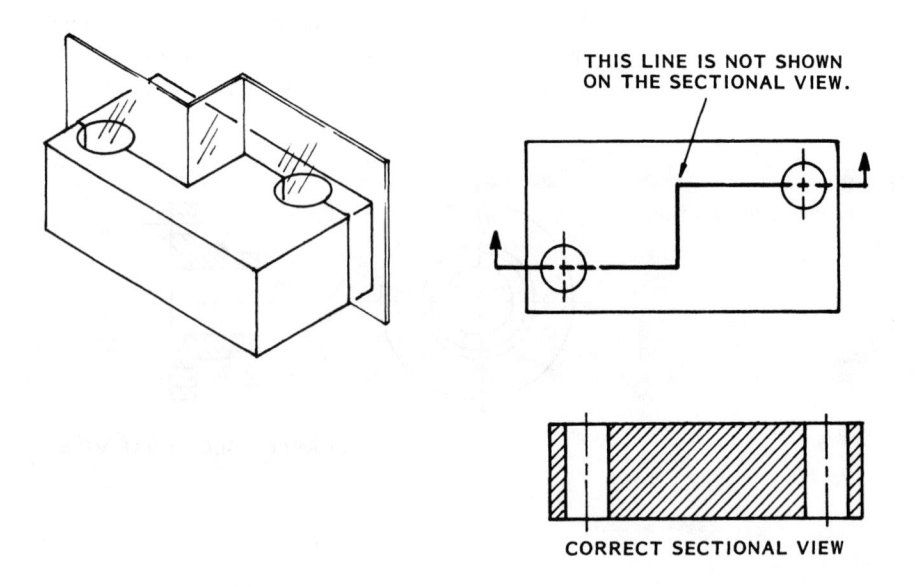

Rotating Features to Show Their Shape

It is clearer to show some shapes rotated to the cutting plane in the sectional view. In Figure 9–11A, the two ribs are moved to the imaginary cutting plane in the sectional view. The front view remains unchanged. In Figure 9–11B the cutting plane is rotated, but the small holes are shown in the sectional view as if the top hole were on a vertical cutting plane. The reason for drawing the sectional view this way is to make the drawing clearer to the person who will be making the part. The true shape of the ribs in Figure 9–11A makes it appear that they do not extend to the circumference of the circular base. In Figure 9–11B the top hole would be shown as if it were not the same distance from the edge of the part. Either of these views could result in the part's being manufactured incorrectly.

FIGURE 9–10
Offset Cutting Plane

FIGURE 9–11
Rotating Features

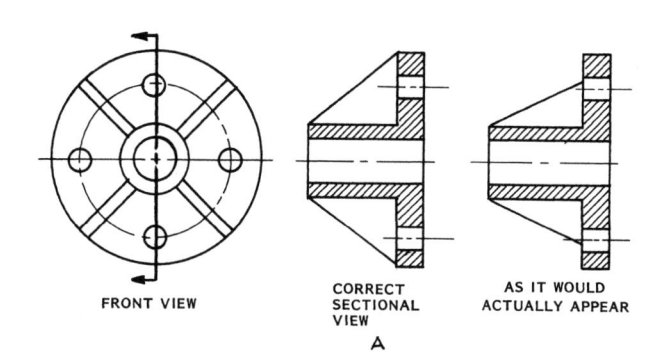

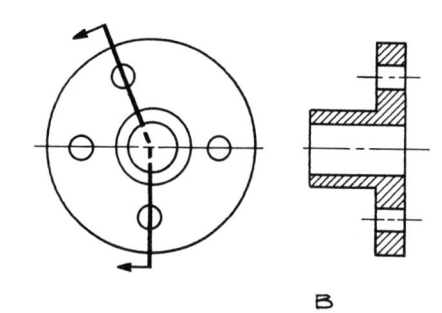

Types of Sectional Views

Two major types of sections are described by the portion of the part that is cut by the cutting plane: the full section and the half section. Other types are partial or broken-out and assembly sections.

Full Sections

In a full section, the cutting plane passes through the entire object. It may be straight or offset if that is necessary to show additional detail. If the cutting plane is offset, as in Figure 9–10, the edges of the offset cut are not shown on the sectional view.

Half Sections

Half sections are used to show both the internal and external shape of an object on the same view (Figure 9–12). Half sections are used when the object is symmetrical and both internal and external views are needed to completely describe the object. The line separating the internal and external features is a center line. The cutting-plane line has an arrow on only one end to show the line of sight to the viewer.

FIGURE 9–12
Half Sections

Sketching Sectional Views 155

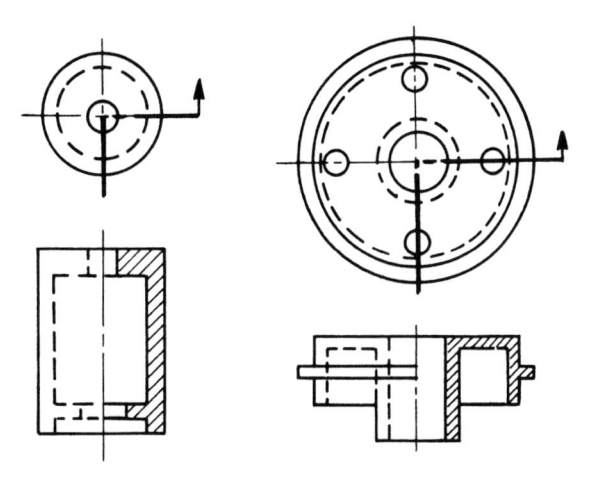

FIGURE 9–13
Partial or Broken-out Sections

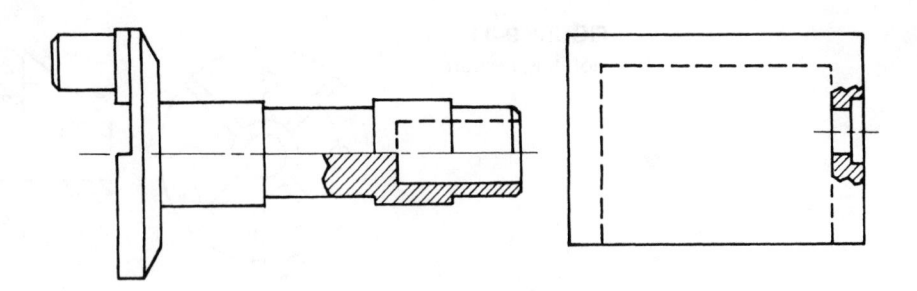

Partial or Broken-out Sections

It is sometimes helpful to show a partial section of a view, often called a *broken-out section*. The break line is shown with straight lines at obtuse angles (Figure 9–13). This allows a particular feature to be shown without sectioning the complete object. Broken-out sections are often used to avoid dimensioning to a hidden line, which is undesirable.

Assembly Sections

Assembly sections are used to show how parts function or fit together in an assembly. The following are features of assembly sections:

> Many of the parts of an assembly are not shown as cut in the sectional view. This makes the drawing much clearer. In general, no fasteners, springs, rods, shafts, or spheres are shown as cut (Figure 9–14).
> Parts that fit together have section lines at opposite 45° angles. When you run out of 45° angles, use 60° angles, then 30° angles.
> Some industries use different section lining for different materials. Often, very thin parts and parts made from rubber are shown solid black.
> Assembly sections must show the parts in such a manner that complete assembly and building instructions may be written.
> Parts lists are often included on the same sheet or on additional sheets so that each individual part may be identified.

Sketching hatch patterns on sectional views will make you appreciate the hatching feature of AutoCAD and AutoCAD LT, which is described in the following chapter. So now, sketch some sectional views.

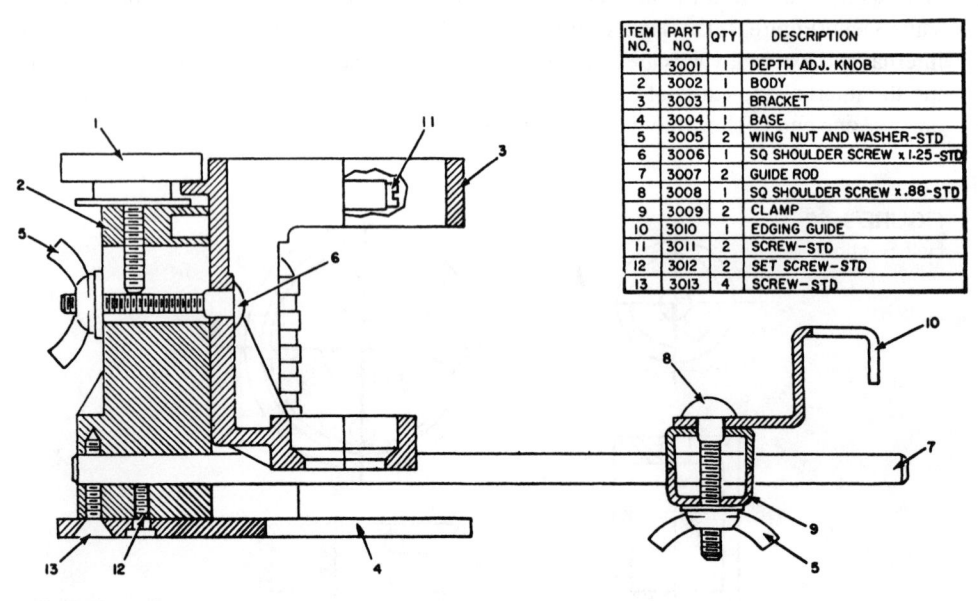

ITEM NO.	PART NO.	QTY	DESCRIPTION
1	3001	1	DEPTH ADJ. KNOB
2	3002	1	BODY
3	3003	1	BRACKET
4	3004	1	BASE
5	3005	2	WING NUT AND WASHER-STD
6	3006	1	SQ SHOULDER SCREW x 1.25-STD
7	3007	2	GUIDE ROD
8	3008	1	SQ SHOULDER SCREW x .88-STD
9	3009	2	CLAMP
10	3010	1	EDGING GUIDE
11	3011	2	SCREW-STD
12	3012	2	SET SCREW-STD
13	3013	4	SCREW-STD

FIGURE 9–14
An Assembly Section

Chapter 9

EXERCISES

EXERCISE 9–1 Complete Exercise 9–1 using the steps described.

Step 1. Remove the sheet labeled Exercise 9–1 from your book.

Step 2. Using the sketching and construction techniques you used in Chapter 7, complete the right-side view as a sectional view. The depth dimensions are shown by the horizontal lines at the top and bottom of the vertical line. The smallest circles on the front view are holes that go all the way through the part.

Sprinkle drafting powder over your drawing before you begin to darken lines.

Be sure to line up all features of the object with the adjacent views.

Make sure that your lines are the correct weight and are of even width and darkness. Try to match the thickness and darkness of the existing lines.

Step 3. Draw 45° section lines in the right-side view approximately $\frac{1}{10}''$ apart. Use the lines shown as an example. These lines should be thin and dark.

Step 4. Draw centerlines in the right-side view.

Step 5. Fill in the title block with your best lettering. Title the drawing SECTION 1.

EXERCISE 9–2 Complete Exercise 9–2 using the steps described.

Step 1. Remove the sheet labeled Exercise 9–2 from your book.

Step 2. Sketch the front and right-side views of Figure 9–15 using the sketching and construction techniques you used in Chapter 7. Complete the right-side view as a sectional view.

Draw full scale.

Sprinkle drafting powder over your drawing before you begin to darken lines.

Be sure to line up all features of the object with the adjacent views.

Make sure that your lines are the correct weight and are of even width and darkness. Try to match the thickness and darkness of the existing lines.

DO NOT SHOW DIMENSIONS. DO SHOW CENTERLINES.

Step 3. Draw 45° section lines in the right-side view approximately $\frac{1}{10}''$ apart. Use the lines shown in Exercise 9–1 as an example. These lines should be thin and dark.

Step 4. Fill in the title block with your best lettering. Title the drawing, SECTION 2.

FIGURE 9–15
Dimensions for Exercise 9–2

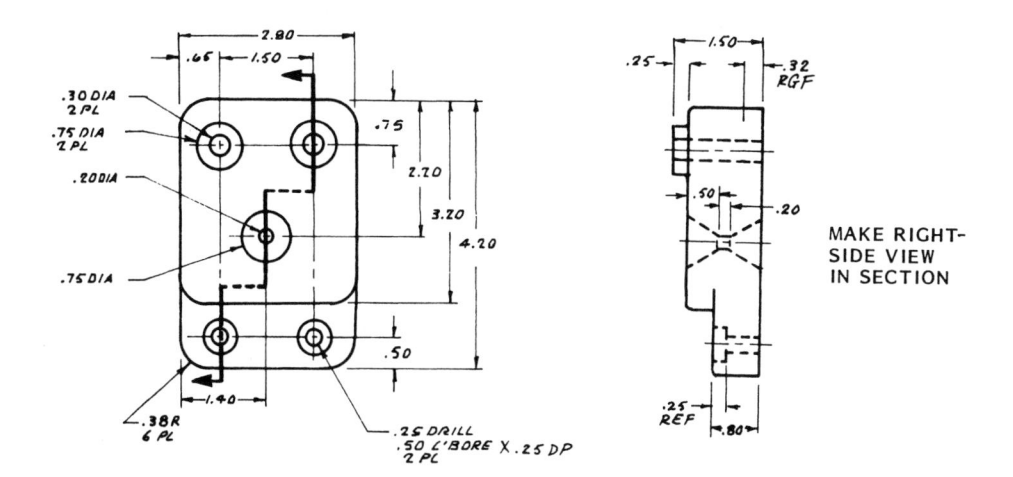

FIGURE 9–16
Dimensions for Exercise 9–3

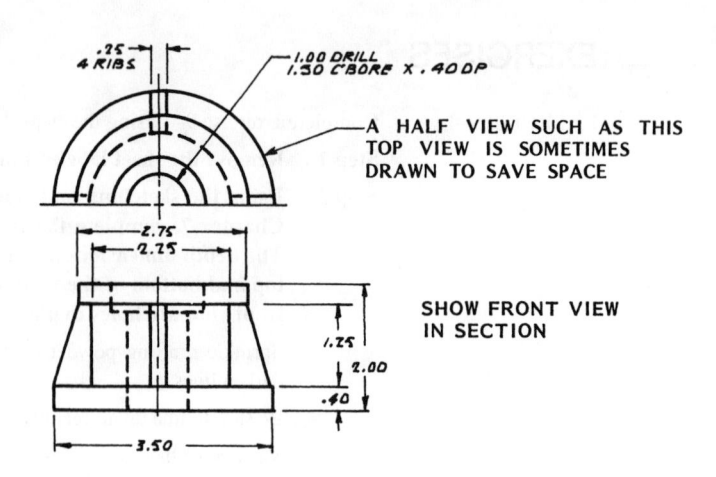

EXERCISE 9–3 Complete Exercise 9–3 using the steps described.

Step 1. **Remove the sheet labeled Exercise 9–3 from your book.**

Step 2. **Sketch the top and front views of Figure 9–16 using the sketching and construction techniques you used in Chapter 7. Sketch circles with a circle template.**

Draw full scale.

Complete the front view as a sectional view.

Sprinkle drafting powder over your drawing before you begin to darken lines.

Be sure to line up all features of the object with the adjacent views.

Make sure that your lines are the correct weight and are of even width and darkness. Try to match the thickness and darkness of the existing lines.

DO NOT SHOW DIMENSIONS. DO SHOW CENTERLINES.

Step 3. **Draw 45° section lines in the front view approximately $\frac{1}{10}''$ apart. Use the lines shown in Exercise 9–1 as an example. These lines should be thin and dark.**

Step 4. **Fill in the title block with your best lettering. Title the drawing SECTION 3.**

EXERCISE 9–4 Complete Exercise 9–4 using the steps described.

Step 1. **Remove the sheet labeled Exercise 9–4 from your book.**

Step 2. **Sketch the front and right-side views of Figure 9–17 using the sketching and construction techniques you used in Chapter 7. Complete the right-side view as a sectional view.**

FIGURE 9–17
Dimensions for Exercise 9–4

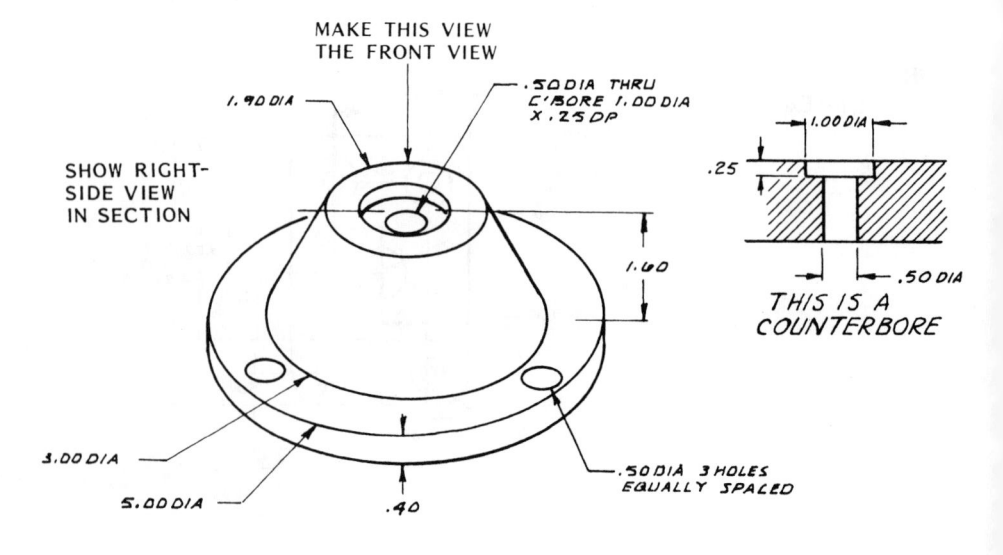

Draw at a scale of $\frac{1}{2}'' = 1''$

Sprinkle drafting powder over your drawing before you begin to darken lines.

Be sure to line up all features of the object with the adjacent views.

Make sure that your lines are the correct weight and are of even width and darkness. Try to match the thickness and darkness of the existing lines.

DO NOT SHOW DIMENSIONS. DO SHOW CENTERLINES.

Step 3. Draw 45° section lines in the right-side view approximately $\frac{1}{10}''$ apart. Use the lines shown in Exercise 9–1 as an example. These lines should be thin and dark.

Step 4. Fill in the title block with your best lettering. Title the drawing SECTION 4.

EXERCISE 9–5   Complete Exercise 9–5 using the steps described.

Step 1. Remove the sheet labeled Exercise 9–5 from your book.

Step 2. Sketch the front and right-side views of Figure 9–18 using the sketching and construction techniques you used in Chapter 7. Complete the right-side view as a sectional view.

Draw at a scale of $\frac{1}{2}'' = 1''$

Sprinkle drafting powder over your drawing before you begin to darken lines.

Be sure to line up all features of the object with the adjacent views.

Make sure that your lines are the correct weight and are of even width and darkness. Try to match the thickness and darkness of the existing lines.

DO NOT SHOW DIMENSIONS. DO SHOW CENTERLINES.

Step 3. Draw 45° section lines in the right-side view approximately $\frac{1}{10}''$ apart. DO NOT SHOW SECTION LINES ON THE RIBS. Use the lines shown in Exercise 9–1 as an example. These lines should be thin and dark.

Step 4. Fill in the title block with your best lettering. Title the drawing SECTION 5.

EXERCISE 9–6   Complete Exercise 9–6 using the steps described.

Step 1. Remove the sheet labeled Exercise 9–6 from your book.

Step 2. Sketch the front and right-side views of Figure 9–19 using the sketching and construction techniques you used in Chapter 7. Complete the front view as a half section. Show the internal shape on the top half of the view.

FIGURE 9–18
Dimensions for Exercise 9–5

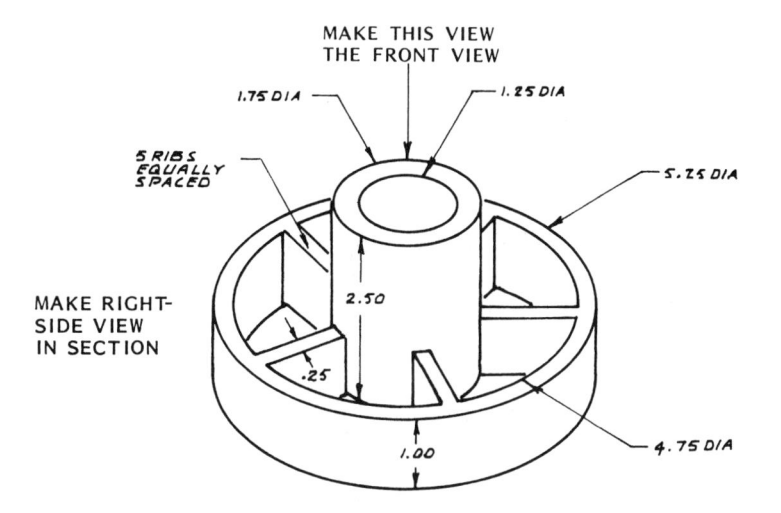

Sketching Sectional Views

FIGURE 9–19
Dimensions for Exercise 9–6

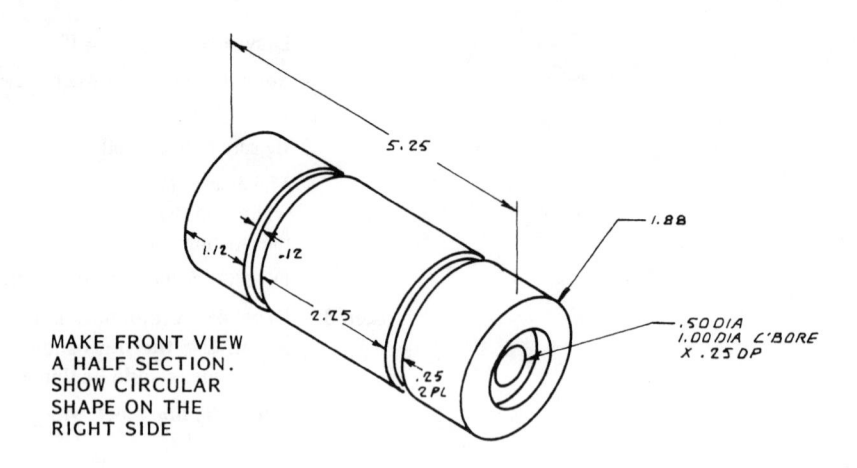

MAKE FRONT VIEW
A HALF SECTION.
SHOW CIRCULAR
SHAPE ON THE
RIGHT SIDE

Draw full scale.

Sprinkle drafting powder over your drawing before you begin to darken lines.

Be sure to line up all features of the object with the adjacent views.

Make sure that your lines are the correct weight and are of even width and darkness. Try to match the thickness and darkness of the existing lines.

DO NOT SHOW DIMENSIONS. DO SHOW CENTERLINES.

Step 3. Draw 45° section lines in the front view approximately $\frac{1}{10}''$ apart. Use the lines shown in Exercise 9–1 as an example. These lines should be thin and dark.

Step 4. Fill in the title block with your best lettering. Title the drawing **SECTION 6**.

EXERCISE 9–7 Complete Exercise 9–7 using the steps described.

Step 1. Remove the sheet labeled Exercise 9–7 from your book.

Step 2. Sketch the front view only of Figure 9–20 as an assembly sectional view. Put all of the parts together as an assembly.

Draw full scale.

Do not cut the shaft.

Sprinkle drafting powder over your drawing before you begin to darken lines.

Make sure that your lines are the correct weight and are of even width and darkness. Try to match the thickness and darkness of the existing lines.

DO NOT SHOW DIMENSIONS.

Step 3. Draw 45° section lines on the bushing approximately $\frac{1}{10}''$ apart. Draw 45° section lines on the hub in the opposite direction. Show the rubber tire solid black.

Step 4. Fill in the title block with your best lettering. Title the drawing **ASSEMBLY SECTION**.

REVIEW QUESTIONS

Circle the best answer.

1. Why are sectional views needed to describe objects?
   a. To describe surface textures
   b. To show complex interior details
   c. To describe the overall shape of the part

FIGURE 9-20
Dimensions for Exercise 9-7

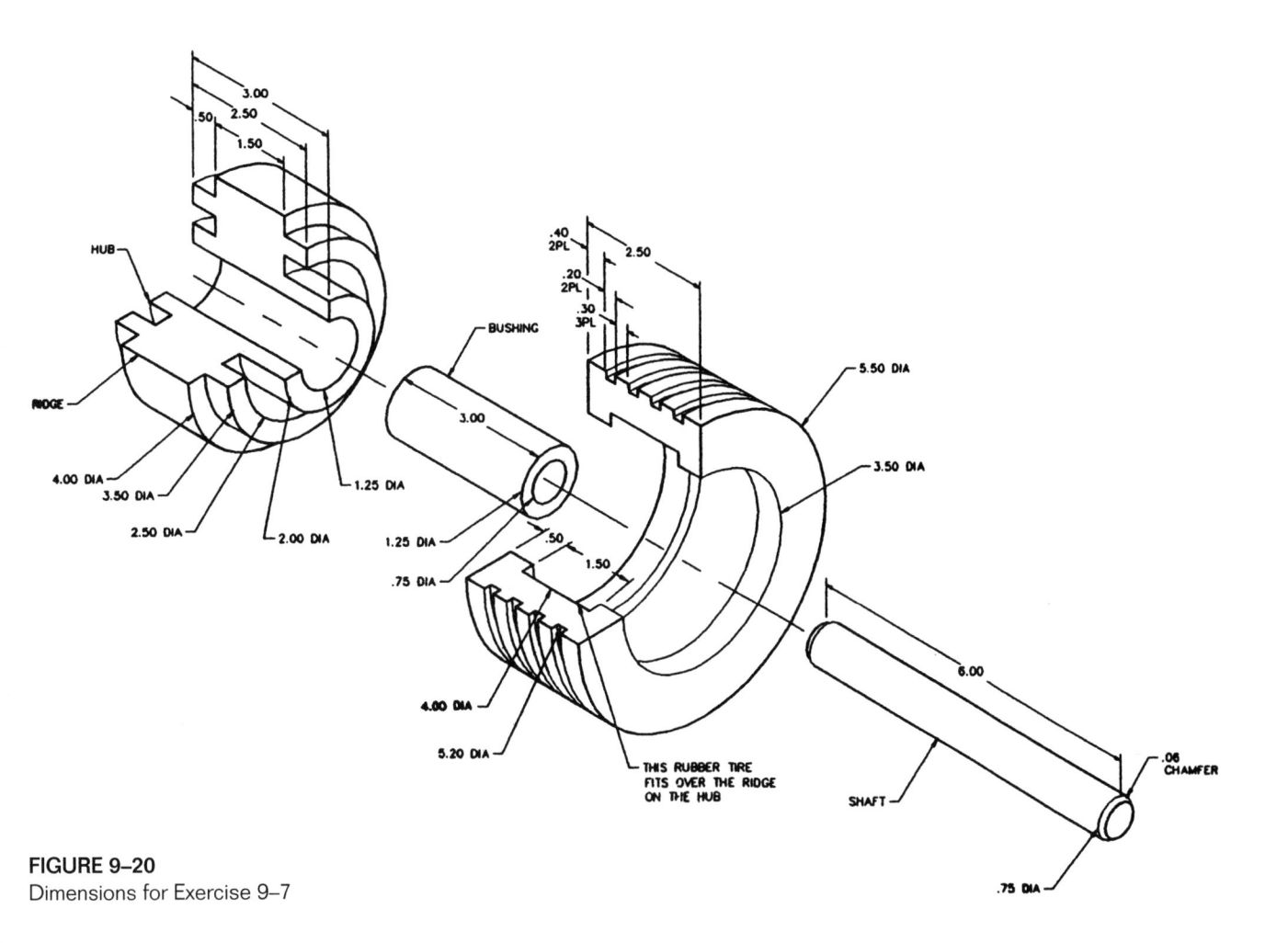

    d. To show different sides of a part
    e. To complete a top view
2. The object is cut by a _____ to describe the sectional view.
    a. Cutting plane
    b. Saw
    c. Knife
    d. String
    e. Blade
3. The solid material cut by the cutting plane is hatched with diagonal lines, usually drawn at
    a. 20°
    b. 15°
    c. 90°
    d. 0°
    e. 45°
4. The line representing the edge view of the cutting plane is called
    a. The edge-view line
    b. The section line
    c. The cut line
    d. The hatch line
    e. The cutting-plane line
5. If the object is cut all the way across, the view is called
    a. A full section
    b. A quarter view
    c. A half section
    d. A partial section
    e. The plan view

Sketching Sectional Views

6. If the cutting plane cuts halfway across a symmetrical object, the resulting sectional view is called
   a. A full section
   b. A quarter view
   c. A half section
   d. A partial section
   e. The plan view
7. In a half-section view, the exterior view and the interior view are separated by
   a. A centerline
   b. A cutting-plane line
   c. A hidden line
   d. An object line
   e. A phantom line
8. Hidden lines in a half-sectioned object show only on the side representing
   a. The interior view
   b. The exterior view
9. What are the advantages of sketching a half section of a symmetrical object rather than a full section?
   a. It shows exterior and interior at the same time.
   b. It saves sketching time.
   c. It helps the observer understand the part better.
   d. It requires fewer views
   e. All the above
10. A cutting plane that offsets to pass through small details of a part creates
    a. An offset section
    b. A quarter section
    c. A half section
    d. A partial section
    e. The plan view
11. What is the name given to the section created when a small portion of the part is broken away to reveal an interior detail?

    _____

12. List three items that are *not* section lined in an assembly section even though the cutting plane cuts right through them.

    _____ _____ _____

13. List two purposes that an assembly section view serves.

    _____

    _____

14. Why are hidden lines not usually shown in sectional views?

    _____

15. How are sectional views labeled when there are four sectional views on the same drawing?

    _____ _____ _____ _____

16. Describe how thick the cutting-plane line should be in relation to the object line.

    _____

17. Describe how thick the section lines should be in relation to the object line.

    _____

18. Why are ribs and spokes not hatched in a sectional view?

    _____

19. What is the purpose of a parts list?

    _____

20. Why are some features rotated in a sectional view?

    _____

10 Making Sectional Views with AutoCAD

OBJECTIVES

After completing this chapter, you will be able to:

□ Make accurate drawings containing hatch patterns using AutoCAD or AutoCAD LT.
□ Use the following commands to produce drawings containing hatch patterns:

Array	Zoom	Fillet
Line	Trim	Chamfer
Osnap	Offset	Hatch
Circle		

□ Answer questions regarding the commands listed above.

HATCHING USING AUTOCAD OR AUTOCAD LT

AutoCAD and AutoCAD LT use the Hatch command to draw section lines on sectional views. The Hatch command allows you to create a boundary containing the hatch pattern by picking individual lines or with the use of a window. It also allows you to pick a single point inside the boundary that is to contain the hatch pattern. In this chapter you will use select objects to create a boundary for the hatch pattern.

HATCH

The Hatch command contains a number of hatch patterns that may be used to fill areas. Because this is an introductory course the details of how to use those patterns in a variety of ways are omitted. Hatch in this chapter requires that the boundary selected be complete and that none of the lines extend outside the area to be hatched. Therefore, temporary lines are drawn for the boundary and erased after the hatching is complete. Also, the hatch patterns are placed on a layer named HATCH so that the pattern remains on a different layer that is assigned a different color. In Exercise 10–1 you will draw a single view and place a hatch pattern on it. In Exercise 10–2 you will draw a front view that contains a cutting-plane line and a right-side view that contains the hatching necessary to make it a sectional view.

EXERCISE 10–1
Making a Sectional Drawing Using Hatch Patterns

Step 1. To begin Exercise 10–1, turn on the computer and start AutoCAD or AutoCAD LT.
Step 2. Open drawing EX10-1 supplied on the disk that came with your book.
Step 3. Use Zoom-All to view the entire drawing area.

Step 4. Use the Line command to draw object lines in the right-side view.

Prompt	Response
Command:	Type: **L**↵
Specify first point:	Type: **END**↵
of	Click: **D1** (Figure 10–1)
Specify next point or [Undo]:	Type: **@4.5<90**↵
Specify next point or [Undo]:	Type: **@.75<180**↵
Specify next point or [Close/Undo]:	Type: **@2.5<270**↵
Specify next point or [Close/Undo]:	Type: **@.75<180**↵
Specify next point or [Close/Undo]:	Type: **END**↵
of	Click: **D2**
Specify next point or [Close/Undo]:	↵
Command:	↵
Specify first point:	Click: **D3** (Be sure D3 is aligned with the top quadrant of the small upper circle.)
Specify next point or [Undo]:	Click: **D4**
Specify next point or [Undo]:	↵
Command:	↵
Specify first point:	Click: **D5** (Be sure D5 is aligned with the bottom quadrant of the small upper circle.)
Specify next point or [Undo]:	Click: **D6**
Specify next point or [Undo]:	↵
Command:	↵
Specify first point:	Click: **D7** (Be sure D7 is aligned with the top quadrant of the small lower circle.)
Specify next point or [Undo]:	Click: **D8**
Specify next point or [Undo]:	↵
Command:	↵
Specify first point:	Click: **D9** (Be sure D9 is aligned with the bottom quadrant of the small lower circle.)

Important: Be sure Snap and Ortho are ON so you can align picked points with the circles in the front view.

FIGURE 10–1

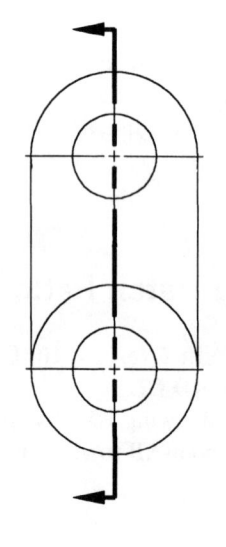

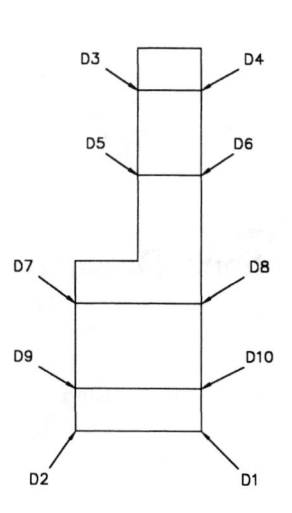

Prompt	Response
Specify next point or [Undo]:	Click: **D10**
Specify next point or [Undo]:	↵

Step 5. Because the Hatch command (select objects) requires a complete boundary with no lines extending outside that boundary, use the Line command to draw boundary lines in the right-side view. Begin by setting a new layer current and a running osnap mode of Endpoint so you do not have to type END↵ for every click.

Prompt	Response
Command:	Type: **-LA**↵
Enter an option [?/Make/Set/New/ON/ OFF/Color/Ltype/LWeight/Plot/ Freeze/Thaw/LOck/Unlock/stAte]:	Type: **S**↵
Enter layer name to make current or <select object>:	Type: **0**↵
Command:	Type: **-OS**↵
Enter list of object snap modes:	Type: **END**↵
Command:	Type: **L**↵
Specify first point:	Click: **D1** (Figure 10–2)
Specify next point or [Undo]:	Click: **D2**
Specify next point or [Undo]:	↵
Command:	↵
Specify first point:	Click: **D3**
Specify next point or [Undo]:	Click: **D4**
Specify next point or [Undo]:	↵
Command:	↵
Specify first point:	Click: **D5**
Specify next point or [Undo]:	Click: **D6**
Specify next point or [Undo]:	↵
Command:	↵
Specify first point:	Click: **D7**
Specify next point or [Undo]:	Click: **D8**
Specify next point or [Undo]:	↵
Command:	↵
Specify first point:	Click: **D9**
Specify next point or [Undo]:	Click: **D10**
Specify next point or [Undo]:	↵
Command:	↵
Specify first point:	Click: **D11**
Specify next point or [Undo]:	Click: **D12**
Specify next point or [Undo]:	↵
Command:	↵
Specify first point:	Click: **D13**
Specify next point or [Undo]:	Click: **D14**
Specify next point or [Undo]:	↵

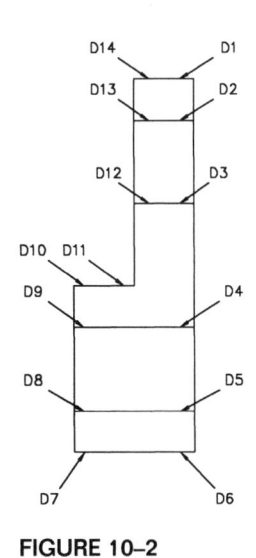

FIGURE 10–2

Step 6. Take off the running osnap mode Endpoint so you will not be picking endpoints without knowing it.

Prompt	Response
Command:	Type: **-OS**↵
Enter list of object snap modes:	Type: **NONE**↵

Step 7. Set the layer HATCH current and use the Hatch command to place crosshatching on the areas where the cutting plane touches.

Prompt	Response
Command:	Type: **-LA**↵
Enter an option [?/Make/Set/New/ON/ OFF/Color/Ltype/LWeight/Plot/Freeze/ Thaw/LOck/Unlock/stAte]:	Type: **S**↵
Enter layer name to make current layer or <select object>:	Type: **HATCH**↵
Command:	Type: **-H**↵
Enter a pattern name or [?/Solid/User defined] <ANSI131>:	Type: **U**↵ (to specify a user-defined pattern—meaning you type the angle and the spacing for the pattern)
Specify angle for crosshatch lines <0>:	Type: **45**↵
Specify spacing between lines <1.0000>:	Type: **.1**↵
Double hatch area? [Yes/No] <N>	↵
Select objects to define hatch boundary or <direct hatch>: Select objects:	Click: **D1** (Figure 10–3)
Specify opposite corner:	Click: **D2**
Select objects:	Click: **D3**
Specify opposite corner:	Click: **D4**
Select objects:	Click: **D5**
Specify opposite corner:	Click: **D6**
Select objects:	↵

FIGURE 10–3

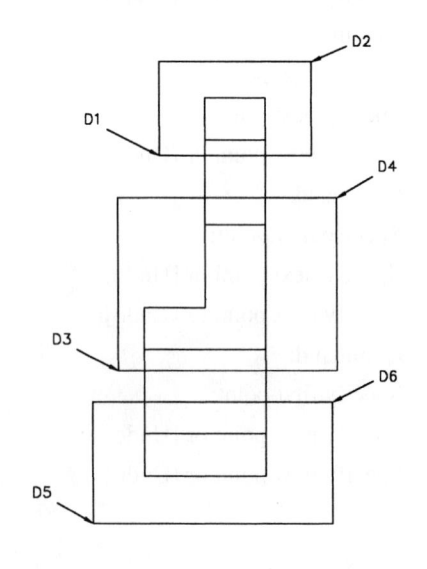

Step 8. Turn off all layers except the 0 layer and erase the boundary lines you drew.

Prompt	Response
Command:	Type: **-LA**↵
Enter an option [?/Make/Set/New/ON/OFF/Color/Ltype/LWeight/Plot/Freeze/Thaw/LOck/Unlock/stAte]:	Type: **OFF**↵
Enter name list of layer(s) to turn off or <select objects>:	Type: *****↵
Really want layer HATCH (the CURRENT layer) off? <N>	Type: **Y**↵
Enter an option [?/Make/Set/New/ON/OFF/Color/Ltype/LWeight/Plot/Freeze/Thaw/LOck/Unlock/stAte]:	Type: **ON**↵
Enter name list of layer(s) to turn ON:	Type: **0**↵
Enter an option [?/Make/Set/New/ON/OFF/Color/Ltype/LWeight/Plot/Freeze/Thaw/LOck/Unlock/stAte]:	↵
Command:	Type: **E**↵
Select objects:	Click: **D1** (Figure 10–4)
Other corner:	Click: **D2**
Select objects:	↵

Step 9. Turn on all layers, set layer CENTER current, and draw center lines in the right-side view.

Prompt	Response
Command:	Type: **-LA**↵
Enter an option [?/Make/Set/New/ON/OFF/Color/Ltype/LWeight/Plot/Freeze/Thaw/LOck/Unlock/stAte]:	Type: **ON**↵
Enter name list of layer(s) to turn ON:	Type: *****↵
Enter an option [?/Make/Set/New/ON/OFF/Color/Ltype/LWeight/Plot/Freeze/Thaw/LOck/Unlock/stAte]:	Type: **S**↵
Enter layer name to make current or <select object>:	Type: **CENTER**↵

FIGURE 10–4

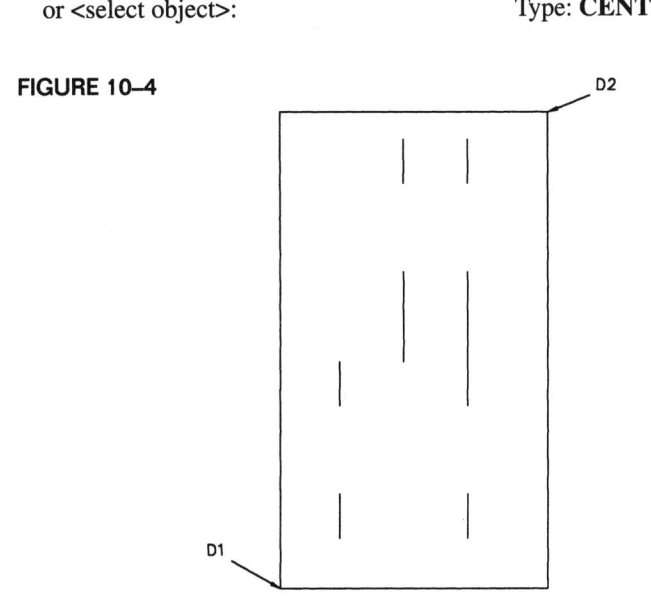

FIGURE 10–5

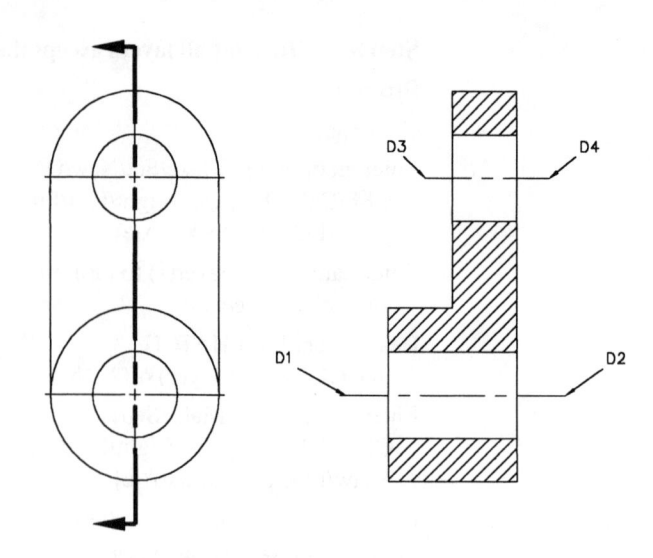

Important: Be sure Snap and Ortho are on so you can align the center lines with the centers of the circles in the front view. Make the center lines in the right-side view approximately the same length as shown in Figure 10–5.

Prompt	Response
Enter an option [?/Make/Set/New/ON/OFF/Color/Ltype/LWeight/Plot/Freeze/Thaw/LOck/Unlock/stAte]:	↵
Command:	Type: **L**↵
Specify first point:	Click: **D1** (Figure 10–5)
Specify next point or [Undo]:	Click: **D2**
Specify next point or [Undo]:	↵
Command:	↵
Specify first point:	Click: **D3**
Specify next point or [Undo]:	Click: **D4**
Specify next point or [Undo]:	↵

Step 10. Set the OBJECT layer current and use the Dtext command to complete the title block as you did for previous exercises. Name the drawing GUIDE.

Step 11. Use the SAVEAS command to save your drawing as EX10-1(your initials) on a floppy disk and again on the hard drive of your computer.

Step 12. Plot or print your drawing full size on an $11'' \times 8\frac{1}{2}''$ sheet.

EXERCISE 10–2
Making a Drawing Containing a Front View with a Cutting Plane, and a Right-Side View in Section from an Isometric Sketch

Step 1. To begin Exercise 10–2, turn on the computer and start AutoCAD or AutoCAD LT.

Step 2. Open drawing EX10-2 supplied on the disk that came with your book.

Step 3. Use Zoom-All to view the entire drawing area.

The dimensions for Exercise 10–2 are taken from the sketch shown in Figure 10–6. The diameters in this sketch use the abbreviation DIA instead of the symbol ∅. You will find both symbols used in drawings you are assigned when you go to work.

Step 4. Use the Polar Array command to complete the front view.

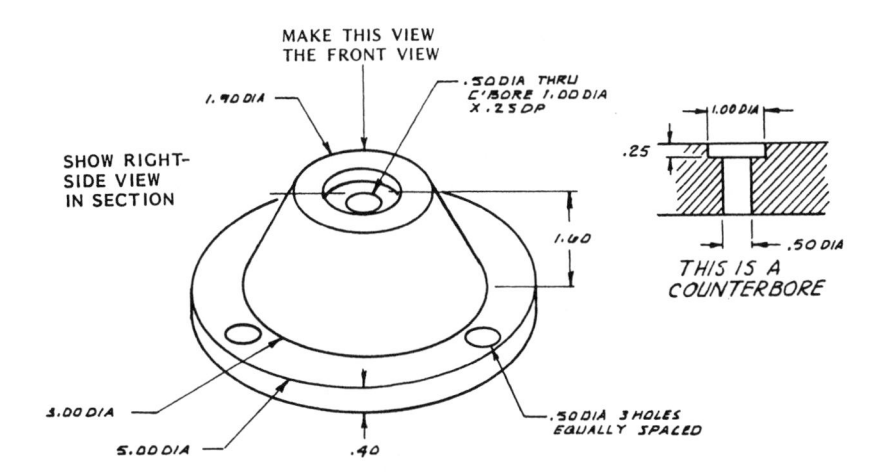

FIGURE 10–6
Dimensions for Exercise 10–2

FIGURE 10–7

Important: Make sure Snap is ON and Ortho is OFF so you can pick the necessary points accurately.

FIGURE 10–8

The Polar Array command allows you to copy objects around a center point. The number of items you specify to be copied must include the item you pick. You may also specify the number of degrees to be included in the array. The default is 360° or a complete circle.

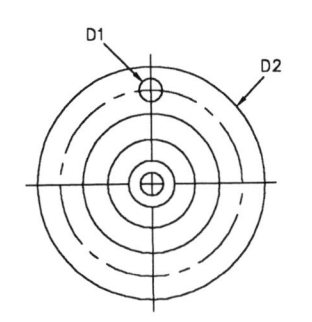

Prompt	Response
Command:	Type: **-AR**↵
Select objects:	Click: **D1** (Figure 10–7)
Select objects:	↵
Enter the type of array [Rectangular/Polar] <R>:	Type: **P**↵
Specify center point of array or [Base]:	Type: **CEN**↵
of	Click: **D2**
Enter the number of items in the array:	Type: **3**↵ (This is the total number of items.)
Specify the angle to fill (+=ccw, −=cw) <360>:	↵
Rotate arrayed objects? [Yes/No] <Y>:	↵ (It does not matter whether you rotate these objects.)

Step 5. **Set layer PHANTOM current and use the Polyline command to draw the cutting-plane line in the front view.**

On your own: Set layer PHANTOM current.

Prompt	Response
Command:	Type: **PL**↵
Specify start point:	Click: **D1** (Figure 10–8)
Specify next point or [Arc/Halfwidth/ Length/Undo/Width]:	Type: **W**↵
Specify starting width <0.0000>:	Type: **.05**↵
Specify ending width <0.0500>:	↵
Specify next point or [Arc/Halfwidth/ Length/Undo/Width]:	Click: **D2** (Make sure Snap is ON and that the pline is aligned with the center of the circle.)
Specify next point or [Arc/Close/ Halfwidth/Length/Undo/Width]:	Click: **D3** (the center of the circle)
Specify next point or [Arc/Close/ Halfwidth/Length/Undo/Width]:	Type: **@1.5<210**↵

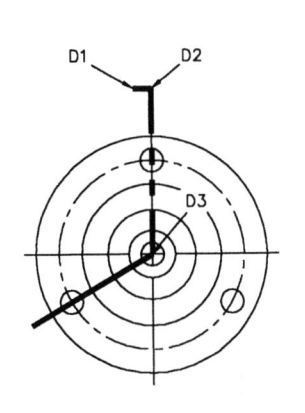

Making Sectional Views with AutoCAD　169

Prompt	Response
Specify next point or [Arc/Close/Halfwidth/Length/Undo/Width]:	↵

Step 6. Set layer OBJECT current and use the Polyline command to draw an arrow on the top end of the cutting-plane line.

On your own: Set layer OBJECT current.

Prompt	Response
Command:	Type: **PL**↵
Specify first point:	Click: **D1** (Figure 10–9)
Specify next point or [Arc/Halfwidth/Length/Undo/Width]:	Type: **W**↵
Specify starting width <0.0000>:	Type: **.15**↵
Specify ending width <0.1500>:	Type: **0**↵
Specify next point or [Arc/Halfwidth/Length/Undo/Width]:	Type: **@.2<180**↵
Specify next point or [Arc/Close/Halfwidth/Length/Undo/Width]:	↵

Step 7. Use the Line command to draw object lines in the right-side view.

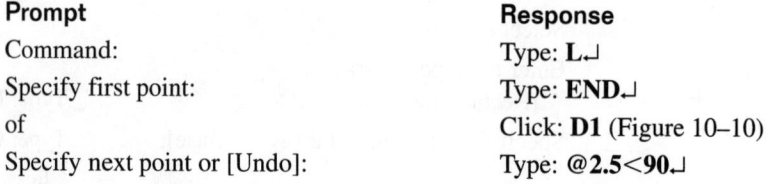

Prompt	Response
Command:	Type: **L**↵
Specify first point:	Type: **END**↵
of	Click: **D1** (Figure 10–10)
Specify next point or [Undo]:	Type: **@2.5<90**↵
Specify next point or [Undo]:	Type: **@.2<180**↵
Specify next point or [Close/Undo]:	Type: **@.5<-90**↵
Specify next point or [Close/Undo]:	Type: **@-.8,-.25**↵
Specify next point or [Close/Undo]:	Type: **@1<-90**↵
Specify next point or [Close/Undo]:	Type: **@.8,-.25**↵
Specify next point or [Close/Undo]:	Type: **@.5<-90**↵
Specify next point or [Close/Undo]:	Type: **END**↵
of	Click: **D2**
Specify next point or [Close/Undo]:	↵

Step 8. Use the Offset command to draw the small holes in the right-side view.

Prompt	Response
Command:	Type: **O**↵
Specify offset distance or [Through]:	Type: **.125**↵
Select object to offset:	Click: **D1** (Figure 10–11)
Specify point on side to offset:	Click: **D2**

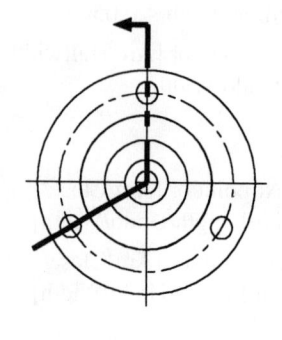

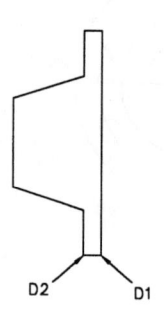

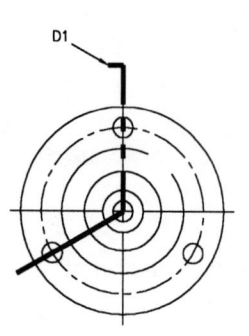

FIGURE 10–9

Note: You may need to zoom a window around the right-side view to make it easier to pick the lines to offset.

FIGURE 10–10

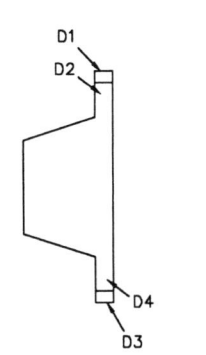

FIGURE 10-11

Important: Make sure Snap and Ortho are ON so you can pick points accurately.

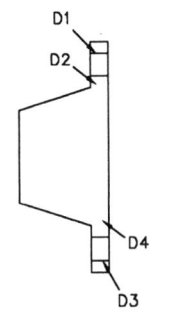

FIGURE 10-12

Prompt	Response
Select object to offset:	Click: **D3**
Specify point on side to offset:	Click: **D4**
Select object to offset:	↵
Command:	↵
Specify offset distance or [Through]:	Type: **.25**↵
Select object to offset:	Click: **D1** (Figure 10–12)
Specify point on side to offset:	Click: **D2**
Select object to offset:	Click: **D3**
Specify point on side to offset:	Click: **D4**
Select object to offset:	↵

Step 9. Use the Line command to draw the counterbore in the right-side view.

Prompt	Response
Command:	Type: **L**↵
Specify first point:	Click: **D1** (Figure 10–13)
Specify next point or [Undo]:	Type: **@.125<0**↵
Specify next point or [Undo]:	Type: **@.5<90**↵
Specify next point or [Close/Undo]:	Type: **@.125<180**↵
Specify next point or [Close/Undo]:	↵
Command:	↵
Specify first point:	Type: **NEA**↵
to	Click: **D2**
Specify next point or [Undo]:	Type: **PER**↵
to	Click: **D3**
Specify next point or [Undo]:	↵
Command:	↵
Specify first point:	Type: **NEA**↵
to	Click: **D4**
Specify next point or [Undo]:	Type: **PER**↵
to	Click: **D5**
Specify next point or [Undo]:	↵

Step 10. Because the Hatch command requires a complete boundary with no lines extending outside that boundary, use the Line command to draw boundary lines in the right-side view. Begin by setting a new layer current and a running osnap mode of Endpoint so you do not have to type END ↵ for every click.

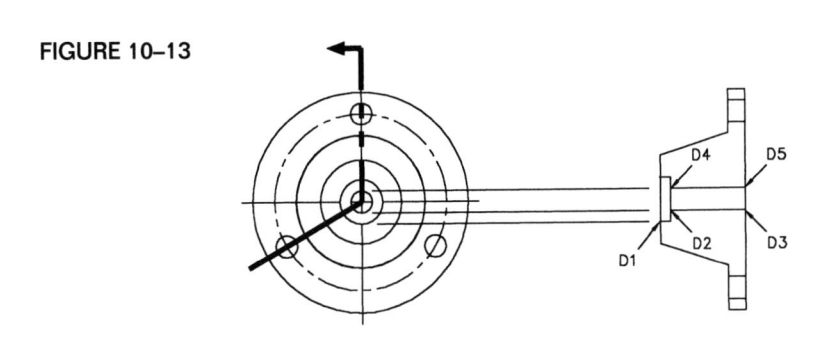

FIGURE 10-13

Making Sectional Views with AutoCAD

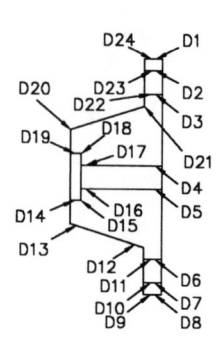

On your own: Set layer current.

Prompt	Response
Command:	Type: **-OSNAP**↵
Object snap modes:	Type: **END**↵

On your own: Zoom a window around the right-side view so you can pick points more easily, then proceed with the Line command.

Prompt	Response
Command:	Type: **L**↵
Specify first point:	Click: **D1** (Figure 10–14)
Specify next point or [Undo]:	Click: **D2**
Specify next point or [Undo]:	↵
Command:	↵
Specify first point:	Click: **D3**
Specify next point or [Undo]:	Click: **D4**
Specify next point or [Undo]:	↵
Command:	↵
Specify first point:	Click: **D5**
Specify next point or [Undo]:	Click: **D6**
Specify next point or [Undo]:	↵
Command:	↵
Specify first point:	Click: **D7**
Specify next point or [Undo]:	Click: **D8**
Specify next point or [Undo]:	↵
Command:	↵
Specify first point:	Click: **D9**
Specify next point or [Undo]:	Click: **D10**
Specify next point or [Undo]:	↵
Command:	↵
Specify first point:	Click: **D11**
Specify next point or [Undo]:	Click: **D12**
Specify next point or [Undo]:	↵
Command:	↵
Specify first point:	Click: **D13**
Specify next point or [Undo]:	Click: **D14**
Specify next point or [Undo]:	↵
Command:	↵
Specify first point:	Click: **D15**
Specify next point or [Undo]:	Click: **D16**
Specify next point or [Undo]:	↵
Command:	↵
Specify first point:	Click: **D17**
Specify next point or [Undo]:	Click: **D18**
Specify next point or [Undo]:	↵
Command:	↵
Specify first point:	Click: **D19**
Specify next point or [Undo]:	Click: **D20**
Specify next point or [Undo]:	↵

FIGURE 10–14

Command:	↵
Specify first point:	Click: **D21**
Specify next point or [Undo]:	Click: **D22**
Specify next point or [Undo]:	↵
Command:	↵
Specify first point:	Click: **D23**
Specify next point or [Undo]:	Click: **D24**
Specify next point or [Undo]:	↵

Step 11. Take off the running osnap mode of Endpoint so you will not be picking endpoints without knowing it.

Prompt	Response
Command:	Type: **-OS**↵
Enter list of object snap modes:	Type: **NONE**↵

Step 12. Set the layer HATCH current, and use the Hatch command to place crosshatching on the areas where the cutting plane touches.

Prompt	Response
Command:	Type: **-H**↵
Enter a pattern name or [?/Solid/User defined] <ANSI131>:	Type: **U**↵ (to specify a user-defined pattern—meaning you type the angle and the spacing for the pattern)
Specify angle for crosshatch lines <0>:	Type: **45**↵
Specify spacing between the lines <1.0000>:	Type: **.1**↵
Double hatch area? [Yes/No] <N>:	↵
Select objects:	Click: **D1** (Figure 10–15)
Specify opposite corner:	Click: **D2**
Select objects:	Click: **D3**
Specify opposite corner:	Click: **D4**
Select objects:	Click: **D5**
Specify opposite corner:	Click: **D6**
Select objects:	Click: **D7**
Specify opposite corner:	Click: **D8**
Select objects:	↵

Step 13. Turn off all layers except the 0 layer, and erase the boundary lines you drew.

Prompt	Response
Command:	Type: **-LA**↵
Enter an option [?/Make/Set/New/ON/ OFF/Color/Ltype/LWeight/Plot/Freeze/ Thaw/LOck/Unlock/stAte]:	Type: **OFF**↵
Enter name list of layer(s) to turn off or <select objects>:	Type: ***** ↵
Really want layer HATCH (the CURRENT layer) off? <N>	Type: **Y**↵
Enter an option [?/Make/Set/New/ON/ OFF/Color/Ltype/LWeight/Plot/Freeze/ Thaw/LOck/Unlock/stAte]:	Type: **ON**↵
Enter name list of layer(s) to turn on:	Type: **0**↵

FIGURE 10–15

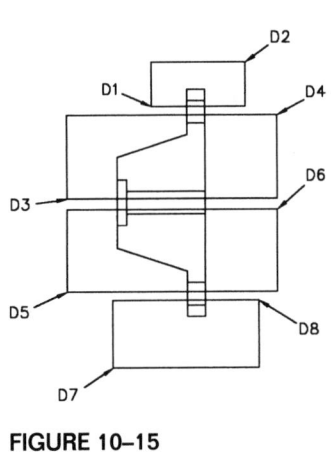

Making Sectional Views with AutoCAD

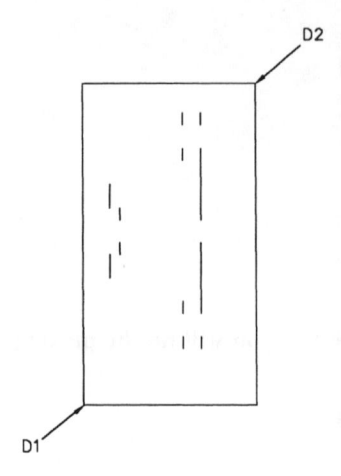

FIGURE 10–16

Prompt	Response
Enter an option [?/Make/Set/New/ON/OFF/Color/Ltype/LWeight/Plot/Freeze/Thaw/LOck/Unlock/stAte]:	↵
Command:	Type: **E**↵
Select objects:	Click: **D1** (Figure 10–16)
Specify opposite corner:	Click: **D2**
Select objects:	↵

Step 14. Turn all layers on, set layer CENTER current, and draw centerlines in the right-side view.

Prompt	Response
Command:	Type: **-LA**↵
Enter an option [?/Make/Set/New/ON/OFF/Color/Ltype/LWeight/Plot/Freeze/Thaw/LOck/Unlock/stAte]:	Type: **ON**↵
Enter a list of layer(s) to turn ON:	Type: ***** ↵
Enter an option [?/Make/Set/New/ON/OFF/Color/Ltype/LWeight/Plot/Freeze/Thaw/LOck/Unlock/stAte]:	Type: **S**↵
Enter layer name to make current or <select object>:	Type: **CENTER**↵
Enter an option [?/Make/Set/New/ON/OFF/Color/Ltype/LWeight/Plot/Freeze/Thaw/LOck/Unlock/stAte]:	↵
Command:	Type: **L**↵
Specify first point:	Click: **D1** (Figure 10–17)
Specify next point or [Undo]:	Click: **D2**
Specify next point or [Undo]:	↵
Command:	↵
Specify first point:	Click: **D3**
Specify next point or [Undo]:	Click: **D4**
Specify next point or [Undo]:	↵
Command:	↵
Specify first point:	Click: **D5**
Specify next point or [Undo]:	Click: **D6**
Specify next point or [Undo]:	↵

Important: Be sure Snap and Ortho are ON so you can align the center lines with the centers of the circles in the front view. Make the center lines in the right-side view approximately the same length as shown in Figure 10–17.

Step 15. Use the Dtext command to complete the title block as you have done previously. Title the drawing FLANGE.

Step 16. Use the SAVEAS command to save your drawing as EX10-2(your initials) on a floppy disk and again on the hard drive of your computer.

Step 17. Plot or print your drawing full size on an 11″ × 8½″ sheet.

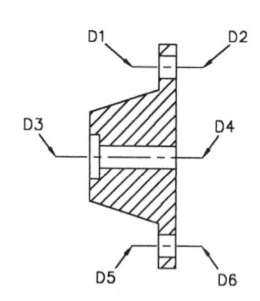

FIGURE 10–17

EXERCISES

EXERCISE 10–1   Complete Exercise 10–1 using steps 1 through 12 described in this chapter.
EXERCISE 10–2   Complete Exercise 10–2 using steps 1 through 17 described in this chapter.

REVIEW QUESTIONS

Circle the best answer.

1. Which of the following will produce a line $3\frac{1}{2}''$ long upward from a point?
   a. 3.5 × 90
   b. @3.5<90
   c. @0<3.5
   d. @3.5<−90
   e. 90<3.5
2. The Hatch command as used in this chapter requires
   a. A complete boundary that must be selected
   b. That you pick a single point within the boundary
   c. That at least two lines proceed outside the boundary
   d. The use of only the uniform hatch pattern
   e. That only a 45° angle be specified
3. The Hatch command also allows you to pick
   a. A complete boundary that must be selected
   b. A single point within the boundary
   c. At least two lines proceed outside the boundary
   d. Only the uniform hatch pattern
   e. Only a 45° angle pattern
4. Which of the following patterns produces continuous, evenly spaced lines?
   a. User defined
   b. MUDST
   c. PAT LINE
   d. U-LINE
   e. LINE
5. Which of the following responses to the "Spacing between lines" prompt will produce Hatch lines $\frac{1}{4}''$ apart?
   a. 1-4
   b. .25
   c. 2.5
   d. 1,4
   e. 25
6. After a hatching command that spaced lines .10 apart has been performed, what is the default response to the "Spacing between lines" prompt (original default was .08)?
   a. 0
   b. .1
   c. .08
   d. .125
   e. .06
7. What do you type to set the width of a polyline at the prompt "specify next point or [Arc/Close/Halfwidth/Length/ Undo/Width];"?
   a. A
   b. H
   c. W
   d. The beginning width of the pline
   e. The beginning width and the ending width separated by a comma
8. What layer used in the exercises in this chapter must be current for hidden lines to be drawn?
   a. HIDDEN
   b. OBJECT
   c. PHANTOM
   d. DASHED
   e. 0
9. What layer used in the exercises in this chapter must be current for phantom lines to be drawn?
   a. HIDDEN
   b. OBJECT
   c. PHANTOM
   d. DASHED
   e. 0

10. If you are using polar array to copy an object four times (total of five objects), what do you type in response to the prompt "Enter the number of items in the array:"?
    a. 0
    b. 4
    c. 5
    d. P
    e. N

11. Which two functions must be ON when you are aligning items between views that have been drawn with Snap ON?

12. What must you type to turn on all layers in response to the prompt "Enter name list of layer(s) to turn on:"

13. Describe how to set a layer current.

14. How can you be sure that only one layer is ON?

15. How do you turn OFF a running osnap mode?

16. Why turn OFF a running osnap mode of Endpoint?

17. Why do you Zoom in closer to an area of the drawing?

18. Describe how to use the Offset command to draw parallel lines $\frac{1}{2}''$ apart.

19. Describe how to draw an arrowhead with PLINE.

20. What is the meaning of "User defined" in the Hatch prompt "Enter a pattern name or [?/Solid/User defined]:"?

11 Sketching Auxiliary Views

OBJECTIVES

After completing this chapter, you will be able to:

☐ Correctly sketch to scale full and partial primary and secondary auxiliary views from two-dimensional or three-dimensional drawings.
☐ Correctly answer questions regarding auxiliary drawings.

USES OF AUXILIARY VIEWS

Many objects can be described with no more than three normal views. Some objects, however, have features that must be described with one or more auxiliary views. An auxiliary view is one that is usually external and describes features that are not on one of the normal planes. In Figure 11–1 notice that the object in 11–1A does not require an auxiliary view because the depth of the slanted surface is shown in the front view. The slot on the object in B is shown on an auxiliary view because it cannot be seen in its true shape in any of the normal views.

Placement on the Page

Auxiliary views are arranged on the page so that they are parallel to the edge view of the slanted surface. Figure 11–2 shows the transparent box again, with the auxiliary plane folded out on the same angle as the slanted surface. All auxiliaries are arranged in this manner. Notice that the projection lines from the front are drawn at 90° to the slanted surface. This is very important in obtaining correct width measurements on the auxiliary view.

Partial Auxiliaries

Most auxiliary views are drawn as partial views. These partial views show the true shape of the slanted surface and nothing else. Confusing hidden surfaces and foreshortened normal surfaces are avoided by using partial views (Figure 11–3). Drawing partial views also takes less time. Notice that the break is shown with straight lines at obtuse angles. Figure

FIGURE 11–1
Surfaces Requiring Auxiliary Views

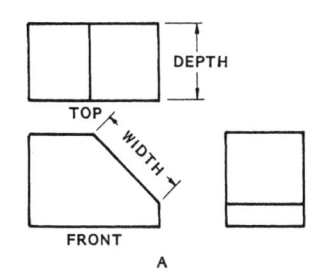

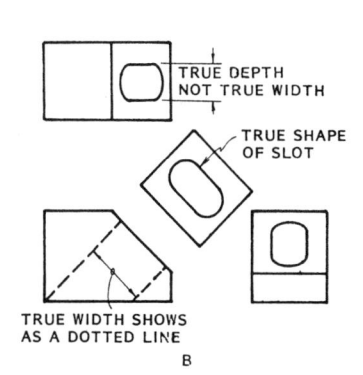

FIGURE 11–2
The Transparent Viewing Box

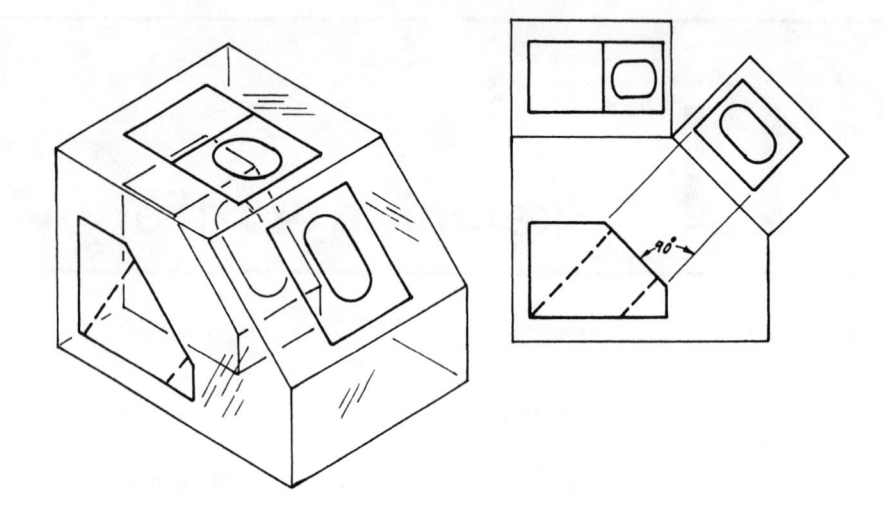

FIGURE 11–3
Partial Views

FULL AUXILIARY

PARTIAL AUXILIARY

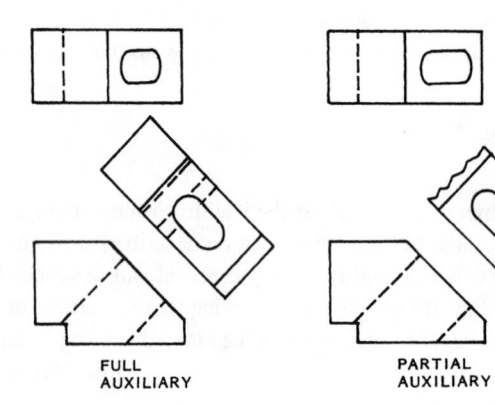

FIGURE 11–4
Complete Object Description with Three Partial Views

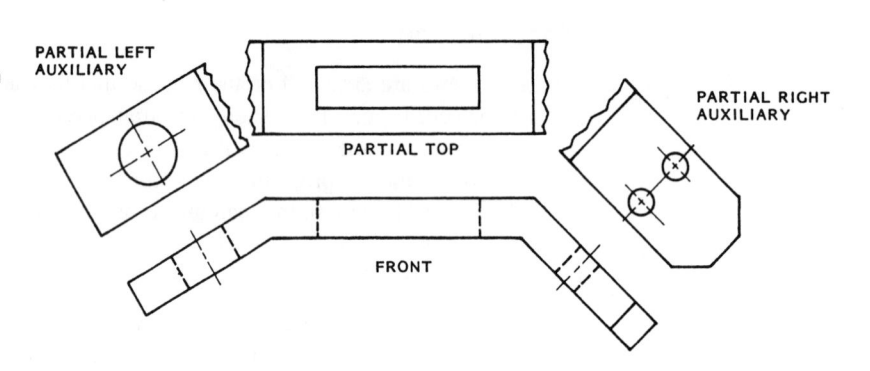

PARTIAL LEFT AUXILIARY

PARTIAL TOP

PARTIAL RIGHT AUXILIARY

FRONT

11–4 shows an object that has been fully described using the front view with partial top, left, and right auxiliaries.

DRAWING A PRIMARY (ALSO CALLED SINGLE) AUXILIARY VIEW FROM A THREE-DIMENSIONAL VIEW

The steps for drawing a primary auxiliary view that shows the inclined surface in the front view (Figure 11–5) are as follows:

Step 1. Draw the front view and as much of the top view as possible. Place this drawing far to the left to allow the right auxiliary to be drawn in the space to the right of the front view.

178   Chapter 11

FIGURE 11-5
Steps in Drawing a Primary Auxiliary View

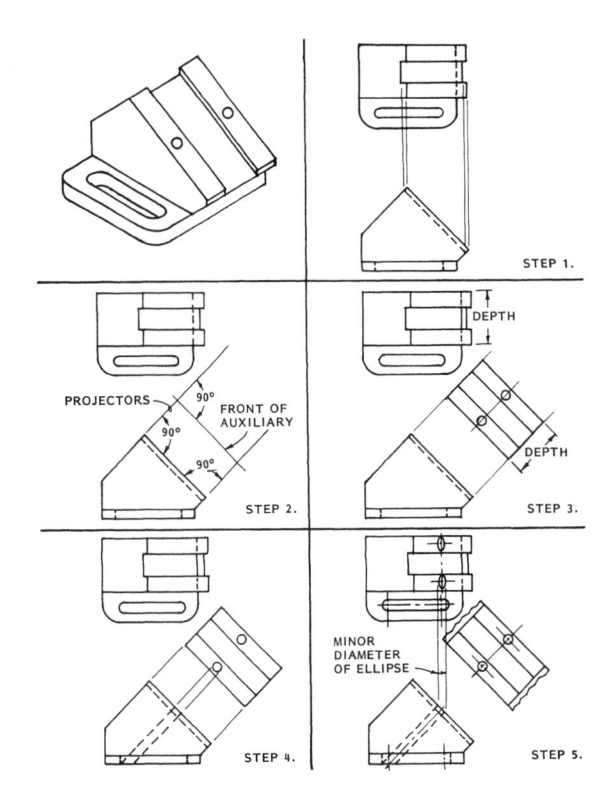

Step 2. Draw lines perpendicular to the slanted surface and locate the front of the auxiliary view at a convenient distance from the other views. These lines are at 90° to the projectors.

Step 3. Mark depth dimensions on the auxiliary view. Take these dimensions from the top view and the three-dimensional view.

Step 4. Project features from the auxiliary view into the front view, and complete the front view.

Step 5. Project features from the front view into the top view, and transfer depth dimensions from the auxiliary view to complete the top view. Notice that the circles appear as ellipses in the top view. The depth of the ellipse is the same as the true diameter of the circle in the auxiliary view, but the minor diameter must be projected from the front view.

DRAWING A SECONDARY (ALSO CALLED DOUBLE) AUXILIARY VIEW FROM A THREE-DIMENSIONAL VIEW

A secondary auxiliary view is used when a surface is not perpendicular to any of the normal views. This view requires that an edge view be drawn before the auxiliary view can be positioned and constructed. It also requires the use of reference planes, which can be thought of as the hinged joints of the transparent box used to describe views.

The steps for drawing a secondary auxiliary view (Figure 11-6) are as follows:

Step 1. Block in the top and front views from given information.

Step 2. Draw two reference planes (RP). One should be parallel to the width dimensions of the front view. The other should be parallel to the edge of the slanted surface, which, when folded out at 90°, will show an edge view of the slanted surface. Locate these two lines at any convenient distance from the views.

Step 3. Project lines at 90° to RP2 from all points in the top view that you want to find in the edge view.

Sketching Auxiliary Views

FIGURE 11–6
Steps in Drawing a Secondary Auxiliary View

180

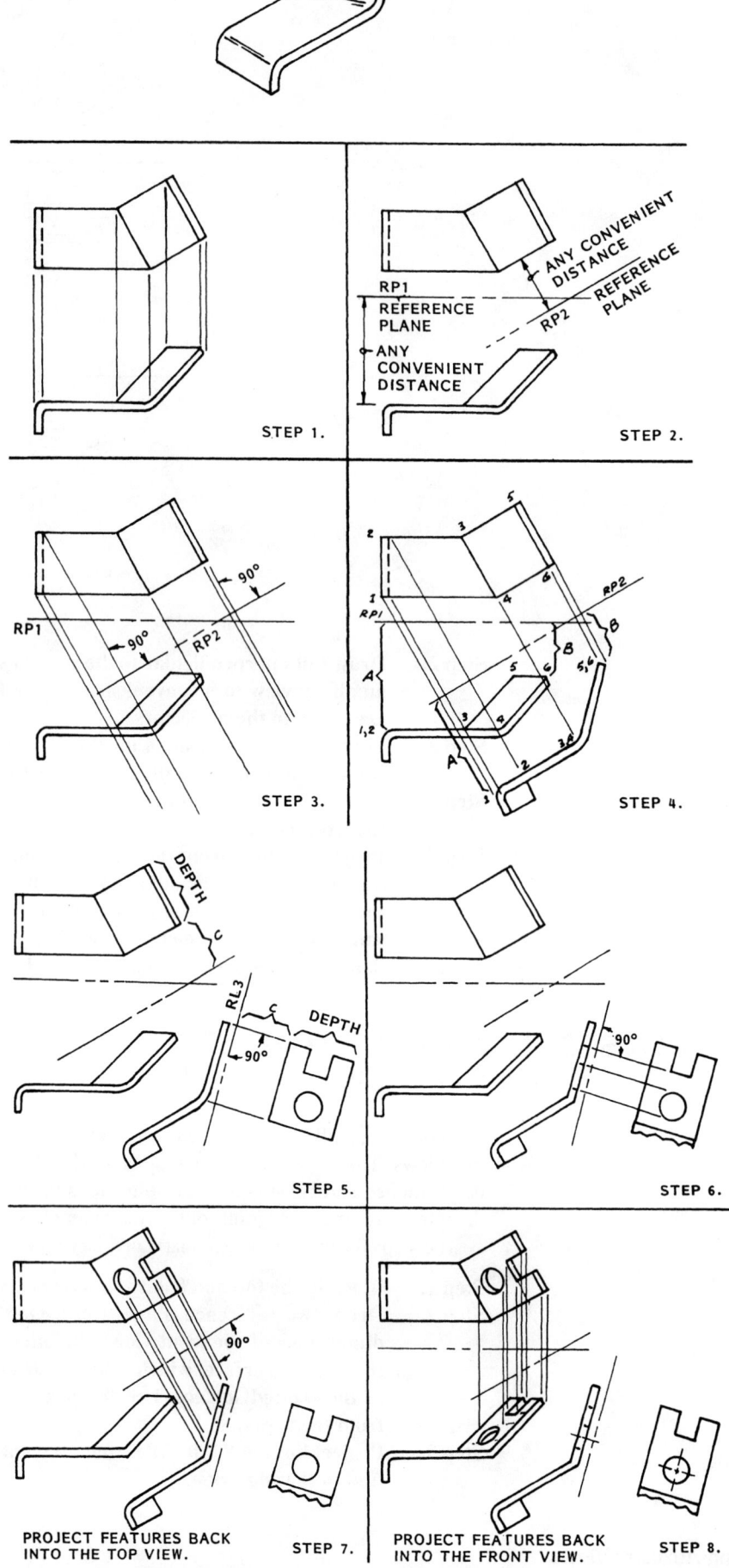

Step 4. Transfer distances by taking the distances from RP1 to the object in the front view and transferring them to along the projectors from RP2 to the object in the auxiliary view. Consider the object sitting in the transparent box at the same distance down from RP1 in both the front and edge views.

Step 5. Draw another reference line at a convenient distance from the edge view, and transfer depth distances from the top view into the auxiliary view. Complete the auxiliary from given information.

Step 6. Project the circle and the slot from the auxiliary view back into the edge view to complete the edge view.

Step 7. Project all features from the edge view into the top view, and transfer depth dimensions from the auxiliary view to complete the top view.

Step 8. Complete the drawing by projecting all features from the top view into the front view. You will have to project the bottom of the notch and the top and bottom quadrants of the circle to the edge of the oblique surface in the top view, then to the edge of the oblique surface in the front view, and then horizontally to the projected lines from the top view.

Study Figures 11–5 and 11–6 and try to visualize the transparent box and the object within it. Although primary auxiliary views are used a great deal and are projected as shown in Figure 11–5, secondary auxiliary views are often not constructed in the manner shown in Figure 11–6. They are often drawn as a view and labeled as View A (for example). An arrow showing the line of sight on one of the normal views would indicate where the view was located. Dimensions for drawing the view are taken from a manufactured part or other specifications.

EXERCISES

EXERCISE 11–1   Complete Exercise 11–1 using the steps described.

NOTE: IF YOU HAVE DIFFICULTY IN CONSTRUCTING THE FIRST AUXILIARY VIEW, READ CHAPTER 12 WHICH EXPLAINS EACH STEP IN DETAIL.

Step 1. Remove the sheet labeled EXERCISE 11–1 from your book.

Step 2. Use construction lines to sketch as much of the front, top, and right-side views of Figure 11–7 as you can using the sketching and construction techniques from Chapter 7. You will have to wait until you complete the auxiliary view to sketch the T-shaped slot in all the other views.

Draw half scale.

Be sure to line up all features of the object with the adjacent views.

FIGURE 11–7
Exercise 11–1

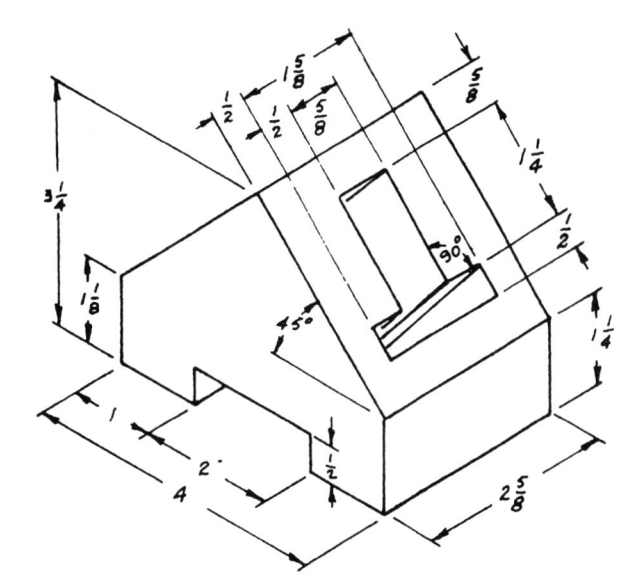

Sketching Auxiliary Views

Step 3. Using construction lines, sketch the auxiliary view in the position indicated. Sketch this view as a partial auxiliary using straight break lines at obtuse angles as shown in Figure 11–3.

Step 4. Project the T-shaped slot into the front view. The depth of this slot is $\frac{3}{4}''$ and will be shown with hidden lines.

Step 5. Project the T-shaped slot into the top and right-side views. Take depth dimensions from either Figure 11–7 or the auxiliary view.

Step 6. Darken all object and hidden lines. Leave all your light construction lines.

Sprinkle drafting powder over your drawing before you begin to darken lines.

Make sure that your lines are the correct weight and are of even width and darkness. Try to match the thickness and darkness of the existing lines.

DO NOT SHOW DIMENSIONS.

Step 7. Fill in the title block with your best lettering. Title the drawing AUXILIARY 1.

EXERCISE 11–2   Remove the sheet labeled EXERCISE 11-2 from your book. Complete Exercise 11–2 using the technique described in Exercise 11–1.

Sketch the full front, partial top, and partial auxiliary views of Figure 11–8. Sketch this drawing full scale.

Fill in the title block with your best lettering and name the drawing AUXILIARY 2.

EXERCISE 11–3   Remove the sheet labeled EXERCISE 11-3 from your book. Complete Exercise 11–3 using the technique described in Exercise 11–1.

Sketch the full front, partial top, and partial auxiliary views of Figure 11–9. Sketch this drawing full scale.

Fill in the title block with your best lettering and name the drawing AUXILIARY 3.

EXERCISE 11–4   Remove the sheet labeled EXERCISE 11-4 from your book. Complete Exercise 11–4 using the technique described in Exercise 11–1.

Sketch the full front, partial top, and partial auxiliary views of Figure 11–10. Sketch this drawing half scale.

Fill in the title block with your best lettering and name the drawing AUXILIARY 4.

Sketch this drawing half scale.

Complete the title block with your best lettering and name the drawing AUXILIARY 4.

FIGURE 11–8
Exercise 11–2

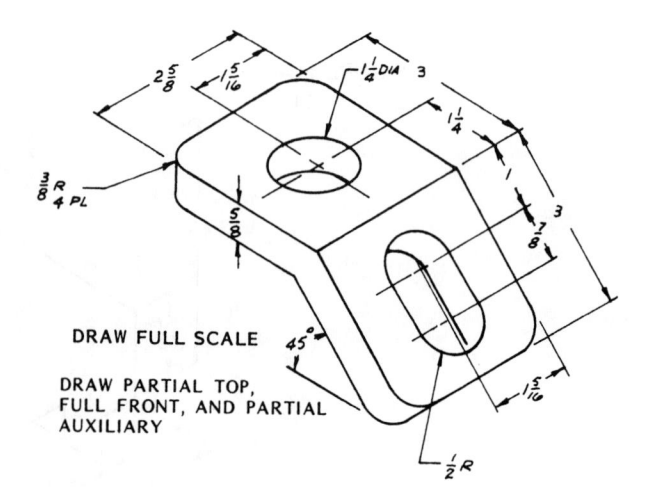

FIGURE 11-9
Exercise 11-3

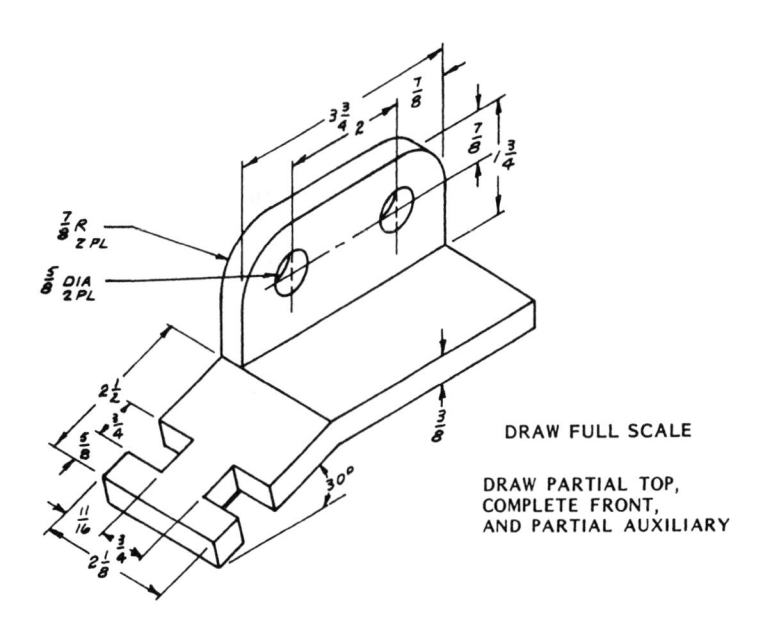

DRAW FULL SCALE

DRAW PARTIAL TOP,
COMPLETE FRONT,
AND PARTIAL AUXILIARY

FIGURE 11-10
Exercise 11-4

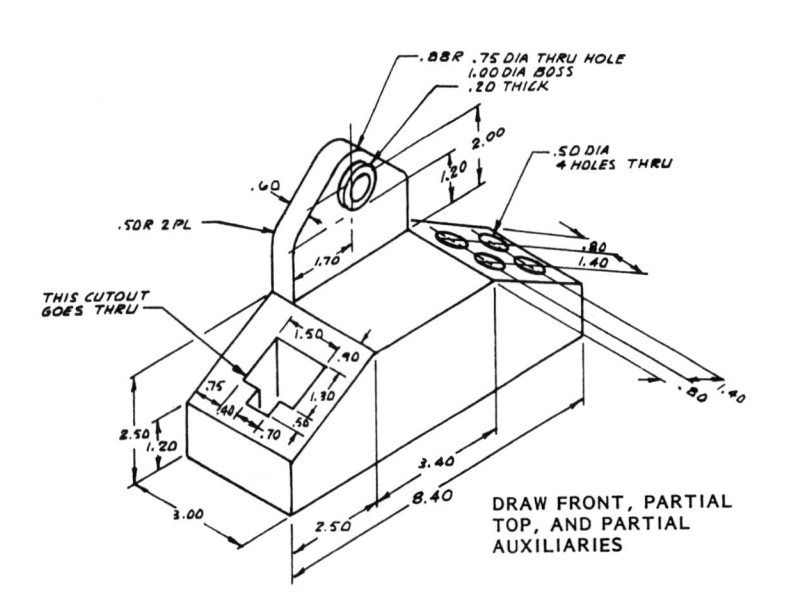

DRAW FRONT, PARTIAL
TOP, AND PARTIAL
AUXILIARIES

EXERCISE 11-5   Complete Exercise 11-5 using the steps described.

   Step 1.   Remove the sheet labeled EXERCISE 11-5 from your book.

   Step 2.   Draw the edge view and the secondary auxiliary view in the positions indicated. Use the drawing and construction techniques described in Exercise 11-1 and Figure 11-6.

   Step 3.   Complete the title block with your best lettering. Title the drawing SECONDARY AUXILIARY.

Sketching Auxiliary Views

REVIEW QUESTIONS

Circle the best answer.

1. Auxiliary views are used to show true size and true shape of
   a. Horizontal surfaces
   b. Vertical surfaces
   c. Slanted surfaces
   d. Normal surfaces
   e. Edge views
2. Surfaces shown true size and true shape in auxiliary views are seen _____ in the normal views.
   a. Foreshortened or as edges
   b. True size and shape
   c. Normal
   d. Seldom
   e. Not at all
3. The line of sight of an auxiliary view is _____ to the slanted surface on an object.
   a. Perpendicular
   b. Slanted
   c. Foreshortened
   d. Parallel
   e. At any angle
4. The reference line from which distances are stepped off in the auxiliary view is _____ to the edge view of the slanted surface.
   a. Perpendicular
   b. Slanted
   c. Foreshortened
   d. Parallel
   e. At any angle
5. The auxiliary view will show a slanted surface
   a. Foreshortened
   b. True size and shape
   c. Normal
   d. Seldom
   e. Not at all
6. Auxiliary views may be projected from
   a. Front views only
   b. Top views only
   c. Right-side views only
   d. Left-side views only
   e. Any of the normal views
7. Partial auxiliaries are used more often than full auxiliaries because
   a. It takes less time to draw a partial auxiliary
   b. A full auxiliary is often confusing to the user
   c. A full auxiliary takes up too much space on a drawing
   d. Full auxiliaries often require too many hidden lines
   e. All the above
8. A secondary auxiliary is a view of
   a. A normal surface
   b. An inclined surface
   c. An oblique surface
   d. A circular surface
   e. The right-side auxiliary
9. A secondary auxiliary must be projected from
   a. An edge view
   b. A normal surface
   c. An inclined surface
   d. Any view
   e. A front view

10. Depth dimensions are shown in
    a. The front view only
    b. The top view only
    c. The right-side view only
    d. The front and right-side views
    e. The top and right-side views

12 Making Auxiliary Views with AutoCAD

OBJECTIVES

After completing this chapter, you will be able to:

☐ Make accurate drawings of auxiliary views with AutoCAD or AutoCAD LT.
☐ Use the commands introduced in previous chapters to make drawings containing auxiliary views.

DRAWING AUXILIARY VIEWS USING AUTOCAD OR AUTOCAD LT

Because you will not use any commands that you have not used in previous chapters to draw auxiliary views, this chapter will proceed directly to the exercises. In this chapter you will make the first two drawings you sketched in Chapter 11. The first drawing will follow the Prompt–Response format. The second drawing will use the approach that requires you to activate commands and use them on your own so that you will become independent of the Prompt–Response method.

EXERCISE 12–1
Making a Drawing Containing Front, Top, Right-Side, and Auxiliary Views

Your final drawing will look similar to the drawing in Figure 12–1.

Step 1. To begin Exercise 12–1, turn on the computer and start AutoCAD or AutoCAD LT.
Step 2. Open drawing EX12-1 supplied on the disk that came with your book.
Step 3. Use Zoom-All to view the entire drawing area.
Step 4. Turn on a running osnap mode of Endpoint and draw the outlines of the top, front, right-side, and auxiliary views. (You will be using the dimensions from Figure 11–7 and drawing at a scale of $\frac{1}{2}''=1''$. All fractions are converted to decimal inches in the following steps.)

Prompt	Response
Command:	Type: **-OS**↵
Enter list of object snap modes:	Type: **END**↵
Command:	Type: **L**↵
Specify first point:	Click: **D1** (Figure 12–2)
Specify next point or [Undo]:	Click: **D2**
Specify next point or [Undo]:	↵
Command:	↵
Specify first point:	Click: **D3**
Specify next point or [Undo]:	Click: **D4**

FIGURE 12-1
Exercise 12-1 Completed

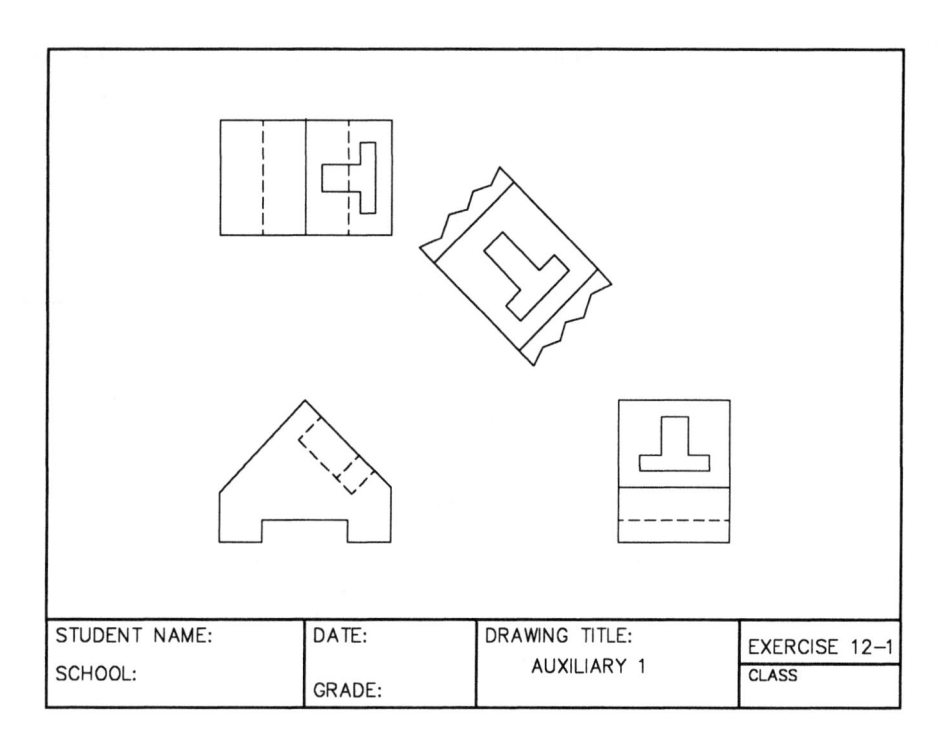

Specify next point or [Undo]:	↵
Command:	↵
Specify first point:	Click: **D5**
Specify next point or [Undo]:	Click: **D6**
Specify next point or [Undo]:	↵
Command:	↵
Specify first point:	Click: **D7**
Specify next point or [Undo]:	Type: **@.5<0**↵
Specify next point or [Undo]:	Type: **@.25<90**↵
Specify next point or [Close/Undo]:	Type: **@1<0**↵
Specify next point or [Close/Undo]:	Type: **@.25<270**↵

FIGURE 12-2
Drawing Outlines of All Views

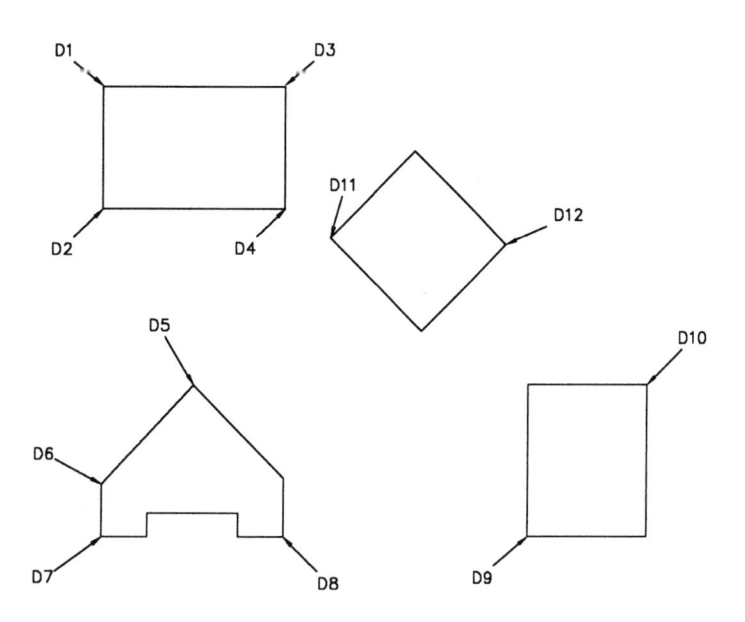

Making Auxiliary Views with AutoCAD

187

Prompt	Response
Specify next point or [Close/Undo]:	Click: **D8**
Specify next point or [Close/Undo]:	↵
Command:	↵
Specify first point:	Click: **D9**
Specify next point or [Undo]:	Type: **@1.625<90**↵
Specify next point or [Undo]:	Click: **D10**
Specify next point or [Close/Undo]:	↵
Command:	↵
Specify first point:	Click: **D11**
Specify next point or [Undo]:	Type: **@1.3125<45**↵
Specify next point or [Undo]:	Click: **D12**
Specify next point or [Close/Undo]:	↵

Important: Be sure you turn Snap and Ortho ON so that your lines will be in the correct location.

Step 5. Turn OFF the running osnap mode of Endpoint and draw the other visible lines in the top and right-side views.

Prompt	Response
Command:	Type: **-OS**↵
Enter list of object snap modes:	Type: **NONE**↵
Command:	Type: **L**↵
Specify first point:	Click: **D1** (Figure 12–3)
Specify next point or [Undo]:	Click: **D2**
Specify next point or [Undo]:	↵
Command:	↵
Specify first point:	Click: **D3**
Specify next point or [Undo]:	Click: **D4**
Specify next point or [Undo]:	↵

Important: Be sure you turn Snap and Ortho ON so that your lines will be in the correct location.

Step 6. Set layer HIDDEN current and draw the other hidden lines in the top and right-side views.

On your own: Set layer HIDDEN current.

Prompt	Response
Command:	Type: **L**↵
Specify first point:	Click: **D1** (Figure 12–4)

FIGURE 12–3
Drawing Other Visible Lines in the Top and Right-Side Views

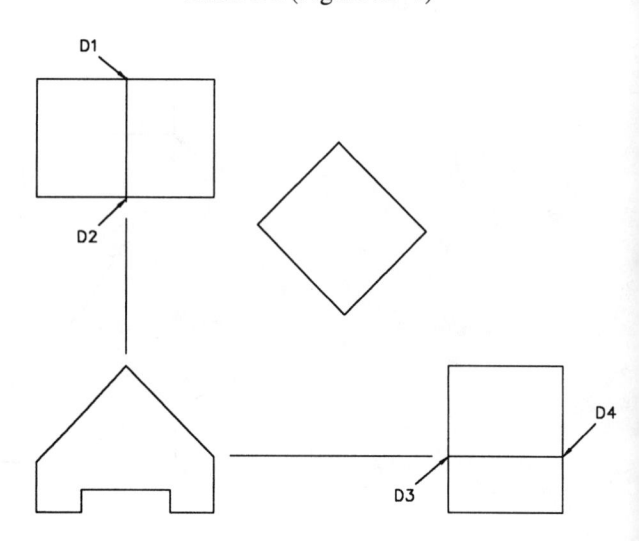

FIGURE 12–4
Adding Hidden Lines in the Top and Right-Side Views

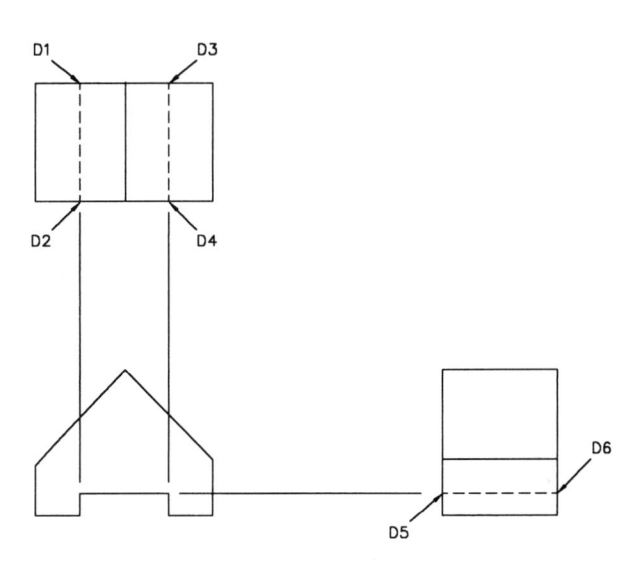

Prompt	Response
Specify next point or [Undo]:	Click: **D2**
Specify next point or [Undo]:	↵
Command:	↵
Specify first point:	Click: **D3**
Specify next point or [Undo]:	Click: **D4**
Specify next point or [Undo]:	↵
Command:	↵
Specify first point:	Click: **D5**
Specify next point or [Undo]:	Click: **D6**
Specify next point or [Undo]:	↵

Step 7. Use the Offset command to draw visible lines in the auxiliary view.

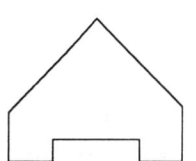

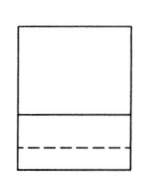

Important: Notice that even though the HIDDEN layer is current, the lines are offset as lines drawn on the OBJECT layer.

Prompt	Response
Command:	Type: **O**↵
Specify offset distance or [Through]:	Type: **.3125**↵
Select object to offset:	Click: **D1** (Figure 12–5)
Specify point on side to offset:	Click: **D2**
Select object to offset:	↵
Command:	↵
Specify offset distance or [Through] <.3125>:	Type: **.625**↵

FIGURE 12–5
Using the Offset Command to Draw Two of the Horizontal Lines of the T-Shaped Slot

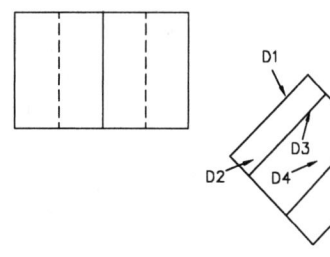

Making Auxiliary Views with AutoCAD

FIGURE 12–6
Using the Offset to Draw All Parallel Lines That Are .25″ Apart

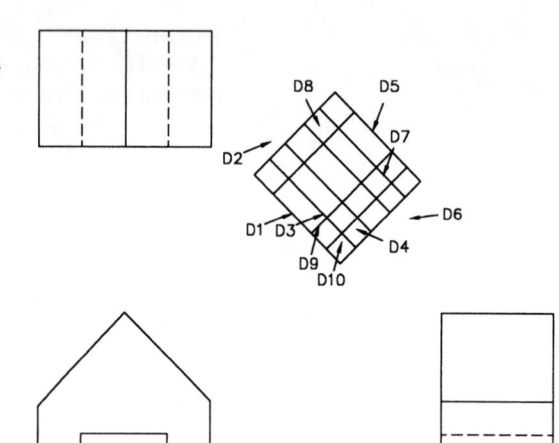

Select object to offset:	Click: **D3**
Specify point on side to offset:	Click: **D4**
Select object to offset:	↵
Command:	↵
Specify offset distance or [Through] <0.6250>:	Type: **.25**↵
Select object to offset:	Click: **D1** (Figure 12–6)
Specify point on side to offset:	Click: **D2**
Select object to offset:	Click: **D3**
Specify point on side to offset:	Click: **D4**
Select object to offset:	Click: **D5**
Specify point on side to offset:	Click: **D6**
Select object to offset:	Click: **D7**
Specify point on side to offset:	Click: **D8**
Select object to offset:	Click: **D9**
Specify point on side to offset:	Click: **D10**
Select object to offset:	↵

Step 8. Use the Trim command to remove unnecessary parts of the offset lines.

Prompt	Response
Command:	Type: **TR**↵
Select cutting edges:	Click: **D1** (Figure 12–7)
Specify opposite corner:	Click: **D2**
Select objects:	↵
Select object to trim or shift-select to extend or [Project/Edge/Undo]:	Click: **D1** (Figure 12–8)
Select object to trim or shift-select to extend or [Project/Edge/Undo]:	Click: **D2**
Select object to trim or shift-select to extend or [Project/Edge/Undo]:	Click: **D3**
Select object to trim or shift-select to extend or [Project/Edge/Undo]:	Click: **D4**
Select object to trim or shift-select to extend or [Project/Edge/Undo]:	Click: **D5**

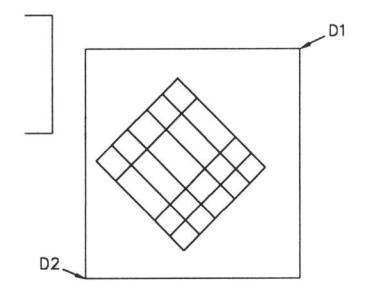

FIGURE 12–7
Selecting Cutting Edges for the Trim Command

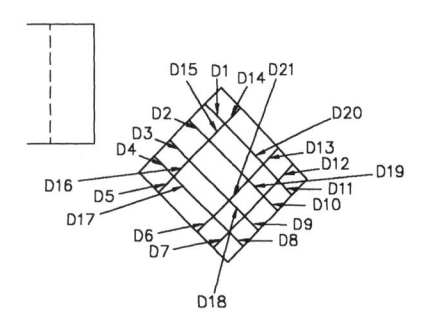

FIGURE 12–8
Selecting Objects to Trim

Prompt	Response
Select object to trim or shift-select to extend or [Project/Edge/Undo]:	Click: **D6**
Select object to trim or shift-select to extend or [Project/Edge/Undo]:	Click: **D7**
Select object to trim or shift-select to extend or [Project/Edge/Undo]:	Click: **D8**
Select object to trim or shift-select to extend or [Project/Edge/Undo]:	Click: **D9**
Select object to trim or shift-select to extend or [Project/Edge/Undo]:	Click: **D10**
Select object to trim or shift-select to extend or [Project/Edge/Undo]:	Click: **D11**
Select object to trim or shift-select to extend or [Project/Edge/Undo]:	Click: **D12**
Select object to trim or shift-select to extend or [Project/Edge/Undo]:	Click: **D13**
Select object to trim or shift-select to extend or [Project/Edge/Undo]:	Click: **D14**
Select object to trim or shift-select to extend or [Project/Edge/Undo]:	Click: **D15**
Select object to trim or shift-select to extend or [Project/Edge/Undo]:	Click: **D16**
Select object to trim or shift-select to extend or [Project/Edge/Undo]:	Click: **D17**
Select object to trim or shift-select to extend or [Project/Edge/Undo]:	Click: **D18**
Select object to trim or shift-select to extend or [Project/Edge/Undo]:	Click: **D19**
Select object to trim or shift-select to extend or [Project/Edge/Undo]:	Click: **D20**
Select object to trim or shift-select to extend or [Project/Edge/Undo]:	Click: **D21**
Select object to trim or shift-select to extend or [Project/Edge/Undo]:	↵

Step 9. **Use the Insert command to add a predefined block that will form the break lines of the partial auxiliary.**

Insert allows you to insert a drawing called a block that is defined on this drawing only. It also allows you to insert any drawing into any other drawing from any disk.

Making Auxiliary Views with AutoCAD

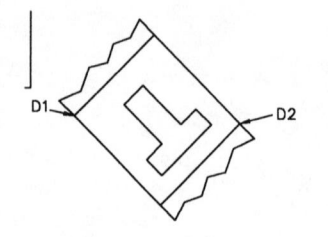

FIGURE 12–9
Inserting the Break Block Twice

Prompt	Response
Command:	Type: **-I**↵
Enter block name or [?]:	Type: **BREAK**↵
Specify insertion point or [Scale...PRotate]:	Type: **END**↵
of	Click: **D1** (Figure 12–9)
Enter X scale factor, specify opposite corner or [Corner/XYZ] <1>:	↵
Enter Y scale factor <use X scale factor>:	↵
Specify rotation angle <0>:	↵
Command:	↵
Enter block name or [?] <BREAK>:	↵
Specify insertion point or [Scale...PRotate]:	Type: **END**↵
of	Click: **D2** (Figure 12–9)
Enter X scale factor, specify opposite corner or [Corner / XYZ] <1>:	↵
Enter Y scale factor <use X scale factor>:	↵
Specify rotation angle <0>:	Type: **180**↵

Step 10. Use the Offset command to draw a line for the bottom of the slot (which you will later change to the HIDDEN layer) and some of the visible lines of the T-shaped slot in the top and right-side views.

Prompt	Response
Command:	Type: **O**↵
Specify offset distance or [Through]:	Type: **.375**↵
Select object to offset:	Click: **D1** (Figure 12–10)
Specify point on side to offset:	Click: **D2**
Select object to offset:	↵
Command:	↵
Specify offset distance or [Through] <0.3750>:	Type: **.25**↵
Select object to offset:	Click: **D3**
Specify point on side to offset:	Click: **D4**
Select object to offset:	Click: **D5**

FIGURE 12–10
Using the Offset Command to Draw Construction Lines in the Front, Top, and Right-Side Views

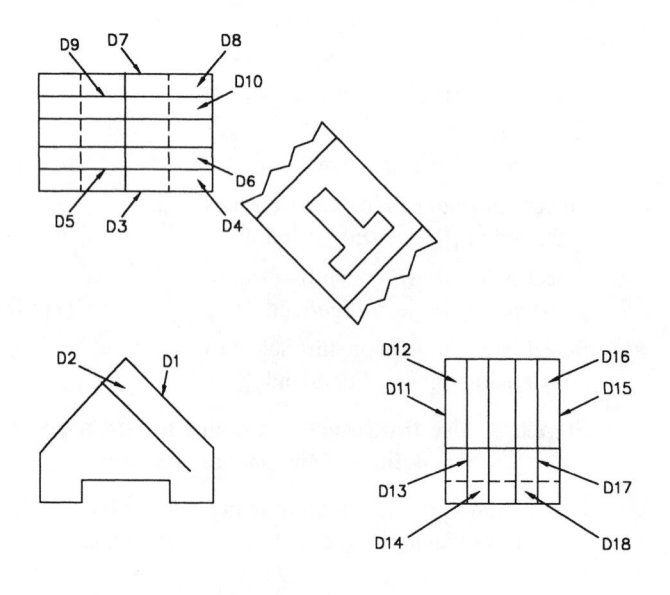

Prompt	Response
Specify point on side to offset:	Click: **D6**
Select object to offset:	Click: **D7**
Specify point on side to offset:	Click: **D8**
Select object to offset:	Click: **D9**
Specify point on side to offset:	Click: **D10**
Select object to offset:	Click: **D11**
Specify point on side to offset:	Click: **D12**
Select object to offset:	Click: **D13**
Specify point on side to offset:	Click: **D14**
Select object to offset:	Click: **D15**
Specify point on side to offset:	Click: **D16**
Select object to offset:	Click: **D17**
Specify point on side to offset:	Click: **D18**
Select object to offset:	↵

Step 11. Set layer OBJECT current and use the Line command to draw perpendicular lines from the auxiliary view to the front view showing the depth of the slot. (You will change those lines to the HIDDEN layer in a later step.)

On your own: Set layer OBJECT current.

Prompt	Response
Command:	Type: **L**↵
Specify start point:	Type: **END**↵
of	Click: **D1** (Figure 12–11)
Specify next point or [Undo]:	Type: **PER**↵ (to make these lines perpendicular)
to	Click: **D2** (any point on this line)
Specify next point or [Undo]:	↵
Command:	↵
Specify first point:	Type: **END**↵
of	Click: **D3**
Specify next point or [Undo]:	Type: **PER**↵
to	Click: **D2** (again)
Specify next point or [Undo]:	↵
Command:	↵
Specify first point:	Type: **END**↵

FIGURE 12–11
Drawing Perpendicular Lines from the Auxiliary View to the Bottom of the Slot

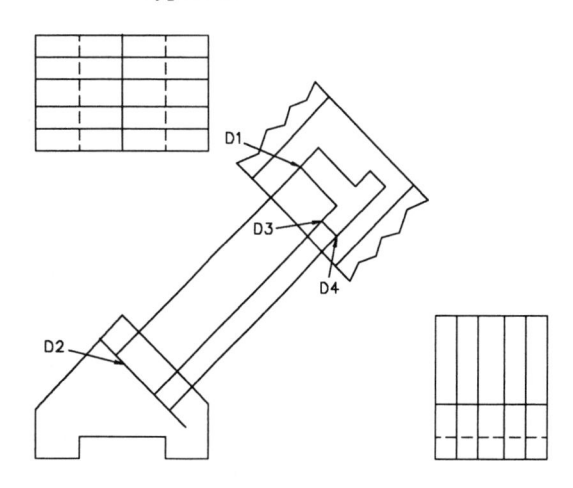

Making Auxiliary Views with AutoCAD

Prompt	Response
of	Click: **D4**
Specify next point or [Undo]:	Type: **PER**↵
to	Click: **D2** (again)
Specify next point or [Undo]:	↵

Step 12. Use the Line command to draw perpendicular lines from the front view to the top and right-side views to show the slot in those views.

Prompt	Response
Command:	Type: **L**↵
Specify first point:	Type: **INT**↵ (to select the intersection of the lines)
of	Click: **D1** (Figure 12–12)
Specify next point or [Undo]:	Type: **PER**↵ (to make these lines perpendicular)
to	Click: **D2** (any point on this line)
Specify next point or [Undo]:	↵
Command:	↵
Specify first point:	Type: **INT**↵
of	Click: **D3**
Specify next point or [Undo]:	Type: **PER**↵
to	Click: **D4**
Specify next point or [Undo]:	↵
Command:	↵
Specify first point:	Type: **INT**↵
of	Click: **D5**
Specify next point or [Undo]:	Type: **PER**↵
to	Click: **D4** (again)
Specify next point or [Undo]:	↵
Command:	↵
Specify first point:	Type: **INT**↵
of	Click: **D1** (Figure 12–12)
Specify next point or [Undo]:	Type: **PER**↵

FIGURE 12–12
Drawing Perpendicular Lines from the Front View to the Right-Side and Top Views

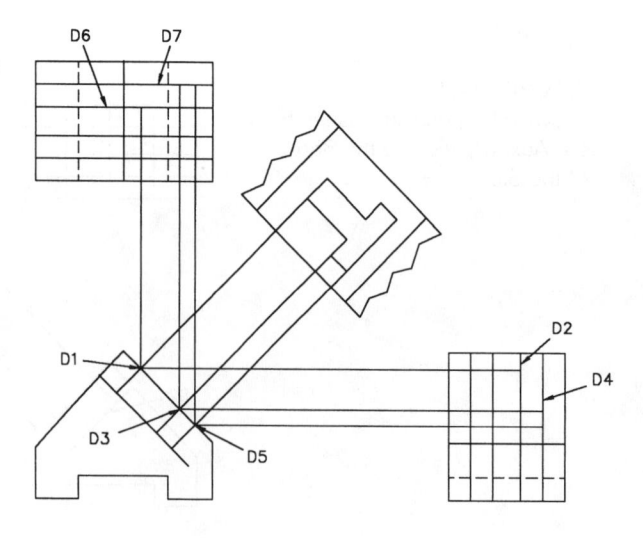

to	Click: **D6** (any point on this line)
Specify next point or [Undo]:	↵
Command:	↵
Specify first point:	Type: **INT**↵
of	Click: **D3**
Specify next point or [Undo]:	Type: **PER**↵
to	Click: **D7**
Specify next point or [Undo]:	↵
Command:	↵
Specify first point:	Type: **INT**↵
of	Click: **D5**
Specify next point or [Undo]:	Type: **PER**↵
to	Click: **D7** (again)
Specify next point or [Undo]:	↵

Step 13. Use the Trim command to remove unnecessary parts of the offset lines in the top view. Begin by zooming a window around the top view so you can pick lines more easily.

Prompt	Response
Command:	Type: **Z**↵
Specify corner of window, enter a scale factor (nX or nXP), or [All/Center/Dynamic/Extents/Previous/Scale/Window] <real time>:	Click: **D1** (Figure 12–13)
Specify opposite corner:	Click: **D2**
Command:	Type: **TR**↵
Select cutting edges:	Click: **D3, D4, D5, D6, D7, and D8** (Figure 12–13)
Select objects:	↵
Select object to trim or shift-select to extend or [Project/Edge/Undo]:	Click: **D3, D4, D5, D6, D7, D8,** (again) **D9, D10, D11, D12, D13, D14, D15, D16, D17, and D18**
Select object to trim or shift-select to extend or [Project/Edge/Undo]:	↵

Step 14. On your own:

Use the Trim command to remove unnecessary parts of the offset lines in the front and right-side views as shown in Figure 12–14. Begin by zooming windows around the views so you can pick lines more easily.

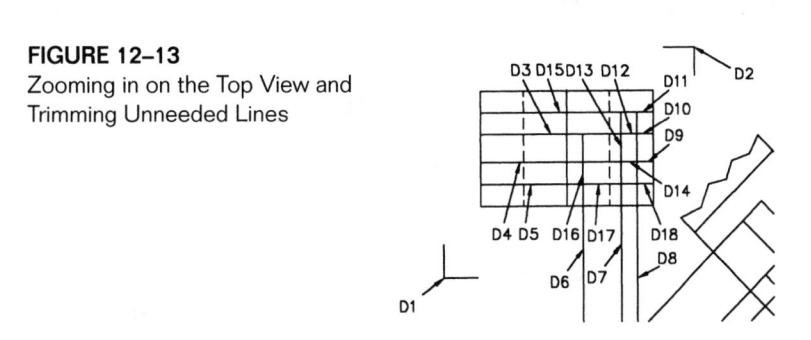

FIGURE 12–13
Zooming in on the Top View and Trimming Unneeded Lines

Making Auxiliary Views with AutoCAD

FIGURE 12–14
Removing Unneeded Lines in the
Right-Side and Front Views Using
the Trim Command

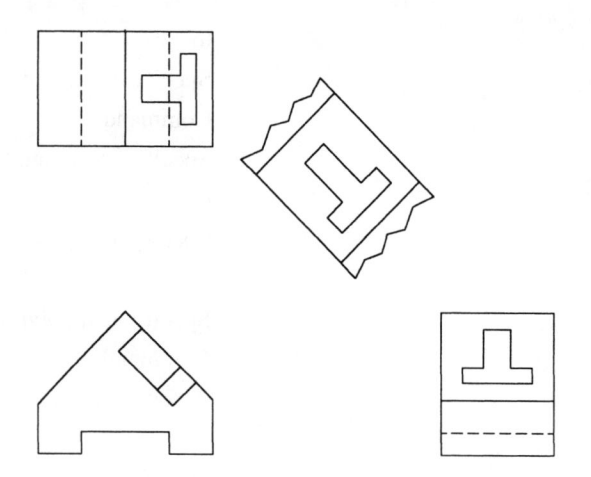

Note: You may also use Properties on the Modify menu to change any property of any AutoCAD object. You may change lines to another layer by clicking on the lines to be changed and then clicking on the new layer in the layer list.

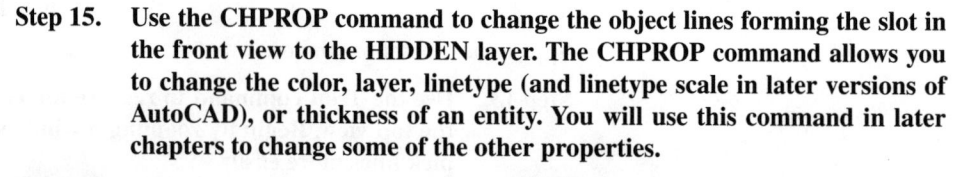

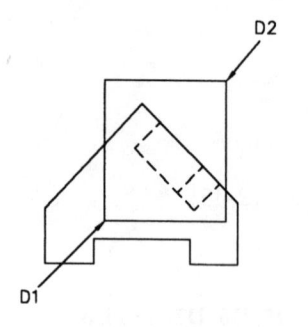

FIGURE 12–15
Using the CHPROP (Change Properties) Command to Change Lines in the Front View to the HIDDEN Layer

Step 15. Use the CHPROP command to change the object lines forming the slot in the front view to the HIDDEN layer. The CHPROP command allows you to change the color, layer, linetype (and linetype scale in later versions of AutoCAD), or thickness of an entity. You will use this command in later chapters to change some of the other properties.

Prompt	Response
Command:	Type: **CHPROP**↵
Select objects:	Click: **D1** (Figure 12–15) (Make sure your window does not include any lines other than those shown in this figure.)
Specify opposite corner:	Click: **D2**
Select objects:	↵
Enter property to change [Color/LAyer/LType/ltScale/LWeight/Thickness]:	Type: **LA**↵
Enter new layer name <OBJECT>:	Type: **HIDDEN**↵
Enter property to change [Color/LAyer/LType/ltScale/LWeight/Thickness]:	↵

The drawing is now complete except for the title block.

Step 16. Set the OBJECT layer current and use the Dtext command to complete the title block as you have done previously. Title the drawing AUXILIARY 1.

Step 17. Use the SAVEAS command to save your drawing as EX12-1(your initials) on a floppy disk and again on the hard drive of your computer.

Step 18. Plot or print your drawing full size on an 11″ × 8½″ sheet.

EXERCISE 12–2
Making a Drawing Containing Front, Partial Top, and Auxiliary Views

Now you will begin to make drawings on your own with suggested commands. The Prompt–Response format will not be used in this exercise. Refer to previous exercises if you need to remember how to use a suggested command.

Your final drawing will look like Figure 12–16.

Step 1. To begin Exercise 12–2, turn on the computer and start AutoCAD or AutoCAD LT.

Step 2. Open drawing EX12-2 supplied on the disk that came with your book. Use the dimensions from Figure 11–9 to make this drawing.

FIGURE 12-16
Exercise 12-2 Completed

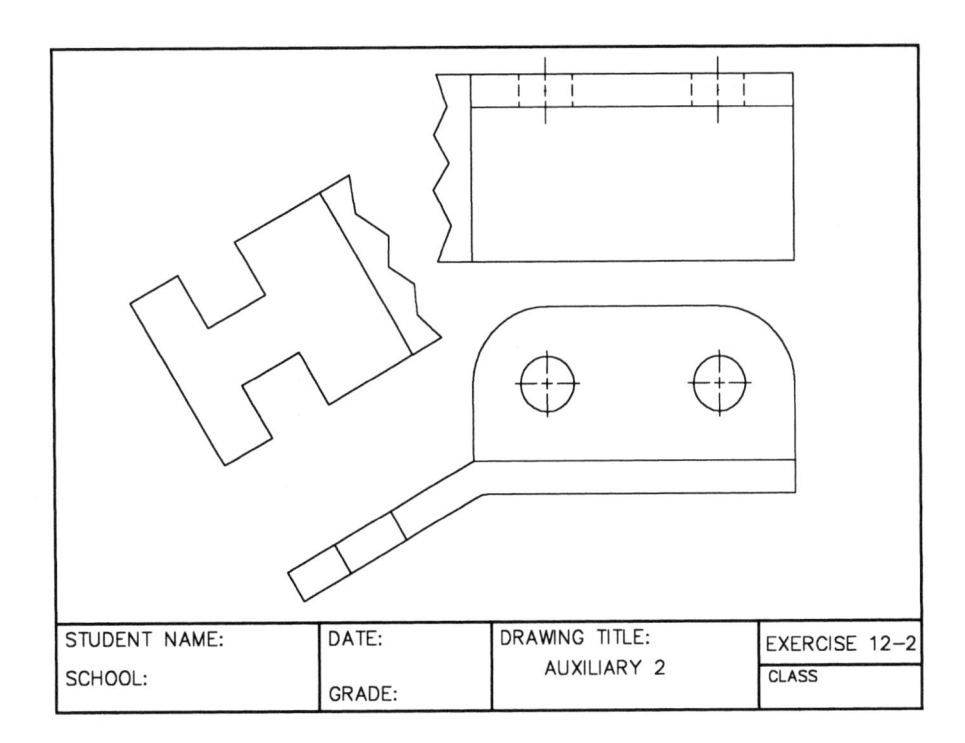

Step 3. Use Zoom-All to view the entire drawing area.
Step 4. Use the Line command starting with D1 (use osnap mode-Endpoint) (Figure 12–17) as the "First point:" to draw the angular line in the front view. You will draw this line 2.5" long at a 210° angle from D1. Use polar coordinates to draw this line.
Step 5. Use the Offset command to offset the two lines in the front view .375" down (Figure 12–18). Use the Fillet command with a radius of 0 to clean up the intersection of the offset lines.
Step 6. Use the Line command to close the ends of the front view; you will have to use osnap modes of Endpoint and Perpendicular to draw accurately (Figure 12–19).

FIGURE 12-17
Drawing a 2½" Line at a 210° Angle from D1

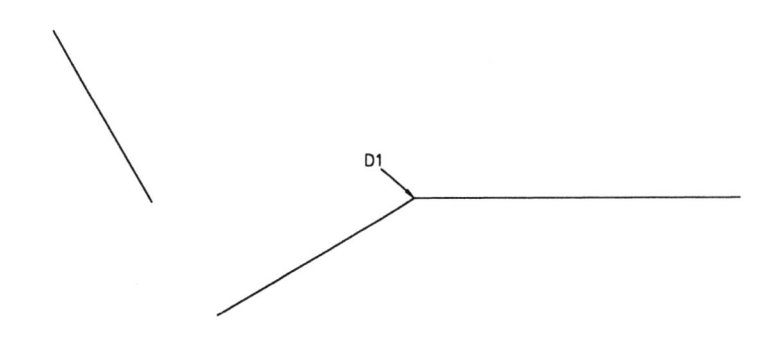

Making Auxiliary Views with AutoCAD

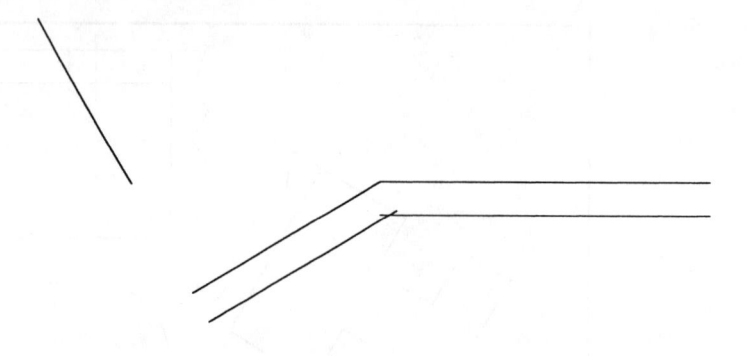

FIGURE 12–18
Using the Offset Command to Offset the Two Lines in the Front View $\frac{3''}{8}$

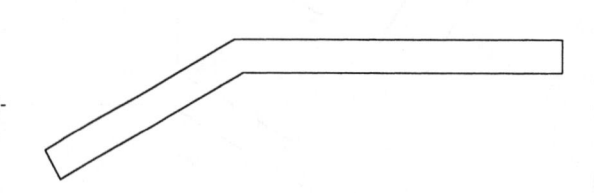

FIGURE 12–19
Using the Line Command with Osnap-Endpoint and -Perpendicular to Close the Front View Ends and Trimming the Intersecting Lines Using the Trim Command

Step 7. Use the Line command to draw the upright part of the front view (Figure 12–20). After you click D1 at the "First point:" prompt, use the following polar coordinates and osnap mode.

@1.75<90

@3.75<0

Type: **END**↵ and Click: **D2**

Step 8. Use the Fillet command to draw the arcs at the top corners. Set the fillet radius at .875. After you have made the fillets, use the Circle command to draw the two .625-diameter circles in the front view. Use Osnap-Center to locate the centers of the circles at the center of the fillets (Figure 12–21).

Step 9. Use the Copy command to copy the angular line from the front view to both ends of the line in the auxiliary view. Then copy the existing line in the auxiliary view to the end of one of the copied lines. Use Osnap-Endpoint for all clicks (Figure 12–22).

Step 10. Use the Offset command to offset lines to form the slots (Figure 12–23). Use the following distances:

Set Offset distance at .625—Select **D1**, side to offset, Click: **D2**

Set Offset distance at .75—Select **D3**, side to offset, Click: **D4**

Set Offset distance at .6875—Select **D5**, side to offset, Click: **D6**, Select **D7**, side to offset, Click: **D8**

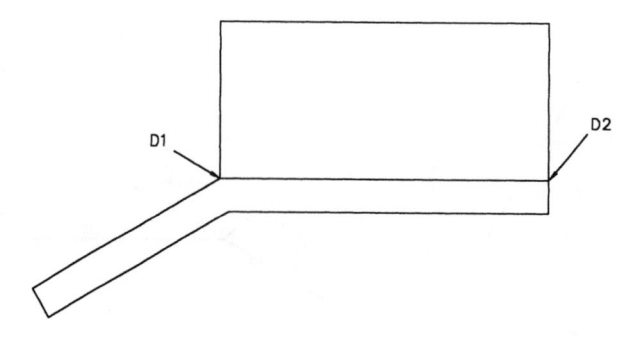

FIGURE 12–20
Using the Line Command to Draw the Upright Part of the Front View

FIGURE 12-21
Using the Fillet Command to Draw $\frac{7}{8}''$ Arcs at the Corners, and Using the Circle Command to Draw $\frac{5}{8}''$ Diameter Circles

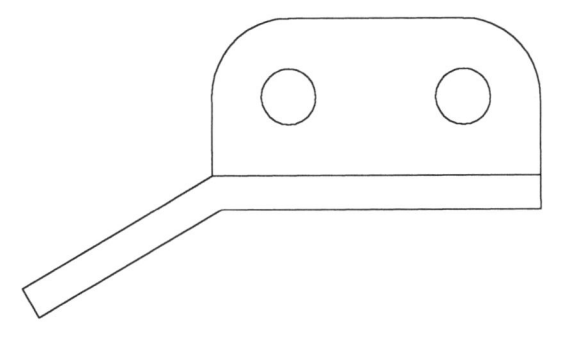

FIGURE 12-22
Using the Copy Command to Draw the Outside Lines of the Auxiliary View

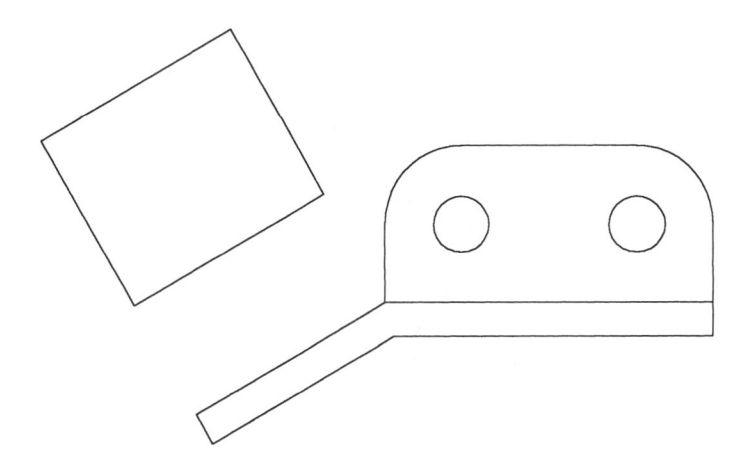

FIGURE 12-23
Using the Offset Command to Draw the Lines Forming the Slots

Step 11. Use the Trim command to remove unnecessary lines, as shown in Figure 12-24.
Step 12. Use the Extend command to extend lines from the auxiliary view slot to the front view. Click: D1 as the boundary edge (Figure 12-25).
Step 13. Use the Trim command to remove lines between the front and auxiliary views as shown in Figure 12-26. Click: D1, D2, and D3 as cutting edges.
Step 14. Use the Insert command to insert a predefined block named **BREAK2**. Click: D1 (Figure 12-27), using Osnap-Endpoint as the insertion point, and accept all other defaults.

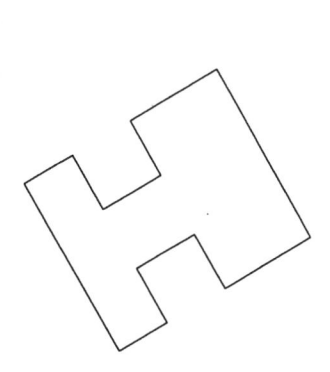

FIGURE 12-24
Using the Trim Command to Remove Unnecessary Line Segments

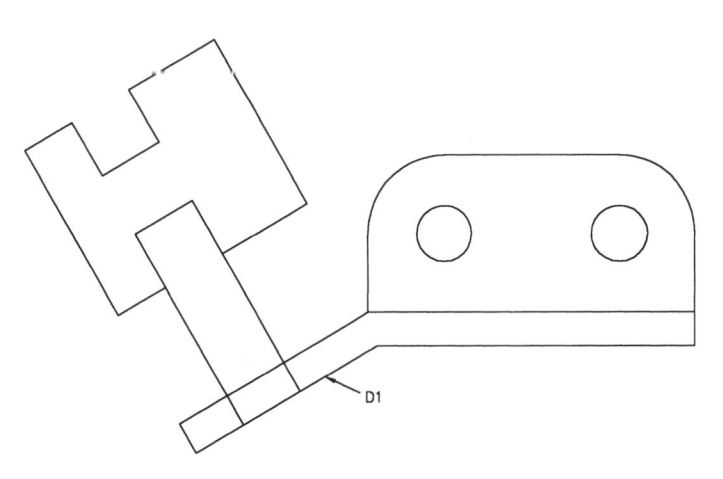

FIGURE 12-25
Using the Extend Command to Extend Lines from the Slot to the Front View

Making Auxiliary Views with AutoCAD

FIGURE 12–26
Using the Trim Command to Remove Lines Between Auxiliary and Front Views

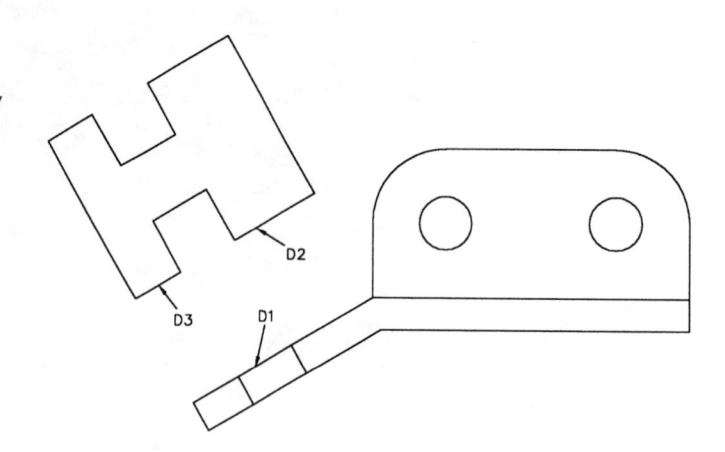

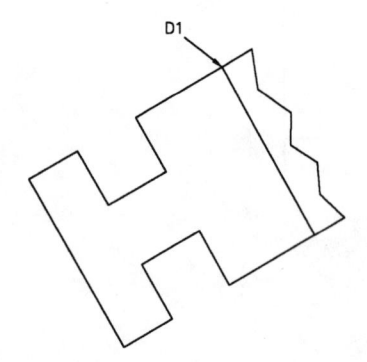

FIGURE 12–27
Using the Insert Command to Insert the Block BREAK 2

Step 15. Use the Offset command to offset the existing horizontal line in the top view .375 down, then offset the same line 2.125 down, as shown in Figure 12–28.

Step 16. Use the Line command to draw the two vertical lines in the top view (Figure 12–29). Use Osnap-Endpoint to click both ends of each line.

Step 17. Use the Insert command to insert a predefined block named BREAK3. Click: D1 (Figure 12–30), using Osnap-Endpoint as the insertion point, and accept all other defaults.

Step 18. Set layer HIDDEN current, activate the Line command and draw hidden lines in the top view to represent the holes. With Snap on, activate the Line command, align the cursor crosshair with the left quadrant of the hole (Figure 12–31), and Click: D1. After you Click: D1, use Osnap-Perpendicular and Click: D2. Draw the other three hidden lines in the top using the same method.

Step 19. Set layer OBJECT current, and use the Dtext command to complete the title block. Title the drawing AUXILIARY 2.

Step 20. Use the SAVEAS command to save your drawing as EX12-2(your initials) on a floppy disk and again on the hard drive of your computer.

Step 21. Plot or print your drawing full size on an 11" × 8½" sheet.

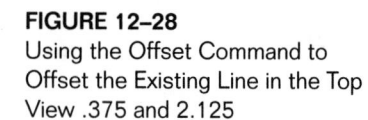

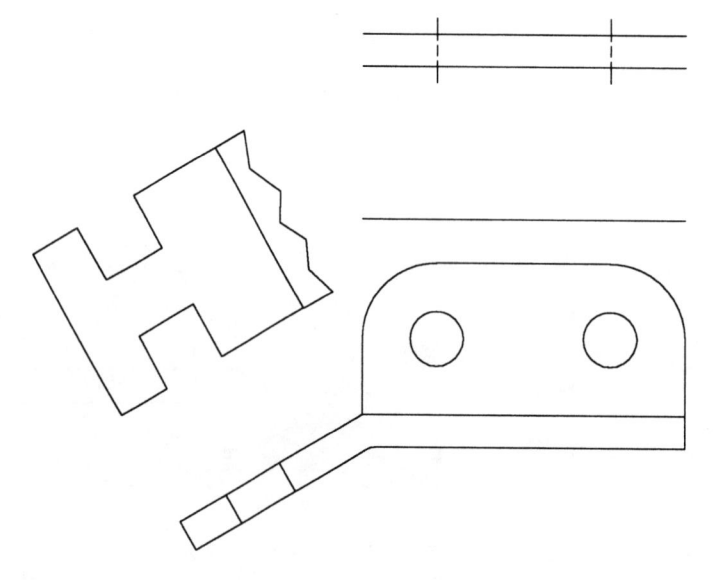

FIGURE 12–28
Using the Offset Command to Offset the Existing Line in the Top View .375 and 2.125

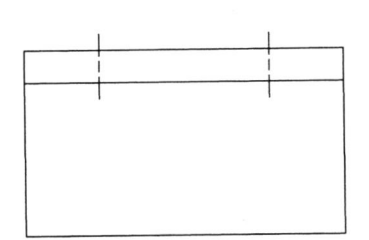

FIGURE 12–29
Using the Line Command to Draw the Two Vertical Lines in the Top View

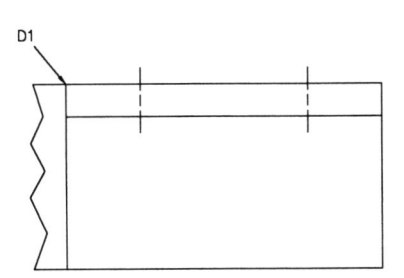

FIGURE 12–30
Using the Insert Command to Insert the Block BREAK 3

FIGURE 12–31
Drawing Hidden Lines in the Top View

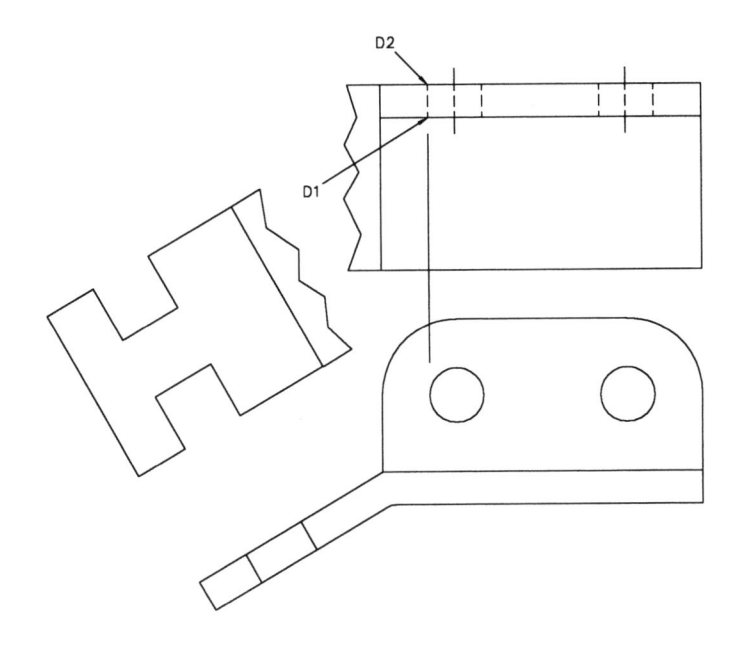

EXERCISES

EXERCISE 12–1 Complete Exercise 12–1 using steps 1 through 18 described in this chapter.
EXERCISE 12–2 Complete Exercise 12–2 using steps 1 through 21 described in this chapter.

REVIEW QUESTIONS

Circle the best answer.

1. To make sure a line is drawn from the exact end of another existing line
   a. Turn Snap ON and draw it
   b. Be sure Ortho is ON
   c. Use Osnap-Endpoint
   d. Move your cursor over the end of the line and click
   e. Use Osnap-Perpendicular
2. To make sure a line is drawn at 90° to an existing line
   a. Turn Snap ON and draw it
   b. Be sure Ortho is ON
   c. Use Osnap-Endpoint
   d. Activate the Line command and Type: 90
   e. Use Osnap-Perpendicular

Making Auxiliary Views with AutoCAD

3. To draw hidden lines as described in this chapter you must
   a. Set a layer current that is assigned the HIDDEN linetype
   b. Draw a continuous line and break it
   c. Set the HIDDEN linetype current
   d. Set linetype scale
   e. Use the Extend command
4. Which of the following commands can be used to change a line from the OBJECT layer to the HIDDEN layer?
   a. Erase
   b. Extend
   c. Hide
   d. Ltscale
   e. Chprop
5. Which of the following commands can be used to draw a line parallel to another existing line?
   a. Extend
   b. Parallel
   c. Chprop
   d. Offset
   e. Trim
6. Which of the following commands can be used to remove a part of an existing line using a cutting edge?
   a. Extend
   b. Erase
   c. Chprop
   d. Offset
   e. Trim
7. Which of the following commands can be used to call up a predefined block on the current drawing?
   a. Extend
   b. Block
   c. Insert
   d. Draw
   e. Wblock
8. If you want to call up a predefined block and rotate it so it is 180° from its defined position
   a. Use an X-scale factor of 2
   b. Use a Y-scale factor of 2
   c. Use a rotation angle of 360
   d. Use a rotation angle of 90
   e. Use a rotation angle of 180
9. If you are using the Trim command and you are having a difficult time picking lines to trim,
   a. Zoom in closer using Zoom-Window
   b. Turn Snap ON
   c. Turn Ortho ON
   d. Keep trying—some lines are difficult to pick
   e. Use the Chprop command
10. To draw a line $2\frac{1}{2}''$ to the right of a point Type:
    a. 2.5<90
    b. 2.5<0
    c. @2.5<90
    d. @2.5<0
    e. @2.5<180

13 Sketching Pictorial Views

OBJECTIVES

After completing this chapter, you will be able to:

- Arrange correctly the three major forms of pictorial drawings in order of difficulty of drawing and shape distortion.
- Correctly sketch oblique figures to scale from orthographic drawings.
- Correctly sketch isometric figures to scale from orthographic drawings.
- Select shapes that are most effectively illustrated in oblique or isometric drawing methods.

PICTORIAL DRAWING FORMS

There are three major pictorial sketching forms: oblique, axonometric, and perspective (Figure 13–1). Each of these methods has merits and drawbacks:

Oblique is usually the easiest to draw but is the most distorted.

Axonometric is more difficult to draw but is less distorted. There are three forms of axonometric drawing: isometric, dimetric, and trimetric. By far the most commonly used is isometric. Isometric is the only axonometric form that will be covered in this book.

Perspective is the most difficult to draw and is usually the least distorted. The perspective method is not covered in this book.

In summary, to give you skills that you can readily use with the least difficulty, only oblique and isometric drawing forms are covered in this book.

FIGURE 13–1
Pictorial Drawing Forms

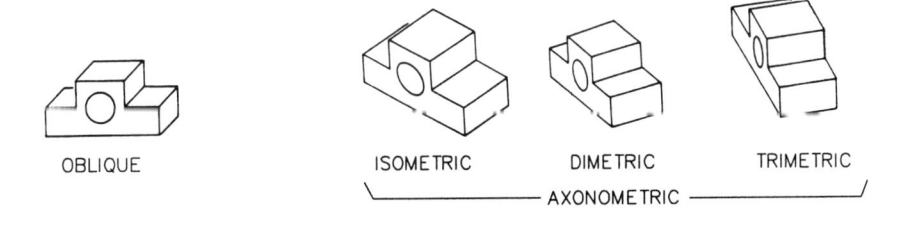

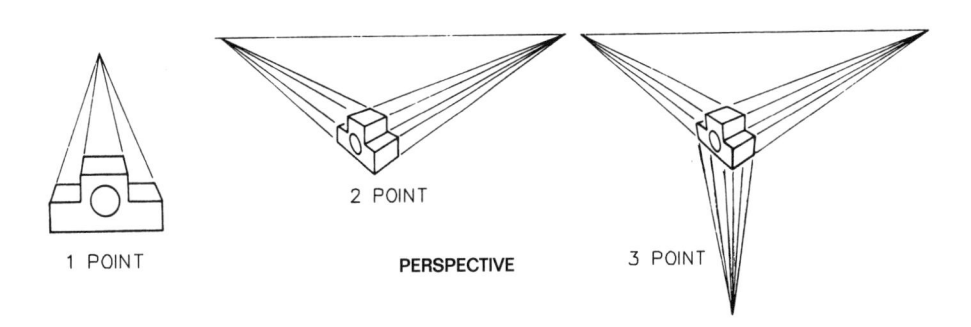

203

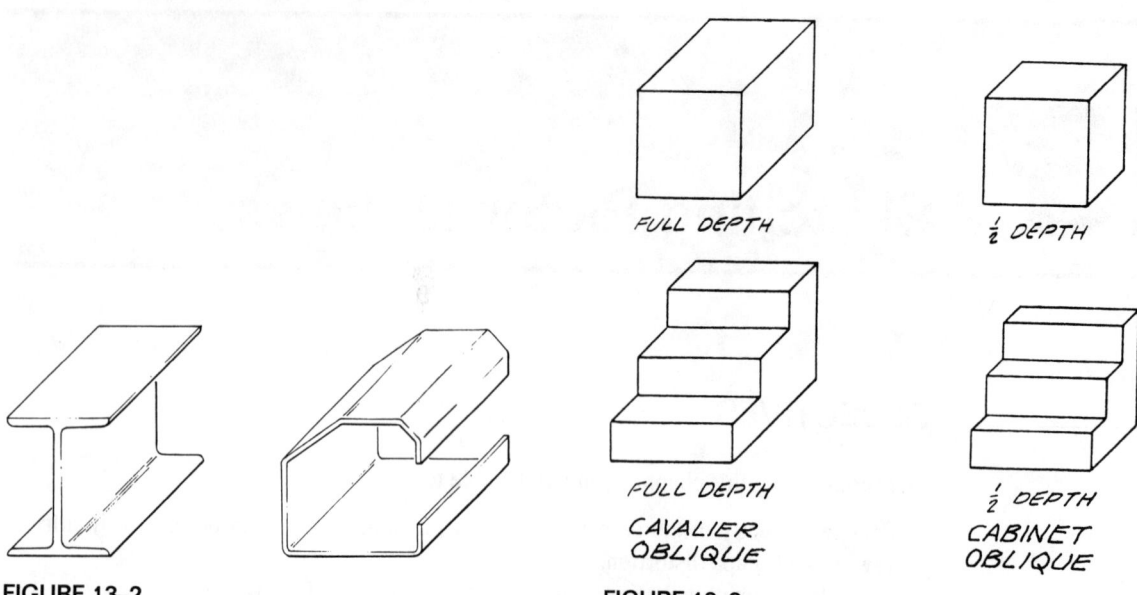

FIGURE 13–2
Objects with Uniform Contour

FIGURE 13–3
Cavalier and Cabinet Oblique

Oblique Drawing

The distortions created by oblique drawing can be an advantage in sketching some objects but are unusable for others. A metal part that has the same cross-sectional shape along its length, for example, looks fine in oblique (Figure 13–2). An advantage of oblique is that it shows the complete two-dimensional shape of the front view, so there is no possibility of confusion.

Oblique drawing simply involves drawing the front view of an object as an orthographic view and then adding depth to it at an angle, usually 45°. To reduce the distorted appearance, the depth is often drawn half its true size. This technique is called *cabinet oblique*. When the depth dimension is full size, the drawing form is called *cavalier oblique* (Figure 13–3). In Exercise 13–1 you will draw four shapes using the cabinet oblique drawing form.

Figure 13–4 shows a technique for sketching ellipses in oblique. Avoid placing round shapes on the depth planes. Place them on the view where they appear as circles. If you have a shape that has round shapes on depth planes, the oblique method is probably not the one to use.

Isometric Drawing

The isometric drawing form is used a great deal because it is easy to draw and looks good for many objects. If you understand isometric well, the other two axonometric forms, dimetric and trimetric, are easy to learn.

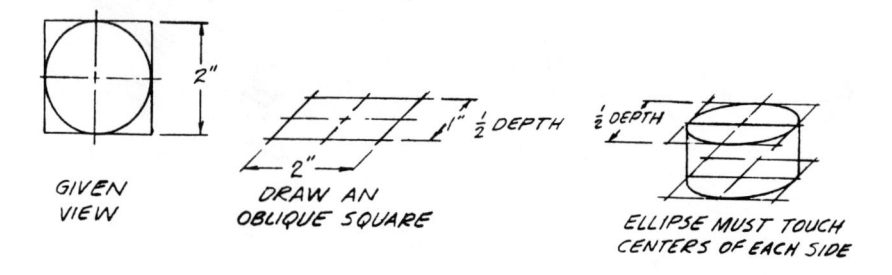

Measurements in isometric are the same in all three dimensions—height, width, and depth. Figure 13–5 shows how to use this method for sketching a simple 4″ × 2″ × 2″ box.

FIGURE 13–4
Sketching Ellipses in Oblique

FIGURE 13–5
Measurements in Isometric

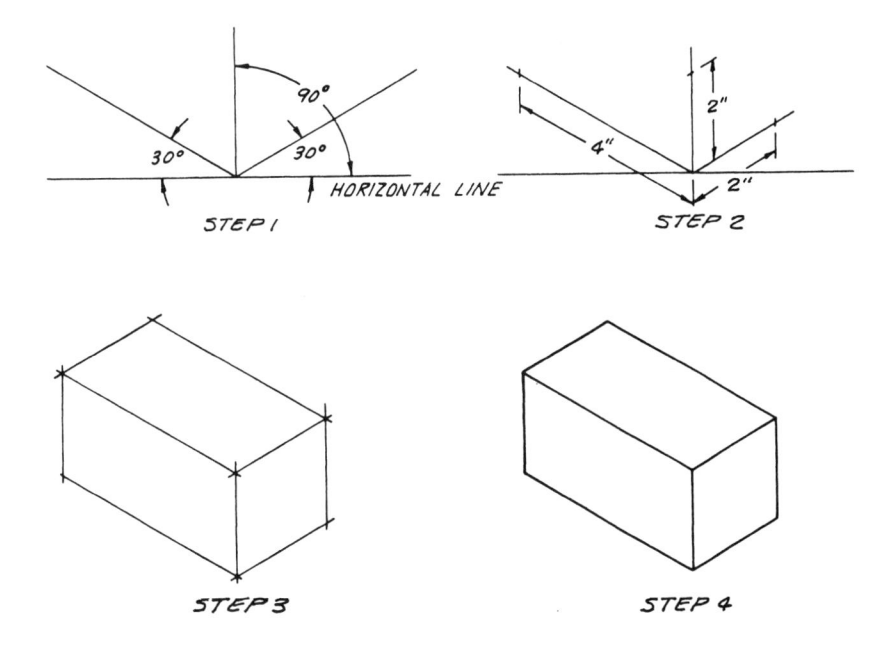

Step 1. Lightly sketch the three isometric axes (or use isometric grid paper): 30° right, 30° left, and 90° to a horizontal line.
Step 2. Measure 4" along one of the 30° lines, 2" along the other 30° line, and 2" on the 90° line.
Step 3. Extend height, width, and depth lines, making sure that all lines are parallel, until they meet. Use your 30–60° triangle to sketch the 30° lines if you are not using isometric grid paper.
Step 4. Darken all object lines to complete the drawing.

This simple example provides you with the basics of isometric drawing. There are some details of construction, which will be explained, but if you can draw Figure 13–5, you have a good understanding of the basic principles. The details of construction that you will need for complex shapes follows.

Angles in Isometric Drawing

Without special tools, no measurements can be made on any lines that are not parallel to one of the isometric axes, and angles cannot be measured directly. Each end of the angle must be located and then joined as shown in Figure 13–6.

FIGURE 13–6
Sketching Angles in Isometric

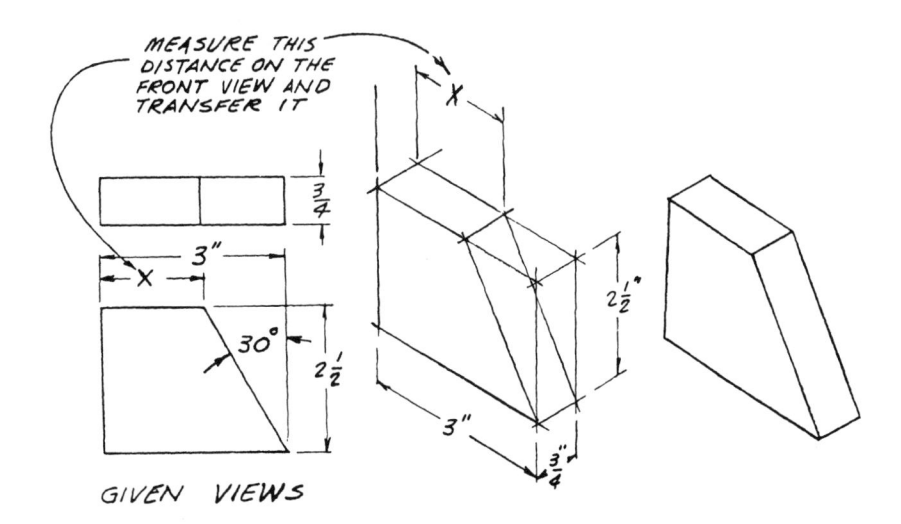

Sketching Pictorial Views

FIGURE 13–7
Positioning Isometric Ellipses

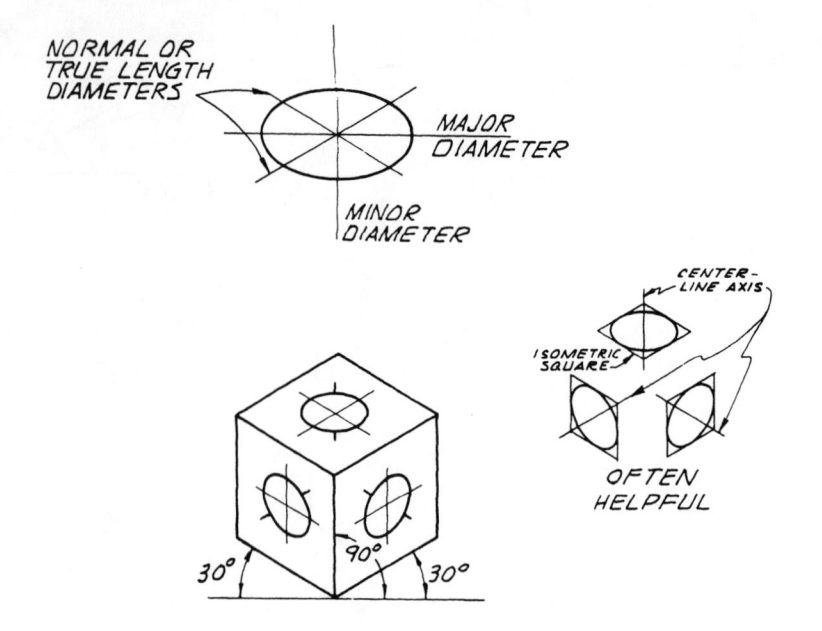

Circles in Isometric Drawing

Circles on the isometric planes appear as ellipses. The angle through which the circle is tilted is 35° 16′. The positions for ellipses on each face of an isometric cube are shown in Figure 13–7. Notice that the minor diameter of the ellipse is always lined up on a centerline that is parallel to one of the isometric axes. Correct positioning of ellipses is very important to the appearance of a drawing and should be carefully studied. It is often helpful for beginners to sketch the centerline axis and the isometric square that the ellipse will fit. In the past the ellipse itself was often drawn with an isometric ellipse template. As you will discover in the next chapter, AutoCAD makes drawing isometric ellipses very easy and almost impossible to turn in the wrong direction. To avoid the use of isometric ellipses, the exercises in this chapter will contain very few ellipses, so you can sketch them without the use of a template.

Cylindrical objects must be drawn so that the centers of their ellipses lie on a centerline, as shown in Figure 13–8. To simplify construction, holes on a curved surface, as hole A in Figure 13–8, are drawn as if they were lying on a flat surface.

Curves in Isometric Drawing

Isometric curves are drawn by locating a number of points (any number that is appropriate for the feature) on the curve in a flat (orthographic) view. Those points are then transferred to the isometric view, using the correct isometric measurements for each point. This process is illustrated in Figure 13–9.

FIGURE 13–8
Sketching Ellipses on a Curved Surface in Isometric

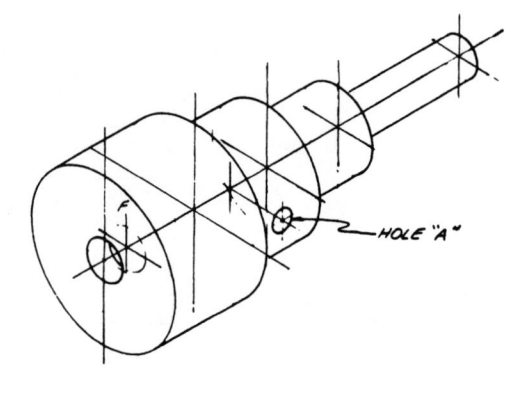

FIGURE 13–9
Sketching Curves in Isometric

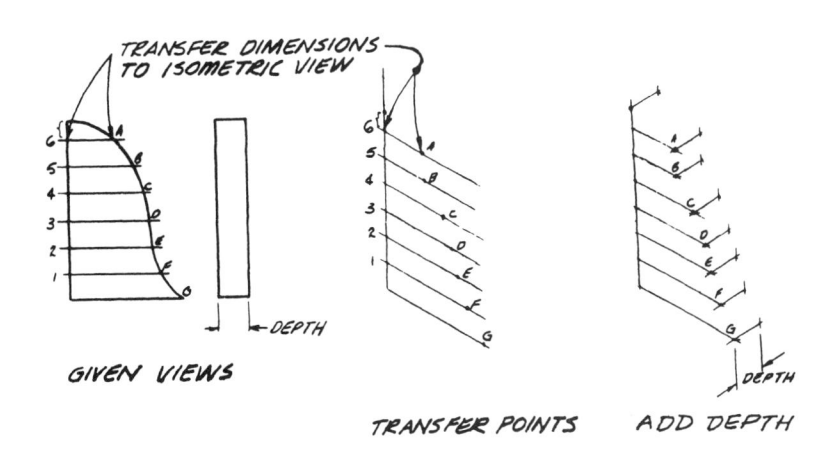

Spheres in Isometric Drawing

A sphere using any drawing method is drawn as a circle. If it is necessary to cut pieces out of it, isometric ellipses are helpful. Notice that the ellipses are drawn so that their minor diameters are lined up on one of the isometric axes (Figure 13–10). A sphere cannot be measured directly in isometric. You must take its measurement from the long diameter of an isometric ellipse. To draw a 1″ sphere in isometric, draw a circle that touches the outside edges of the major diameter of a 1″ diameter ellipse. It will measure about $\frac{1}{4}$″ more, or $1\frac{1}{4}$″ in diameter.

General Guidelines for Sketching in Isometric

Some general guidelines for drawing in isometric are as follows:

> For complex shapes it is often helpful to draw a box around the object and measure from the corners of the box to locate features (Figure 13–11).
> Locate the centers of all holes and curves and draw those first (Figure 13–12).

FIGURE 13–10
Isometric Shapes

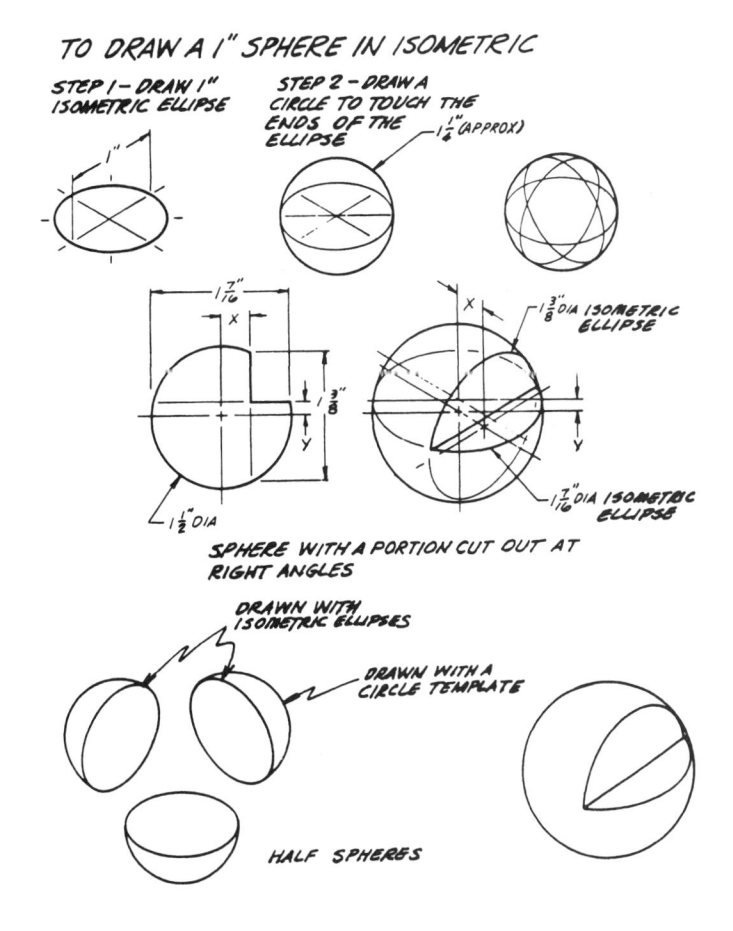

Sketching Pictorial Views

207

FIGURE 13–11
Sketching Within a Box

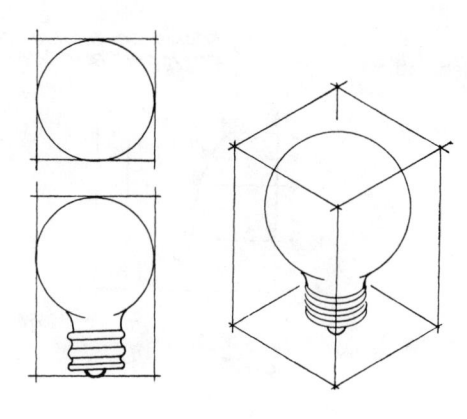

FIGURE 13–12
Locating Centers of Ellipses First

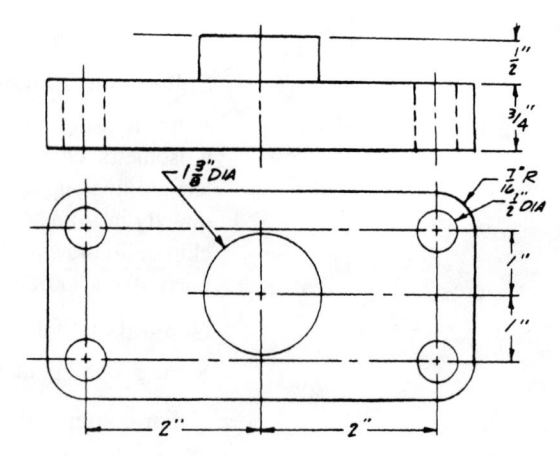

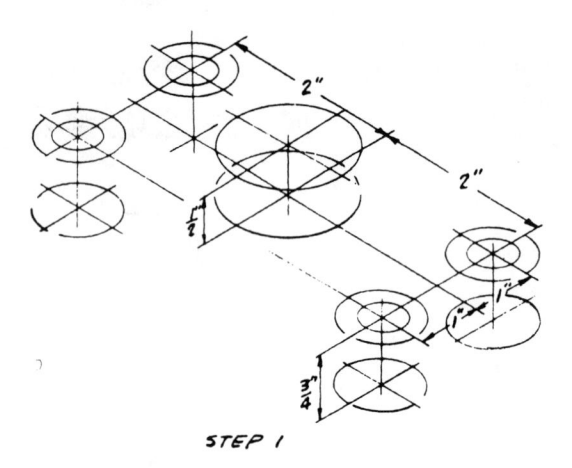

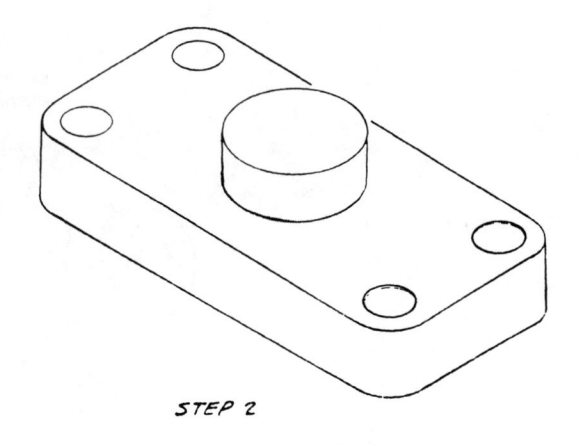

STEP 1

STEP 2

208　　Chapter 13

FIGURE 13–13
Cutaway Example

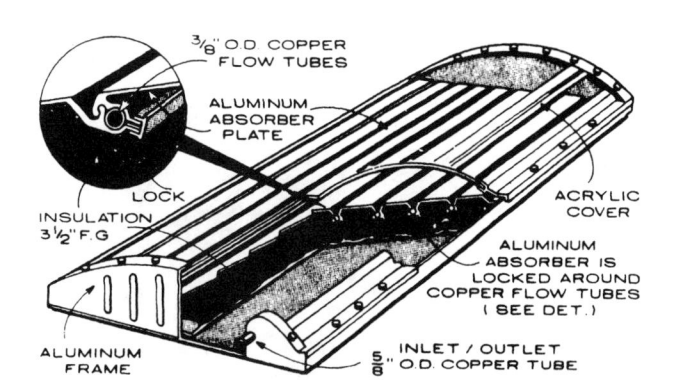

Be sure to locate points and features in all three dimensions: height, width, and depth.

You cannot measure angles in isometric. Locate the ends of the angular line, using height, width, and depth dimensions, as was shown in Figure 13–6.

Be sure to position ellipses correctly, as was shown in Figure 13–7.

For cylindrical objects, work from a centerline, and locate ellipse centers on that line, as was shown in Figure 13–8.

For curves and irregular shapes, establish some points on the orthographic views and transfer them to the isometric view, as was shown in Figure 13–9.

Simplify the intersections of cylinders and holes, as was shown in Figure 13–8.

Cutaway Drawings

Cutaway drawings are among the most impressive technical drawings. They are used in many different types of publications such as advertising pieces, specification sheets, and repair manuals. Examples of cutaway drawings are shown in Figures 13–13 through 13–15.

Although cutaways are impressive and can take a great deal of time if the assembly has many parts, they are not difficult to draw. The main things to remember when drawing cutaways are to complete one part at a time and to make sure parts are in the correct position. The drawing shown in Figure 13–16 is an excellent example of a complex cutaway. If you study a few individual parts for a moment, you will quickly see that there is nothing that you have not drawn or could not draw if you knew the shape and size of the part. There are

FIGURE 13–14
Cutaway Example

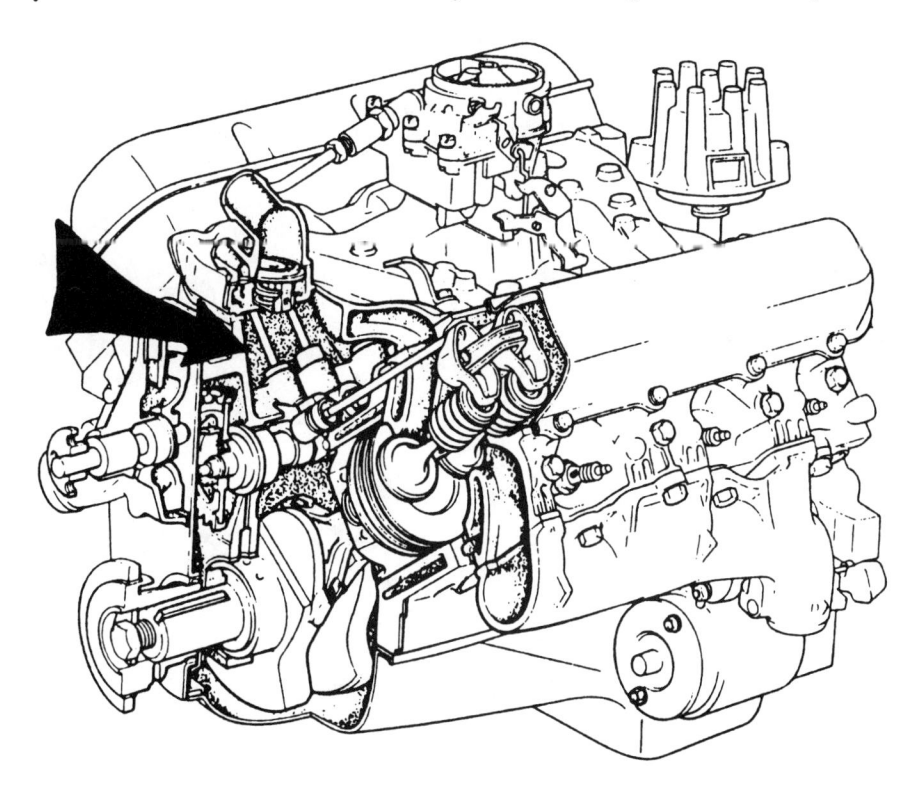

Sketching Pictorial Views

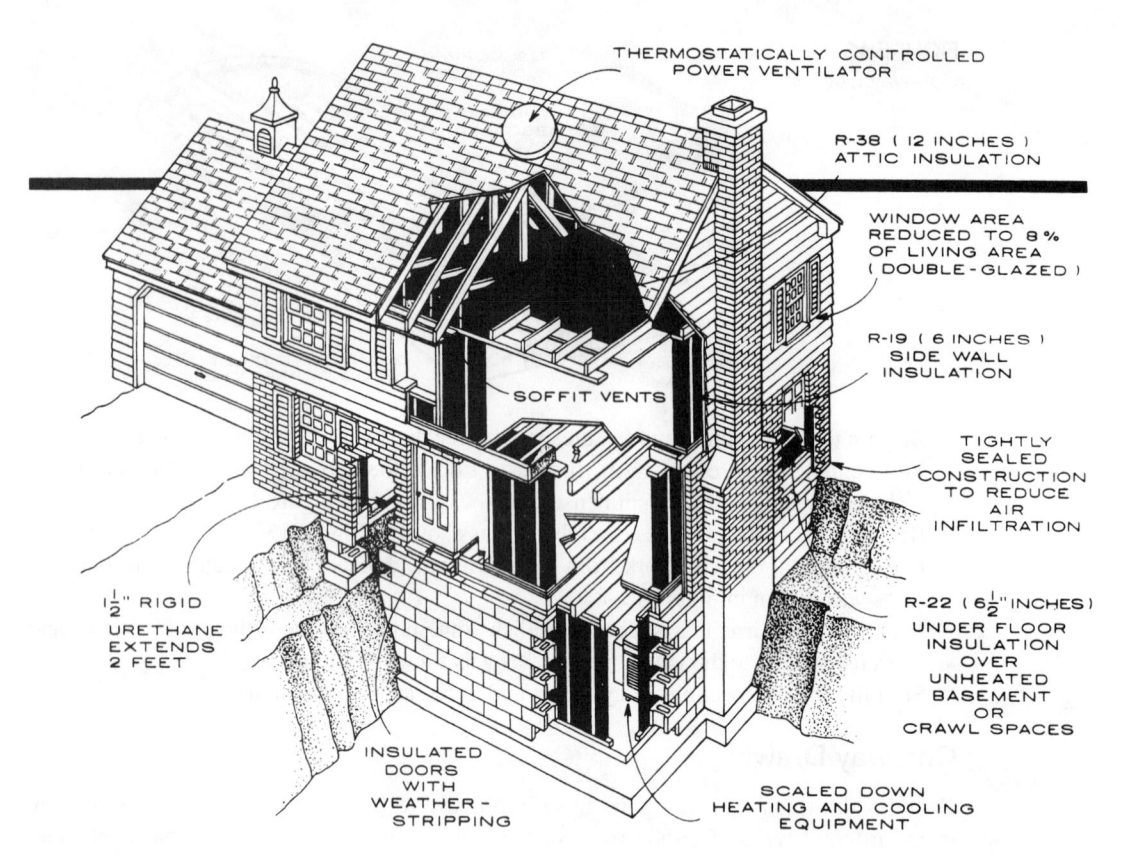

FIGURE 13–15
Cutaway Example

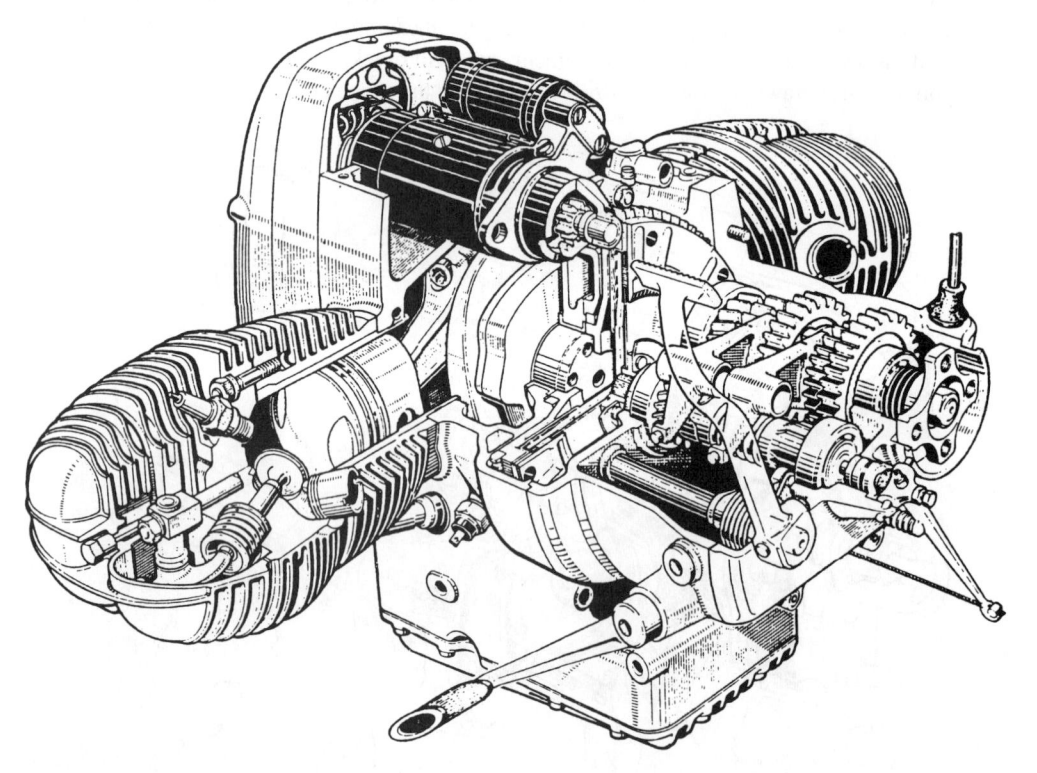

FIGURE 13–16
Complex Cutaway

FIGURE 13–17
Steps in Drawing Cutaways

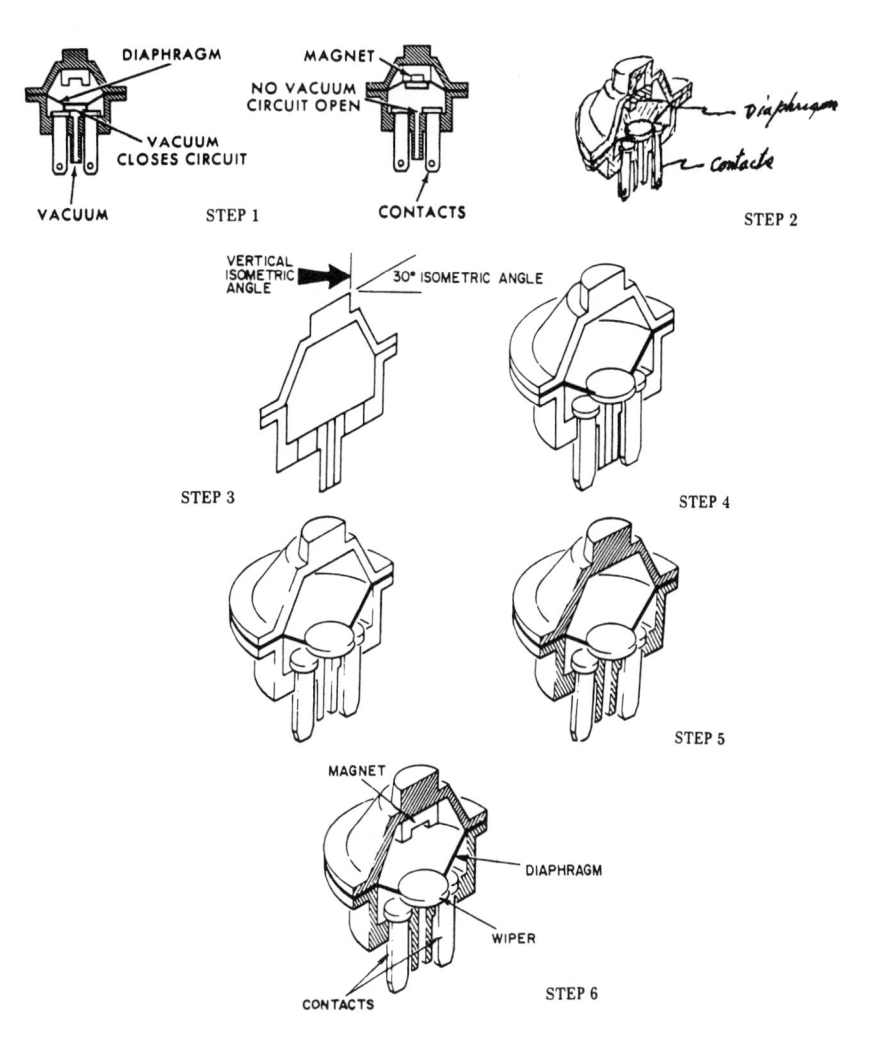

a few rules and details, however, that will help you with drawing cutaways. The following steps, illustrated in Figure 13–17 cover those rules and details.

General Guidelines for Cutaway Drawings

Step 1. Determine the surfaces to be cut.
Step 2. Make a rough freehand sketch.
Step 3. Draw the cut surfaces.
Step 4. Draw the full shapes.
Step 5. Put shading on the cut surfaces.
Step 6. Add callouts.

We shall examine each step in detail.

Determine the surfaces to be cut:

Find out exactly what the drawing is trying to show. When you know the parts to be shown and what position they should be in, you can determine which surfaces should be cut and how deep the cut should be. If the drawing in Figure 13–17 were to be shown as a cutaway, for example, and your instructions were to show how the vacuum switch works, a cut completely across the upper and lower housing through the magnet and the diaphragm should be used. To show less would hide some of the internal parts that reveal how the switch functions.

Make a rough freehand sketch:

Making a rough freehand sketch allows you to decide which isometric position to use and to make the correct decision about which surfaces to cut. In Figure 13–17, the sketch confirms that a full cut across the part is needed to show both contacts. It also indicates that the vertical isometric axis is a good position for the drawing because it shows how the part functions.

Sketching Pictorial Views

211

Draw the cut surfaces:

It is usually best to draw the cut surfaces first. Sometimes this cannot be done before some of the parts that are not cut must be drawn. You will discover that as you proceed through the drawing. In Figure 13–17, notice that the cut surfaces were drawn just as they appear on the orthographic views, except that isometric angles have now taken the place of horizontal and vertical lines.

Draw the full shapes:

Draw the full shapes and complete the uncut parts of the cut surfaces. You must be sure that the parts are shown in their correct positions, which you can determine by measuring. If they are not in the correct positions and still do not show clearly what is intended, do not hesitate to distort dimensions or to make additional cuts, so long as parts do not get too far out of proportion.

Put shading on the cut surfaces:

In Figure 13–17, shading lines were used on the cut surfaces. Notice that the shading lines are at steep angles of approximately 60° in opposite directions. It is important to make these lines different from isometric angles, because they can be easily confused with object lines.

Add callouts:

Notice that the lettering in the callout is lined up horizontally. The blunt end of the leader points to the center of the line of lettering, and the arrow end breaks the part about 1/16″. Keep leaders short and with a fan arrangement or with the same angle. Lines of lettering can be centered or arranged flush right or left, whichever is convenient or specified by your customer. Be sure to follow a consistent pattern, and be sure that all callouts are clear and easily read.

Further Details

If you must cut a flat plane in a cutaway drawing, use obtuse angles for the cut, as shown in Figure 13–18. This allows you to show uniform thickness easily by using the opposite isometric angle and does not distract from the drawing, as extremely jagged cuts do.

Fasteners, shafts, and spheres are shown much more clearly if they are not cut on the drawing. Generally, nothing should be cut in a cutaway that is not necessary to show what is intended. Figure 13–19 is a good example of the selection process.

If you like this kind of drawing, study cutaways in books and magazines to decide what looks good and what does not. Not all the drawings you see printed in books and magazines are good ones.

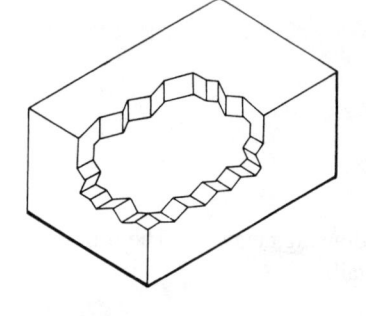

FIGURE 13–18
Using Obtuse Angles for Cut Surfaces

FIGURE 13–19
Do Not Cut Shafts, Fasteners, Gears, or Anything Else That Does Not Aid in Understanding

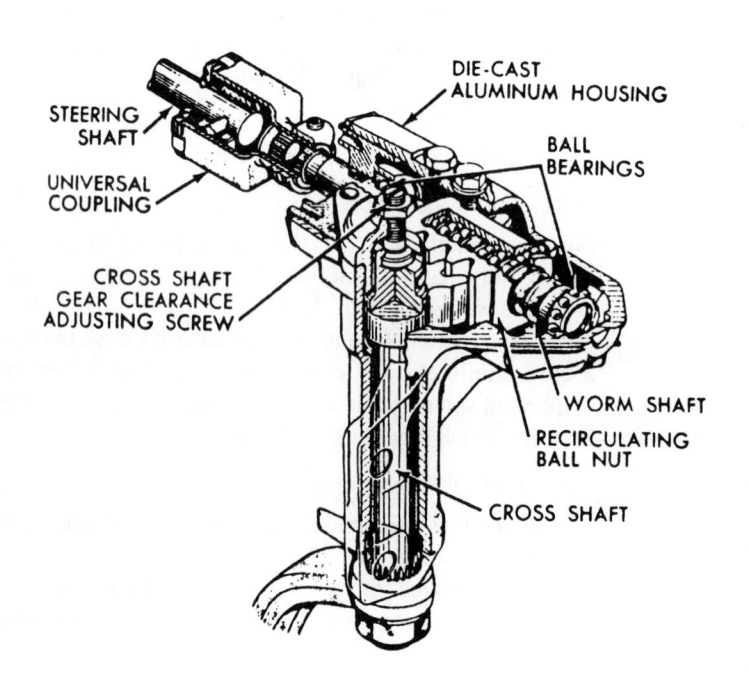

EXERCISES

EXERCISE 13–1 Use the following instructions to sketch the final step shown in Figures 13–20, 13–21, 13–22, and 13–23, on the sheet provided using the cabinet oblique drawing form. Make sure your lines are dense and the same width as object lines. Your drawing will show only the final step in each of these figures.

Remove the sheet labeled EXERCISE 13–1 from your book.

A. **Step 1.** Sketch the front view of the object in Figure 13–20 using the dimensions shown in the given views. The lower left corner of the front view is in the upper left of the exercise sheet.

  Step 2. Extend light construction lines from the front view at a 45° angle to show the depth. Make sure the dimension is half the true depth (in this case, $\frac{1}{2}''$).

  Step 3. Complete the illustration by connecting the points to form the back surface, and darken all object lines.

B. **Step 1.** Sketch front surface A of Figure 13–21 using the dimensions shown in the given views. The lower left corner of this drawing is in the upper right of the exercise sheet.

  Step 2. Lightly sketch lines D1 and D2 at a 45° angle. Make D1 and D2 half their true length. Sketch front surface B using the dimensions shown in the given views.

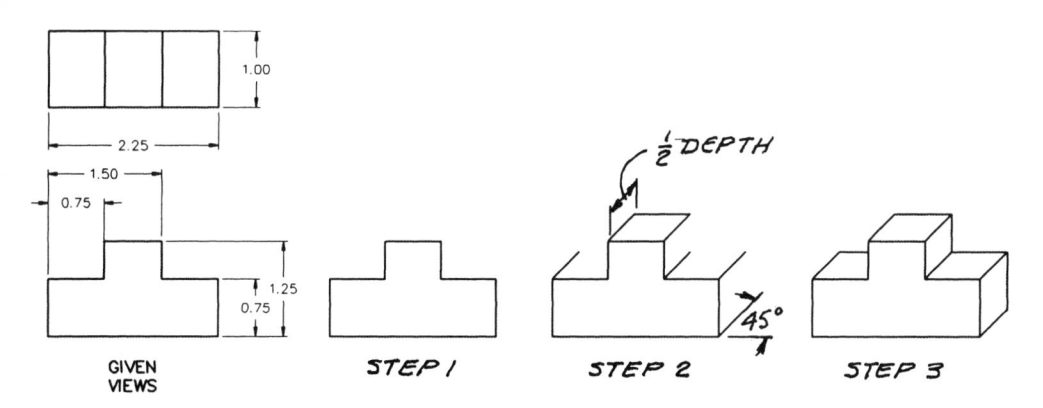

FIGURE 13–20
Exercise 13–1A

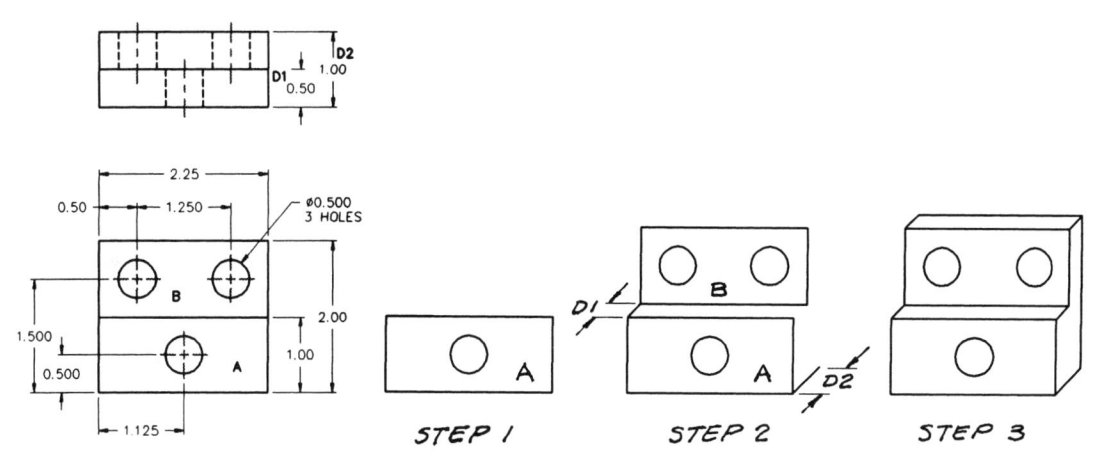

FIGURE 13–21
Exercise 13–1B

Sketching Pictorial Views

FIGURE 13–22
Exercise 13–1C

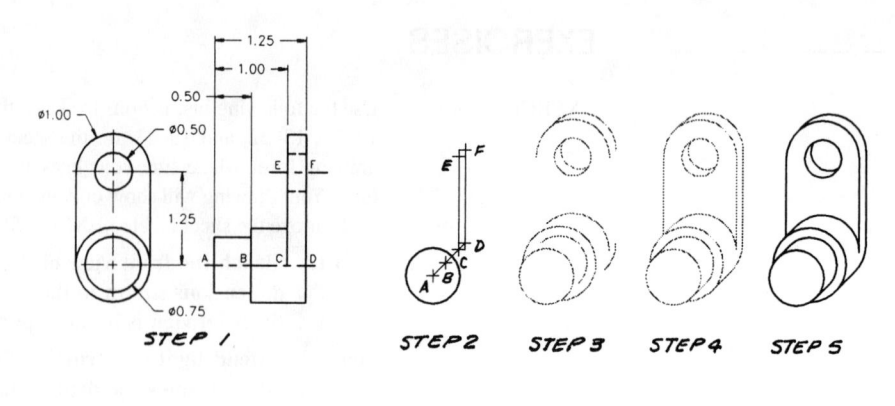

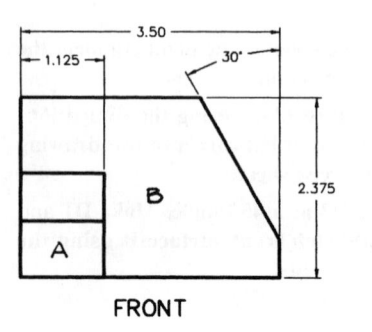

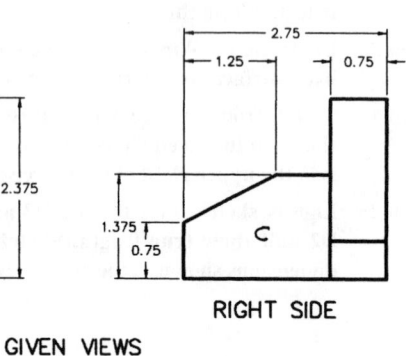

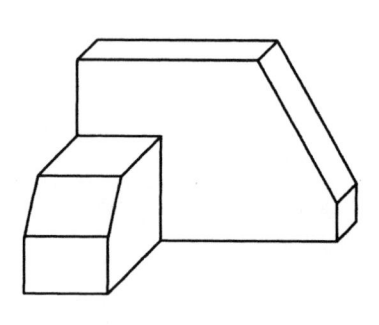

FIGURE 13–23
Exercise 13–1D

Step 3. Darken the other depth lines, and darken all object lines to complete the illustration.

C. Step 1. Look at the given views in Figure 13–22 to decide which circles are on the same centerlines.

Step 2. Locate the centers for all circles. Sketch a light 45° construction line from the center of the circle on the exercise sheet. The center of this circle is point A.

Measure half the distance from A to B and mark point B on the construction line. Do the same for points C and D.

From points C and D sketch light construction lines straight up. Take the distance from C to E on the given views and mark point E on the vertical construction line.

From point E sketch a light 45° construction line to intersect with the vertical line from point D. This locates point F.

Step 3. Lightly sketch the parts of the circles that will show. You can tell which ones will show by beginning with the circle at point A and working backward.

Step 4. Locate the tangent points by drawing light 45° and vertical construction lines. On this drawing, the tangent points are where straight lines meet the circles. You need to find the tangent points so that your lines have a nice, smooth flow and so that none of the circles are flattened.

Step 5. Darken all lines that show to complete the drawing.

FIGURE 13–24
Exercise 13–2A

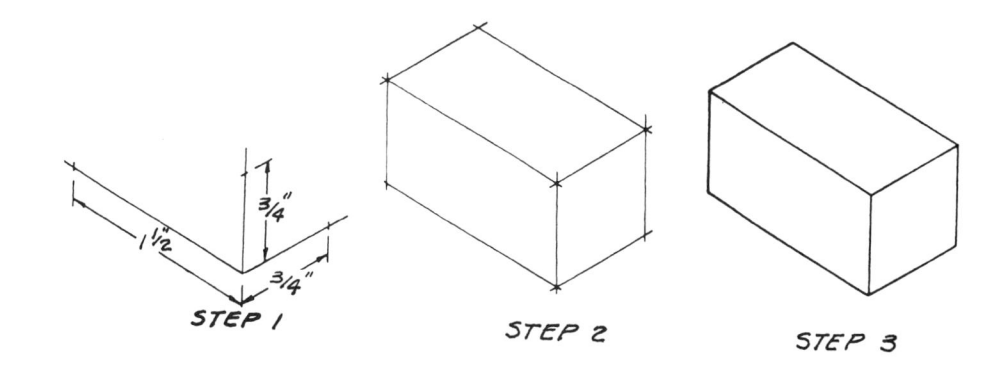

STEP 1   *STEP 2*   *STEP 3*

D. **On your own: Complete the oblique view on the right side of Figure 13–23 at half scale. Surface A of this drawing is located in the lower right of the exercise sheet. The angle on surface B can be measured directly $\frac{1}{2}''$ up from its lower right corner. The angle on surface C cannot be measured. You must locate the ends of the line and connect those points to form the angle. Darken all lines after you have completed the construction.**

Complete the title block with your best lettering and name the drawing CABINET OBLIQUE SKETCHES.

EXERCISE 13–2 Use the following instructions to sketch the final step shown in Figures 13–24 and 13–25 on the sheet provided using the isometric drawing form. Make sure your lines are dense and the same width as object lines. Your drawing will show only the final step in each of these figures.

Remove the sheet labeled EXERCISE 13–2 from your book.

A. Step 1. Mark $\frac{3}{4}''$ from the lower left corner on the 30° axis to the right, $\frac{3}{4}''$ on the vertical axis, and $1\frac{1}{2}''$ on the 30° axis to the left. The lower left corner of the drawing in Figure 13–24 is on the left side of the exercise sheet.

Step 2. Using light construction lines, extend height, width, and depth lines until they meet, and sketch parallel lines as needed.

Step 3. Darken the object lines.

B. Step 1. Using light construction lines, sketch an isometric box measuring 2" on the isometric axis to the right, $2\frac{1}{2}''$ on the isometric axis to the left, and $1\frac{1}{2}''$ on the vertical axis. Each mark on the 3D view represents $\frac{1}{4}''$. The lower right corner of the box is on the right side of the exercise sheet.

Step 2. Use light construction lines to sketch the general shape of the front view of Figure 13–25 as shown.

FIGURE 13–25
Exercise 13–2B

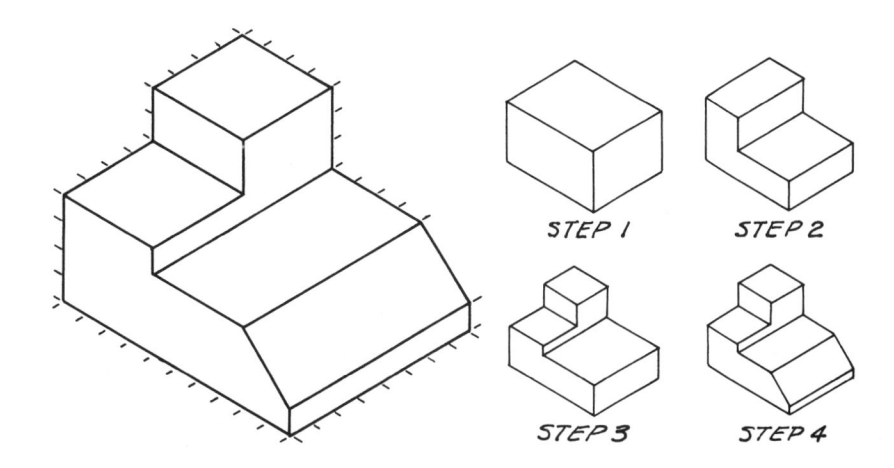

STEP 1   *STEP 2*   *STEP 3*   *STEP 4*

Sketching Pictorial Views

FIGURE 13–26
Exercise 13–3A

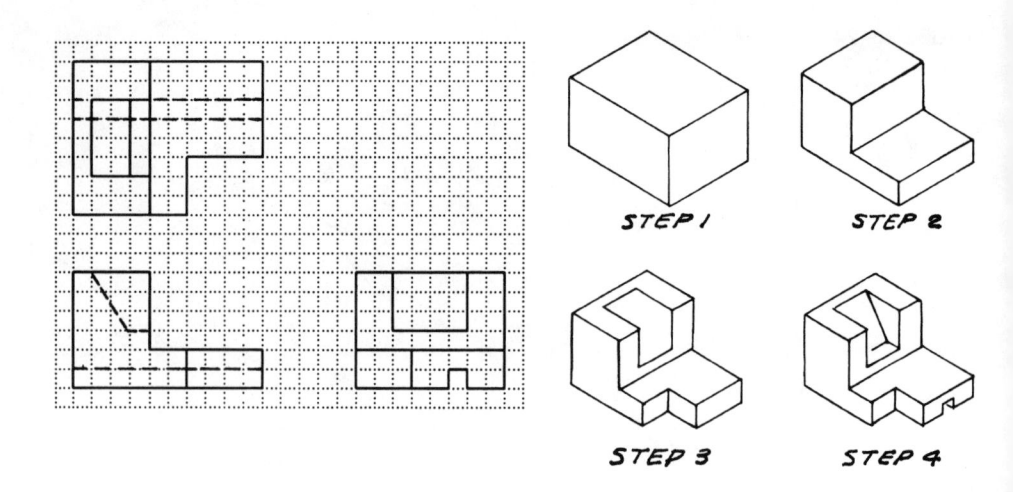

STEP 1 STEP 2
STEP 3 STEP 4

Step 3. Use light construction lines to sketch the notch in the upper surface shown in the right-side view.

Step 4. Locate the ends of the angular shape and complete the shape as shown. Darken all object lines to complete the illustration.

Complete the title block with your best lettering and name the drawing ISOMETRIC SKETCHES NO. 1.

EXERCISE 13–3 Use the following instructions to sketch the final step shown in Figure 13–26 and an isometric sketch of Figure 13–27 on the sheet labeled EXERCISE 13–3 in your book using the isometric drawing form. Make sure your lines are dense and the same width as object lines. Your drawing will show only the final step in Figure 13–26.

A. Step 1. Using light construction lines sketch an isometric box measuring 2" on the isometric axis to the right, $2\frac{1}{2}"$ on the isometric axis to the left, and $1\frac{1}{2}"$ on the vertical axis. Each mark on the orthographic views represents $\frac{1}{4}"$.

Step 2. Use light construction lines to sketch the general shape of the front view of Figure 13–26 as shown.

Step 3. Use light construction lines to sketch the top and right side of the notch in the upper surface. Sketch the notch in the base as shown.

Step 4. Complete the shape of the notch in the upper surface and sketch the .25 slot in the bottom of the right side of the base as shown. Darken all lines to complete the drawing.

B. On your own: Make an isometric sketch from the top, front, and right-side orthographic views shown in Figure 13–27. You will have to make isometric squares to help you sketch the ellipses forming the holes and the fillets at the corners of the base as shown in Figure 13–7. Use the same method that you used for the other sketches in Exercises 13–1 and 13–2.

FIGURE 13–27
Exercise 13–3B

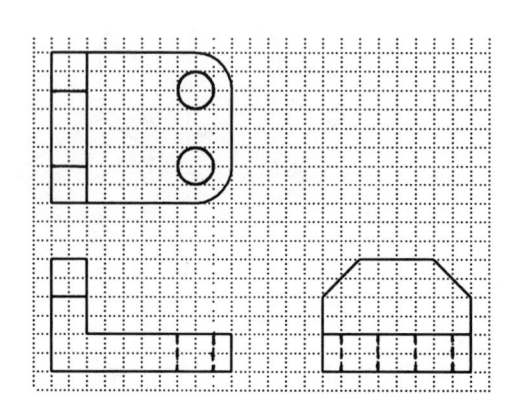

FIGURE 13–28
Exercise 13–4
Full Scale

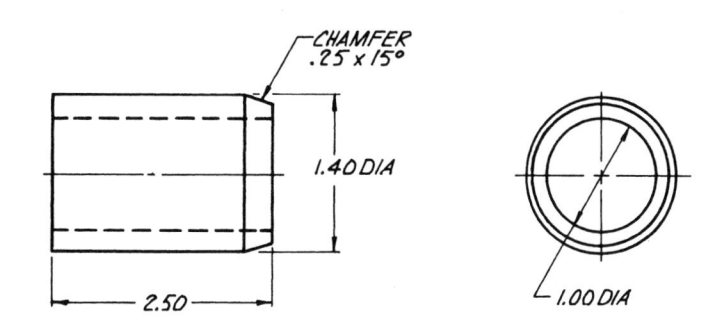

Complete the title block with your best lettering and name the drawing ISOMETRIC SKETCHES NO. 2.

EXERCISE 13–4 Make a CABINET OBLIQUE sketch of Figure 13–28 on the sheet labeled EXERCISE 13–4 in your book.

Draw the figure full scale using the dimensions shown, and put the circular shapes on the front so they can be drawn as circles. Center the drawing in the sheet by using light construction lines until you are sure the drawing is centered. You will have to draw an orthographic view of the chamfered end so you can determine the diameter of the 15° chamfer.

Be sure to make the depth half of the 2.50″ dimension. Do not show any dimensions.

Complete the title block with your best lettering and name the drawing SLEEVE.

EXERCISE 13–5 Make a CABINET OBLIQUE sketch of Figure 13–29 on the sheet labeled EXERCISE 13–5 in your book.

Draw the figure full scale using the dimensions shown, and put the circular shapes on the front so they can be drawn as circles. Center the drawing in the sheet by using light construction lines until you are sure the drawing is centered.

Be sure to make the depth half of the $1\frac{1}{4}''$ dimension. Do not show any dimensions.

Complete the title block with your best lettering and name the drawing BASE PLATE.

EXERCISE 13–6 Make an ISOMETRIC sketch of Figure 13–30 on the sheet labeled EXERCISE 13–6 in your book.

Draw the figure $\frac{3}{4}$ scale using the dimensions shown. Center the drawing in the sheet by using light construction lines until you are sure the drawing is centered.

Do not show any dimensions.

Complete the title block with your best lettering and name the drawing HOLDER.

FIGURE 13–29
Exercise 13–5
Full Scale

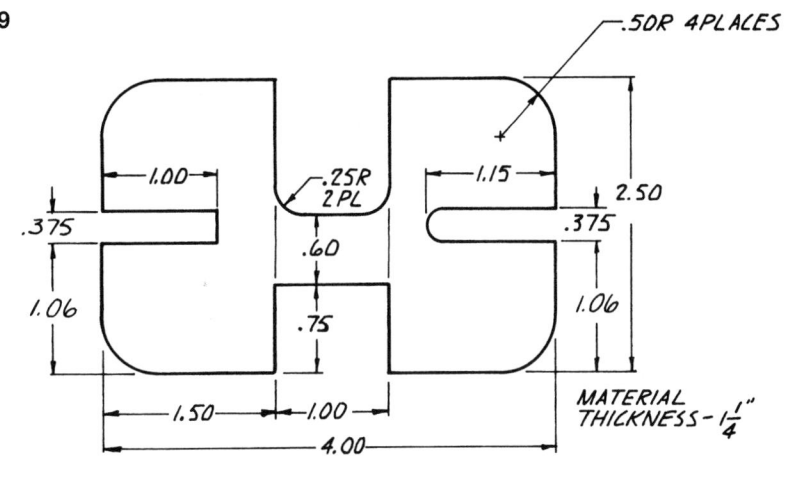

Sketching Pictorial Views

FIGURE 13–30
Exercise 13–6
Scale: $\frac{3}{4}'' = 1''$

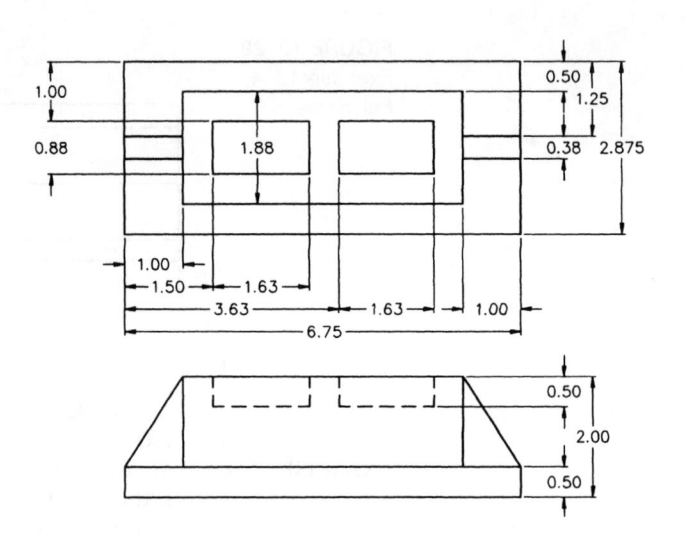

EXERCISE 13–7   Make an ISOMETRIC sketch of Figure 13–31 on the sheet labeled EXERCISE 13–7 in your book.
  Draw the figure at a scale of $1''=1'$ using the dimensions shown. Center the drawing in the sheet by using light construction lines until you are sure the drawing is centered. You will have to make isometric squares to help you sketch the ellipses forming the holes as shown in Figure 13–7.
  Do not show any dimensions.
  Complete the title block with your best lettering and name the drawing STOP.

EXERCISE 13–8   Make an ISOMETRIC sketch of Figure 13–32 on the sheet labeled EXERCISE 13–8 in your book.
  Draw the figure at a scale of $\frac{3}{8}''=1'$ using the dimensions shown. Center the sketch of the entire reception area furniture in the sheet by using light construction lines until you are sure the drawing is centered.
  Do not show any dimensions.
  Complete the title block with your best lettering and name the drawing RECEPTION AREA FURNITURE.

FIGURE 13–31
Exercise 13–7
Scale: $1'' = 1'$

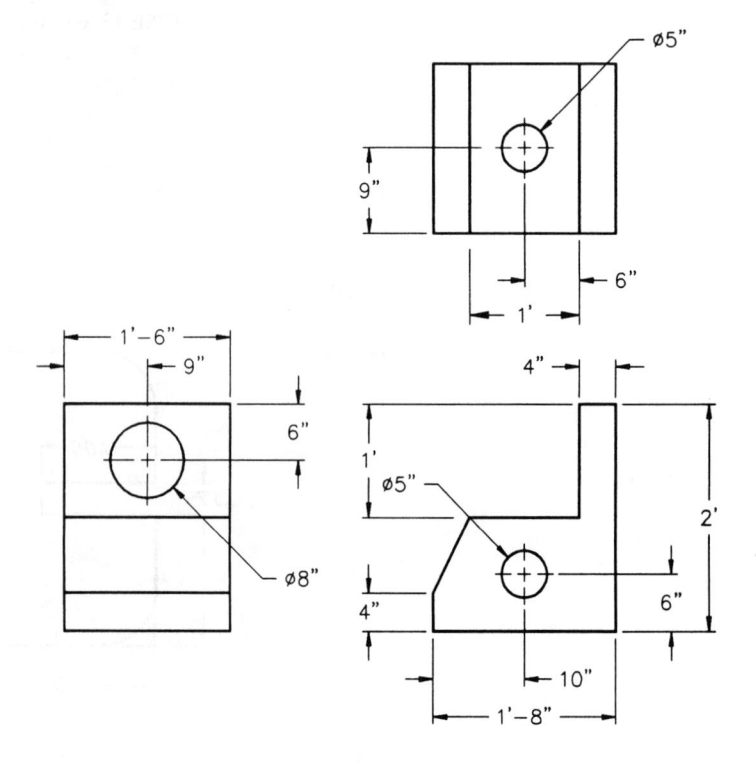

FIGURE 13–32
Exercise 13–8
Scale $\frac{3"}{8} = 1'$

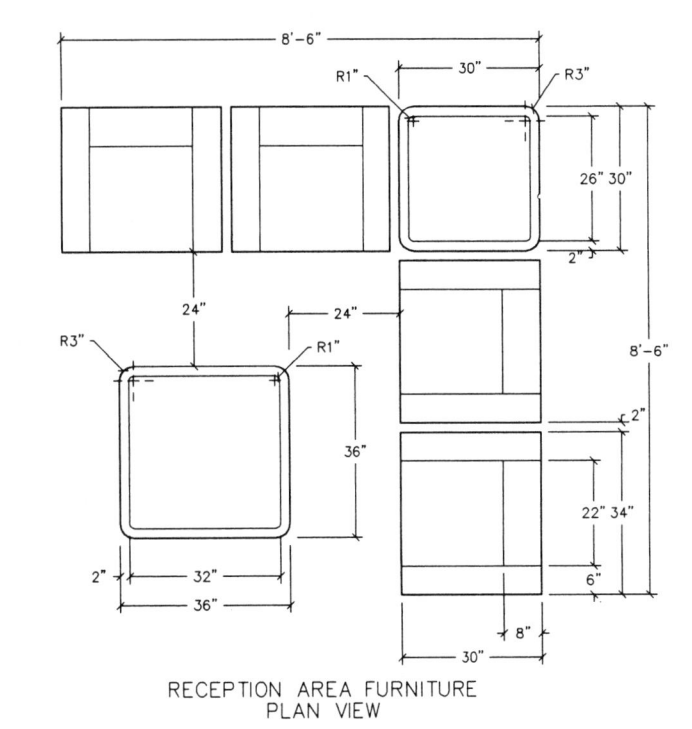

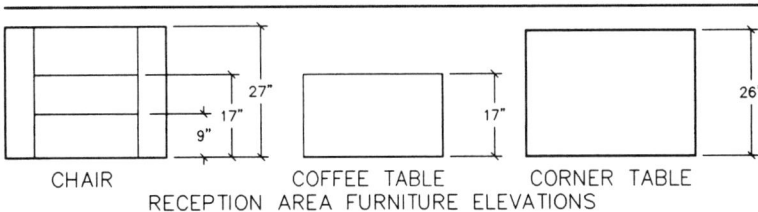

RECEPTION AREA FURNITURE
PLAN VIEW

CHAIR COFFEE TABLE CORNER TABLE
RECEPTION AREA FURNITURE ELEVATIONS

EXERCISE 13–9 Make an ISOMETRIC sketch of Figure 13–33 on the sheet labeled EXERCISE 13–9 in your book.

Draw the figure at a scale of $\frac{3"}{8} = 1"$ using the dimensions shown. Center the drawing in the sheet by using light construction lines until you are sure the drawing is centered.

Do not show any dimensions.

Show the fasteners holding detail A onto the bracket as if they were screwed in holding detail A in place. You will see only the head of the screw and a washer under it. Approximate the size of the screw and washer.

Show only the isometric view.

Complete the title block with your best lettering and name the drawing CONDENSER BRACKET.

EXERCISE 13–10 Make an ISOMETRIC sketch of Figure 13–34 on the sheet labeled EXERCISE 13–10 in your book.

Draw the figure at a scale of $\frac{1"}{4} = 1"$ using the dimensions shown. Center the drawing in the sheet by using light construction lines until you are sure the drawing is centered.

Do not show any dimensions.

Show the three pieces in the approximate positions shown in Figure 13–34.

Complete the title block with your best lettering and name the drawing HOLDER.

EXERCISE 13–11 Make an ISOMETRIC cutaway sketch of Figure 13–35 on the sheet labeled EXERCISE 13–11 in your book.

Draw the figure full scale using the dimensions shown. Center the drawing in the sheet by using light construction lines until you are sure the drawing is centered.

Sketching Pictorial Views

FIGURE 13–33
Exercise 13–9
Scale $\frac{3}{8}'' = 1''$

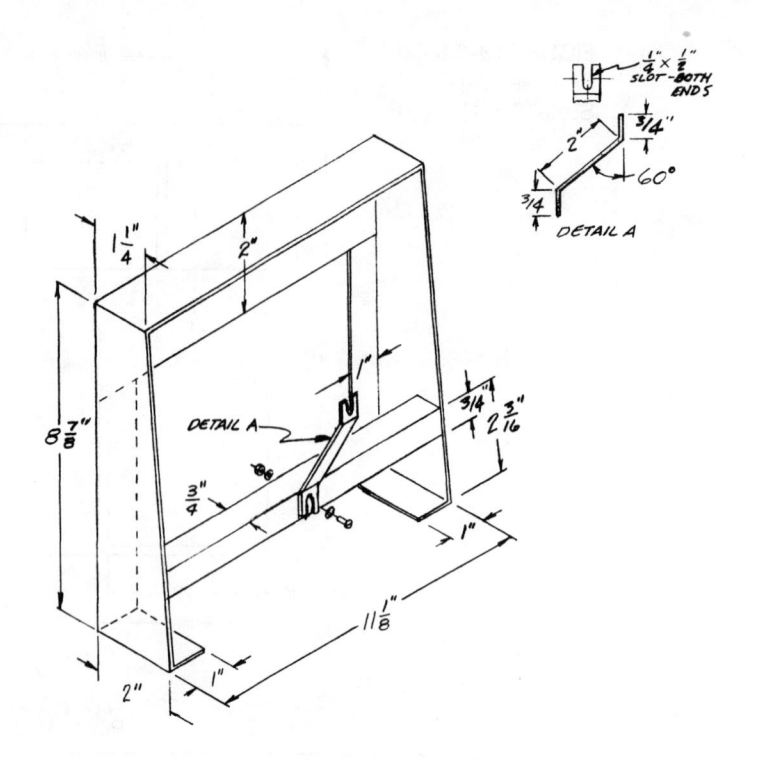

Make the cut with obtuse angle break lines. Notice that the thickness of all three cut walls is the same.
Do not show any dimensions.
Complete the title block with your best lettering and name the drawing CUTAWAY.

FIGURE 13–34
Exercise 13–10
Scale: $\frac{1}{4}'' = 1''$

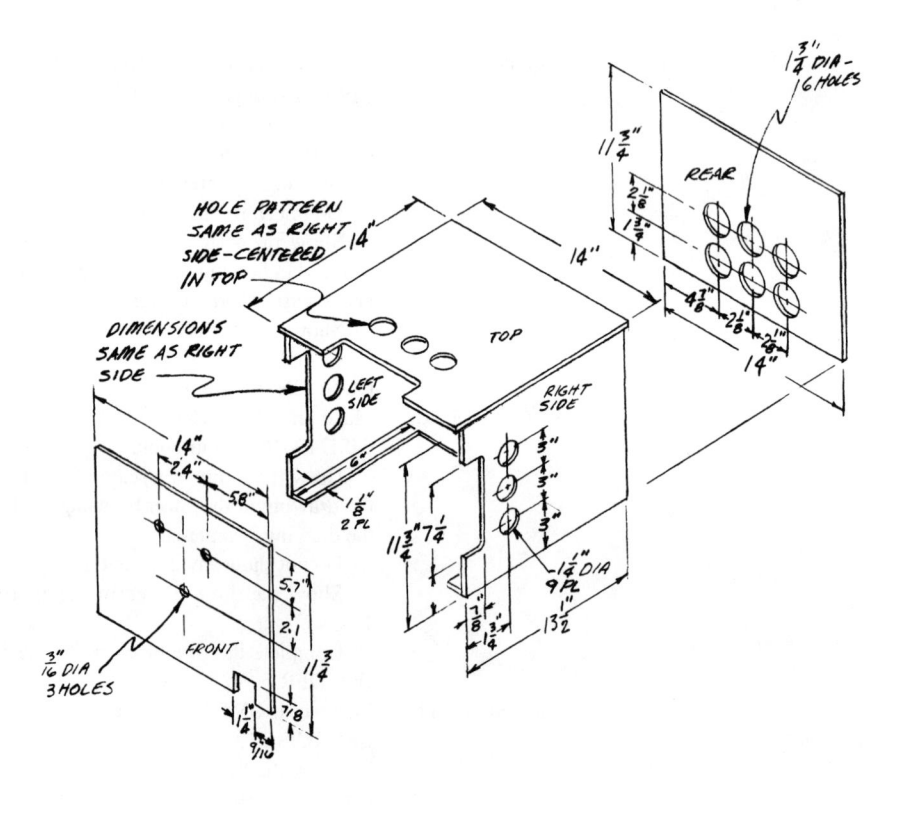

FIGURE 13–35
Exercise 13–11
Full Scale

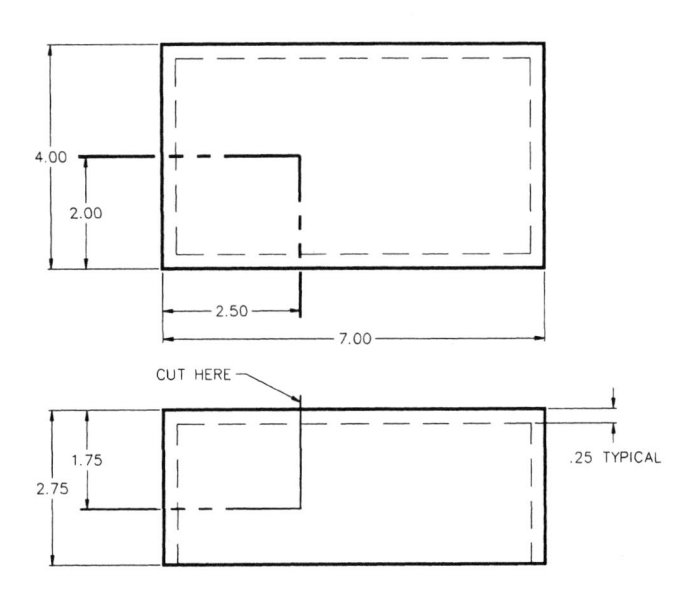

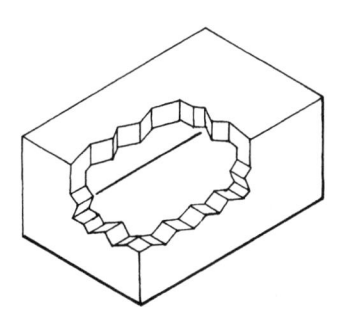

REVIEW QUESTIONS

Circle the best answer.

1. Which of the following is usually the least distorted type of pictorial but the most difficult to draw?
   a. Cabinet oblique
   b. Cavalier oblique
   c. Isometric
   d. Dimetric
   e. Two-point perspective
2. Holes or curves should be placed on receding surfaces in oblique drawing.
   a. True
   b. False
3. The receding oblique surfaces are usually drawn at
   a. 45°
   b. 30°
   c. 60°
   d. 75°
   e. 15°
4. To reduce distortion, receding oblique surfaces are often drawn at half scale.
   a. True
   b. False
5. An isometric gives equal presentations of
   a. Front, top, and right-side views
   b. Front and left-side views
   c. Top and right-side views
   d. Left and right-side views
   e. None of the above

Sketching Pictorial Views

6. On isometric drawings_____ are used.
   a. Isometric measurements
   b. Pictorial measurements
   c. Oblique measurements
   d. True-length measurements
   e. Measurements are not the same on all axes.
7. The three isometric axes are
   a. 30° left, 60° right, and vertical
   b. 30° left, 45° right, and vertical
   c. 30° left, 30° right, and vertical
   d. 60° left, 30° right, and vertical
   e. 45° left, 30° right, and vertical
8. Isometric lines are lines that are parallel to
   a. A level line
   b. The front view
   c. The right-side view
   d. The left-side view
   e. Isometric axes
9. Spheres in isometric drawing appear as
   a. Circles
   b. Ellipses
   c. Straight lines
   d. Irregular curves
   e. It depends on how the sphere is viewed
10. Cylindrical objects in isometric drawing are best drawn by locating centers on
    a. A box around the surface
    b. An outside surface
    c. A centerline or centerlines
    d. The edges of preceding ellipses
    e. Their quadrants
11. Angles are measured in isometric drawing by adding the angle of the isometric axis to the angle to be drawn.
    a. True
    b. False
12. An isometric cutaway drawing is best described as
    a. An isometric sectional view
    b. A perspective
    c. An exploded view
    d. An orthographic sectional view
    e. An oblique view
13. Which of the following should be done first in making an isometric cutaway drawing?
    a. Make a freehand sketch.
    b. Shade the cut surfaces.
    c. Draw the cut surfaces in isometric.
    d. Draw the uncut surfaces in isometric.
    e. Place callouts on the drawing.
14. Which of the following is usually cut on a cutaway?
    a. Cylinder walls
    b. Bearings
    c. Shafts
    d. Spheres
    e. Fasteners

14 Making Isometric Views with AutoCAD

OBJECTIVES

After completing this chapter, you will be able to

☐ Make isometric drawings to scale from two-dimensional drawings using AutoCAD or AutoCAD LT.
☐ Correctly answer questions regarding the following commands and features:
Snap Style
Ellipse
Toggle
Isoplane

ISOMETRIC DRAWING SETTINGS

Isometric drawings can be done quickly and easily using AutoCAD or AutoCAD LT software. Once the Grid and Snap settings are properly made, the drawing itself proceeds with little difficulty. The three isometric axes are 30° left, 30° right, and vertical. All measurements are made full scale on all three axes. AutoCAD has three positions that make drawing in isometric almost foolproof. They are called isoplanes: top, right, and left. You can toggle from one to another by pressing the Ctrl and E keys at the same time or pressing the function key F5.

We will begin isometric drawing with four simple shapes (Figure 14–1):

An isometric box
An isometric cube with an ellipse on each of the three isoplanes
An isometric view of a part with an inclined surface
An isometric cylindrical shape

EXERCISE 14–1
Drawing Four Isometric Shapes Using the AutoCAD Isometric Grid

Step 1. To begin Exercise 14–1, turn on the computer and start AutoCAD or AutoCAD LT.
Step 2. Open drawing EX14-1 supplied on the disk that came with your book.
Step 3. Use Zoom-All to view the entire drawing area.
Step 4. Set the Snap for isometric, as follows.

Prompt	Response
Command:	Type: SN↵
Specify snap spacing or [ON/OFF/Aspect/ Rotate/Style/Type] <0.125>:	Type: S↵

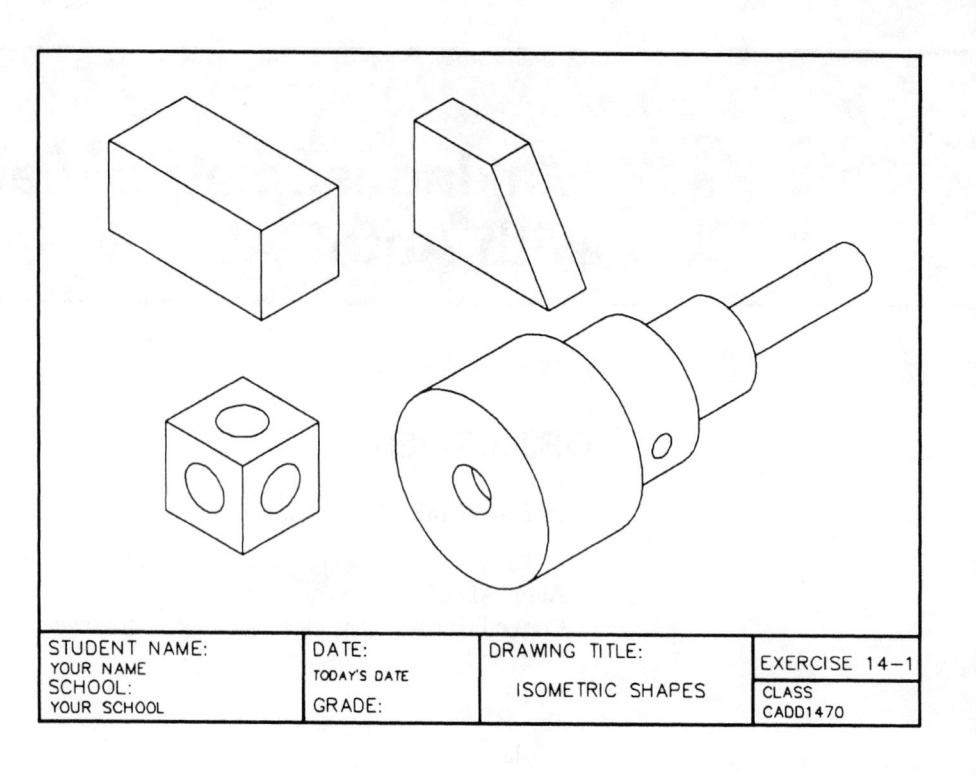

FIGURE 14–1
Exercise 14–1 Completed

| STUDENT NAME: YOUR NAME SCHOOL: YOUR SCHOOL | DATE: TODAY'S DATE GRADE: | DRAWING TITLE: ISOMETRIC SHAPES | EXERCISE 14–1 CLASS CADD1470 |

Prompt	Response
Enter snap grid style [Standard/Isometric] <S>:	Type: **I** ↵
Specify vertical spacing<0.125>:	Type: **.125**↵

    The isometric grid should appear. You may now get control of the plane (or surface) on which you will be working. To do this press the function key F5 so that you toggle between the left, right, and top isometric planes. Now you are ready to draw the shape shown in the upper left corner of Figure 14–1. Follow this sequence:

Draw an isometric rectangle measuring 1″ × 2″ × 1″ (Figure 14–2):

Step 5.   **Draw the left plane of an isometric rectangle measuring 1″ × 2″ × 1″.**

Prompt	Response
Command:	Press: **the Ctrl and E keys at the same time (or the function key F5) until you see** <Isoplane Left>
Command:	Type: **L**↵
Specify first point:	Type: **END**↵
of	Click: **D1** (Figure 14–2)
Specify next point or [Undo]:	Type: **@2<150**↵
Specify next point or [Undo]:	Type: **@1<90**↵
Specify next point or [Close/Undo]:	Type: **@2<330**↵
Specify next point or [Close/Undo]:	↵

You have drawn the left plane. You may leave the isometric plane the same to finish the rectangle, or you may toggle to the right plane.

Step 6.   **Draw the right plane.**

Prompt	Response
Command:	Type: **L**↵

FIGURE 14-2

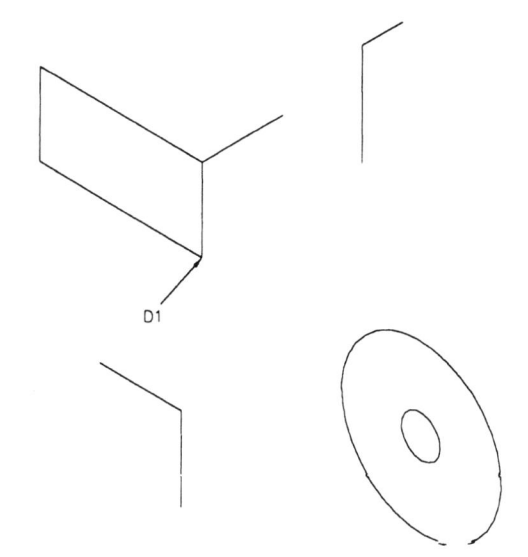

Specify first point:	Type: **END**↵
of	Click: **D1** (Figure 14-3)
Specify next point or [Undo]:	Type: **@1<30**↵
Specify next point or [Undo]:	Type: **@1<90**↵
Specify next point or [Close/Undo]:	↵

Step 7.   Draw the top plane.

Prompt	Response
Command:	↵
Specify first point:	Click: **D2** (Figure 14-3)
Specify next point or [Undo]:	Type: **@2<150**↵
Specify next point or [Undo]:	Type: **@1<210**↵
Specify next point or [Close/Undo]:	↵

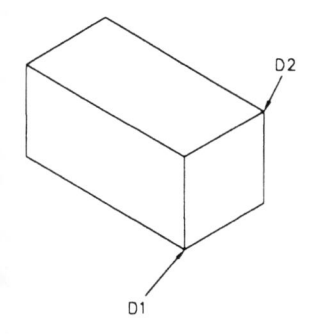

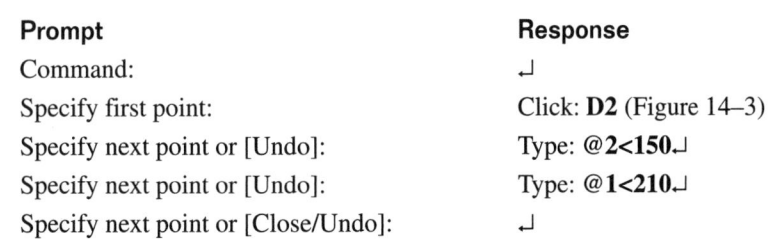

FIGURE 14-3

You have now drawn your first isometric rectangle using AutoCAD.

The next part of the exercise is to draw a 1" isometric cube with a .5"-diameter ellipse on each surface. This will introduce you to the standard isometric planes with which you must be very familiar to be successful with isometric drawing.

Draw a 1" isometric cube with a 1" ellipse on each surface:

Step 8.   Draw the right plane.

Use Ctrl-E or Press: **F5** to toggle to the right isoplane:

Prompt	Response
Command:	Type: **L**↵
Specify first point:	Type: **END**↵
of	Click: **D1** (Figure 14-4)
Specify next point or [Undo]:	Type: **@1<210**↵
Specify next point or [Undo]:	Type: **@1<90**↵
Specify next point or [Close/Undo]:	Type: **@1<30**↵
Specify next point or [Close/Undo]:	↵

Step 9.   Draw the top plane and the left plane.

Now do something a little different.

Making Isometric Views with AutoCAD

FIGURE 14-4

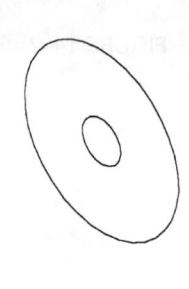

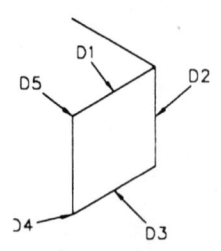

FIGURE 14-5

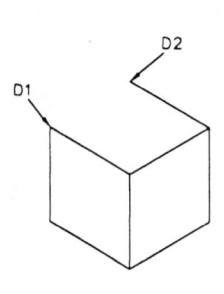

FIGURE 14-6

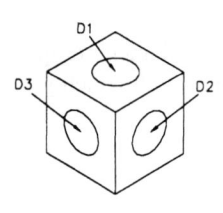

FIGURE 14-7

Prompt	Response
Command:	Type: **MI**↵
Select objects:	Click: **D1, D2, D3** (Figure 14–5)
Select objects:	↵
Specify first point of mirror line:	Click: **D4**
Specify second point of mirror line:	Click: **D5**
Delete source objects? [Yes/No] <No>:	↵
Command:	Type: **L**↵
Specify first point:	Click: **D1** (Figure 14–6)
Specify next point or [Undo]:	Click: **D2**
Specify next point or [Undo]:	↵

Now you are ready to draw ellipses in these cube surfaces.

Use Ctrl-E (or the function key F5) to toggle to the top isoplane:

Step 10. Draw ellipses.

Prompt	Response
Command:	Type: **EL**↵
Specify axis endpoint of ellipse or [Arc/Center/Isocircle]:	Type: **I**↵
Specify center of isocircle:	Click: **D1** (Figure 14–7)
Specify radius of isocircle or [Diameter]:	Type: **.25** ↵ (The default is radius.)

Now use Ctrl-E (or the function key F5) to toggle to the right isometric plane:

Prompt	Response
Command:	↵
Specify axis endpoint of ellipse or [Arc/Center/Isocircle]:	Type: **I**↵
Specify center of isocircle:	Click: **D2**
Specify radius of isocircle or [Diameter]:	Type: **.25** ↵

Now toggle to the left isometric plane:

Prompt	Response
Command:	↵
Specify axis endpoint of ellipse or [Arc/Center/Isocircle]:	Type: **I**↵
Specify center of isocircle:	Click: **D3**
Specify radius of isocircle or [Diameter]:	Type: **.25** ↵

Take a few minutes to study the position of the ellipses on the isometric planes. These positions are the same for all normal (perpendicular) surfaces and must not be rotated in

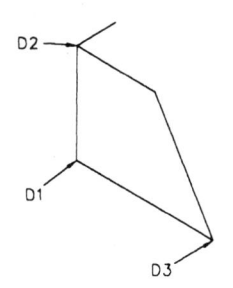

FIGURE 14–8

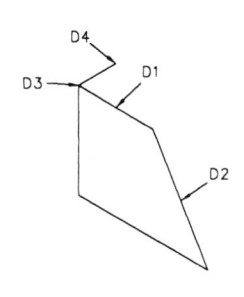

FIGURE 14–9

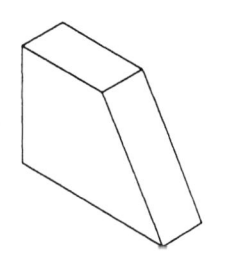

FIGURE 14–10

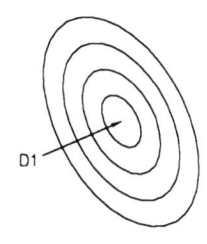

FIGURE 14–11

any direction. Remember where Figure 14–7 is in case you need to return to it to refresh your memory later.

Now draw a shape that has an inclined surface (Figure 14–1, the shape at the top right):

Be sure that Snap is ON, or use Osnap-Endpoint to click points.

Step 11. Use the Line command to draw the right surface of the shape.

Prompt	Response
Command:	Type: **L**↵
Specify first point:	Click: **D1** (Figure 14–8)
Specify next point or [Undo]:	Type: **@1.75<-30**↵
Specify next point or [Undo]:	↵
Command:	↵
Specify first point:	Click: **D2** (Figure 14–8)
Specify next point or [Undo]:	Type: **@1<-30**↵
Specify next point or [Undo]:	Click: **D3**
Specify next point or [Close/Undo]:	↵

Step 12. Use the Copy command to draw the two lines to show the thickness of the part, and use the Line command to connect the two planes.

Prompt	Response
Command:	Type: **CP**↵
Select objects:	Click: **D1 and D2** (Figure 14–9)
Select objects:	↵
Specify base point or displacement or [Multiple]:	Click: **D3**
Specify second point of displacement or <use first point as displacement>:	Click: **D4**

Step 13. On your own: Draw lines connecting the two planes as shown in Figure 14–10.

The final figure in this exercise is a cylindrical shape having four different size cylinders and two holes.

Toggle to the left isoplane:

Step 14. Draw two ellipses in the same location as the two existing ellipses.

Prompt	Response
Command:	Type: **EL**↵
Specify axis endpoint of ellipse or [Arc/Center/Isocircle]:	Type: **I**↵
Specify center of isocircle:	Click: **D1** (Figure 14–11)
Specify radius of isocircle or [Diameter]:	Type: **.75**↵ (The default is radius.)
Command:	↵
Specify axis endpoint of ellipse or [Arc/Center/Isocircle]:	Type: **I**↵
Specify center of isocircle:	Click: **D1** (again)
Specify radius of isocircle or [Diameter]:	Type: **.5**↵

Step 15. Use the Copy and Move commands to draw ellipses in their correct locations.

Making Isometric Views with AutoCAD

FIGURE 14–12

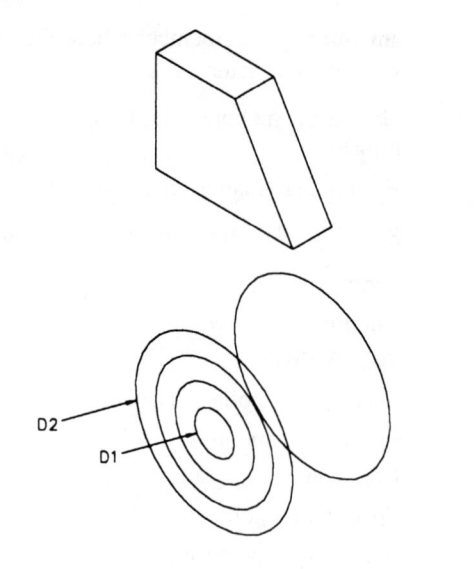

Prompt	Response
Command:	Type: **CP**↵
Select objects:	Click: **D1** (Figure 14–12)
Select objects:	↵
Specify base point or displacement or [Multiple]:	Click: **any point**
Specify second point of displacement or <use first point as displacement>:	Type: **@5<30**↵
Command:	↵
Select objects:	Click: **D2** (Figure 14–12)
Select objects:	↵
Specify base point or displacement or [Multiple]:	Click: **any point**
Specify second point of displacement or <use first point as displacement>:	Type: **@1.25<30**↵
Command:	Type: **M**↵

FIGURE 14–13

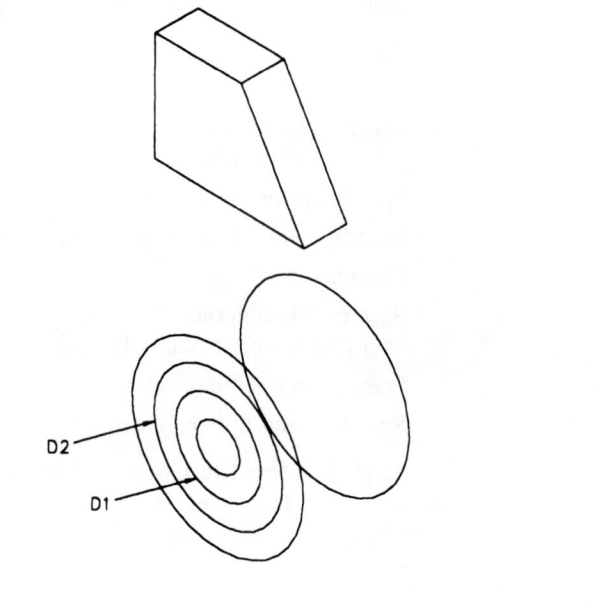

228   Chapter 14

FIGURE 14-14

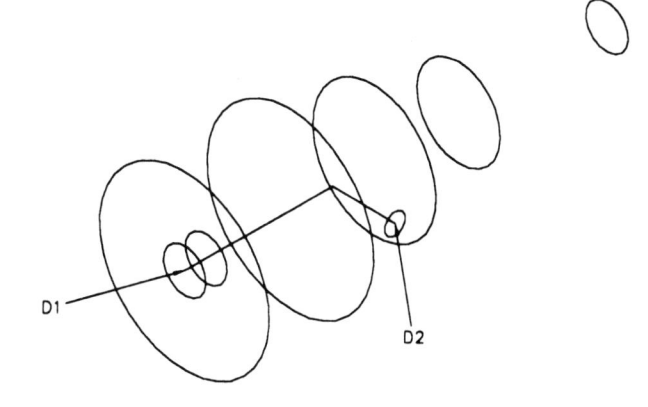

Select objects:	Click: **D1** (Figure 14–13)
Select objects:	↵
Specify base point or displacement or [Multiple]:	Click: **any point**
Specify second point of displacement or <use first point as displacement>:	Type: **@3.25<30**↵
Command:	↵
Select objects:	Click: **D2** (Figure 14–13)
Select objects:	↵
Specify base point or displacement or [Multiple]:	Click: **any point**
Specify second point of displacement or <use first point as displacement>:	Type: **@2.25<30**↵

Step 16. Use the Line command to draw a construction line to locate the small hole at right angles to the cylinder and draw it using the Ellipse command.

Prompt	Response
Command:	Type: **L**↵
Specify first point:	Click: **D1** (Figure 14–14)
Specify next point or [Undo]:	Type: **@1.75<30**↵
Specify next point or [Undo]:	Type: **@.75<330**↵
Specify next point or [Close/Undo]:	↵

Use Ctrl-F (or the control key F5) to toggle to the right isoplane:

Prompt	Response
Command:	Type: **EL**↵
Specify axis endpoint of ellipse or [Arc/Center/Isocircle]:	Type: **I**↵
Specify center of isocircle:	Click: **D2** (Figure 14–14)
Specify radius of isocircle or [Diameter]:	Type: **.125**↵

Step 17. On your own:
Use the Line command with Ortho ON to draw lines tangent to the ellipses as shown in Figure 14–15. Use Osnap-Tangent, -Endpoint, or -Quadrant to pick tangent points. (One of the three should work. If one of these three osnap modes doesn't work, use Nearest as a last resort.)

Allow lines on the three smaller ellipses to extend past the larger ellipses when you pick the second point of the line, and use the Trim command to get rid of the excess.

Making Isometric Views with AutoCAD

FIGURE 14–15

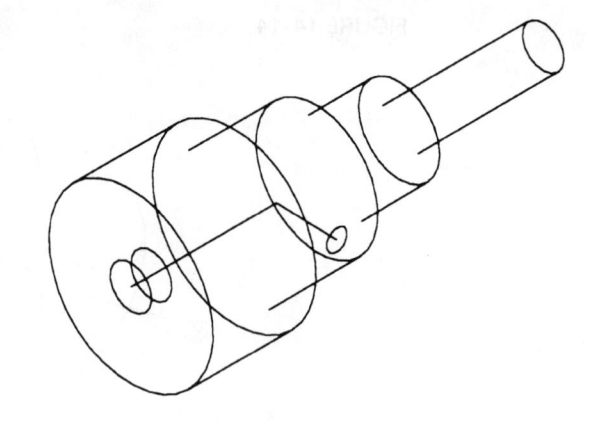

FIGURE 14–16

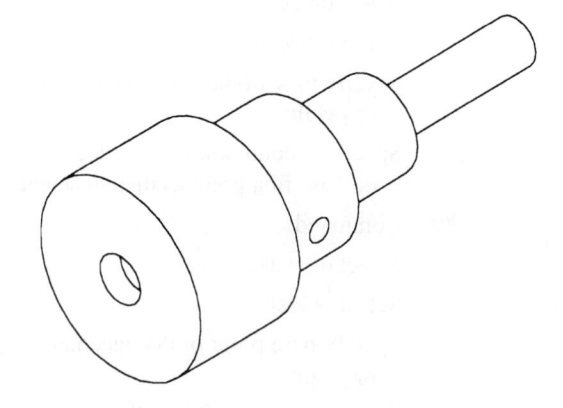

Use the Trim command to trim the unnecessary parts of the ellipses so the drawing appears as shown in Figure 14–16. Click the straight tangent lines as the cutting edges, press ↵, then click the unneeded back halves of the ellipses as the objects to trim. Click the small hole on the front surface as the cutting edge, press ↵, then click the ellipse forming the bottom of the hole as the object to trim.

Use the Erase command to erase all construction lines.

Step 18. Use the Dtext command to complete the title block as you did for previous exercises. **Name the drawing ISOMETRIC SHAPES.**

Step 19. **Use the SAVEAS command to save your drawing as EX14-1(your initials) on a floppy disk and again on the hard drive of your computer.**

Step 20. **Plot or print your drawing full size on an 11″ × 8½″ sheet.**

EXERCISE

EXERCISE 14–1   Complete Exercise 14–1 using steps 1 through 20 described in this chapter.

REVIEW QUESTIONS

It is suggested that you complete this test at the computer after completing Exercise 14–1. Circle the correct answer.

1. The isometric grid is obtained from which of the following commands?
    a. Elev
    b. Line
    c. Snap
    d. 3-D
    e. Grid

2. Which of the following is used to toggle from one isoplane to another?
   a. F3
   b. F5
   c. F8
   d. F7
   e. F9
3. When a rectangle is being drawn, what happens when the C key is struck as a response to the prompt: "to point"?
   a. The rectangle is copied.
   b. The rectangle is changed to layer C.
   c. The rectangle is opened.
   d. The rectangle is closed.
   e. The rectangle is erased.
4. In response to the prompt: "Line: Specify first point," which of the following starts a line from the endpoint of the last line drawn?
   a. ↵
   b. F6
   c. Ctrl-E
   d. REPEAT
   e. Ctrl-C
5. An ellipse that appears in the following position would have been drawn in?
   a. Left isoplane
   b. Right isoplane
   c. Top isoplane
   d. Either the top or the left isoplane
   e. None of the above
6. An ellipse that appears in the following position would have been drawn in?
   a. Left isoplane
   b. Right isoplane
   c. Top isoplane
   d. Either the top or the left isoplane
   e. None of the above
7. An ellipse that appears in the following position would have been drawn in?
   a. Left isoplane
   b. Right isoplane
   c. Top isoplane
   d. Either the top or the left isoplane
   e. None of the above
8. Which of the following is the same as $-30°$?
   a. 60°
   b. 150°
   c. 210°
   d. 330°
   e. 270°
9. Which of the following commands has a feature labeled "multiple"?
   a. Copy
   b. Change
   c. Line
   d. Array
   e. Trim
10. Which of the following isoplanes will not allow vertical lines to be drawn with a mouse when Ortho is ON?
    a. Left
    b. Right
    c. Top
    d. All will allow vertical lines to be drawn.
    e. None will allow vertical lines to be drawn.

11. What response must be made to the "SNAP-style" prompt in order for an isometric grid to appear?

12. Which function key is used to toggle between isoplanes?

13. Which two keys are used to toggle from one isoplane to another?

14. How are the three isoplanes labeled?

15. Write the correct syntax (letters and numbers) to draw a line 5.25″ long at an angle 30° upward to the right.

16. Write the correct sequence of keystrokes to draw the right side of the isometric rectangle. Start at the lower left corner and draw the first line upward at a 30° angle to the right.

2.50
4.00

17. Which command is used to produce a mirror image of an object?

15 Sketching Dimensions

OBJECTIVES

After completing this chapter, you will be able to

☐ Correctly sketch dimension lines, extension lines, centerlines, arrowheads, and leaders.
☐ Correctly dimension rectangles, angles, circles, and arcs.
☐ Use the unidirectional and aligned systems for dimensioning.
☐ Correctly select and place dimensions.
☐ Dimension hole patterns for accurate manufacturing.
☐ Use current symbols for diameter and radius.

INTRODUCTION

Most engineering and architectural drawings are used for manufacturing, assembling, or constructing a product. The person who builds or assembles the product must know not only the shape of the object but also its exact size, where features are located, what materials are to be used, and how the product is to be finished.

Although the CAD operator should know basic manufacturing and construction processes so that correct instructions can be issued to manufacturing people through dimensions and notes, most of the problems of basic dimensioning are covered by a few simple rules. You will have little difficulty if you follow these rules carefully and apply them knowing that someone must actually make the object from your drawing or sketch.

STANDARD DIMENSIONING PRACTICES

Each part of the dimensioning process has specific rules governing it. These parts include:

Lines, symbols, and abbreviations
Size and location dimensions
Drawing to scale
Placement of dimensions
Overdimensioning
Aligned and unidirectional systems of dimensioning
Dimensioning features
Notes
Datum dimensioning
Tabular dimensioning

Lines, Symbols, and Abbreviations

The four types of lines used in dimensioning are the extension line, dimension line, centerline, and the leader. All four lines are drawn thin and dark.

FIGURE 15–1
Extension and Dimension Lines

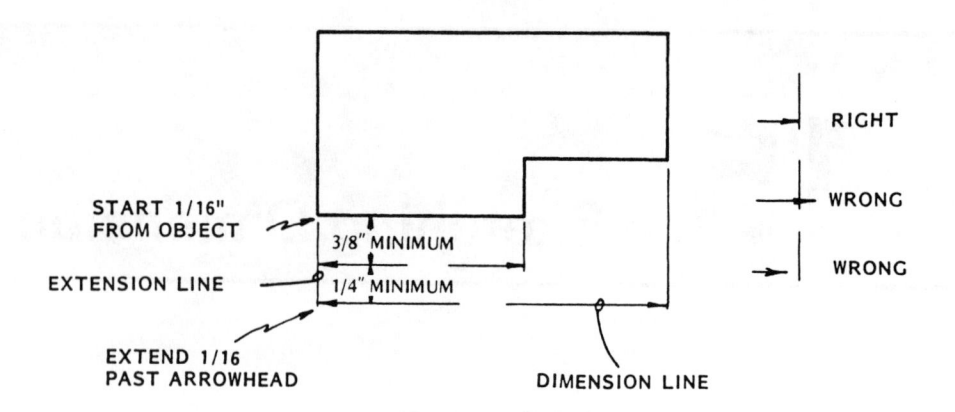

The symbols for diameter and radius are described in this chapter, as are several of the most commonly used abbreviations, namely, PL, TYP, CSK, C'BORE, and S'FACE.

Extension Line

The extension line (Figure 15–1) extends from the object, with a gap of about $\frac{1}{16}''$ next to the object, and goes to about $\frac{1}{16}''$ beyond the arrowhead. Leave a gap where extension lines cross other extension lines or dimension lines.

Dimension Line

The dimension line (Figure 15–1) has an arrowhead or other symbol such as a slash, called a *tick mark*, at each end to show where the measurement is made. The arrowhead just touches the extension line. The arrowhead is approximately three times as long as it is wide. The wings of the arrowhead are straight and very close to the shaft. A gap is left near the middle for the dimension text. In architectural drawings the dimension line is often unbroken and the text is placed above a horizontal dimension or to the left of a vertical one. On small drawings such as the ones in this chapter, dimension lines are spaced $\frac{1}{2}''$ from the object and $\frac{3}{8}''$ apart. The minimum of $\frac{3}{8}''$ and $\frac{1}{4}''$ shown in Figure 15–1 is used where space is very limited. On larger drawings the spacing can be greater, but the first dimension is always placed farther from the object lines of the drawing than the spacing between dimensions.

Centerline

Centerlines (Figure 15–2) are used to show the centers of symmetrical features and are used in place of extension lines for locating holes and other round features. Make centerlines end about $\frac{1}{16}''$ to $\frac{1}{8}''$ outside the hole or feature.

Leader

A leader (Figure 15–3) is a thin solid line that leads from a note or dimension and ends with an arrowhead touching the part.

Leaders are straight, inclined lines (never vertical or horizontal) that are usually drawn at 45°, 60°, or 30° angles but may be drawn at any convenient angle. A short horizontal

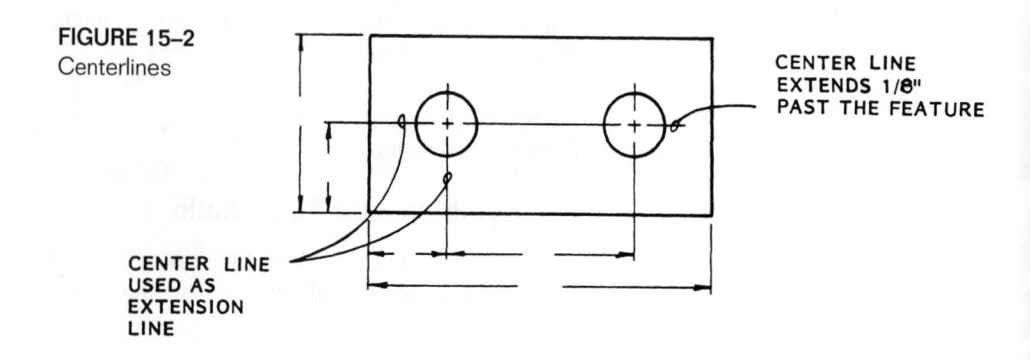

FIGURE 15–2
Centerlines

FIGURE 15-3
Leaders

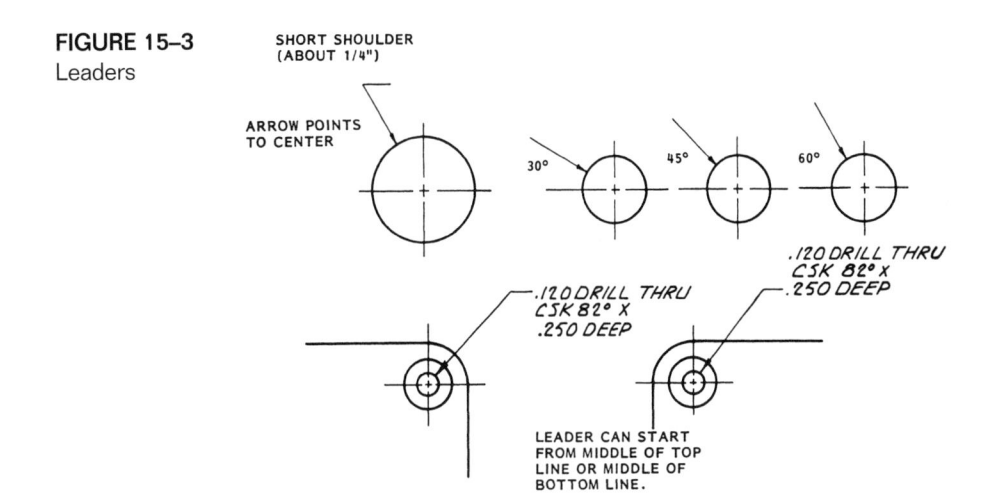

shoulder extends out from midheight of the lettering of the note accompanying the leader. Leaders may extend from either the center of the beginning line or the ending line of the note, not from the inside lines.

Symbols and Abbreviations

The following standard symbols and abbreviations are used in this chapter:

ϕ	Diameter
R	Radius
PL	Places (as in "the dimension occurs in four places")
TYP	Typical (as in "this thickness is the same or typical throughout the part")
CSK	Countersink
C'BORE	Counterbore
S'FACE	Spotface
DP	Deep

Size and Location Dimensions

There are two types of dimensions: those that show the size of a feature and those that show the location of the feature (Figure 15–4). Location dimensions are taken from a machined (finished) edge or the center of a hole whenever possible.

FIGURE 15-4
Size and Location Dimensions

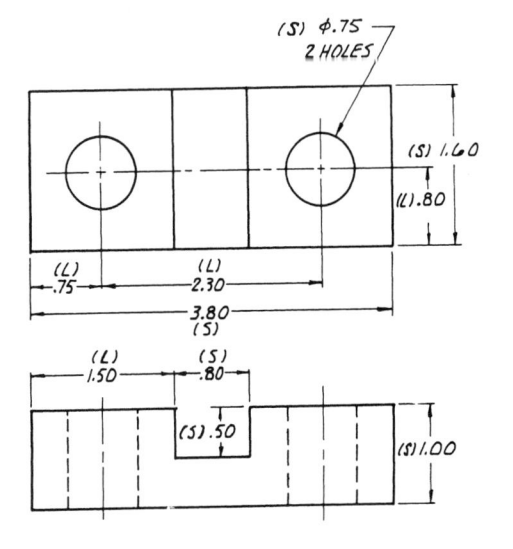

Sketching Dimensions

235

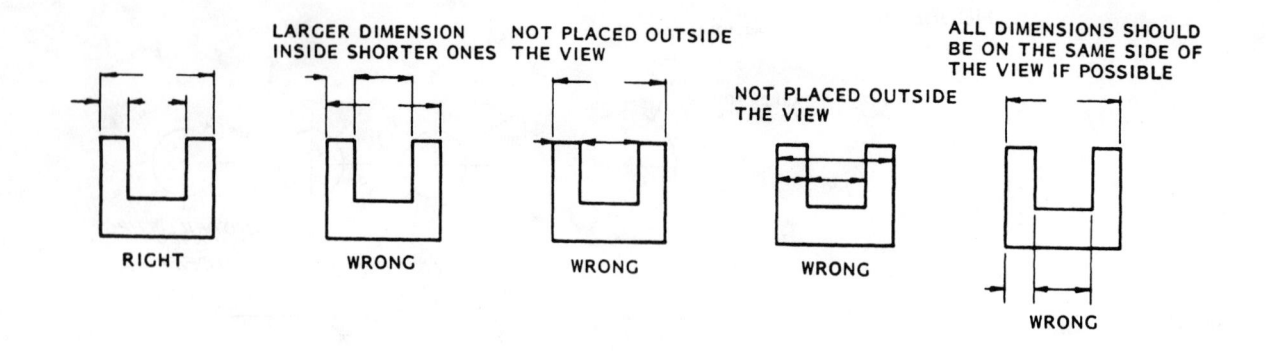

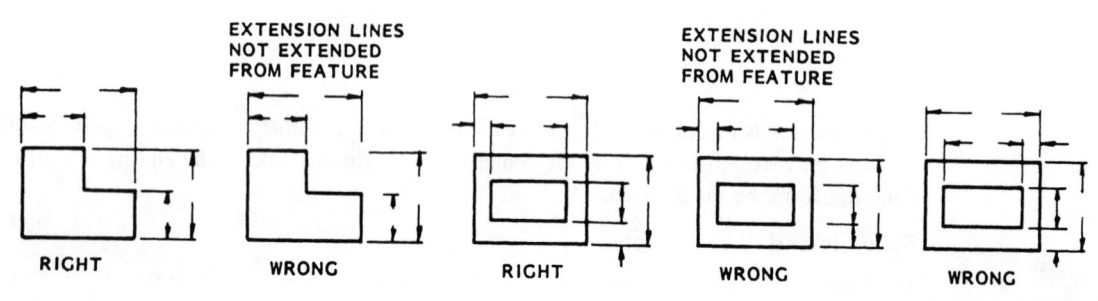

FIGURE 15–5
Correct and Incorrect Placement of Dimensions

Drawing to a Scale

Drawings and sketches are often made to a scale. This scale is shown on the drawing, usually in the title block. If a minor change is to be made in one of the dimensions, a wavy line may be placed under the dimension to show the dimension is not to scale. Regardless of the scale used, dimensions are always those of the part to be manufactured, not the size of that part on the drawing.

Placement of Dimensions

Examples of correct and incorrect placement of dimension lines are shown in Figure 15–5. The shortest dimensions are closest to the object outline.

Dimension lines should not cross extension lines, which happens when the shorter dimensions are placed outside. (This is sometimes unavoidable, however.) Be aware that it is acceptable for extension lines to cross each other.

A dimension line should never coincide with or form a continuation of any line of the drawing.

Avoid crossing dimension lines with other dimension lines whenever possible. In general, avoid dimensioning to hidden lines (Figure 15–6). Place dimensions where the feature is seen as a solid line. (A section drawing may be necessary.)

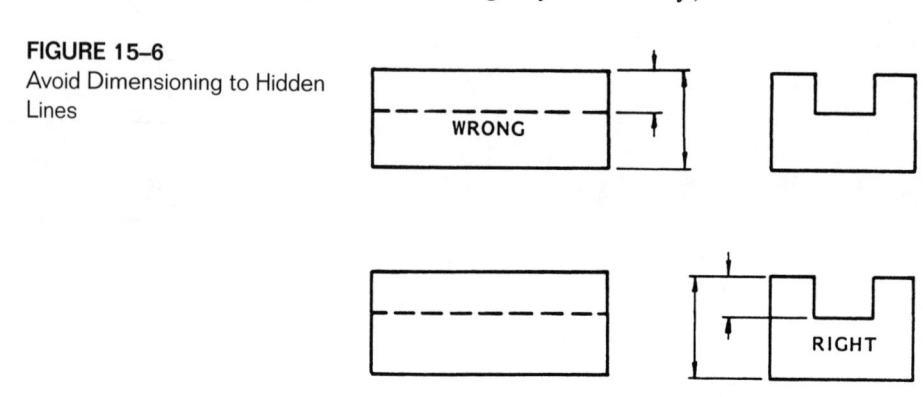

FIGURE 15–6
Avoid Dimensioning to Hidden Lines

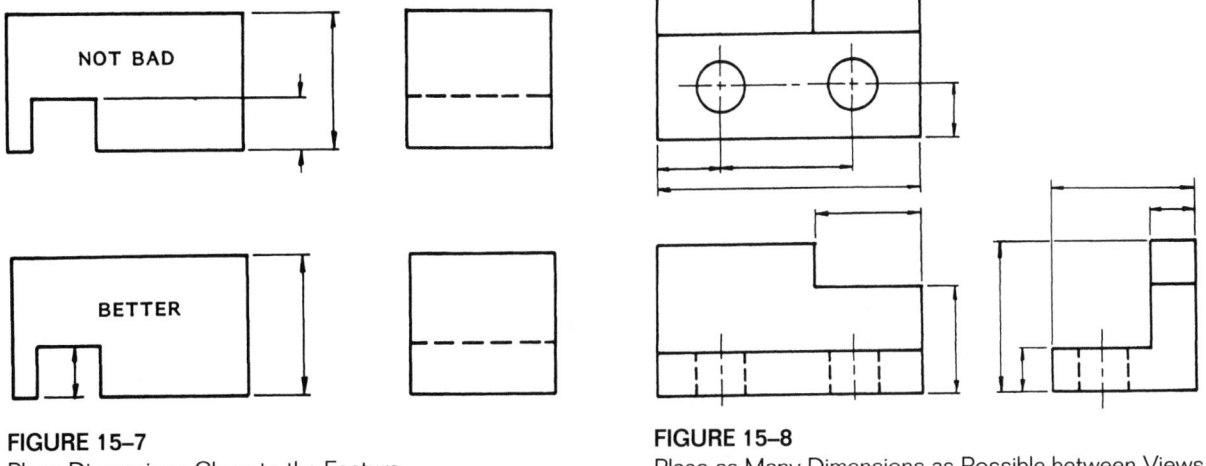

FIGURE 15–7
Place Dimensions Close to the Feature

FIGURE 15–8
Place as Many Dimensions as Possible between Views

If possible, place dimensions in close relationship to the feature shown (Figure 15–7). Place as many of the dimensions as possible between the views and on the view that shows the shape of the feature best (Figure 15–8).

If the drawing contains fractional dimensions that run into one another, stagger the dimensions to place one above the other so that both are easily read.

Overdimensioning

The same dimension should never be shown twice on the same drawing, and tolerance buildup should be avoided (Figure 15–9). Tolerances are the variations allowed in all

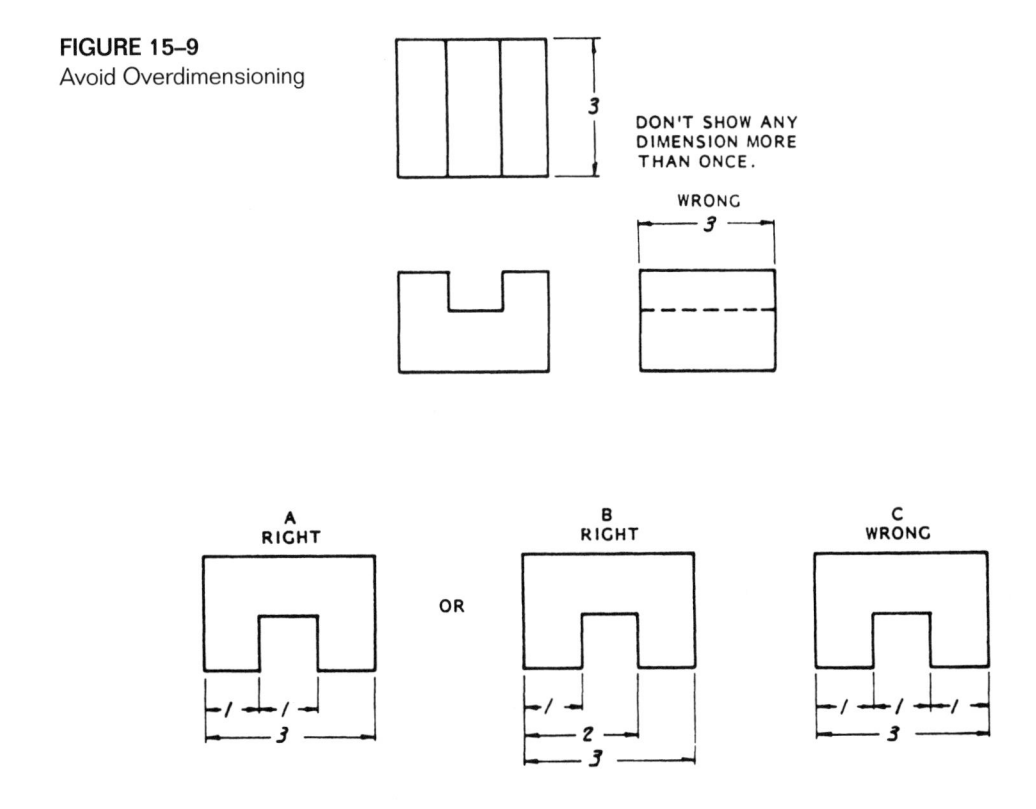

FIGURE 15–9
Avoid Overdimensioning

DON'T SHOW ANY DIMENSION MORE THAN ONCE.

WRONG

A RIGHT
OR
B RIGHT
C WRONG

WHY? ASSUME THAT ALL DIMENSIONS ON THE PART CAN BE 1/16" TOO BIG AND THE PART WILL STILL FUNCTION CORRECTLY. A AND B HAVE NO DIMENSIONS THAT ARE TOO BIG. C IS WRONG BECAUSE THE THREE 1" DIMENSIONS COULD RESULT IN THE PART BEING 3/16" TOO BIG. THE 3" DIMENSION IS IN CONFLICT WITH THE THREE 1" DIMENSIONS.

Sketching Dimensions

FIGURE 15–10
Unidirectional System

FIGURE 15–11
Aligned System

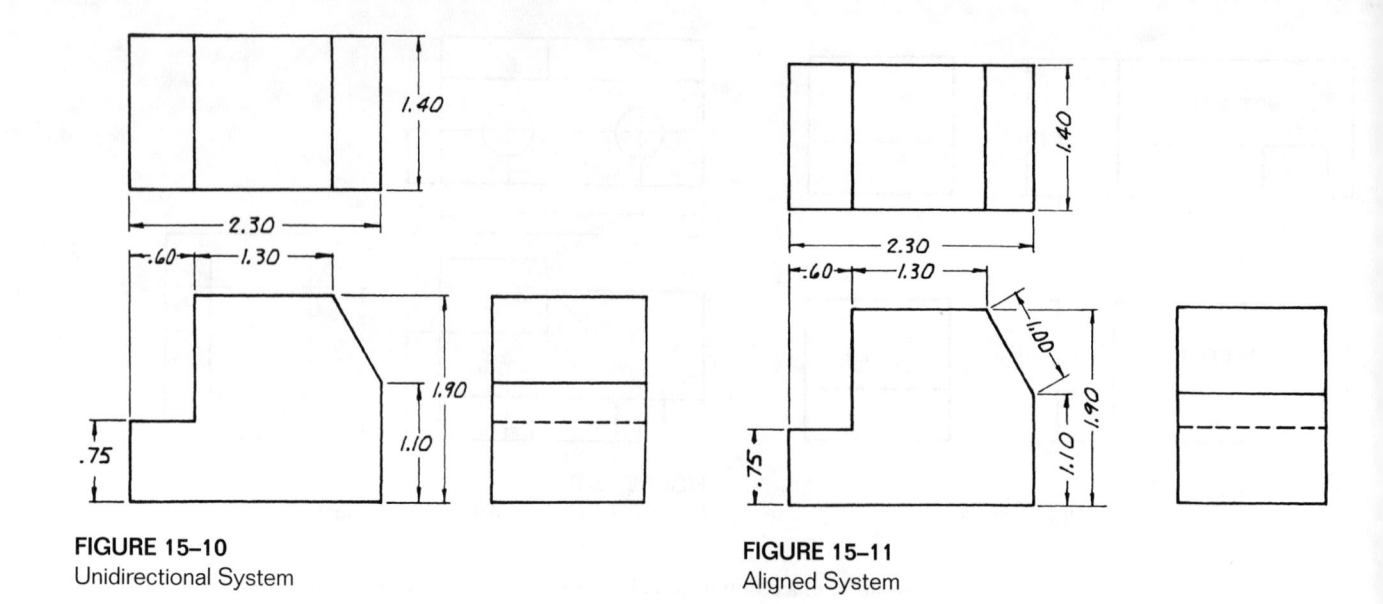

dimensions of a manufactured part. If a part has a tolerance of ±.005, a 1.000″ dimension could measure 1.005 to .995 and still be accepted by an inspector. Figure 15–9 shows how tolerances can accumulate.

Unidirectional and Aligned Systems of Dimensioning

Both the aligned and the unidirectional systems of dimensioning have been widely used in industry. Each system has advantages and disadvantages, but many manufacturing companies use the unidirectional system.

Unidirectional System

In the unidirectional system (Figure 15–10), the dimensions are placed to read from the bottom of the drawing, and fraction bars (if any) are parallel with the bottom of the drawing.

Aligned System

In the aligned system of dimensioning (Figure 15–11), the dimensions are placed parallel to the lines of the drawing, and they are read from the bottom or right side of the drawing.

Dimensions and notes with leaders are aligned with the bottom of the drawing (placed in a horizontal position) in both systems.

Dimensioning Features

Angles

Angles are dimensioned in degrees (Figure 15–12). All the methods of dimensioning angles shown in the figure are acceptable. You will find occasion to use all of them if you dimension many angles.

Circles

There are two common types of circular features: holes and cylinders (Figure 15–13). Holes are dimensioned on the view in which the hole is seen as a circle. Cylinders are dimensioned on the view where the cylinder is seen as a rectangle.

A circular center line (called a bolt circle) is used to locate holes from the center of a cylindrical piece. The words DRILL, REAM, PUNCH, and other similar process notes are not normally used because they tell the machinist which process to use to make the hole. Many companies prefer to tell the machinist what size the feature should be and let that person select the process. The people who manufacture the part are usually in a much better position to know the best process to use.

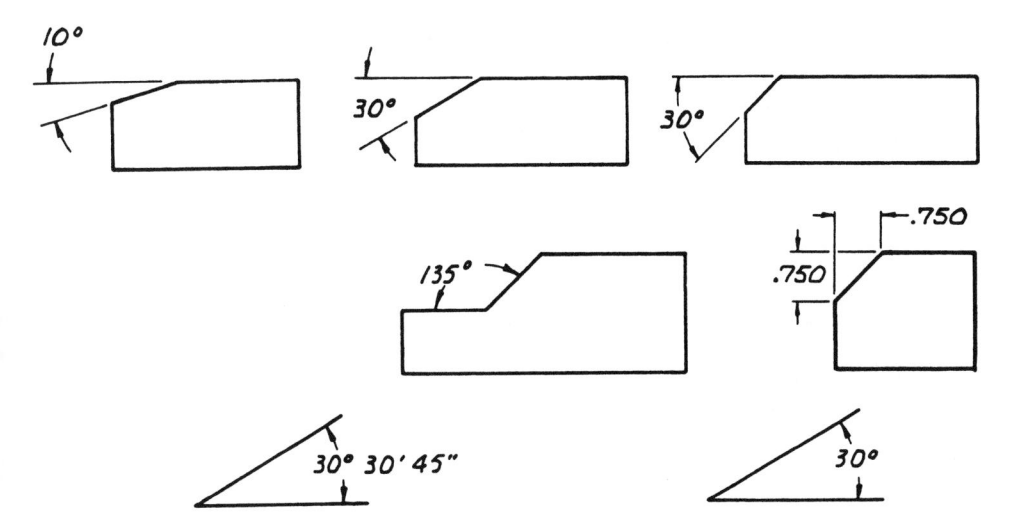

FIGURE 15–12
Dimensioning Angles

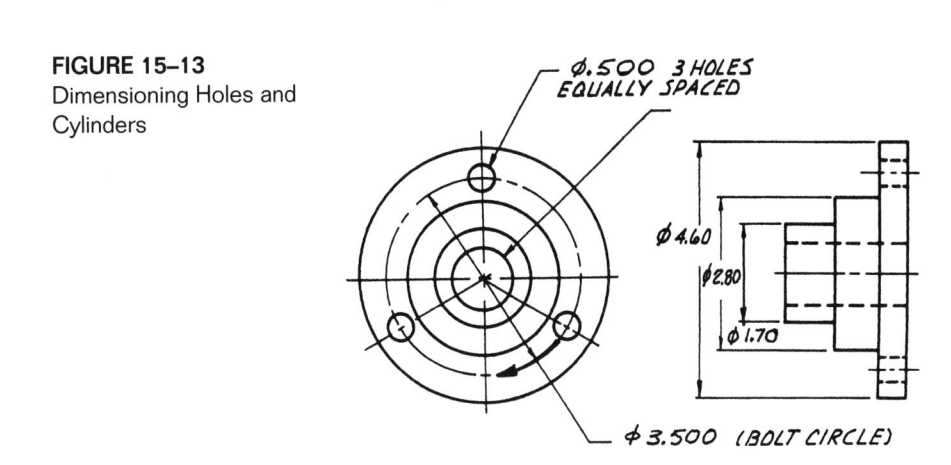

FIGURE 15–13
Dimensioning Holes and Cylinders

Radius

A radius is dimensioned in the view in which its true shape is seen (Figure 15–14). The abbreviation R is used for radius. If a large radius cannot be located because of space, a false center can be located.

Countersink, Counterbore, and Spotface

The countersink, counterbore, and spotface features are dimensioned by use of a note and leader (Figure 15–15).

The countersink can be dimensioned by using an included angle between opposite faces of a diameter, or the diameter of the countersink and its depth, but not both.

The counterbore shows the diameter of the primary hole, the diameter of the counterbore, and the depth of the counterbore. The depth of the counterbore is necessary in all cases.

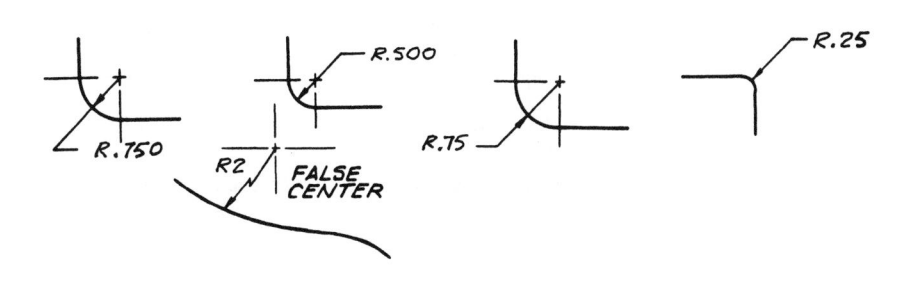

FIGURE 15–14
Dimensioning Radii

Sketching Dimensions

FIGURE 15–15
Dimensioning Countersink, Counterbore, and Spotface

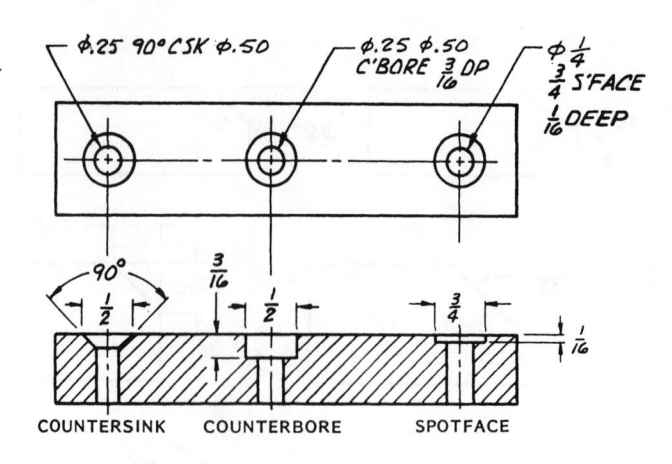

The spotface shows the diameter of the primary hole, the diameter of the spotface, and its depth. The depth of the spotface is made just deep enough to provide a smooth surface for a bolt head, nut, or similar fasteners.

Internal and External Surfaces with Rounded Ends

Slots can be punched (if the metal is thin enough) or drilled. Figure 15–16 shows dimensioning practices for slots and for external surfaces with rounded ends. The overall dimension at C is an overall dimension that is not really necessary. It can be shown as a reference dimension, which is a means of telling the person who will make the part the approximate length. The reference dimension, which is not used in measuring the part for accuracy, is followed by the abbreviation REF.

Special Features

Figure 15–17 shows the methods used to dimension several special features. When you encounter a feature that appears to need a special method of dimensioning, look up the method in this text or some other reference such as ASMEY14.5M1994 Dimensioning and Tolerancing Standard. Do not guess about how to dimension features, especially as a beginner.

Notes

Two types of notes are shown on drawings: general and specific. General notes pertain to the whole drawing, and specific notes apply to one feature on the drawing. These notes

FIGURE 15–16
Internal and External Features with Rounded Ends

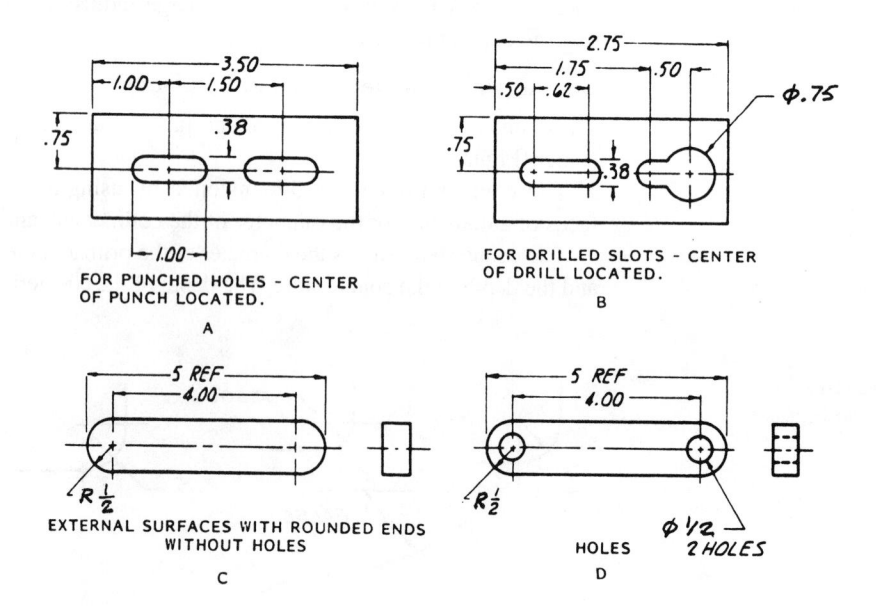

240

Chapter 15

FIGURE 15–17
Special Features

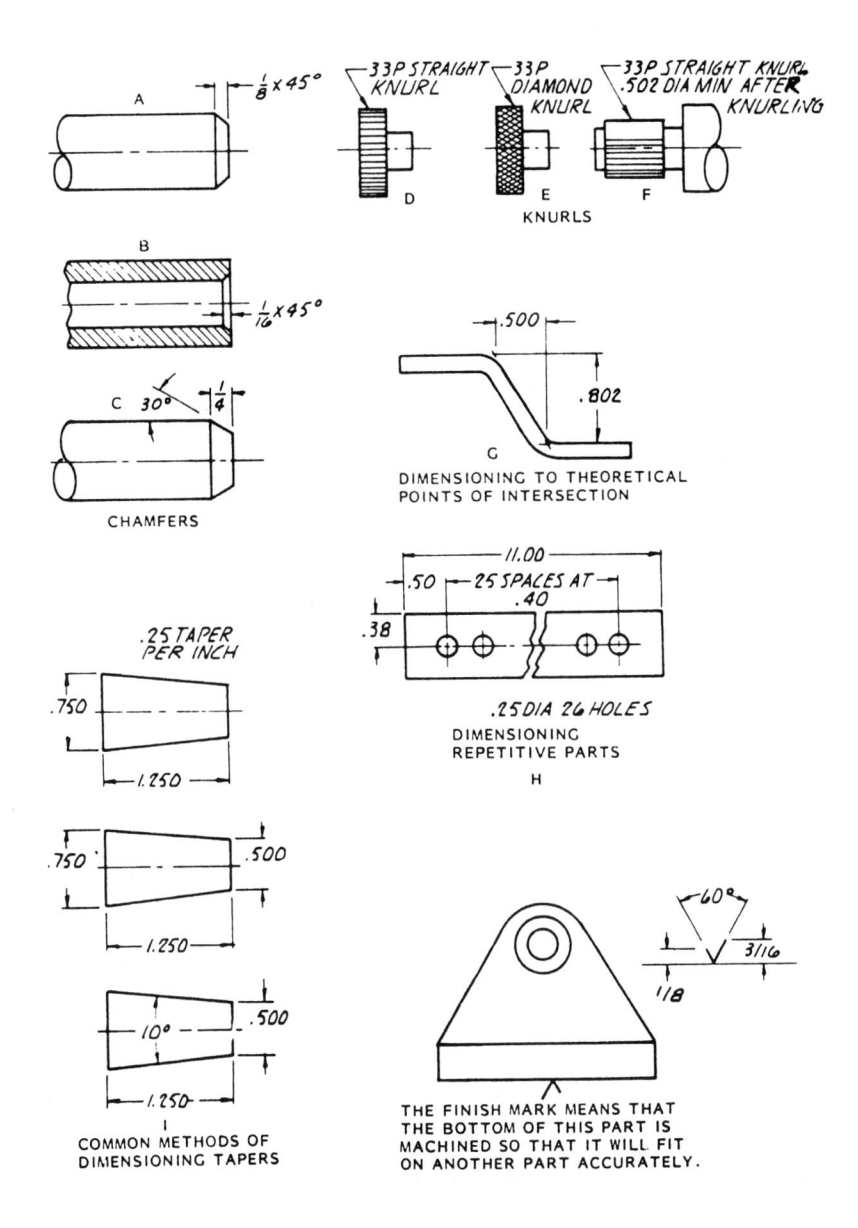

are usually located on the drawing in a specific place that is preprinted with the drawing format. If there is no specific place for notes, they can be placed anywhere on the field of the drawing where they will be uncrowded and can be read easily. Figure 15–18 shows examples of specific and general notes.

SPECIFIC:
.125 DIA – 6 HOLES
.062R – 4 PLACES
MARK PART NUMBER WITH PERMANENT BLACK INK.
WELD ALL AROUND TWO CORNERS.
4 RIBS EQUALLY SPACED
3 SCREWS EVENLY SPACED

CLEARANCE CUT IN UPPER FLANK OF CHANNEL TO FIT AT TIME OF INSTALLATION.
TERMINAL NUMBERING FOR SWITCH SL IS FOR REFERENCE ONLY.
TOP OF PIECE IS TO PARALLEL WITHIN .005 OF THE SHIP MOUNT.
PARTIAL REFERENCE DESIGNATIONS ARE SHOWN PREFIX THE PART DESIGNATION WITH SUBASSEMBLY DESIGNATION.

GENERAL:
FINISH ALL OVER
ALL GEARS ARE 48 PITCH
ALL RESISTORS ARE 1/4 WATT
ALL DIMENSIONS ARE IN INCHES
DEBURR AND REMOVE ALL SHARP EDGES WITH .03R MINIMUM.
UNLESS OTHERWISE NOTED ALL RADII ARE .125.

FIGURE 15–18
Notes

Sketching Dimensions

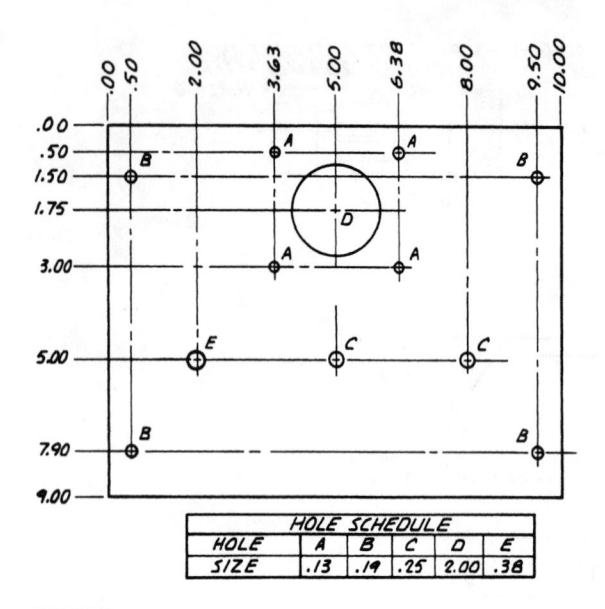

FIGURE 15-19
Datum Dimensioning

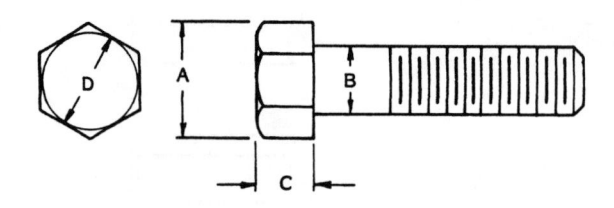

FIGURE 15-20
Tabular Dimensioning

Datum Dimensioning

A datum is an origin from which measurements are made. Figure 15-19 shows a drawing that used a datum dimensioning system to locate holes. The datum is the upper left corner, marked .00. All measurements are made from the datum. This drawing also has a hole schedule, which tells the sizes of all the holes. An identifying letter is placed next to each hole.

Tabular Dimensioning

Some drawings can be used for several sizes of an object that have the same appearance. Figure 15-20 is an example of tabular dimensioning.

EXERCISES

EXERCISE 15-1   Sketch the dimensions on the sheet in the back of the book labeled EXERCISE 15-1. Refer to Figures 16-2 and 16-5 if necessary to determine what dimensions are necessary and where they should be placed.

    Use 3-place decimal dimensions.
    Leave $\frac{1}{2}''$ (two grid marks) between the drawing and the first dimension.
    Leave $\frac{3}{8}''$ (one and a half grid marks) between dimension lines.
    Use good line weights and be consistent with letter size, arrowheads, and extension line offset from the part and extension line extension past the arrowhead.
    Complete the title block with your best lettering, and name the drawing DIMENSIONING SKETCH 1.

EXERCISE 15-2   Sketch top, front, and right-side views of Figure 15-21 full scale, and dimension it using good dimensioning practice as described in this chapter and Exercise 15-1 above. Use the sheet marked EXERCISE 15-2.

    Use 2-place decimal dimensions for all dimensions.
    Complete the title block with your best lettering, and name the drawing DIMENSIONING SKETCH 2.

EXERCISE 15-3   Sketch top, front, and right-side views of Figure 15-22 full scale, and dimension it using good dimensioning practice as described in this chapter and Exercise 15-1 above. Use the sheet labeled EXERCISE 15-3 in the back of the book.

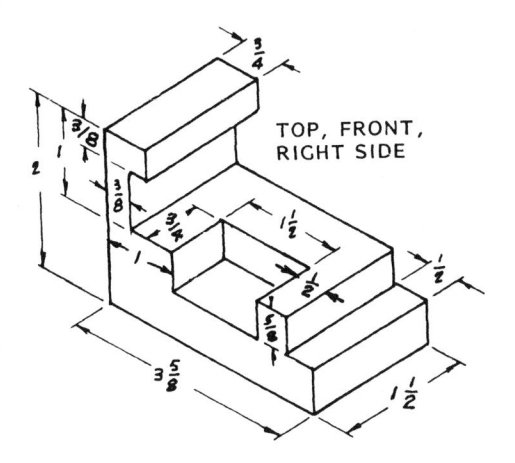

FIGURE 15–21
Exercise 15–2

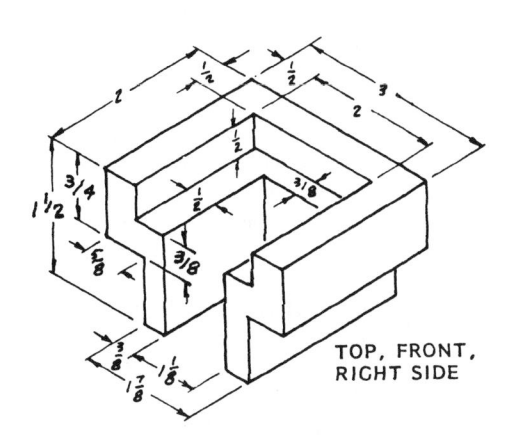

FIGURE 15–22
Exercise 15–3

Use 2-place decimal dimensions for all dimensions.

Complete the title block with your best lettering, and name the drawing DIMENSIONING SKETCH 3.

EXERCISE 15–4 Sketch the dimensions on the sheet labeled EXERCISE 15–4 in the back of the book. Use good dimensioning practices as described in this chapter and Exercise 15–1 above.

Use 2-place decimal dimensions for all dimensions except the distance between holes on the right-side view. Use a 3-place decimal for that dimension because it assumes a smaller tolerance, which is necessary for hole patterns.

If you do not know the decimal equivalent of any fraction, divide the top number by the bottom number of the fraction to get the decimal.

You will have to measure this drawing and choose your own dimensions based on rules described in this chapter. Refer to Figures 15–4 and 15–8.

Complete the title block with your best lettering and name the drawing DIMENSIONING SKETCH 4.

REVIEW QUESTIONS

Circle the best answer.

1. How much space should be left between the dimension line arrowhead and the extension line to which it points?
    a. $\frac{1}{2}''$
    b. $\frac{3}{8}''$
    c. $\frac{1}{32}''$
    d. $\frac{1}{16}''$
    e. None

2. What proportion should the length of arrowheads be to their width?
    a. 3 to 1
    b. 4 to 1
    c. 2 to 1
    d. 1 to 1
    e. Proportion does not matter in this case.

3. How much space should be left between the object and the first dimension line on the small drawings in the exercises in this chapter.
    a. $\frac{1}{4}''$
    b. $\frac{3}{8}''$
    c. $\frac{1}{2}''$
    d. $\frac{3}{4}''$
    e. Any spacing is OK.

Sketching Dimensions

4. Is it permissible for dimension lines to cross extension lines or other dimension lines?
   a. Yes in all cases
   b. Never
   c. Avoid if possible
   d. No, leave off the crossing dimension
   e. This will never happen if the dimensioning is carefully planned.
5. Where is the best place to show dimensions?
   a. Inside the views
   b. Each dimension on every view
   c. Outside and between the views
   d. As close as possible to the view
   e. On top and front views only
6. If more than one dimension containing a fraction is given on one side of a view, what should be done to make the fractions more legible?
   a. Make the fractions twice the normal size.
   b. Stagger the numbers containing the fractions.
   c. Line up the numbers containing the fractions.
   d. Make the fractions half the normal size.
   e. None of the above
7. Cylindrical objects are dimensioned in the rectangular view with
   a. Extension lines and dimension lines
   b. A leader and a note
   c. Dimension lines only
   d. A leader and a centerline
   e. Cylindrical objects are not dimensioned in the rectangular view
8. Which of these angles would be best for drawing a leader?
   a. 0°
   b. 45°
   c. 90°
   d. 180°
   e. All are equally good.
9. Small holes are dimensioned in the circular view with
   a. Extension lines and dimension lines
   b. A leader and a note
   c. Dimension lines only
   d. A leader and a centerline
   e. Holes are not dimensioned in the circular view.
10. Location dimensions to locate the center of a hole should be taken from
    a. Any surface on the object
    b. Finished surfaces or other holes
    c. Unfinished surfaces
    d. The left side of the part
    e. The right side of the part
11. A circular centerline used to locate holes from the center of a cylindrical piece is known as
    a. A bolt circle
    b. A cylindrical radius
    c. A circular diameter
    d. A drilling center
    e. None of the above
12. If an object has all fillets and rounds of the same size, how should they be dimensioned?
    a. Separately
    b. Not at all
    c. With a general note
    d. Dimension only one
    e. Such will never be the case
13. The major difference between a spotface and a counterbore is
    a. The diameter of the primary hole
    b. The diameter of the counterbore
    c. The angle between faces of a countersink
    d. The depth of the counterbore or spotface
    e. The finished surface

14. The note "3 HOLES EQUALLY SPACED" is an example of a
    a. Specific note
    b. General note
15. The note "UNLESS OTHERWISE NOTED ALL RADII ARE .125" is an example of a
    a. Specific note
    b. General note

16 Dimensioning with AutoCAD

OBJECTIVES

After completing this chapter, you will be able to

☐ Use the dimensioning commands to dimension full-scale drawings of
   Architectural floor plans
   Mechanical parts using decimal parts of an inch
☐ Correctly answer questions regarding dimensioning commands.

DIMENSIONING

Up to this point you have made several different types of drawings. For the manufacturing and construction drawings to be used to manufacture a part or to construct a building, dimensions must be added. Adding dimensions to a drawing manually is a very time consuming process. With AutoCAD, adding dimensions is much easier. In addition, the AutoCAD dimensioning process verifies the accuracy of the drawing. The associative dimensioning feature of AutoCAD also allows a part size to be changed or corrected, automatically changing the dimension with it.

In this chapter you will dimension some drawings full size, so the dimensioning procedure is relatively simple. Pay particular attention to where you place the dimension lines in relation to the drawing and other dimension lines. Although AutoCAD dimensioning has many settings and other features, this chapter does not cover those. This chapter will give you a good idea of how dimensioning is done when all the settings have been made.

EXERCISE 16-1
Dimensioning a Mechanical Part

Step 1. To begin Exercise 16–1, turn on the computer and start AutoCAD or AutoCAD LT.
Step 2. Open drawing EX16-1 supplied on the disk that came with your book.
Step 3. Use Zoom-All to view the entire drawing area.
Step 4. Use the Dimensioning Mode to draw center lines in the circle in the front view.

Note: The dimensioning prompts among versions of AutoCAD are slightly different. You may need to experiment with dimensioning commands a little if the prompts are not quite the same as shown in this chapter.

Prompt	Response
Command:	Type: **DIM**↵
Dim:	Type: **CEN**↵
Select arc or circle:	Click: **any point on the circumference of the circle in the front view**
Dim:	Type: **DIMDEC**↵
Enter new value for DIMDEC <4>:	Type: **3**↵ (This sets the number of places to the right of the decimal.)

Be sure Snap is ON and Ortho is OFF before you start.

Step 5. Continue using the Dimensioning Mode to draw all horizontal dimensions

Prompt	Response
Dim:	Type: **HOR**↵

FIGURE 16–1

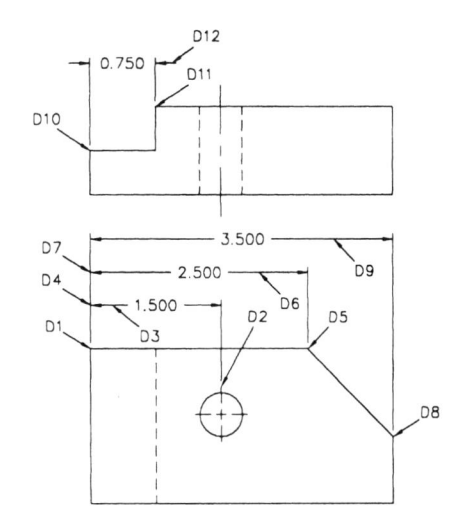

Prompt	Response
Specify first extension line origin or <select object>:	Click: **D1** (Figure 16–1)
Specify second extension line origin:	Click: **D2**
Specify dimension line location or [MText/Text/Angle]:	Click: **D3** ($\frac{1}{2}''$ from the top line of the part).⏎
Enter dimension text <1.500>:	⏎
Dim:	⏎
Specify first extension line origin or <select object>:	Click: **D4** (Figure 16–1)
Specify second extension line origin:	Click: **D5**
Specify dimension line location or [MText/Text/Angle]:	Click: **D6** ($\frac{3}{8}''$ from the first dimension line).⏎
Enter dimension text <2.500>:	⏎
Dim:	⏎
Specify first extension line origin or <select object>:	Click: **D7**
Specify second extension line origin:	Click: **D8**
Specify dimension line location or [MText/Text/Angle]:	Click: **D9** ⏎ ($\frac{3}{8}''$ from the second dimension)
Enter dimension text <3.500>:	⏎
Dim:	⏎
Specify first extension line origin or <select object>:	Click: **D10**
Specify second extension line origin:	Click: **D11**
Specify dimension line location or [MText/Text/Angle]:	Click: **D12** ($\frac{1}{2}''$ from the part)
Enter dimension text <0.750>:	⏎

Step 6. Use the Dimensioning Mode to draw all vertical dimensions.

Prompt	Response
Dim:	Type: **VER**⏎

FIGURE 16–2

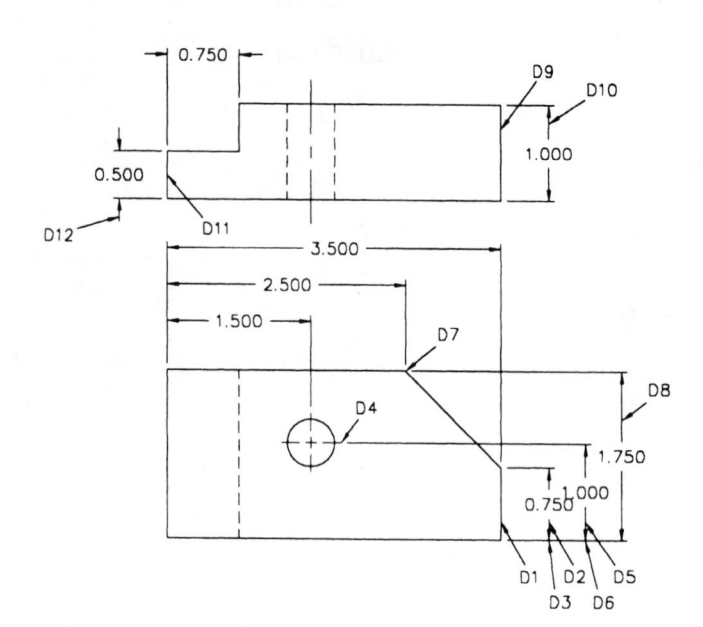

Prompt	Response
Specify first extension line origin or <select object>:	↵
Select object to dimension:	Click: **D1** (Figure 16–2)
Specify dimension line location or [MText/Text/Angle]:	Click: **D2** ($\frac{1}{2}''$ from the right side of the part) ↵
Enter dimension text <0.750>:	↵
Dim:	↵
Specify first extension line origin or <select object>:	Click: **D3** (Figure 16–2)
Specify second extension line origin:	Click: **D4**
Specify dimension line location or [MText/Text/Angle]:	Click: **D5** ($\frac{3}{8}''$ from the first vertical dimension line) ↵
Enter dimension text <1.000>:	↵
Dim:	↵
Specify first extension line origin or <select object>:	Click: **D6**
Specify second extension line origin:	Click: **D7**
Specify dimension line location or [MText/Text/Angle]:	Click: **D8** ↵ ($\frac{3}{8}''$ from the second vertical dimension)
Enter dimension text <1.750>:	↵
Dim:	↵
Specify first extension line origin or <select object>:	↵
Select object to dimension:	Click: **D9**
Specify dimension line location or [MText/Text/Angle]:	Click: **D10** ↵ ($\frac{1}{2}''$ from the part)
Enter dimension text <1.000>:	↵
Dim:	↵

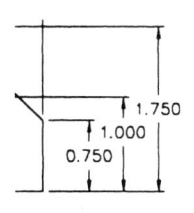

FIGURE 16–3

FIGURE 16–4

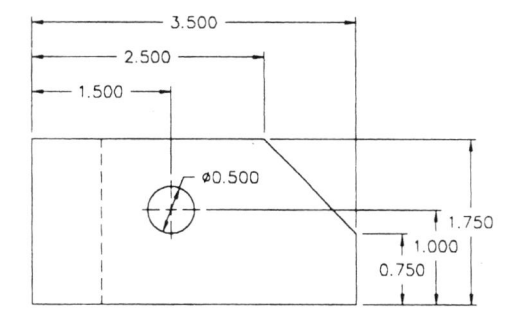

Prompt	Response
Specify first extension line origin or <select object>:	↵
Select object to dimension:	Click: **D11**
Specify dimension line location or [MText/Text/Angle]:	Click: **D12** ↵
Enter dimension text <0.500>:	↵

Step 7. Use the Dimtedit command from the Dim: prompt to move some vertical dimension text so it does not overlap.

Prompt	Response
Dim:	Type: **TEDIT**↵
Select dimension:	Click: **the 1.000 vertical dimension in the front view**
Specify new location for dimension text or [Left/Right/Home/Angle]:	Click: **a new position for the 1.000 text as shown in Figure 16–3**

Step 8. Use the Dimensioning Mode to dimension the circle diameter.

Prompt	Response
Dim:	Type: **DIA**↵
Select arc or circle:	Click: **any point on the circumference of the circle in the front view**
Enter dimension text <0.500>:	↵
Specify dimension line location or [MText/Text/Angle]:	Click: **a point so the diameter is shown in the approximate position shown in Figure 16–4**

Step 9. Use the Tedit command from the Dim: prompt to move the diameter text off the object to a new location.

Prompt	Response
Dim:	Type: **TEDIT**↵
Select dimension:	Click: **the 0.500-diameter dimension in the front view**
Specify new locatoin for dimension text or [Left/Right/Home/Angle]:	Click: **a new position for the 0.500-diameter text as shown in Figure 16–5**

Step 10. Use the Dtext command to complete the title block as you did for previous exercises. Name the drawing DIMENSIONING.

Dimensioning with AutoCAD

FIGURE 16–5

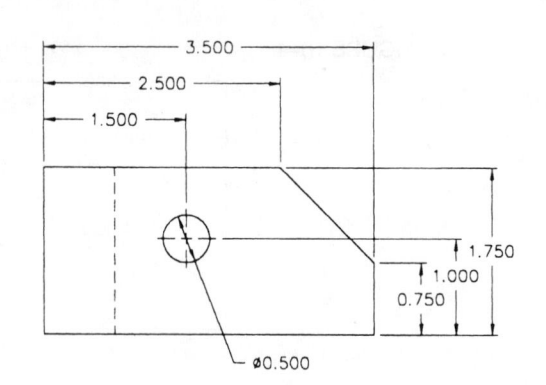

EXERCISE 16–2
Dimensioning an Architectural Floor Plan

Step 11. Use the SAVEAS command to save your drawing as EX16-1(your initials) on a floppy disk and again on the hard drive of your computer.
Step 12. Plot or print your drawing full size on an 11" × 8½" sheet.

Step 1. To begin Exercise 16–2, turn on the computer and start AutoCAD or AutoCAD LT
Step 2. Open drawing EX16-2 supplied on the disk that came with your book.
Step 3. Use Zoom-All to view the entire drawing area.
Step 4. Use the Dimensioning Mode to draw all horizontal dimensions on the top of the floor plan.

Important: Be sure Snap is ON and Ortho is OFF when you pick points.

Prompt	Response
Command:	Type: **DIM**↵
Dim:	Type: **DIMUNIT**↵
Current value<2> New Value:	Type: **4**↵ (to set architectural units)
Dim:	Type: **HOR**↵
Specify first extension line origin or <select object>:	Click: **D1** (Figure 16–6)
Specify second extension line origin:	Click: **D2**
Specify dimension line location or [MText/Text/Angle]:	Click: **D3** (2 grid marks from the top line of the floor plan)
Enter dimension text <5'-6">:	↵
Dim:	Type: **CO**↵ (for continue dimensioning)

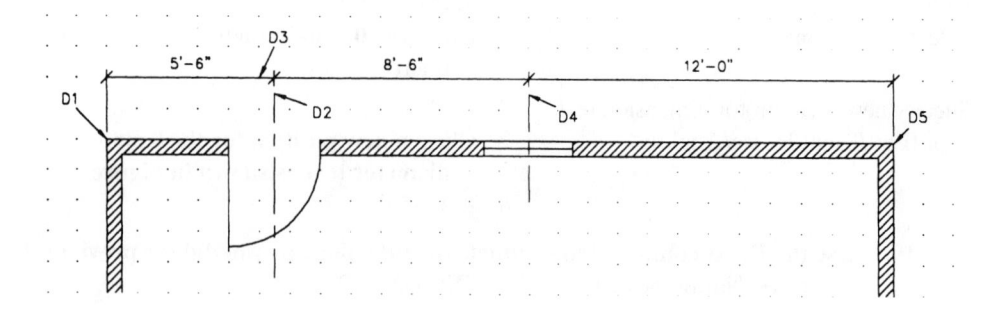

FIGURE 16–6

250 Chapter 16

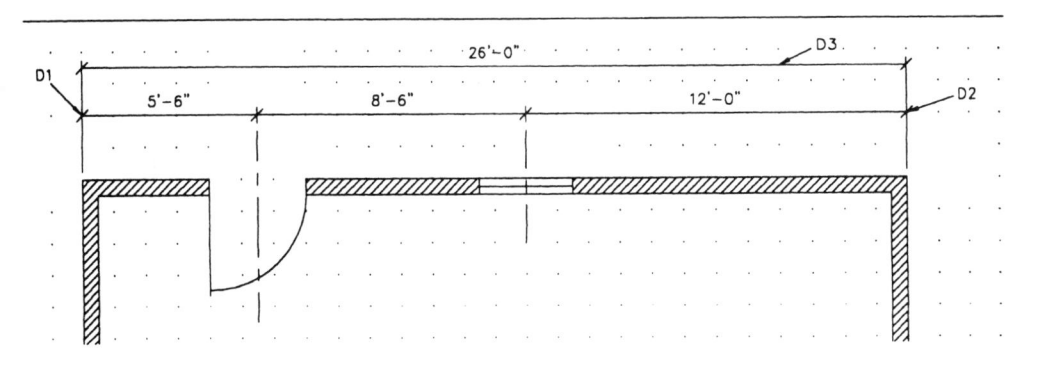

FIGURE 16–7

Prompt	Response
Specify second extension line origin:	Click: **D4**
Enter dimension text <8'-6">:	↵
Dim:	↵
Specify second extension line origin:	Click: **D5**
Enter dimension text <12'-0">:	↵
Dim:	Type: **HOR**↵
Specify first extension line origin or <select object>:	Click: **D1** (Figure 16–7)
Specify second extension line origin:	Click: **D2**
Specify dimension line location or [MText/Text/Angle]:	Click: **D3** (1½ grid marks from the first dimension line)
Enter dimension text <26'-0">:	↵

Step 5. **On your own:**
Draw dimensions at the bottom of the floor plan as shown in Figure 16–8.

Step 6. **Use the Dimensioning Mode to draw all vertical dimensions.**

Prompt	Response
Dim:	Type: **VER**↵
Specify first extension line origin or <select object>:	Click: **D1** (Figure 16–9)
Specify second extension line origin:	Click: **D2**
Specify dimension line location or [MText/Text/Angle]:	Click: **D3** (2 grid marks to the left of the wall)
Enter dimension text <9'-0">:	↵
Dim:	Type: **CO**↵

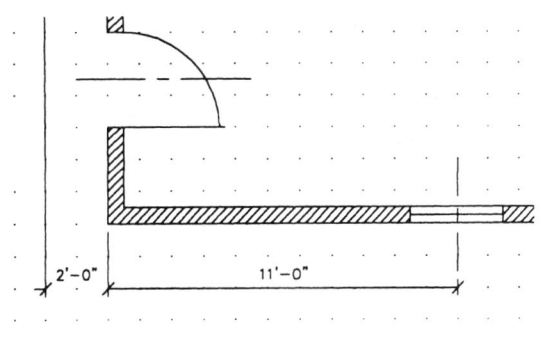

FIGURE 16–8

Dimensioning with AutoCAD

251

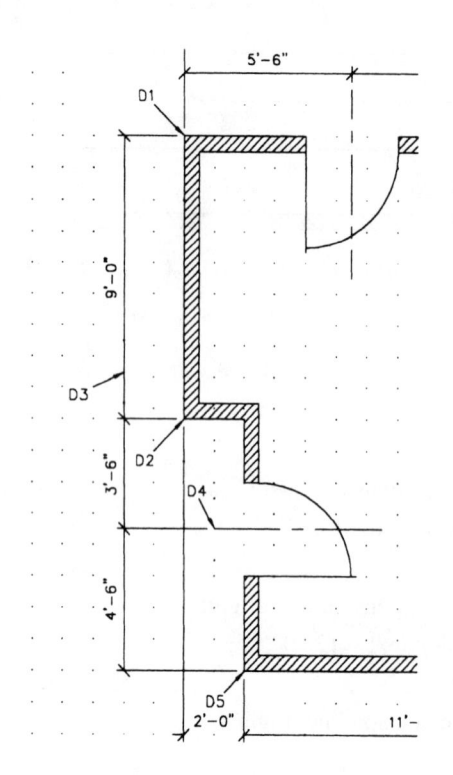

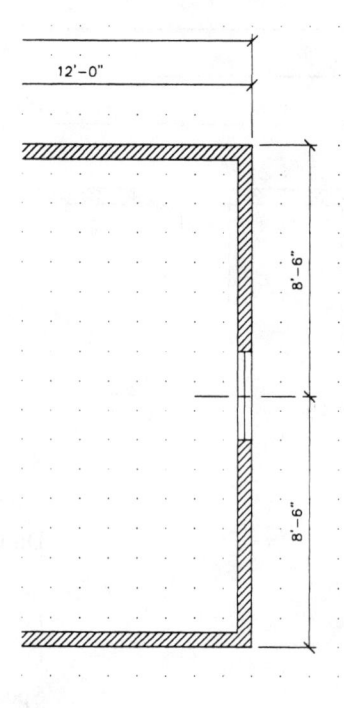

FIGURE 16–9

FIGURE 16–10

Prompt	Response
Specify second extension line origin:	Click: **D4**
Enter dimension text <3'-6">:	↵
Dim:	↵
Specify second extension line origin:	Click: **D5**
Enter dimension text <4'-6">:	↵

FIGURE 16–11

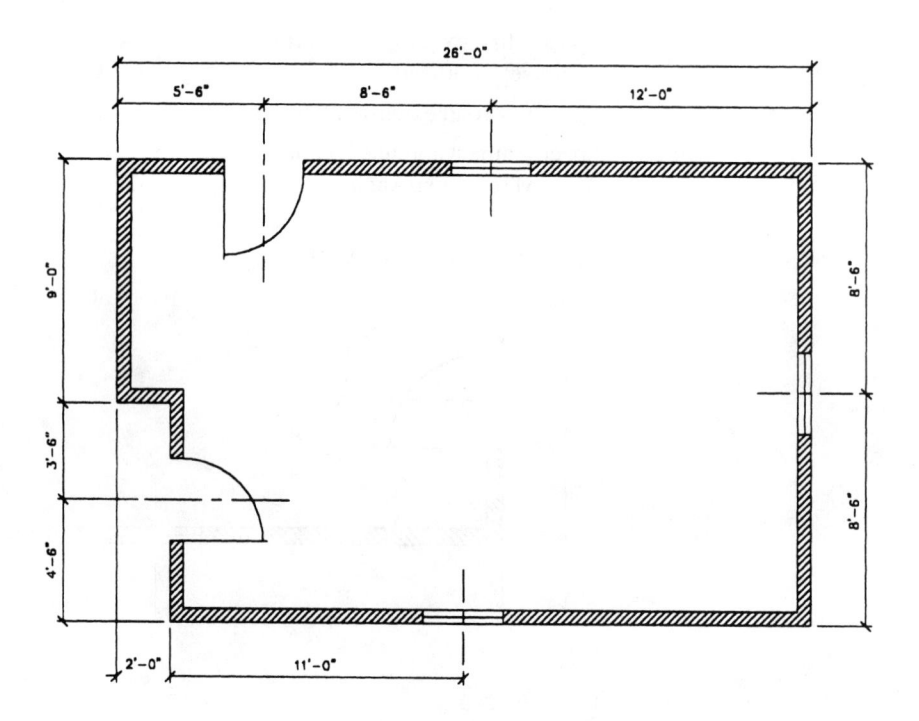

252 Chapter 16

Dim: ↵

Step 7. **On your own:**
Draw dimensions on the right side of the floor plan as shown in Figure 16–10. Your final floor plan should appear as shown in Figure 16–11.

Step 8. **Use the Dtext command to complete the title block as you did for previous exercises. Name the drawing FLOOR PLAN WITH DIMENSIONS.**

Step 9. **Use the SAVEAS command to save your drawing as EX16-2(your initials) on a floppy disk and again on the hard drive of your computer.**

Step 10. **Plot or print your drawing at a scale of 1 plotted inch = 48 drawing units ($\frac{1}{4}''=1'$) on an 11" × $8\frac{1}{2}''$ sheet. You will have to rotate the plot 90°.**

EXERCISES

EXERCISE 16–1   Complete Exercise 16–1 using steps 1 through 12 described in this chapter.
EXERCISE 16–2   Complete Exercise 16–2 using steps 1 through 10 described in this chapter.

REVIEW QUESTIONS

1. What three letters must be typed to draw a horizontal dimension from the Dim: prompt?

2. What three letters must be typed to draw a vertical dimension from the Dim: prompt?

3. In what position does the CONTINUE dimensioning command place dimensions?

4. What three letters must be typed from the Dim: prompt to draw a diameter dimension for a circle?

5. For what purpose was the Tedit command used in this chapter?

17 Sketching Threads and Fasteners

OBJECTIVES

After completing this chapter, you will be able to

☐ Correctly sketch simplified schematic and detailed drawings of threads to commonly accepted industrial standards from written descriptions.
☐ Sketch and identify screws, bolts, rivets, and nuts.

INTRODUCTION

Threads, fasteners, and springs are features that almost every CAD operator will draw at one time or another. Almost every assembly of two or more parts uses some type of fastener (Figure 17–1).

Thread Terms

Thread terms must be understood before the correct specifications can be written. These terms, as illustrated in Figure 17–2 (see p. 257), are as follows:

Root diameter   The smallest diameter of the thread (also called minor diameter).
Crest diameter   The largest diameter of the thread (also called major diameter).
Pitch diameter   The point between the crest and the root at which the widths of the thread and the space between threads are the same.
Pitch   The distance from a point on a thread to a point in the same location on the next thread, such as crest to crest or root to root.

Thread Forms

The standard thread forms, shown in Figure 17–3 (see p. 257), are used as follows:

Sharp V   Used in a fastener where friction is required, as in a set screw.
Sharp V, unified   Used in such fasteners as screws, bolts, and nuts.
Square, acme, and buttress   Used to transmit motion and power, as in gears.
Knuckle   Used for cheap, loose-tolerance fastening, such as twist-off bottle caps and light bulbs.

THREAD SPECIFICATION FOR ENGLISH UNITS

Figure 17–4 (see p. 257) shows the parts of a thread specification. This information usually is shown directly on a drawing or in a parts list. You will probably need to know how to specify threads for both English and metric units. The parts of the specification are as follows:

Size

The size given is the crest or major diameter of the thread. Size may be given as a whole number, such as 10, as a fraction, such as $\frac{1}{4}$, or as a decimal, such as .250. Tables of thread and fastener sizes are located in the back of this book.

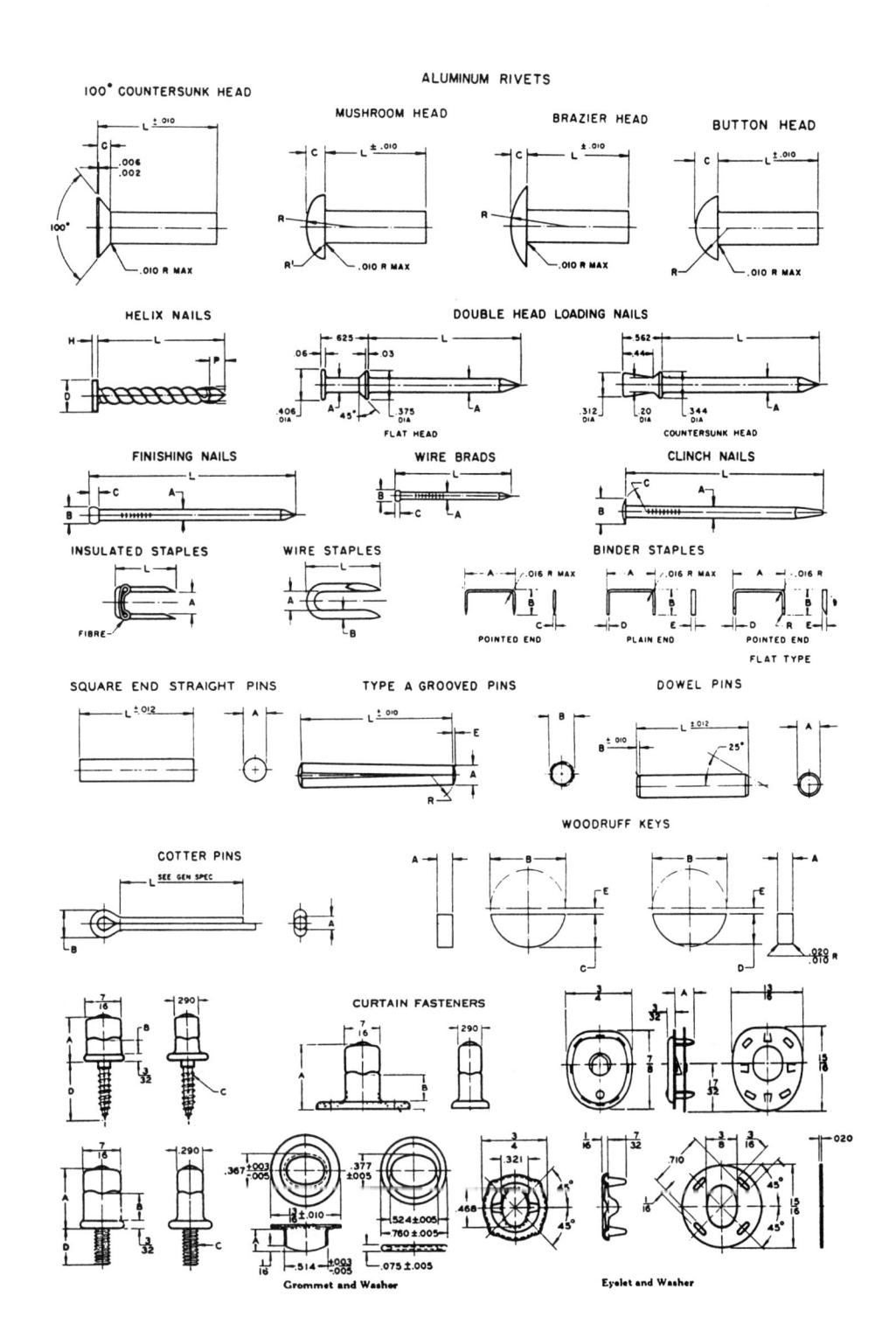

FIGURE 17-1
Sheet 1 of 2
Fasteners

Sketching Threads and Fasteners

FIGURE 17–1
Sheet 2 of 2
Fasteners

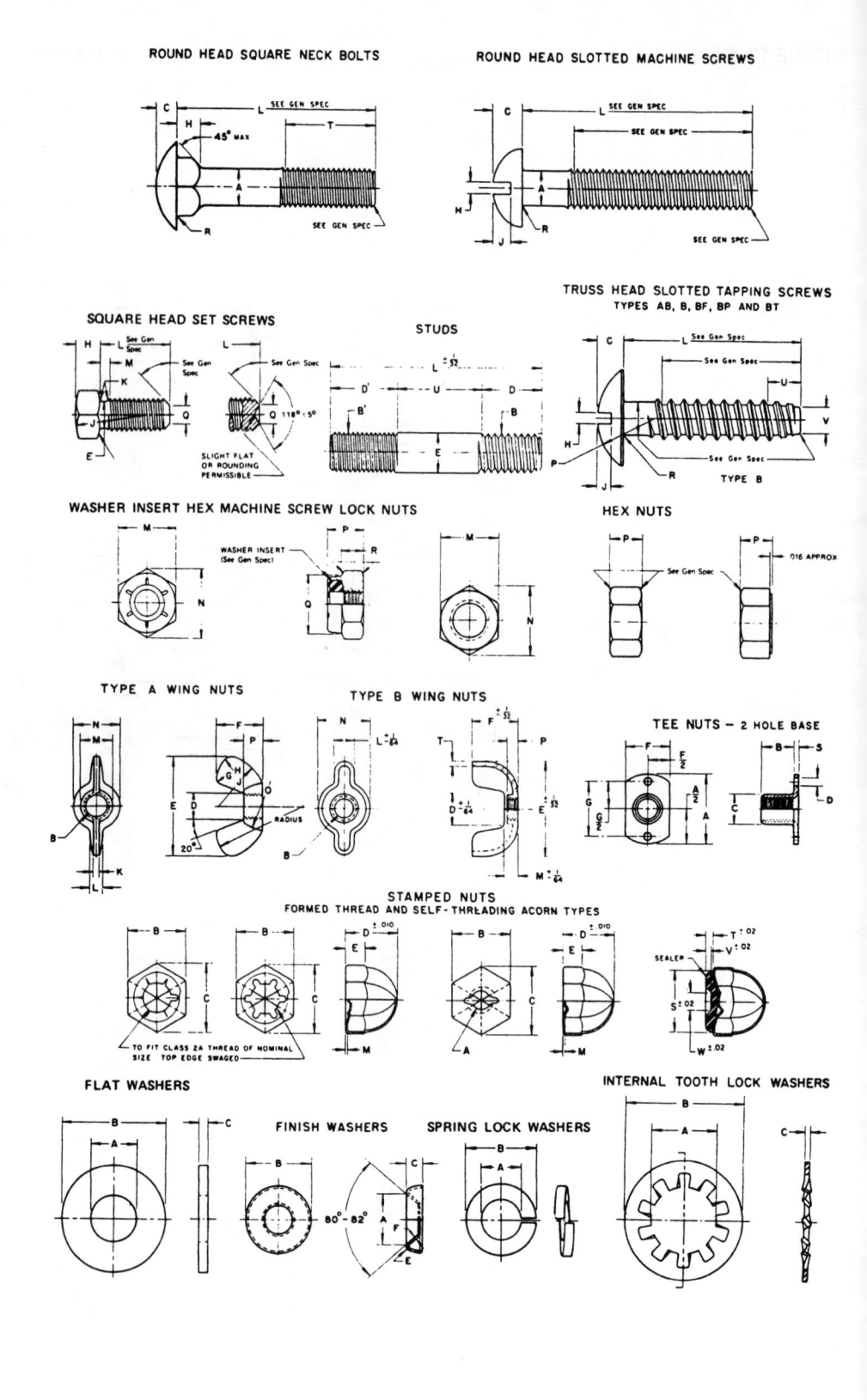

FIGURE 17–2
Thread Terms

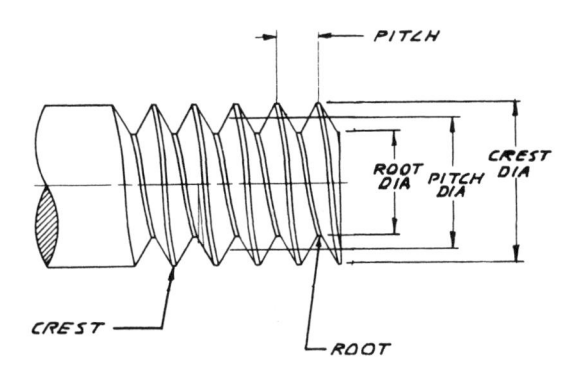

FIGURE 17–3
Thread Forms

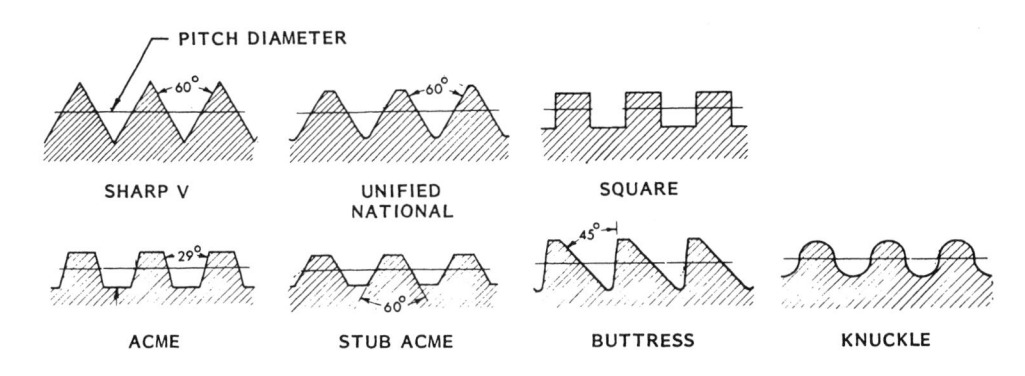

FIGURE 17–4
Standard Thread Specification

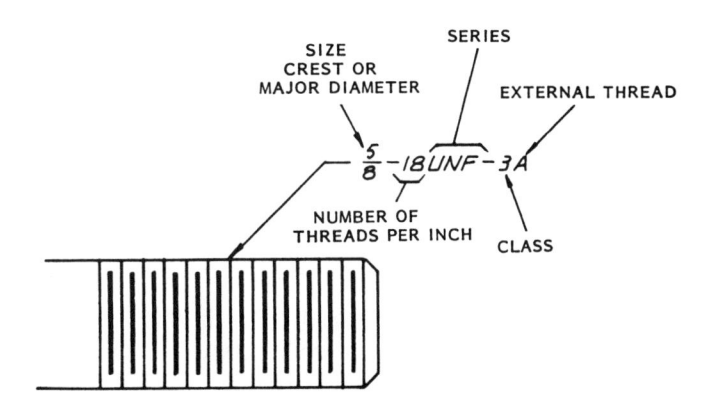

Number of Threads per Inch

This measurement is taken by placing a scale on a screw thread and counting the number of threads per inch.

Thread Series

The thread series is the form of the thread. There are four unified thread series, which were standardized in the United States, Great Britian, and Canada during World War II: Unified National Coarse (UNC), Unified National Fine (UNF), Unified National Extra Fine (UNEF), and Constant Pitch (UN). There are other series of threads, but these are the most commonly used.

Unified National Coarse (UNC) These threads are used on screws, nuts, and bolts and for threads in soft metal or plastic castings in which tolerances are not critical.

Unified National Fine (UNF) These threads are used for bolts, screws, and nuts that require a better fit, such as standard fasteners used on automobile bodies and hardware.

Unified National Extra Fine (UNEF) These are used for thin-wall tubes, nuts, and similar applications in which the length of the thread is relatively short.

Constant Pitch (UN) These are used for special-purpose applications, often large diameters, such as pipe threads.

Sketching Threads and Fasteners

FIGURE 17–5
Complete Thread Specification with Countersink

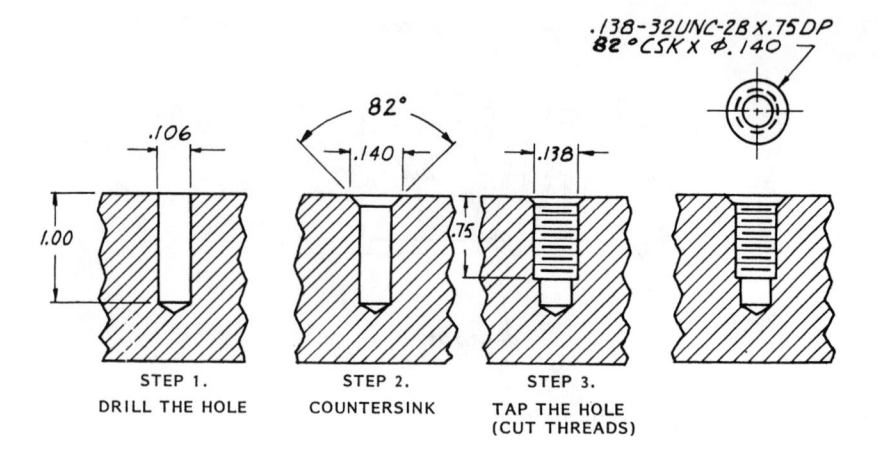

Thread Class

Thread class is determined by the amount of tolerance allowed in fitting together two parts, such as a nut and a bolt. The three thread classes are as follows:

Class 1 This is a loose fit to allow for the cheapest manufacturing methods and to allow parts to be screwed together easily and quickly.
Class 2 This fit is used in mass production of screws, nuts, and bolts that are found at most hardware and automotive parts suppliers.
Class 3 This is a very close tolerance fit, used where a great deal of stress and vibration occurs, such as in aircraft.

External or Internal Thread

External thread, as on a screw, is indicated by the letter A. Internal thread, as on a nut, is indicated by the letter B.

Sample Specification

Figure 17–5 shows a complete specification for a thread in a countersunk hole. This figure also shows the manufacturing processes used to make this thread.

Step 1.   Drill a hole to the depth of the thread plus .25.
Step 2.   Make the countersink.
Step 3.   Cut threads in the hole (this is known as tapping the hole). These threads are:

> .138 in diameter
> 32 threads per inch
> Unified National Coarse series
> class 2
> internal
> .75″ deep

This hole also has an 82° countersink at its top that has a diameter of .140.

THREAD SPECIFICATION FOR METRIC THREADS

The metric thread series is similar to the unified thread series. The complete thread specification for the metric series is shown in Figure 17–6. The letter M identifies the thread as a metric thread. The number 2 is the size of the major diameter of the thread in millimeters. The decimal .40 is the pitch of the thread in millimeters (distance from crest to crest). The pitch is shown on fine threads but not on coarse threads. A coarse thread would be specified simply as M2. The notation 6g is the grade of the thread, which is similar to the class of a unified thread.

FIGURE 17-6
Metric Thread Specification

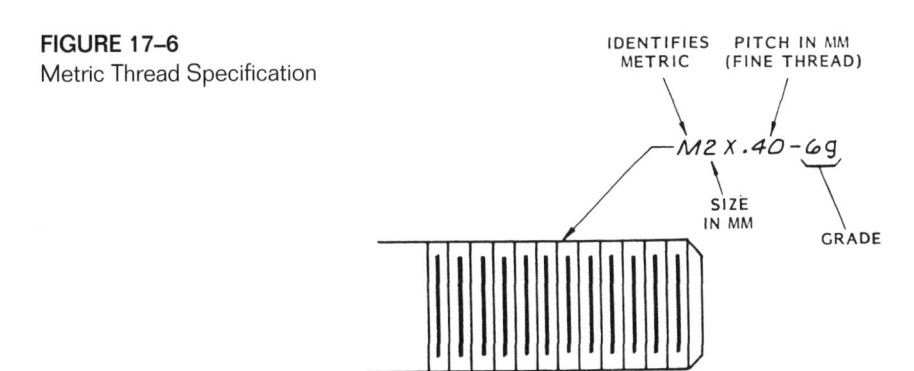

SYMBOLS FOR DRAWING THREADS

Threads may be drawn in three different ways:

Detailed—This is the most realistic symbol.
Schematic—This is the most commonly used symbol.
Simplified—This is the easiest to sketch.

Unless there is a special requirement, the number of threads per inch is ignored in drawing the detailed and schematic symbols. The appropriate symbols and sizes used to draw threads are shown in Figure 17-7. The sizes can vary slightly, depending on personal taste or company specification. On large drawings, the sizes become proportionately larger. Notice that the line showing the root on the schematic symbol is thicker than the crest lines.

TYPES OF FASTENERS

Although there are many types of fasteners, the following are the ones you will encounter most often.

Cap Screws

A cap screw is a fastener that holds two parts together (Figure 17-8). It goes through a clearance hole in one part and screws into a threaded hole in the other part. In this case it does not have a nut.

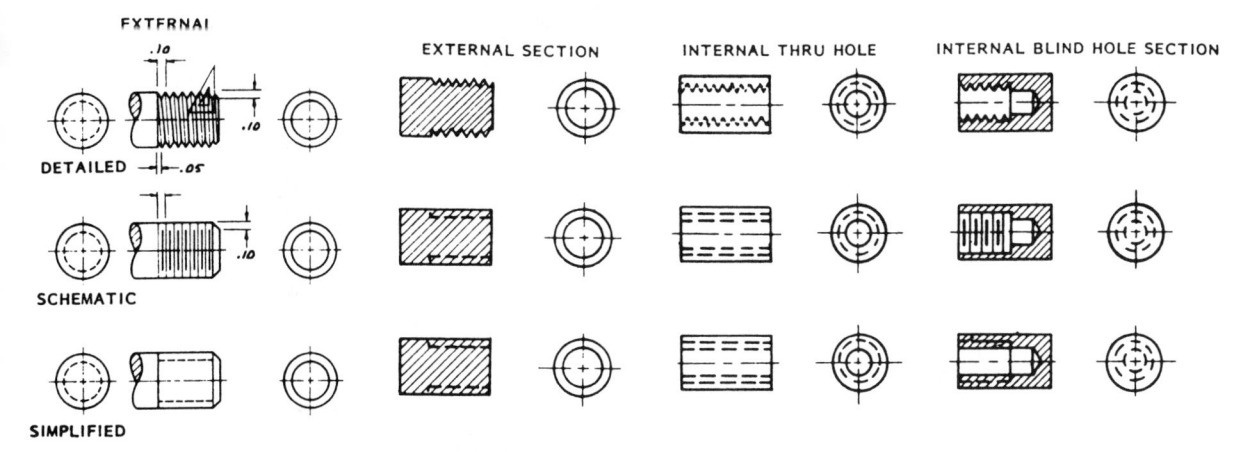

FIGURE 17-7
Thread Symbols

Sketching Threads and Fasteners

FIGURE 17-8
Cap Screws

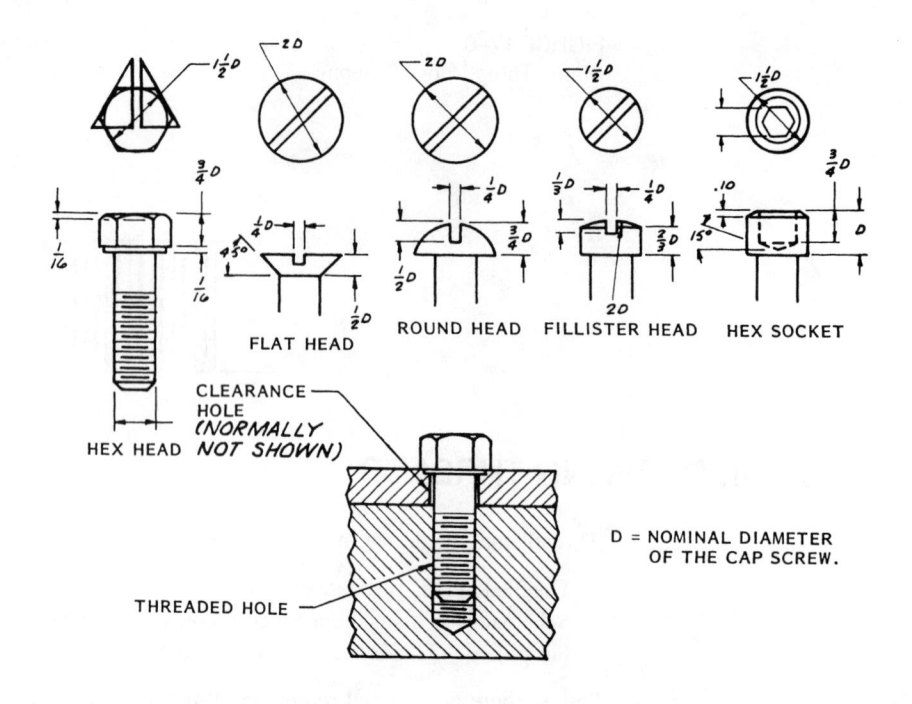

Bolts

A bolt holds two or more parts together. It goes through clearance holes in the parts and holds them together with a nut on the end opposite the head. The front view of the common types of bolt heads looks better if the head is drawn across corners, as shown in Figure 17–9.

Studs

A stud is a screw with threads on both ends and no head (Figure 17–10). It screws into a threaded hole, and then one or more parts are placed over the stud. The parts are held together with a nut on the other threaded end.

FIGURE 17-9
Bolts

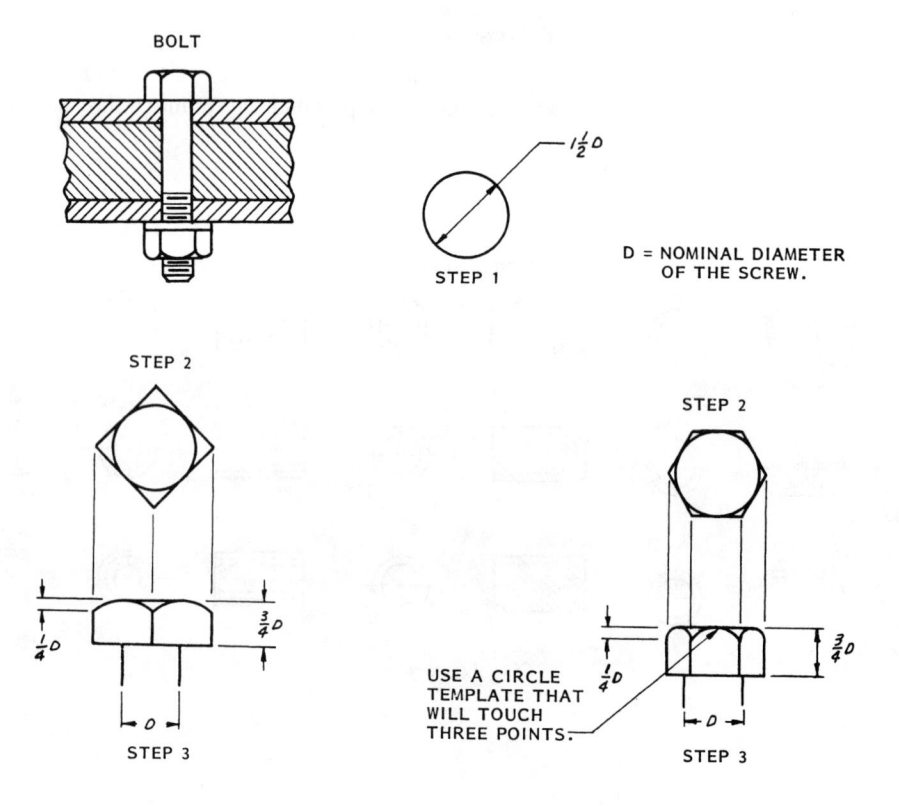

Chapter 17

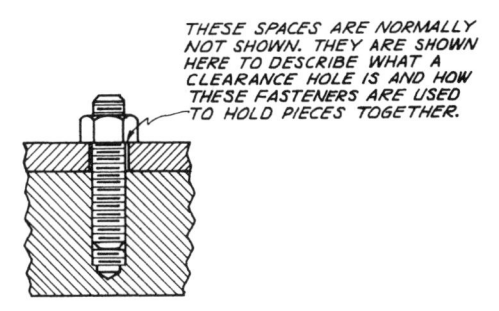

FIGURE 17–10
Stud

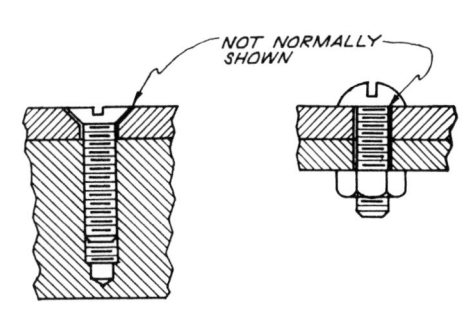

FIGURE 17–11
Machine Screw

Machine Screws

A machine screw is similar to a bolt and a cap screw and often has a nut (Figure 17–11).

Setscrews

A setscrew is used to keep a part from moving in a hole. Figure 17–12 shows several types of setscrews.

Washers

Washers are used with screws, bolts, and nuts for several purposes. They are used as spacers, to protect a surface, to distribute a torque load over a wider surface, to fit over a hole that is too big for the fastener, and to keep a fastener from turning. There are many different types of washers, the two most common being the flat washer and the split or lock washer (Figure 17–13).

Nuts

There are many different types of nuts. Each type has a special application and is the best to use for that particular purpose. Manufacturers' catalogs showing available nuts are a

FIGURE 17–12
Setscrews

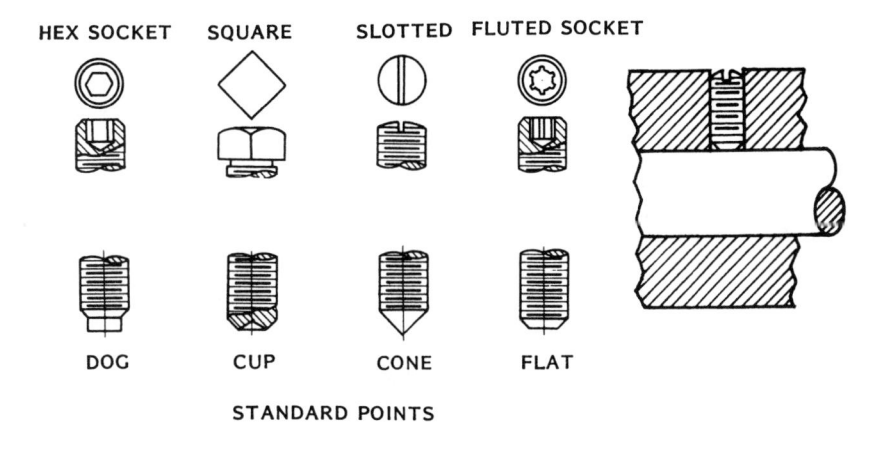

FIGURE 17–13
Washers

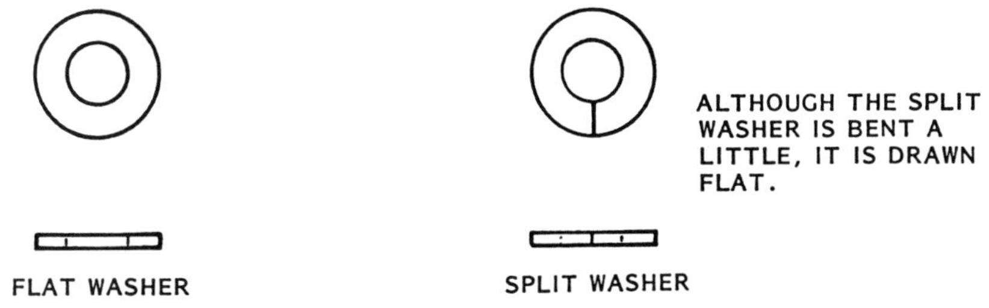

Sketching Threads and Fasteners

FIGURE 17-14
Nuts

TRACK BOLT   SQUARE   MACHINE SCREW   HEX JAM

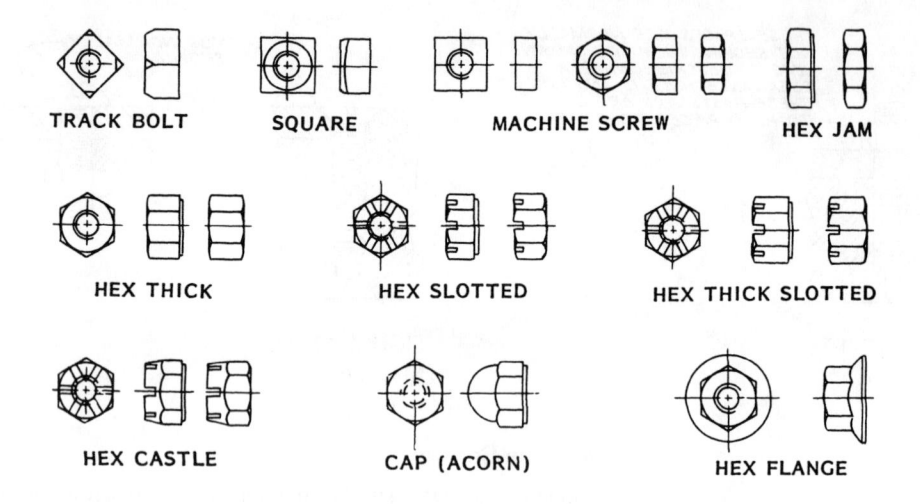

HEX THICK   HEX SLOTTED   HEX THICK SLOTTED

HEX CASTLE   CAP (ACORN)   HEX FLANGE

standard reference in mechanical engineering drawing rooms. A CAD designer should become familiar with the types that apply to the products manufactured by his or her company and should know which are stocked in the shop. Figure 17–14 shows several commonly used nuts.

Rivets

There are two general types of rivets: standard and blind. These two types and the symbols used for them are shown in Figure 17–15.

Standard This type of rivet has a head on one side and is then deformed (or bucked) on the other side to lock it in place.

Blind This rivet is used in places where a tool cannot be used to deform the end. The rivet is placed in the hole and a mandrel or screw in the rivet is used to deform the end.

Keys and Pins

Keys and pins are used to keep parts from turning inside other parts. Some common keys and pins are shown in Figure 17–16. Keys fit into slots that are called keyways on the hub and keyseats on the shaft.

Springs

Although they are not commonly thought of as fasteners, springs are often used to retain parts or to maintain close fits. The three types of springs—extension, compression, and torsion—are shown in Figure 17–17 (see p. 265).

SKETCHING A THREAD

Figure 17–18 (see p. 265) shows a procedure for drawing schematic thread symbols for a hex head (six-sided head) bolt.

Step 1.   **Draw the bolt or screw head and body to specifications.**
Step 2.   **Draw the thread length to specifications.**
Step 3.   **Draw thread crests and guidelines for roots.**
Step 4.   **Draw thread roots thicker than crests.**

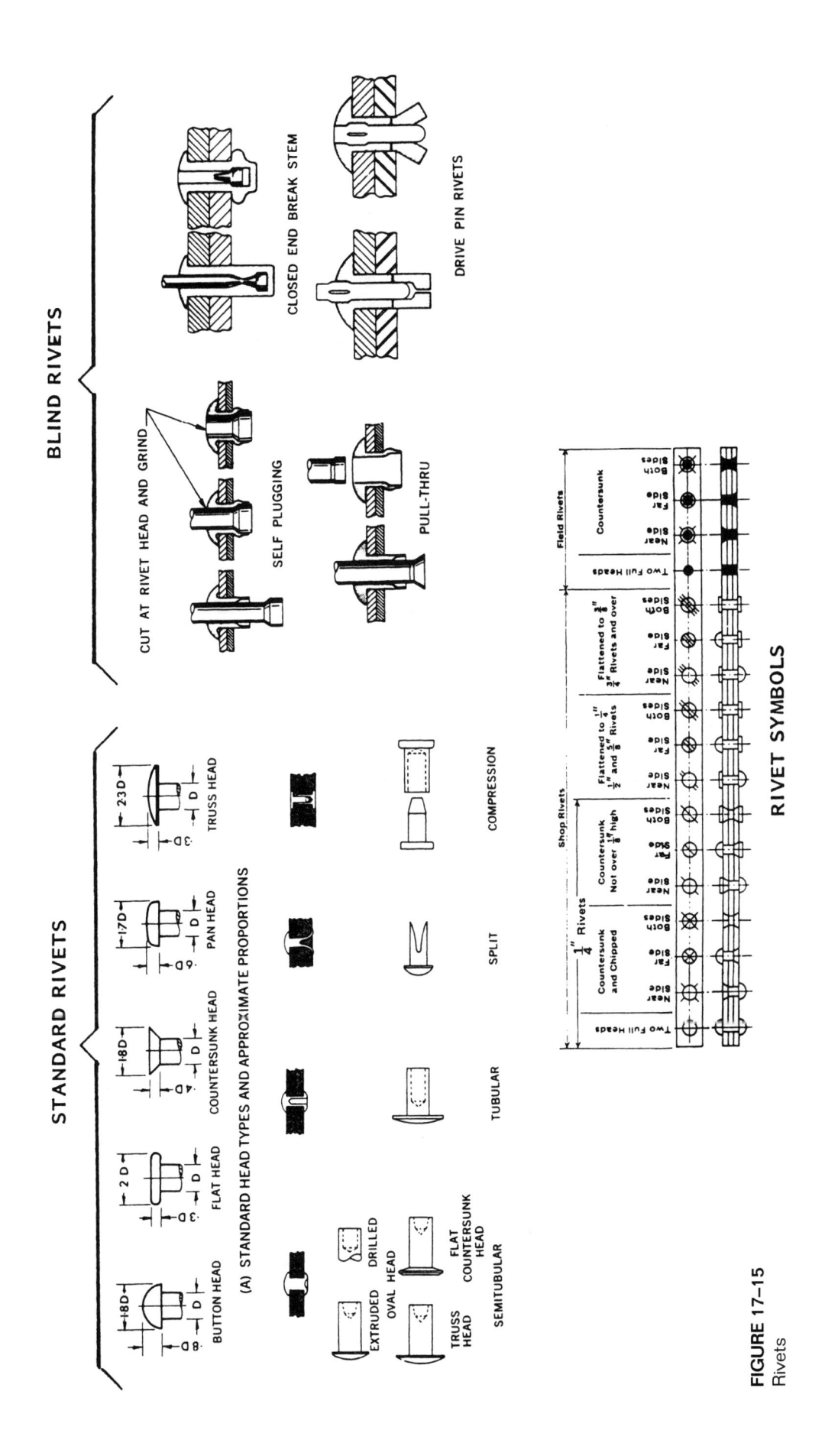

FIGURE 17-15
Rivets

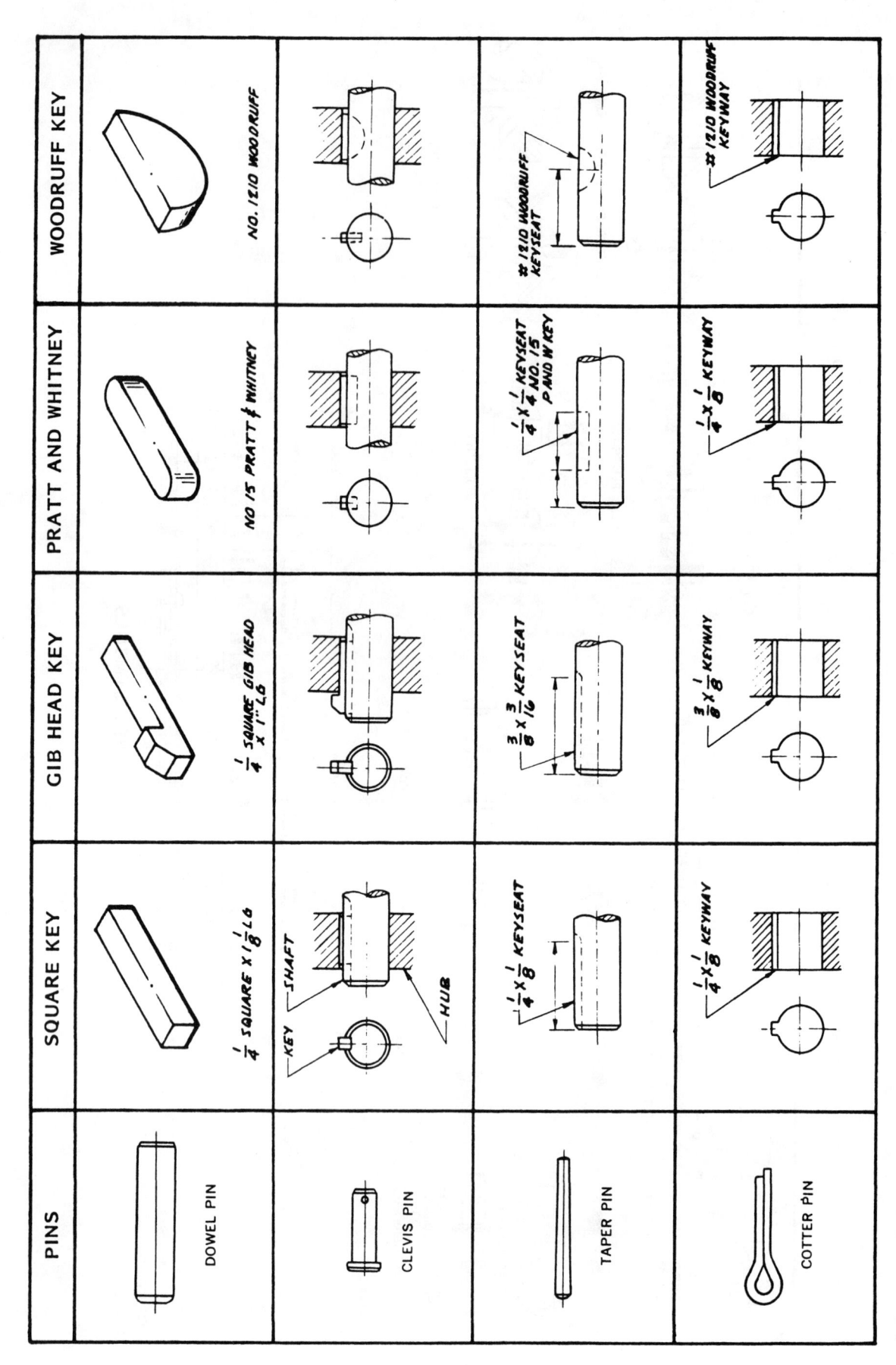

FIGURE 17-16
Keys and Pins

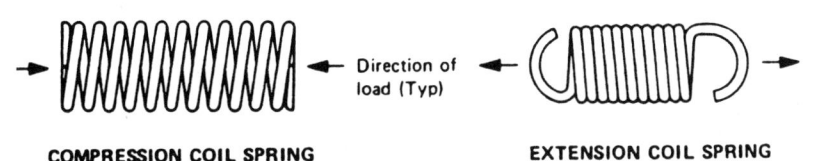

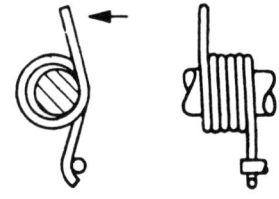

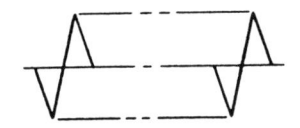

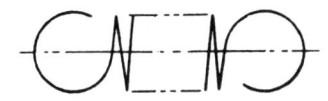

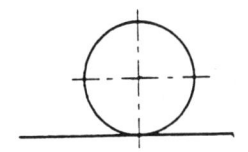

FIGURE 17-17
Springs

FIGURE 17-18
Drawing a Thread

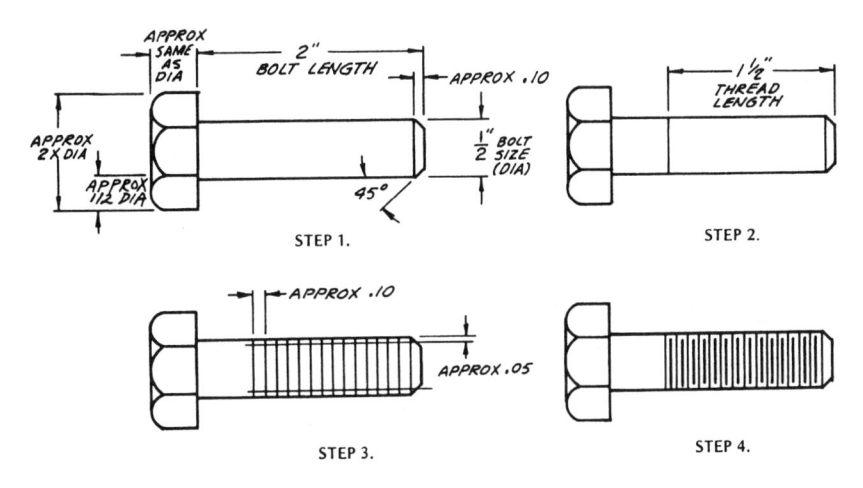

EXERCISE

EXERCISE 17-1 Sketch the required fasteners on the sheet labeled Exercise 17-1 where indicated on Figure 17-19, and letter neatly the complete specification for the thread of the fastener in the dotted guidelines. Refer to tables in the appendix for the numbers 8 and 10 size threads. Use the required symbols as shown in Figures 17-7 and 17-15. You will have no specifications for the rivets, just the symbols from Figure 17-13. Complete the title block information with your best lettering, and turn the assignment in to your instructor.

REVIEW QUESTIONS

Circle the best answer.

1. The largest diameter on a thread is the
    a. Root diameter
    b. Crest diameter
    c. Pitch diameter
    d. Drill diameter
2. A thread type that is used for common screws is
    a. Knuckle
    b. Sharp V
    c. Unified
    d. Square

Sketching Threads and Fasteners

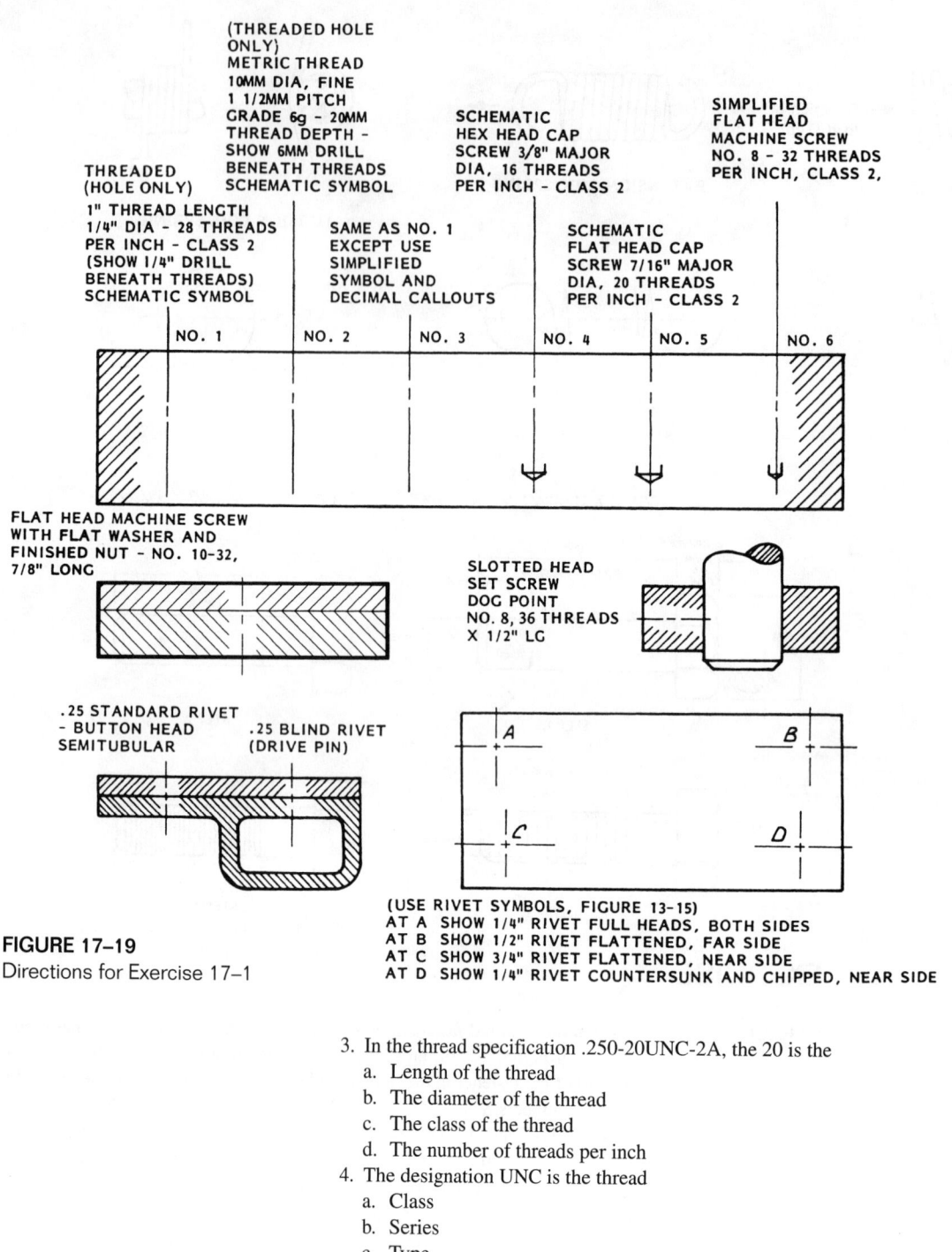

FIGURE 17-19
Directions for Exercise 17-1

3. In the thread specification .250-20UNC-2A, the 20 is the
   a. Length of the thread
   b. The diameter of the thread
   c. The class of the thread
   d. The number of threads per inch
4. The designation UNC is the thread
   a. Class
   b. Series
   c. Type
   d. Diameter
5. The closest tolerance fit is in class
   a. 1
   b. 2
   c. 3
   d. 4
6. An internal thread is shown with the letter
   a. A
   b. B
   c. C
   d. D

7. A coarse thread with the designation M2 has a
   a. Diameter of $\frac{2}{10}''$
   b. Length of 2"
   c. Diameter of 2 mm
   d. Length of 20 mm
8. The most realistic thread symbol is
   a. Detailed
   b. Schematic
   c. Simplified
   d. Metric
9. A bolt goes through a clearance hole in one part and into a threaded hole in the second part.
   a. True
   b. False
10. A type of rivet that does not have an end that can be bucked or deformed is
    a. Standard
    b. Blind
    c. Flanged
    d. Mandrel

18 AutoCAD Drawings of Fasteners Using Blocks and Attributes

OBJECTIVES

After completing this chapter, you will be able to

- Make a block and save it on the current drawing.
- Make a wblock and save it on a floppy disk so it can be used on other drawings.
- Insert predefined blocks to create an assembly drawing.
- Insert a format containing attributes around the drawing.

INTRODUCTION

In every company that uses AutoCAD many of the most commonly drawn symbols or details are made into drawings that can be used on new drawings. The commands that save these drawings in a form so they can be used easily are Block and Wblock. These blocks or wblocks may contain text items that can be changed as the block is inserted into a new drawing. These text items are called *attributes*. As you will see in this chapter, not all blocks contain attributes.

Block

The Block command saves the selected part of the drawing on the current drawing's database. It allows you to name the block and to specify an insertion point that is the point at which the block is located when it is inserted. The block can be used on the current drawing only. You will create a block and insert it on a drawing in Exercise 18–1. You will also use several predefined blocks to draw a complex assembly and insert a format containing attributes around it in Exercise 18–2.

Wblock

The Wblock command allows you to save the selected part of the drawing as a drawing on a floppy disk or in a directory on the hard drive of your computer. A wblock is the same as any other drawing except its insertion point may be selected, as with the Block command. Any AutoCAD drawing may be inserted into any other AutoCAD drawing of the same or compatible versions. You will create a wblock in Exercise 18–1 that can be used on later drawings.

Attributes

Attributes can accompany any block or wblock if they are defined when the block is created. An attribute is text that is changed as the block is inserted. This text may also be extracted and arranged into a useful form such as a parts list. For example, in a large building containing furniture with attributes, the attributes could be extracted so that you would know how many desks, chairs, and computers are located on each floor and what their part numbers, colors, and descriptions were. In Exercise 18–2 you will insert a format containing attributes around a drawing. In this case the attributes are used to make sure that you place the date, school name, assignment number, and your name in the correct location with the least amount of time and effort.

Insert

The Insert command allows you to insert any AutoCAD drawing into any other AutoCAD drawing that is of a compatible version. You can insert a Release 14 drawing into a Release 2002, for example, but you cannot insert a Release 2002 drawing into a Release 14 drawing. Insert also allows you to insert blocks that have been defined on the current drawing's database into the drawing. Insert allows you to specify exactly where you want to place the inserted drawing or block, and it also allows you to specify the scale of the drawing. You can even specify different X,Y, and Z scale factors if necessary. Unless you specify otherwise, the inserted drawing comes in as a single object. If you want to move, erase, or otherwise edit the block, you must explode it first. You will not want to explode blocks with attributes in most cases because the attributes revert to tags. Tags are the names assigned to the attributes so they can be extracted and are not what you see or need on the drawing. You will use the Insert command in Exercises 18–1 and 18–2 many times.

EXERCISE 18–1
Making and Inserting Blocks and Wblocks

Your final drawing will look like the drawing in Figure 18–1.

Step 1. To begin Exercise 18–1, turn on the computer and start AutoCAD or AutoCAD LT.

Step 2. Open drawing EX18-1 supplied on the disk that came with your book.

Step 3. Use Zoom-All to view the entire drawing area.

Step 4. On you own:
With Snap ON, use the Circle command to draw a 1"-diameter circle in the approximate center of the drawing. Be sure it is a 1" diameter ($\frac{1}{2}$" radius).

Step 5. Draw a hexagon around the circle.

FIGURE 18–1
Exercise 18–1 Completed

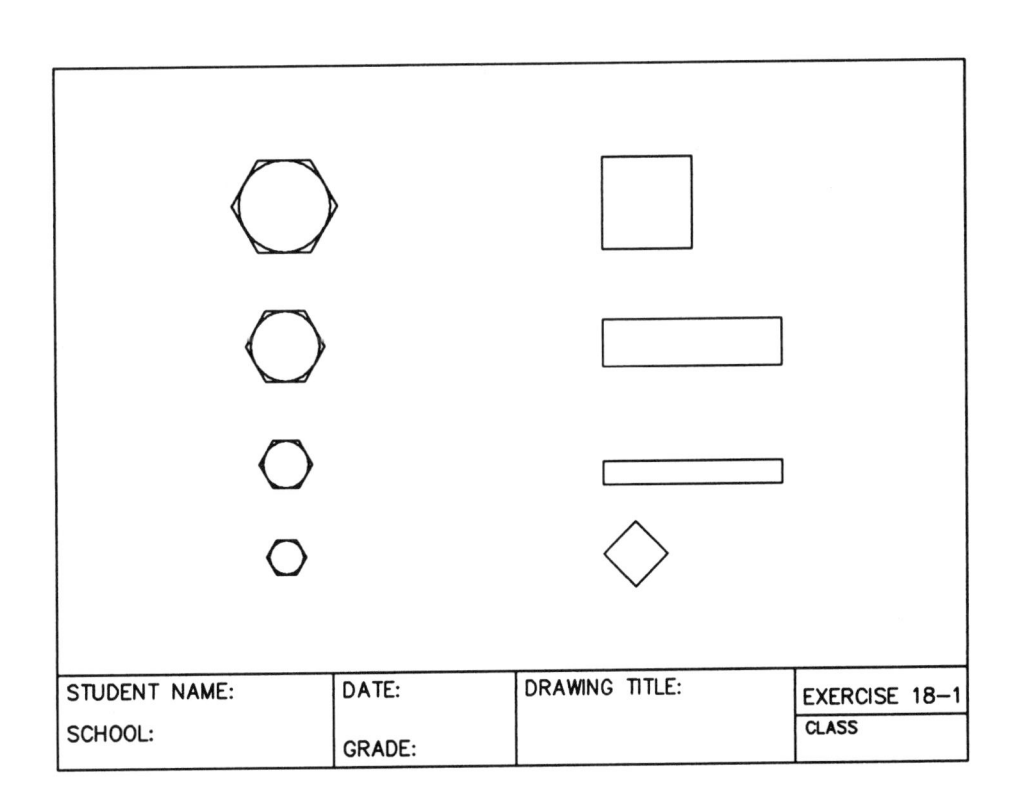

AutoCAD Drawings of Fasteners Using Blocks and Attributes

Prompt	Response
Command:	Type: **POL**↵
Enter number of sides <4>:	Type: **6**↵
Specify Center of polygon or [Edge]:	Type: **CEN**↵
of	Click: **any point on the circle circumference**
Enter an option [Inscribed in circle/ Circumscribed about circle] <I>:	Type: **C**↵
Specify radius of circle:	Type: **.5**↵

You have now drawn the top view of a hexagonal head bolt that can be used for the top view of any size of hexagonal head bolt if it is blocked.

Step 6. Block the hexagon and the circle.

Prompt	Response
Command:	Type: **-B**↵
Enter block name (or ?):	Type: **HEX-HD**↵
Specify insertion base point:	Type: **CEN**↵
of	Click: **any point on the circumference of the circle**
Select objects:	Click: **the circle and the polygon**
Select objects:	↵

The block disappears.

Step 7. Insert the top view of the hex head bolt at scales of 1, .750, .500, and .375.

Prompt	Response
Command:	Type: **-I**↵
Enter block name (or ?):	Type: **HEX-HD**↵
Specify insertion point or [Scale...PRotate]:	Type: **3,6.5**↵
Enter X scale factor, specify opposite corner or [Corner/XYZ] <1>:	↵
Enter Y scale factor <use X scale factor>:	↵
Specify rotation angle <0>:	↵
Command:	↵
Enter block name (or ?) <HEX-HD>:	↵
Specify insertion point or [Scale...PRotate]:	Click: **a point $1\frac{1}{2}''$ down from the center of the first HEX-HD**
Enter X scale factor, specify opposite corner or [Corner/XYZ] <1>:	Type: **.75**↵
Enter Y scale factor <use X scale factor>:	↵
Specify rotation angle <0>:	↵
Command:	↵
Enter block name (or ?) <HEX-HD>:	↵
Specify insertion point or [Scale...PRotate]:	Click: **a point $1\frac{1}{4}''$ down from the center of the second HEX-HD**
Enter X scale factor, specify opposite corner or [Corner/XYZ] <1>:	Type: **.5**↵
Enter Y scale factor <use X scale factor>:	↵
Specify rotation angle <0>:	↵

Command:	↵
Enter block name (or ?) <HEX-HD>:	↵
Specify insertion point or [Scale...PRotate]:	Click: **a point 1" down from the center of the third HEX-HD**
Enter X scale factor, specify opposite corner or [Corner/XYZ] <1>:	Type: **.375**↵
Enter Y scale factor <use X scale factor>:	↵
Specify rotation angle <0>:	↵

Step 8. Turn off dialog boxes so the prompts can be read from the command line.

Prompt	Response
Command:	Type: **FILEDIA**↵
Enter new value for FILEDIA <1>:	Type: **0**↵

Step 9. Wblock the HEX-HD block so it can be used on a later drawing.

Prompt	Response
Command:	Type: **-W**↵
Enter name of output file:	Type: **A:HEX**↵ (Be sure you have a floppy disk in the A drive with a label on it.)
Enter name of existing block or [=(block=output file)*(whole drawing)] <define new drawing>:	Type: **HEX-HD**↵

Because the HEX-HD block had already been defined on the drawing, the Wblock command did not ask for an insertion point or for you to select objects.

The HEX drawing is now on your floppy disk.

Step 10. On your own:
Draw a 1" square in an open space on your drawing.

Step 11. Wblock the 1" square to the floppy disk in your A drive.

Prompt	Response
Command:	Type: **-W**↵
Enter name of output file:	Type: **A:SQUARE**↵ (Be sure you have a floppy disk in the A drive with a label on it.)
Enter name of existing block or [------]<define new drawing>:	↵ (There is no predefined block on your drawing for the square.)
Specify insertion base point:	Type: **END**↵
of	Click: **the upper left corner of your square**
Select objects:	Click: **select the square**

Step 12. Insert the SQUARE drawing into the current drawing with scales of 1, X=2 Y=.5, X=2 Y=.25, and .25 rotated.

Prompt	Response
Command:	Type: **-I**↵
Enter block name (or ?) <HEX-HD>:	Type: **A:SQUARE**↵
Specify insertion point or [Scale...PRotate]:	Type: **6.5,7**↵

Prompt	Response
Enter X scale factor, specify opposite corner or [Corner/XYZ] <1>:	↵
Enter Y scale factor <use X scale factor>:	↵
Specify rotation angle <0>:	↵
Command:	↵
Enter block name (or ?) <SQUARE>:	↵
Specify insertion point or [Scale...PRotate]:	Click: **a point $\frac{3}{4}''$ down from the lower left corner of the first SQUARE**
Enter X scale factor, specify opposite corner or [Corner/XYZ] <1>:	Type: **2.**↵
Enter Y scale factor <use X scale factor>:	Type: **.5.**↵
Specify rotation angle <0>:	↵
Command:	↵
Enter block name (or ?) <SQUARE>:	↵
Specify insertion point or [Scale...PRotate]:	Click: **a point 1″ down from the lower left corner of the second SQUARE**
Enter X scale factor, specify opposite corner or [Corner/XYZ] <1>:	Type: **2.**↵
Enter Y scale factor <use X scale factor>:	Type: **.25.**↵
Specify rotation angle <0>:	↵
Command:	↵
Enter block name (or ?) <SQUARE>:	↵
Specify insertion point or [Scale...PRotate]:	Click: **a point $\frac{3}{4}''$ down from the lower left corner of the third SQUARE**
Enter X scale factor, specify opposite corner or [Corner/XYZ] <1>:	Type: **.25.**↵
Enter Y scale factor <use X scale factor>:	↵
Specify rotation angle <0>:	Type: **45.**↵

Step 13. On your own:
Turn on dialog boxes (Type: **FILEDIA**↵, then Type: **1**↵)
Use the Dtext command to complete the title block with .12 high letters.
Use the SAVEAS command to save your drawing as EX18-1(your initials) on a floppy disk and again on the hard drive of your computer.
Plot or print your drawing full size on an 11″ × 8½″ sheet.

EXERCISE 18–2
Making a Drawing Containing Predefined Blocks of Fasteners and a Format

Your final drawing will look like the drawing in Figure 18–2.

INSERTING THE FORMAT

Step 1. To begin Exercise 18–2, turn on the computer and start AutoCAD or AutoCAD LT.

Step 2. Open drawing EX18-2 supplied on the disk that came with your book.

FIGURE 18–2
Exercise 18–2 Completed

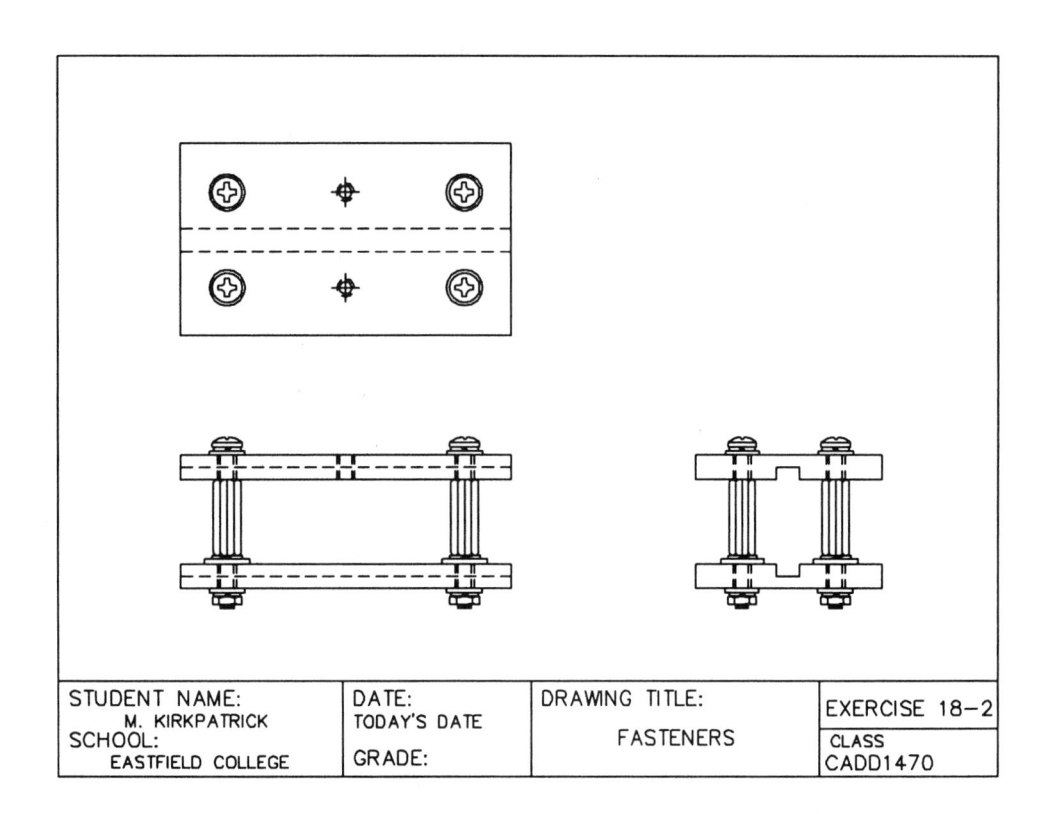

STUDENT NAME: M. KIRKPATRICK SCHOOL: EASTFIELD COLLEGE	DATE: TODAY'S DATE GRADE:	DRAWING TITLE: FASTENERS	EXERCISE 18–2 CLASS CADD1470

Step 3. Use Zoom-All to view the entire drawing area.
Step 4. Insert a predefined format that contains attributes.

Prompt	Response
Command:	Type: **-I↵**
Enter block name (or ?):	Type: **TB↵**
Specify insertion point or [Scale...PRotate]:	Type: **0,0↵**
Enter X scale factor, specify opposite corner or [Corner/XYZ] <1>:	↵
Enter Y scale factor <use X scale factor>:	↵
Specify rotation angle <0>:	↵
TYPE THE NAME OF YOUR SCHOOL <XX>:	Type: **your school name in all caps.↵**
TYPE YOUR FIRST INITIAL AND LAST NAME <XX>:	Type: **your first initial and last name in all caps.↵**
TYPE TODAY'S DATE<XX>:	Type: **today's date.↵**
TYPE THE LETTERS AND NUMBERS OF YOUR CLASS NAME <XX>:	Type: **the numbers and letters identifying your class.↵**
TYPE THE EXERCISE NUMBER <XX>:	Type: **EX18-2.↵**
TYPE THE TITLE OF YOUR DRAWING <XX>:	Type: **FASTENERS.↵**

DRAWING THE TOP VIEW

Step 5. **On your own:**
Set the Object layer current.

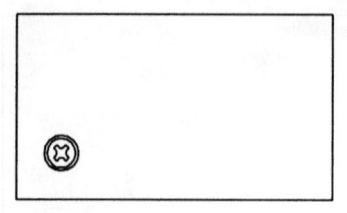

FIGURE 18–3
Inserting the Predefined Block PHIL and Draw the Top View of the Washer

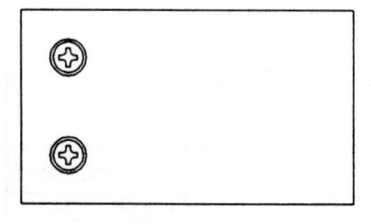

FIGURE 18–4
Mirroring the Phillips Head Block and Washer

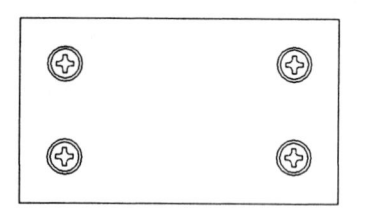

FIGURE 18–5
Mirroring Both Phillips Head Blocks and Washers to the Right Side of the Front View

Step 6. Draw the outline of the top view.

Prompt	Response
Command:	Type: **L**↵
Specify first point:	Type: **1.75,5**↵
Specify next point or [Undo]:	Type: **@3.5,0**↵
Specify next point or [Undo]:	Type: **@0,2**↵
Specify next point or [Close/Undo]:	Type: **@-3.5,0**↵
Specify next point or [Close/Undo]:	Type: **C**↵

Step 7. Insert a predefined block of the top view of a Phillips head screw .500 to the right and .500 up from the lower left corner of the top view (Figure 18–3), and draw the top view of the washer under the screw.

Prompt	Response
Command:	Type: **ID**↵
Specify point:	Type: **END**↵
of	Click: **the lower left corner of your top view**
Command:	Type: **-I**↵
Enter block name (or ?)<TB>:	Type: **PHIL**↵
Specify insertion point or [Scale...PRotate]:	Type: **@.5,.5**↵
Enter X scale factor, specify opposite corner or [Corner/XYZ] <1>:	↵
Enter Y scale factor <use X scale factor>:	↵
Specify rotation angle <0>:	Type: **45**↵
Command:	Type: **C**↵
Specify Center point for circle or [3P/2P/Ttr(tan tan radius)]:	Type: **INS**↵ (the insertion point of the PHIL block)
of	Click: **any point on the PHIL block**
Specify radius of circle or [Diameter]:	Type: **.19**↵

Step 8. On your own:

1. Use the Mirror command to copy the Phillips head block and washer to the location shown in Figure 18–4. Use the midpoints of the vertical lines as the mirror line.
2. Use the Mirror command to copy the two Phillips head blocks and washers to the right side of the top view, as shown in Figure 18–5. Use the midpoints of the horizontal lines as the mirror line.

Step 9. Insert a predefined block of a no. 10 threaded hole. (A no. 10 indicates the size of a hole for a no. 10 screw.)

Prompt	Response
Command:	Type: **ID**↵
Specify point:	Type: **END**↵
of	Click: **the lower left corner of your top view**
Command:	Type: **-I**↵
Enter block name (or ?)<PHIL>:	Type: **THREAD10**↵

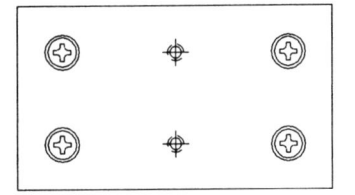

FIGURE 18-6
Copying the Thread Hole Symbol 1″ Up

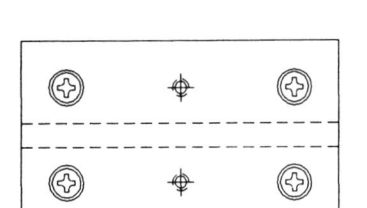

FIGURE 18-7
Drawing Hidden Lines in the Top View

Specify insertion point or [Scale...PRotate]:	Type: **@1.75,.5**↵
Enter X scale factor, specify opposite corner or [Corner/XYZ]:	↵
Enter Y scale factor (default=X) <use X scale factor>:	↵
Specify rotation angle <0>:	↵

Step 10.   On your own:
1. Copy the threaded hole 1″ up (Figure 18–6). Any point will do for the base point. Use relative coordinates when AutoCAD asks for "Second point of displacement," @0,1.
2. Set the Hid layer current.

Step 11.   With the Hid layer current draw hidden lines in the top view (Figure 18–7).

Prompt	Response
Command:	Type: **ID**↵
Specify point:	Type: **END**↵
of	Click: **the lower left corner of the top view**
Command:	Type: **L**↵
Specify first point:	Type: **@0,.875**↵
Specify next point or [Undo]:	Type: **PER**↵
to	Click: **the right vertical line of the top view**
To point:	↵
Command:	Type: **O**↵
Specify offset distance or [Through] <Through>:	Type: **.25**↵
Select object to offset:	Click: **the hidden line just drawn**
Specify point on side to offset:	Click: **any point above the existing hidden line**

DRAWING THE FRONT VIEW

Step 12.   On your own:
Set layer Object current.

Step 13.   With layer Object current draw the outline of the front view (Figure 18–8).

FIGURE 18-8
Drawing the Outline and Hidden Lines of the Front View

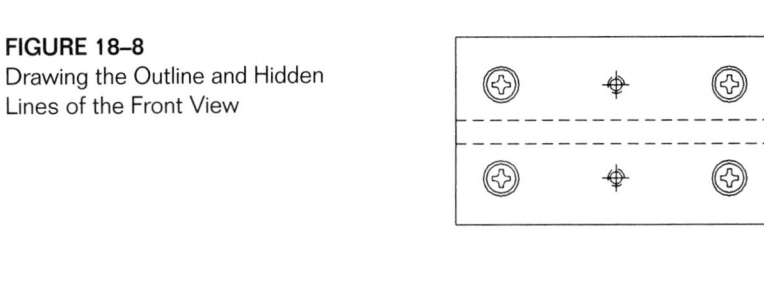

AutoCAD Drawings of Fasteners Using Blocks and Attributes

Prompt	Response
Command:	Type: **L**↵
Specify first point:	Type: **1.75,3.625**↵
Specify next point or [Undo]:	Type: **@3.5,0**↵
Specify next point or [Undo]:	Type: **@0,.25**↵
Specify next point or [Close/Undo]:	Type: **@-3.5,0**↵
Specify next point or [Close/Undo]:	Type: **C**↵

Step 14. Offset one of the hidden lines in the top view through the midpoint of one of the short vertical lines in the front view.

Prompt	Response
Command:	Type: **O**↵
Specify offset distance or [Through] <0.2500>:	Type: **T**↵
Select object to offset or <exit>:	Click: **one of the hidden lines in the top view**
Specify through point:	Type: **MID**↵
of	Click: **D1** (Figure 18–8)

Step 15. Copy the rectangle and hidden line to form the bottom shape of the front view.

Prompt	Response
Command:	Type: **CP**↵
Select objects:	**Use a window to select the rectangle and hidden line in the front view.**
Select objects:	↵
Specify base point or displacement or [Multiple]:	Click: **any point**
Specify second point of displacement or <use first point as displacement>:	Type: **@0,-1.125**↵

Step 16. Insert predefined blocks named STANDOFF, SIDE-PH, and SIDE-NUT to show the fasteners in the front view (**Figure 18–9**).

Prompt	Response
Command:	Type: **ID**↵
Specify point:	Type: **END**↵
of	Click: **D1** (Figure 18–9)
Command:	Type: **-I**↵
Enter block name (or ?)<THREAD10>:	Type: **STANDOFF**↵
Specify insertion point or [Scale...PRotate]:	Type: **@.5,0**↵
Enter X scale factor, specify opposite corner or [Corner/XYZ] <1>:	↵
Enter Y scale factor <use X scale factor>:	↵
Specify rotation angle <0>:	↵

FIGURE 18–9
Inserting and Copying the Fastener Blocks

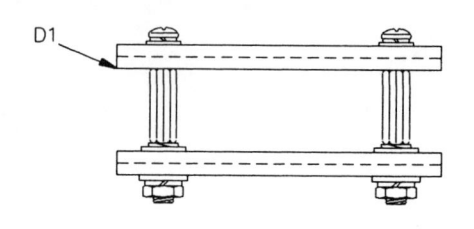

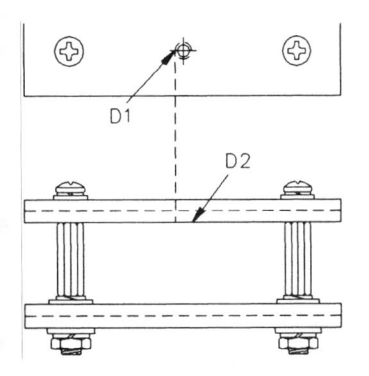

FIGURE 18–10
Drawing a Hidden Line for the Threaded Hole

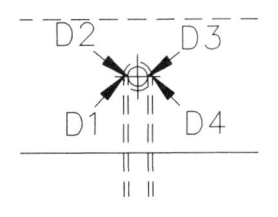

FIGURE 18–11
Copying the Hidden Line to Show Threads

Prompt	Response
Command:	↵
Enter block name (or ?)<STANDOFF>:	Type: **SIDE-PH**↵
Specify insertion point or [Scale...PRotate]:	Type: **@0,.25**↵ (These coordinates are from the last point picked, the insertion point of the STANDOFF block.)
Enter X scale factor, specify opposite corner or [Corner/XYZ] <1>:	↵
Enter Y scale factor <use X scale factor>:	↵
Specify rotation angle <0>:	↵
Command:	↵
Enter block name (or ?)<SIDE-PH>:	Type: **SIDE-NUT**↵
Specify insertion point or [Scale...PRotate]:	Type: **@0,-1.375**↵ (These coordinates are from the last point picked, the insertion point of the SIDE-PH block.)
Enter X scale factor, specify opposite corner or [Corner/XYZ] <1>:	↵
Enter Y scale factor <use X scale factor>:	↵
Specify rotation angle <0>:	↵

Step 17. On your own:
Use the Mirror command to copy the entire screw assembly to the right side of the front view. Use the midpoint of the top line of the rectangle as the first point of the mirror line and the midpoint of the bottom line of the rectangle as the second point.
Set layer Hid current.

Step 18. With layer Hid current draw hidden lines for the threaded holes and fasteners in the front view.

Prompt	Response
Command:	Type: **L**↵
Specify first point:	Type: **QUA**↵
of	Click: **D1** (Figure 18–10)
Specify next point or [Undo]:	Type: **PER**↵
to	Click: **D2** (any point on this line)
Specify next point or [Undo]:	↵
Command:	**Zoom a window around the threaded hole in the top view** (Figure 18–11).
Command:	Type: **CP**↵
Select objects:	Click: **the hidden line just drawn**
Select objects:	↵
Specify base point or displacement or [Multiple]:	Type: **M**↵
Specify base point:	Type: **QUA**↵
of	Click: **D1** (Figure 18–11)
Specify second point of displacement or <use first point as displacement>:	Type: **QUA**↵
of	Click: **D2**
Specify second point of displacement or <use first point as displacement>:	Type: **QUA**↵
of	Click: **D3**

AutoCAD Drawings of Fasteners Using Blocks and Attributes

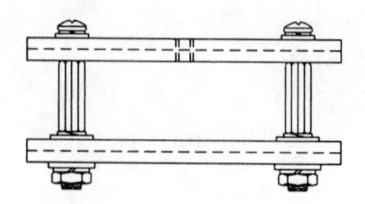

FIGURE 18–12
Trimming the Hidden Lines

FIGURE 18–13
Copying the Hidden Lines to Show the Screws

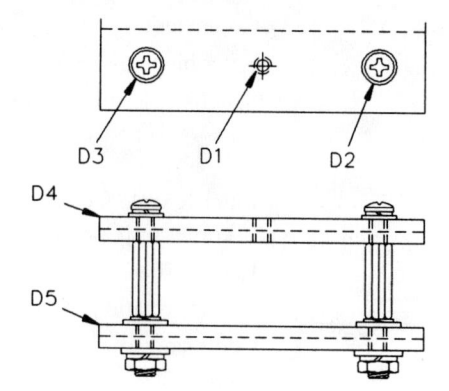

Prompt	Response
Specify second point of displacement or <use first point as displacement>:	Type: **QUA**↵
of	Click: **D4**
Specify second point of displacement or <use first point as displacement>:	↵
Command:	Type: **Z**↵
[All/Center/ Dynamic/Extents/Previous/ Scale/Window] <real time>:	Type: **P**↵
Command:	Type: **TR**↵
Select cutting edges:	
Select objects:	Click: **the top line of the top rectangle in the front view**
Select cutting edges:	
Select objects:	↵
Select object to trim or shift-select to extend or [Project/Edge/Undo]:	Click: **the part of the hidden line that is above the cutting edge** (Figure 18–12)

Select object to trim or shift-select to extend or [Project/Edge/Undo]: ↵

Step 19. Copy the four hidden lines to show the screws in the top plate.

Prompt	Response
Command:	Type: **CP**↵
Select objects:	**Use a window to select the four hidden lines.**
Select objects:	↵
Specify base point or displacement or [Multiple]:	Type: **M**↵
Specify base point:	Type: **CEN**↵
of	Click: **D1** (Figure 18–13)
Specify second point of displacement or <use first point as displacement>:	Type: **CEN**↵
of	Click: **D2**
Specify second point of displacement or <use first point as displacement>:	Type: **CEN**↵

Prompt	Response
of	Click: **D3**
Specify second point of displacement or <use first point as displacement>:	↵

Step 20. Copy eight of the hidden lines to show the screws in the bottom plate.

Prompt	Response
Command:	↵
Select objects:	**Use a window to select the four hidden lines under the screw head on the left side.**
Select objects:	**Use a window to select the four hidden lines under the screw head on the right side.**
Select objects:	↵
Specify base point or displacement or [Multiple]:	Type: **END**↵
of	Click: **D4** (Figure 18–13)
Specify second point of displacement or <use first point as displacement>:	Type: **END**↵
of	Click: **D5**

DRAWING THE RIGHT-SIDE VIEW

Step 21. On your own:
Set layer Object current.

Step 22. Use the line command to draw the right-side view of the top plate.

Prompt	Response
Command:	Type: **L**↵
Specify first point:	Type: **7.375,3.625**↵
Specify next point or [Undo]:	Type: **@.875,0**↵
Specify next point or [Undo]:	Type: **@0,.125**↵
Specify next point or [Close/Undo]:	Type: **@.25,0**↵
Specify next point or [Close/Undo]:	Type: **@0,-.125**↵
Specify next point or [Close/Undo]:	Type: **@.875,0**↵
Specify next point or [Close/Undo]:	Type: **@0,.25**↵
Specify next point or [Close/Undo]:	Type: **@-2,0**↵
Specify next point or [Close/Undo]:	**C**↵

Step 23. Use the Copy command to copy the top plate down and rotate it to form the bottom plate.

Prompt	Response
Command:	Type: **CP**↵
Select objects:	**Use a Window to select the top plate.**
Select objects:	↵
Specify base point or displacement [Multiple]:	Click: **any point**
Specify second point of displacement or <use first point as displacement>:	Type: **@0,-1.375**↵
Command:	Type: **RO**↵

AutoCAD Drawings of Fasteners Using Blocks and Attributes

FIGURE 18–14
Copying the Top Plate and Rotating the Copy to Form the Bottom Plate

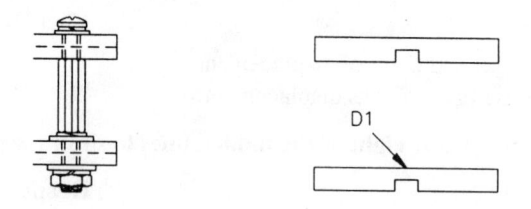

FIGURE 18–15
Copying the Screw, Standoff, Nut, and Washers to the Right-Side View

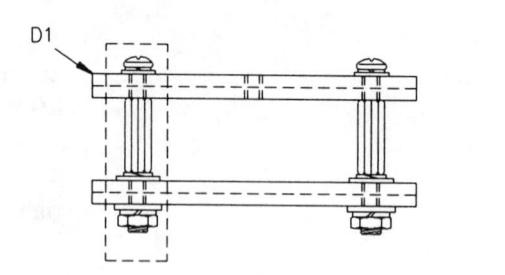

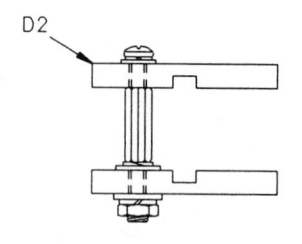

Prompt	Response
Select objects:	Use a window to select the copy of the top plate.
Select objects:	↵
Specify base point:	Type: **MID**↵
of	Click: **D1** (Figure 18–14)
Specify Rotation angle or [Reference]:	Type: **180**↵

Step 24. **Use the Copy command to copy the screw, standoff, nut, and washers from the left side of the front view to the right-side view.** (Because the screw is located .500 to the right and .500 up from the lower left corner of the top view you can use D1 and D2 in Figure 18–15 to locate the copy.)

Prompt	Response
Command:	Type: **CP**↵
Select objects:	Use a Window as shown in Figure 18–15 to select the objects.
Select objects:	↵
Specify base point or displacement or [Multiple]:	Click: **D1** (Figure 18–15)
Specify second point of displacement or <use first point as displacement>:	Click: **D2**

Step 25. **On your own:**

1. Use the Mirror command to copy the screw, standoff, nut, and washers to the right side of the right-side view.

   Use a window to select the objects.

   Use Osnap-Midpoint to select D1 (Figure 18–16) for the first point on the mirror line.

   Use Osnap-Midpoint to select D2 (Figure 18–16) for the second point on the mirror line.

2. Save the drawing in two places.

3. Print the drawing at 1=1 (Plotted inches = Drawing Units) on an 11″ × 8.5″ sheet.

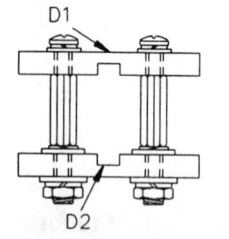

FIGURE 18–16
Using the Mirror Command to Copy the Screw, Standoff, Nut, and Washers to the Right Side of the Side View

EXERCISES

EXERCISE 18–1   Complete Exercise 18–1 using steps 1 through 13 described in this chapter.
EXERCISE 18–2   Complete Exercise 18–2 using steps 1 through 25 described in this chapter.

REVIEW QUESTIONS

1. Which command used in this chapter creates a block that can be used on the current drawing only?
   a. Block
   b. Wblock
   c. Insert
   d. Attribute
2. Which command used in this chapter creates a block that can be used on any drawing?
   a. Block
   b. Wblock
   c. Insert
   d. Attribute
3. All blocks contain attributes.
   a. True
   b. False
4. To change an inserted block into separate objects which of the following commands is used?
   a. Insert
   b. Block
   c. Trim
   d. Explode
5. If you want a block to be twice the size that it was when it was created, which of the following is correct?
   a. X scale factor = 2, Y scale factor = 1
   b. X scale factor = 2, Y scale factor = X
   c. X scale factor = 1, Y scale factor = 2
   d. X scale factor = .5, Y scale factor = X
6. If you want a block to be twice the length and half the height it was when it was created, which of the following is correct?
   a. X scale factor = 2, Y scale factor = 1
   b. X scale factor = .5, Y scale factor = 2
   c. X scale factor = 1, Y scale factor = 2
   d. X scale factor = 2, Y scale factor = .5
7. Blocks containing attributes should not be exploded because
   a. The attributes revert to tags
   b. The attributes disappear
   c. Blocks containing attributes cannot be exploded
   d. No block should ever be exploded.
8. Which command should be used to draw a hexagon?
   a. Hexagon
   b. Rectangle
   c. Polyline
   d. Polygon
9. Which of the following is the correct drawing name if you want to Wblock a drawing with the name HEX to a floppy disk in the A drive?
   a. A:HEX
   b. HEX
   c. HEX-A:
   d. A-HEX
10. Which command will identify a point to AutoCAD so that a block can be inserted a specified distance from that point?
    a. Insert
    b. Osnap-Insertion
    c. ID
    d. Wblock

19 Sketching Development Drawings

OBJECTIVES

When you have completed this chapter, you will be able to

- Sketch accurate parallel line development drawings of rectangles and cylinders.
- Sketch accurate radial line development drawings of pyramids and cones.
- Sketch an accurate development drawings of a warped surface, using triangulation.

INTRODUCTION

A *development drawing* is a flat pattern of a sheet metal (or other material) part. The concepts needed to make accurate developments can be used in many different types of drawing. Development drawings are not dimensioned and do not have any hidden lines (Figure 19–1). There are three methods for making developments: parallel line, radial line, and triangulation. Figure 19–2 shows examples of the three methods.

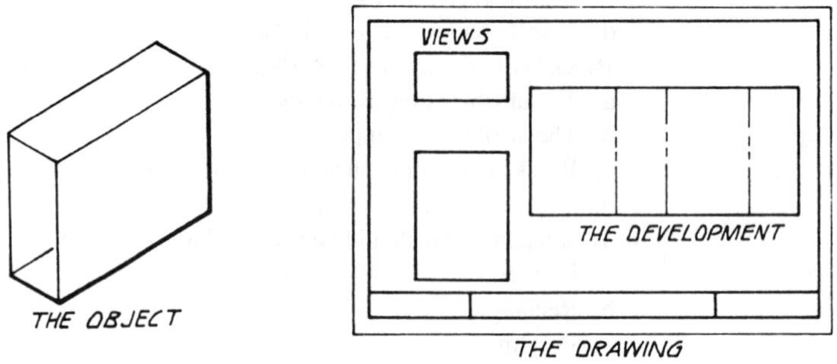

FIGURE 19–1
A Development Drawing

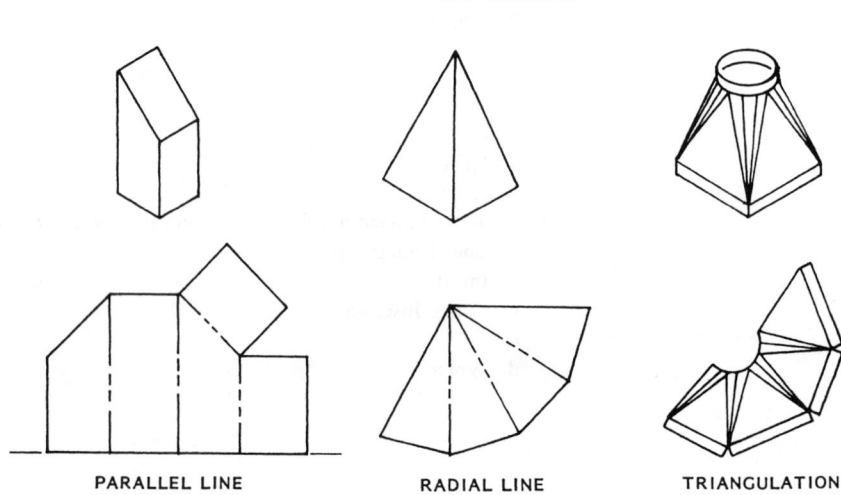

FIGURE 19–2
Methods of Development

FIGURE 19-3
The Stretch-out Line

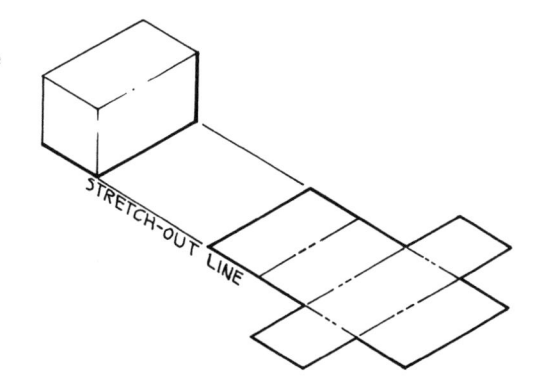

Parallel Line Development

A simple example of parallel line development is the flat pattern of a box (Figure 19-3). The stretch-out line shown is used as a reference line for horizontal and vertical dimensions. Use very light lines for construction, and draw the pattern using standard object line weight.

Parallel Line Method

To draw the development of the truncated prism shown in Figure 19-4 (truncate means to shorten by cutting off a part), follow these steps:

Step 1. **Draw top and front views and number the corners.**
Step 2. **Draw a stretch-out line and mark off the length of each side on it. (Lengths of the sides are found in the top view.)**

Sketching Development Drawings

FIGURE 19–4
Parallel Line Development of a Truncated Prism

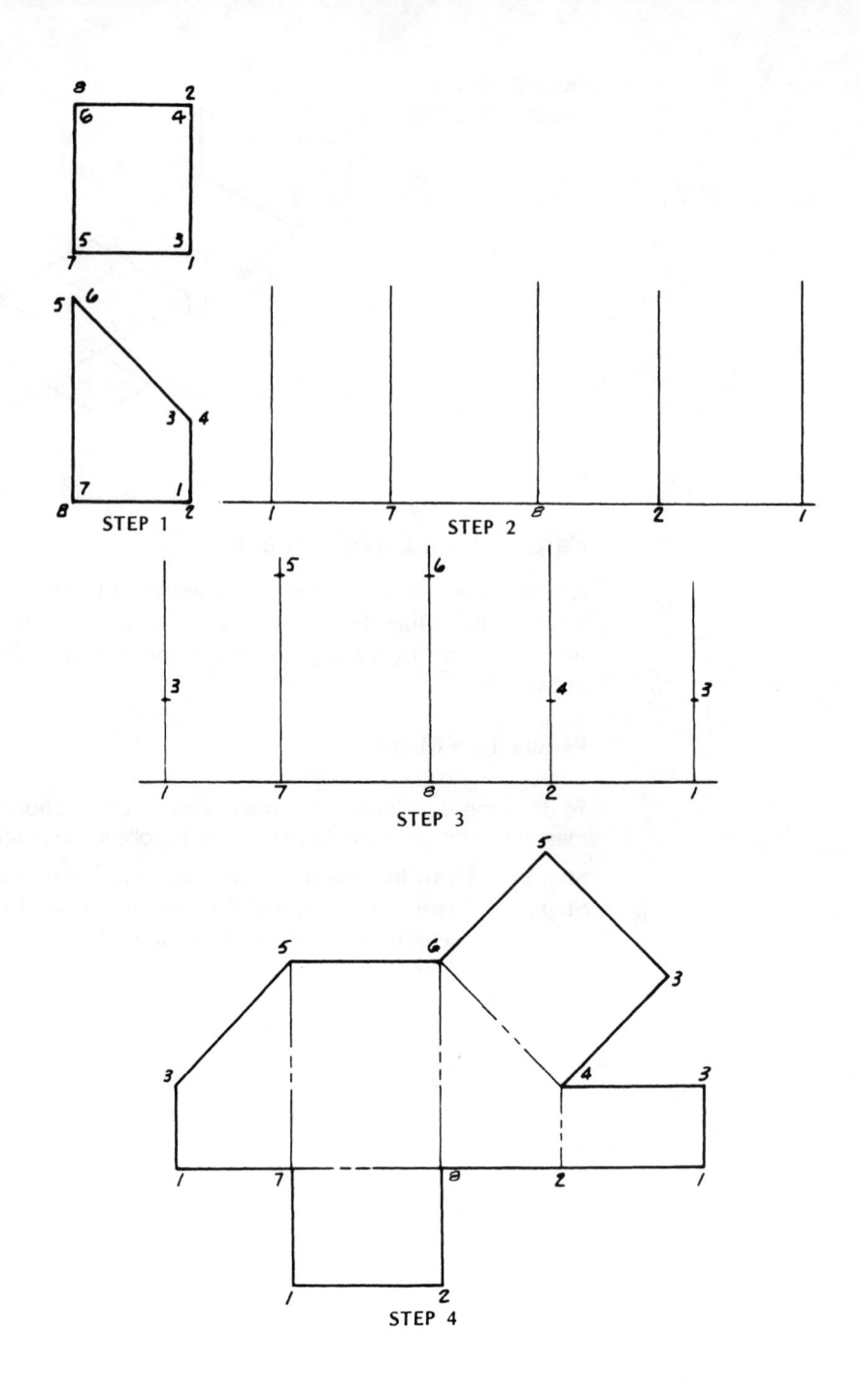

Step 3. Draw vertical lines showing the bends at each intersection, and mark off heights on them.

Step 4. Connect the points to form the sides and add bottom and top pieces. (The depths of the top and bottom pieces are taken from the top view.)

284

Chapter 19

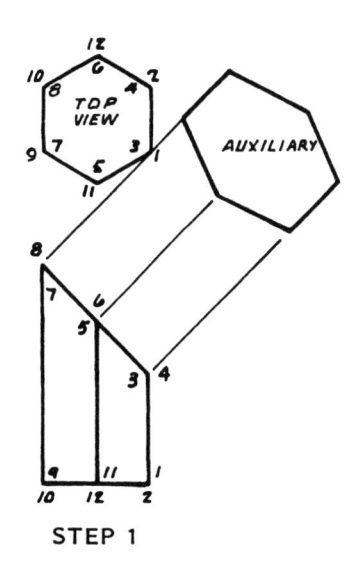

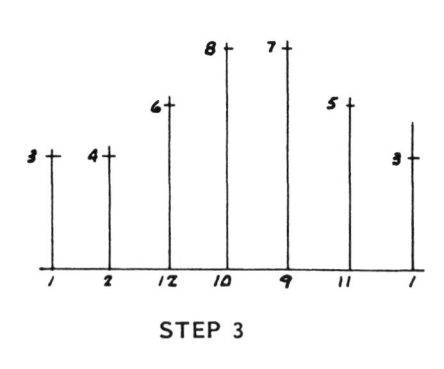

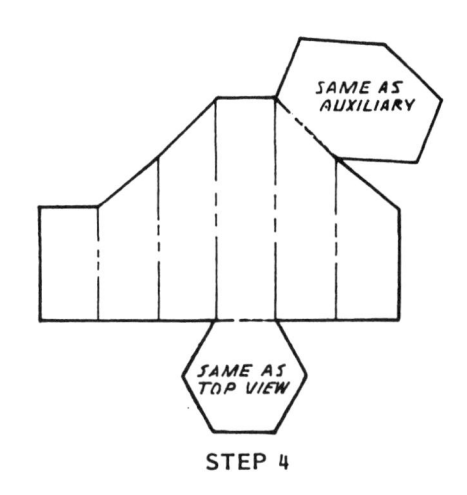

FIGURE 19–5
Parallel Line Development of a Truncated Polygon Prism

To draw the development of the truncated polygon prism shown in Figure 19-5, follow these steps:

Step 1. Draw top and front views and number the corners.
Step 2. Draw a stretch-out line and mark off the length of each side on it. (Lengths of the sides are found in the top view.)
Step 3. Draw vertical lines showing the bends at each intersection, and mark off heights on them.
Step 4. Connect the points to form the sides and add bottom and top pieces. The bottom piece is the same as the top view of the object. The top of the slanted surface must be found by drawing an auxiliary view.

Sketching Development Drawings

FIGURE 19–6
Parallel Line Development of a Cylinder

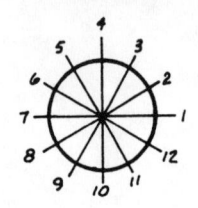

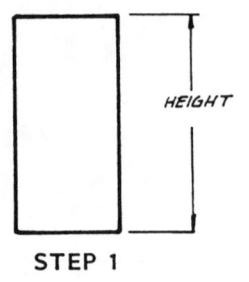

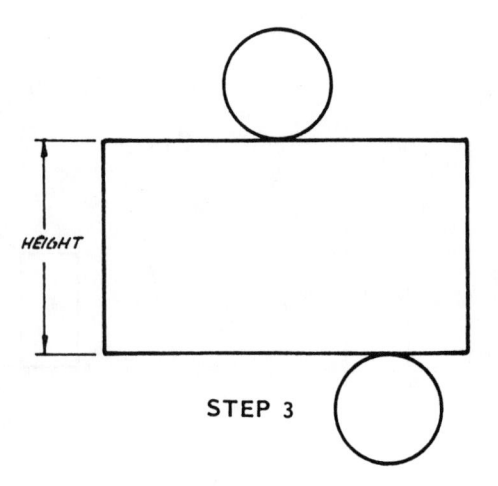

To draw the development of the cylinder shown in Figure 19–6, follow these steps:

Step 1. Draw top and front views. Divide the circumference (the outside boundary of the circle) of the top view into any number of equal parts (using 30° angles works well). The smaller the units are, the greater the accuracy of the part will be. Number the points on the circumference.

Step 2. Draw a stretch-out line and mark off the lengths of the circumference on it.

Step 3. Draw vertical lines at the ends of the circumference stretch-out line, mark off the height, and draw the top edge of the cylinder.

Step 4. Add a top and a bottom to complete the development.

To draw the development of the truncated cylinder shown in Figure 19–7, follow these steps:

Step 1. Draw top and front views. Divide the circumference (the outside boundary of the circle) of the top view into any number of equal parts (using 30° angles works well). The smaller the units are, the greater the accuracy of the part will be. Number the points on the circumference.

Step 2. Project the points, using vertical lines from the top view, onto the slanted surface in the front view. Number the points on the slanted surface.

Step 3. Draw the stretch-out line. Mark off the parts of the circumference of the circle on it and number them. (Take these measurements from the top view.) Draw vertical lines from each point on the circumference. Project the height of each part from the front view onto the vertical lines.

Step 4. Develop an auxiliary view of the slanted surface. This will be the top surface of the truncated cylinder.

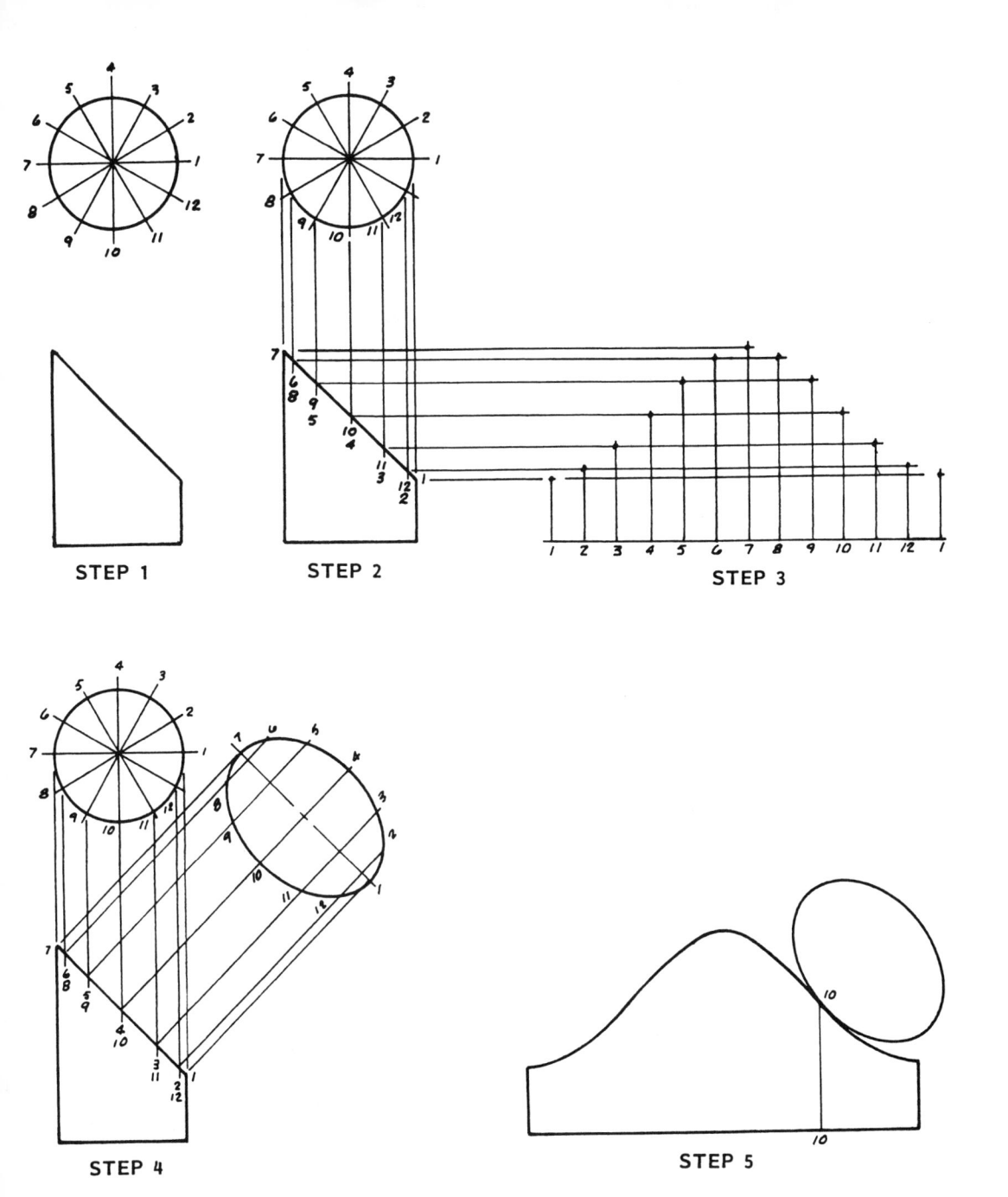

FIGURE 19-7
Parallel Line Development of a Truncated Cylinder

Step 5. Draw a smooth curve through the points, and attach the top surface at point 10, the center of the cylinder.

Radial Line Development

The radial line development method is used for such objects as pyramids and cones in which the sides of the object meet at a common point. Radial line developments are usually drawn as a series of triangles meeting at a common point.

Radial Line Method

To draw the development of the pyramid in Figure 19–8 follow these steps:

Step 1. Draw front and top views and number the corners. Find the true length (AB) of the slanted sides by revolving one of the lines to a centerline and projecting it into the front view. (This will give you the same measurement as drawing an auxiliary view of one of the sides.)

Sketching Development Drawings

287

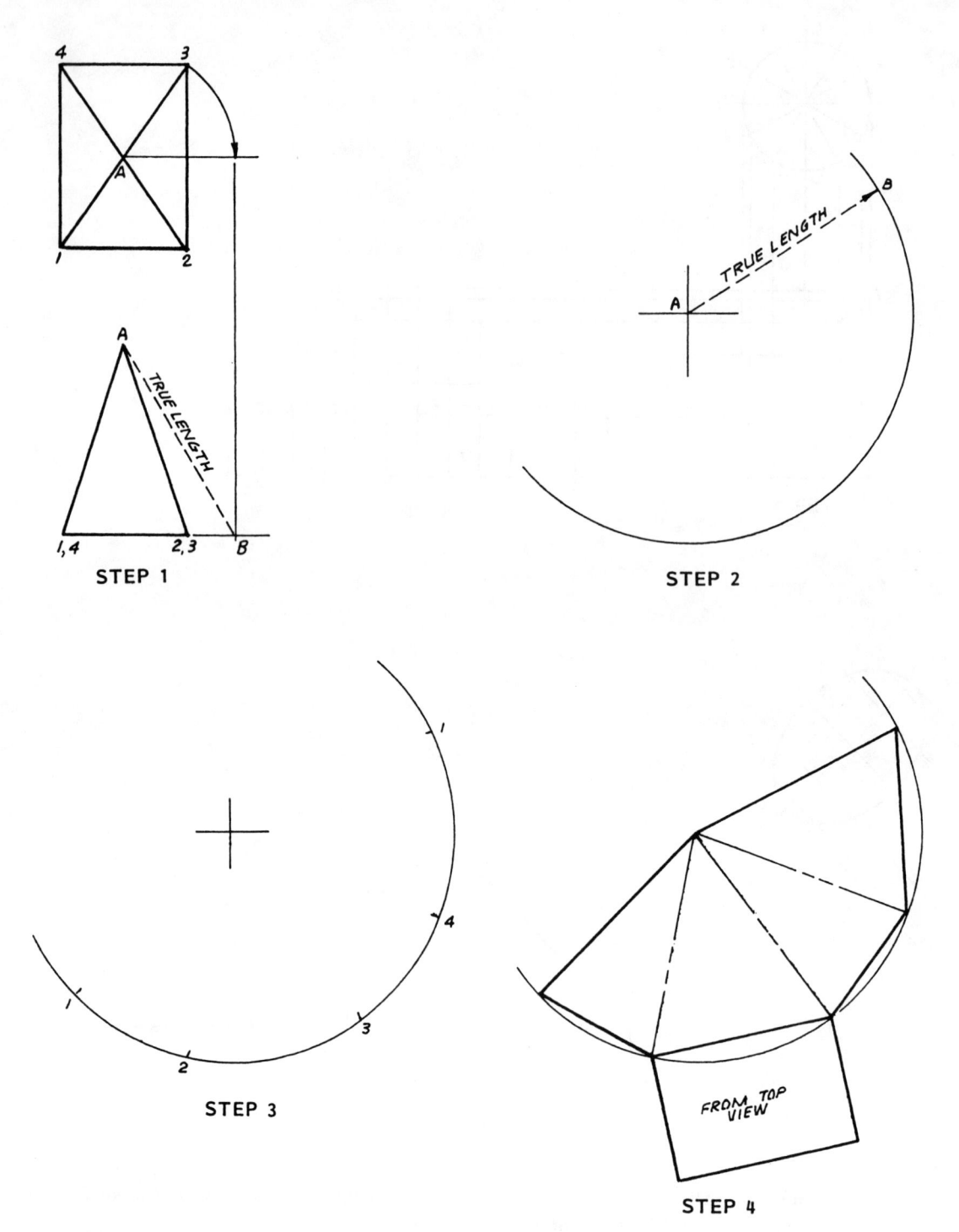

FIGURE 19–8
Radial Line Development of a Pyramid

Step 2. Using line AB from the front view as a radius, draw an arc.
Step 3. Using the measurements of the base (taken from the top view), mark off the bottom edges of the pyramid on the radial stretch-out line. Connect the numbered points.
Step 4. Connect all the numbered points to point A.
Step 5. Add the bottom to complete the development.

To draw the development of the truncated pyramid in Figure 19–9, follow these steps:

Step 1. Draw front and top views. Draw light construction lines to form a complete pyramid. Find the true length (AB) of the slanted sides of the pyramid by revolving one of the lines to a center line and projecting it into the front view.

Step 2. Find the true length (AC) of the distance from the tip of the pyramid to points 1 and 2 by revolving point 2 to the center line and projecting it onto a horizontal line through point 2 in the front view. Find the true length of line AD in the same manner.

Step 3. Using line AB from the front view, draw an arc.

Step 4. Mark off the bottom edges of the pyramid on the radial stretch-out line and connect those points.

Step 5. Using the radius AC from step 2, locate points 1 and 2 by swinging an arc through construction lines connecting points 5 and 6 to point A. Locate points 3 and 4 using the radius AD.

Step 6. Connect the plotted points to form the sides of the pyramid.

Step 7. Construct an auxiliary view to find the true shape of the slanted surface that forms the top surface.

Step 8. Add top and bottom surfaces to allow the part to be manufactured as easily as possible and to conserve the greatest amount of material.

To draw the development of the truncated cone in Figure 19–10, follow these steps:

Step 1. Draw front and side views of the cone and extend the sides of the cone until they meet.

Step 2. Divide the front view into any number of equal parts. Number the points around the circumference of the circle and project these points into the right-side view. Also draw lines from these points in the right-side view to point A as shown in step 4.

Step 3. Using point A as a center and the true length of the side of the cone as a radius, draw the radial stretch-out line. Mark off the length of each part of the circumference on the stretch-out line. Number these points and draw lines from them to point A.

Step 4. Draw vertical lines from the intersections on the slanted surface to a true-length line (the one on the bottom is the most convenient). Using point A as a center and the intersection on the true length line as a radius, swing arcs to locate points on the top edge of the cone. Connect all the points to form the sides of the cone.

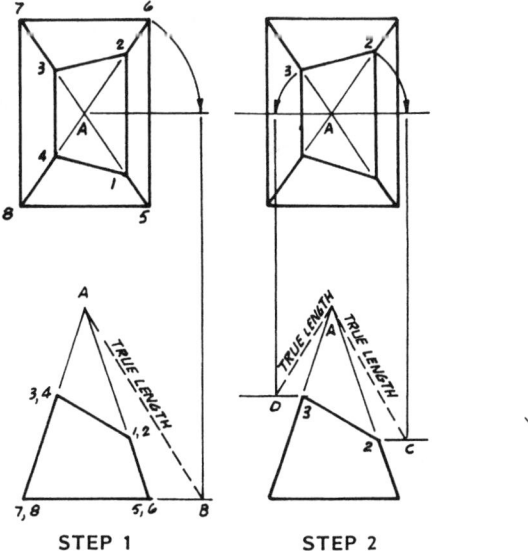

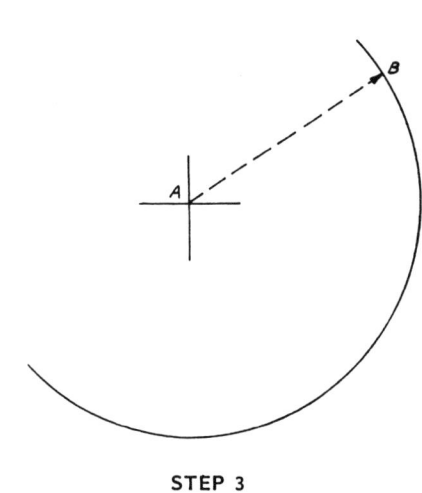

FIGURE 19–9
(Sheet 1 of 2)
Radial Line Development of a Truncated Pyramid

Sketching Development Drawings

STEP 4

STEP 5

STEP 6

STEP 7

TRUE SHAPE

TRUE SHAPE OF AUXILIARY

FROM TOP VIEW

STEP 8

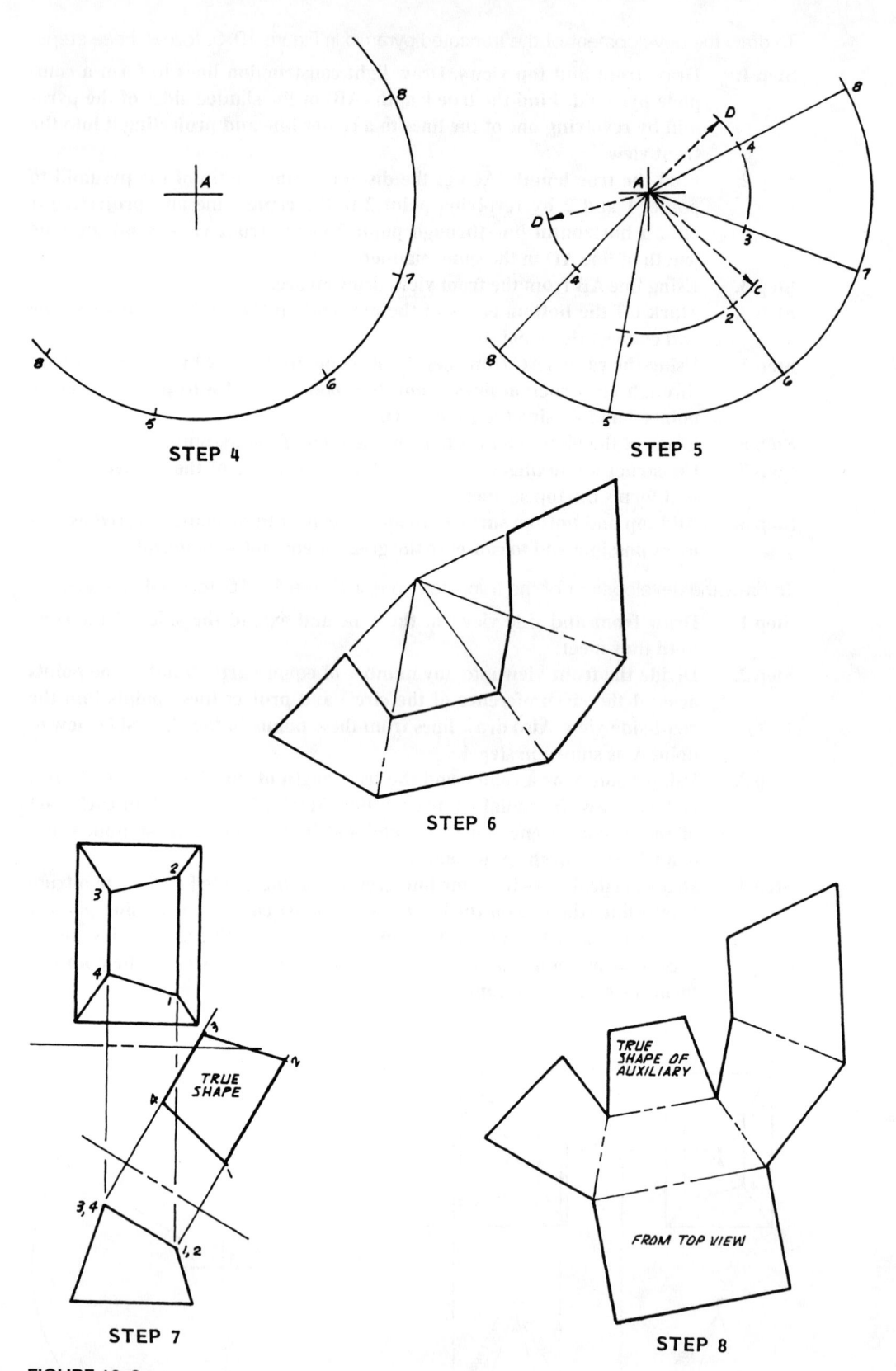

FIGURE 19–9
(Sheet 2 of 2)

290

Chapter 19

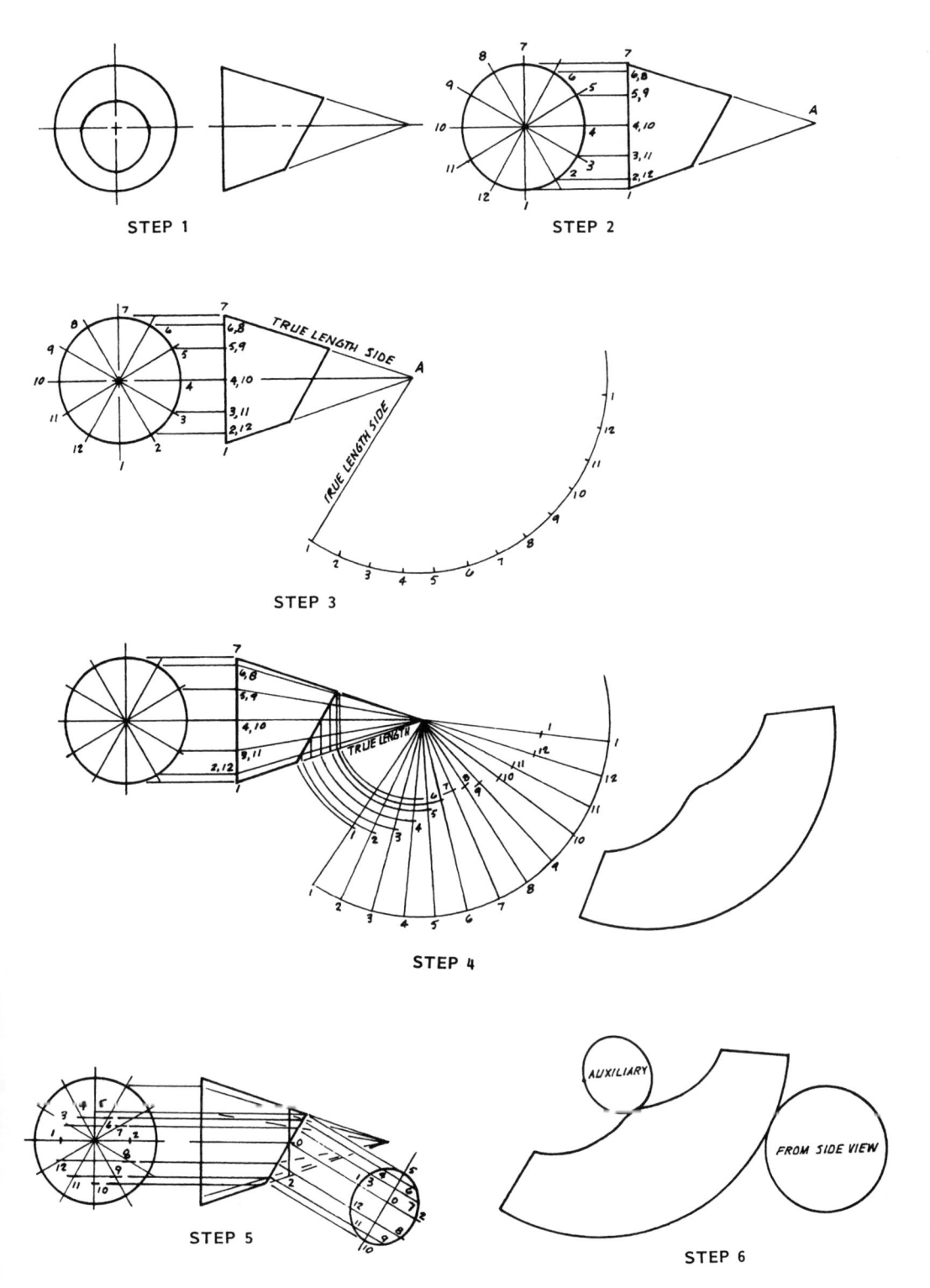

FIGURE 19–10
Radial Line Development of a Truncated Cone

Step 5. Construct an auxiliary view to find the true shape of the top surface.
Step 6. Add the top and bottom surfaces to complete the development.

Triangulation Development

Triangulation is a method of dividing an area into a series of triangles to form a development. It is similar to radial line development, except that it is used for shapes that are not

Sketching Development Drawings   291

uniform cones or pyramids but combinations of these. Triangulation is often used to form sheet metal transition pieces, such as those that connect a rectangular heating air duct with a round one.

Triangulation Method

To develop the transition piece shown in Figure 19-11, follow these steps:

Step 1. Draw front and top views of the object. Divide the circle in the top view into an equal number of segments and number the points on the circumference.

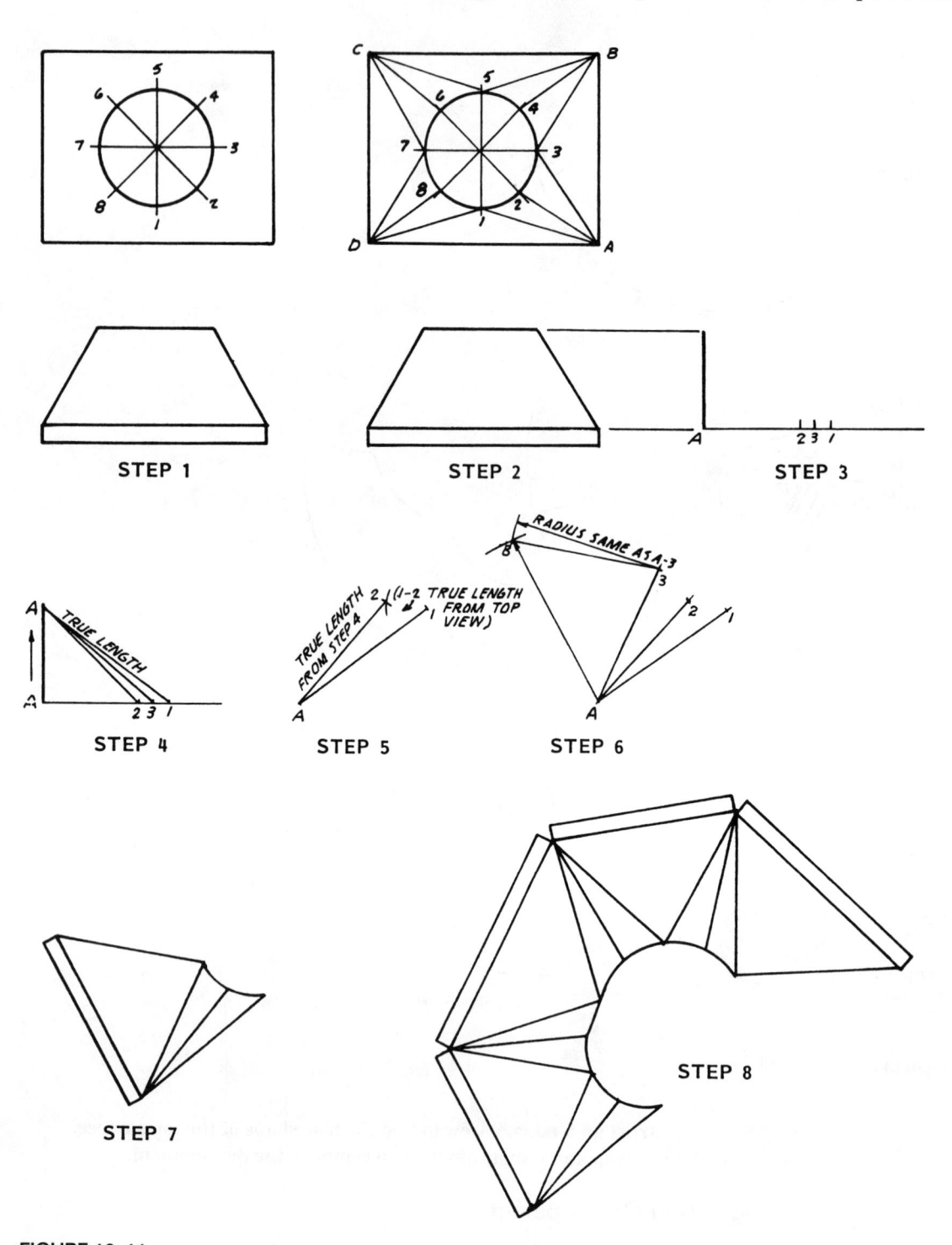

FIGURE 19-11
Triangulation Development of a Transition Piece

292 Chapter 19

Step 2. Draw lines from the numbered points to the corners A, B, C, and D in the top and front views.

Step 3. Draw a true-length triangle by using the height of the transition piece as a vertical line. Make the height meet a horizontal baseline. On the baseline, mark off distances A-1, A-2, and A-3 taken from the top view.

Step 4. Move the corner A up to the top of the triangle and draw straight lines from the new corner A to points 1, 2, and 3. These are true-length lines, which can now be used in the development of the flat pattern. Lines A-1, A-2, and A-3 are the same length as the lines coming from corners B, C, and D.

Step 5. Mark off a line A-1 in a convenient position. From point 1, swing an arc with radius 1-2, taken from the top view. From point A, swing an arc with radius A-2, taken from the true length triangle in step 4. The intersection of these two arcs forms a triangle with line A-1 to complete one segment of the development.

Step 6. Draw the triangle A-3-2, using the method just described.

Then, draw the large triangle: from point 3, swing an arc with radius B-3 (same length as A-3), taken from the true-length triangle. From point A, swing an arc with radius A-B, taken from the top view. The intersection of these two arcs forms a triangle with line A-3.

Step 7. Darken the lines to form one major segment of the development and add the base leg, using the height from the front view and the side A-B.

Step 8. Complete the development, using the method in steps 4 through 7.

EXERCISES

EXERCISES 19–1 to 19–4   Select any two of the objects shown in Figures 19–12 through 19–15 and make a development drawing of each one using the exercise sheets provided at the back of the book. Place only one drawing on a sheet.

EXERCISES 19–5 to 19–8   Use the parallel line method to develop Figures 19–16 through 19–19. Place each drawing on the exercise sheet provided at the back of the book. Remember to draw top and bottom surfaces. After you have made the development, cut one of the patterns out of thin cardboard, such as a manila folder, and fold it into the required shape. Use cardboard tabs and glue or masking tape to hold it together.

EXERCISES 19–9 to 19–11   Use the radial line method to develop Figures 19–20 through 19–22. Place each drawing on the exercise sheet provided at the back of the book. Remember to draw top and bottom surfaces.

EXERCISE 19–12   Use the triangulation method to develop Figure 19–23. Place the drawing on the exercise sheet provided at the back of the book. Do not put a top or bottom on this transition piece.

FIGURE 19–12
Exercise 19–1

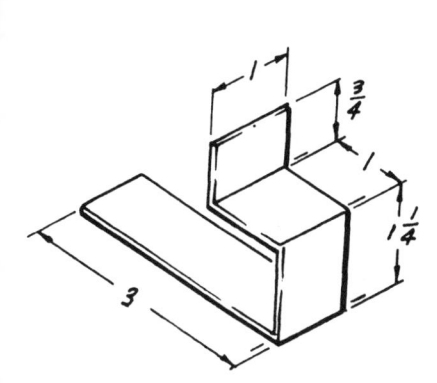

FIGURE 19–13
Exercise 19–2

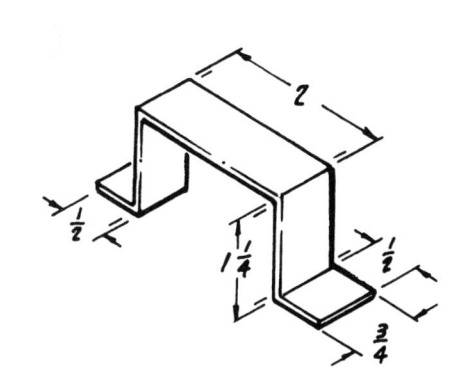

FIGURE 19–14
Exercise 19–3

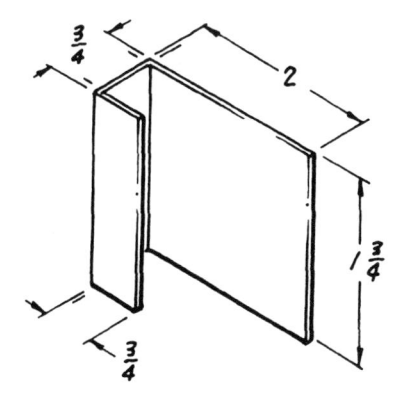

FIGURE 19–15
Exercise 19–4

Sketching Development Drawings

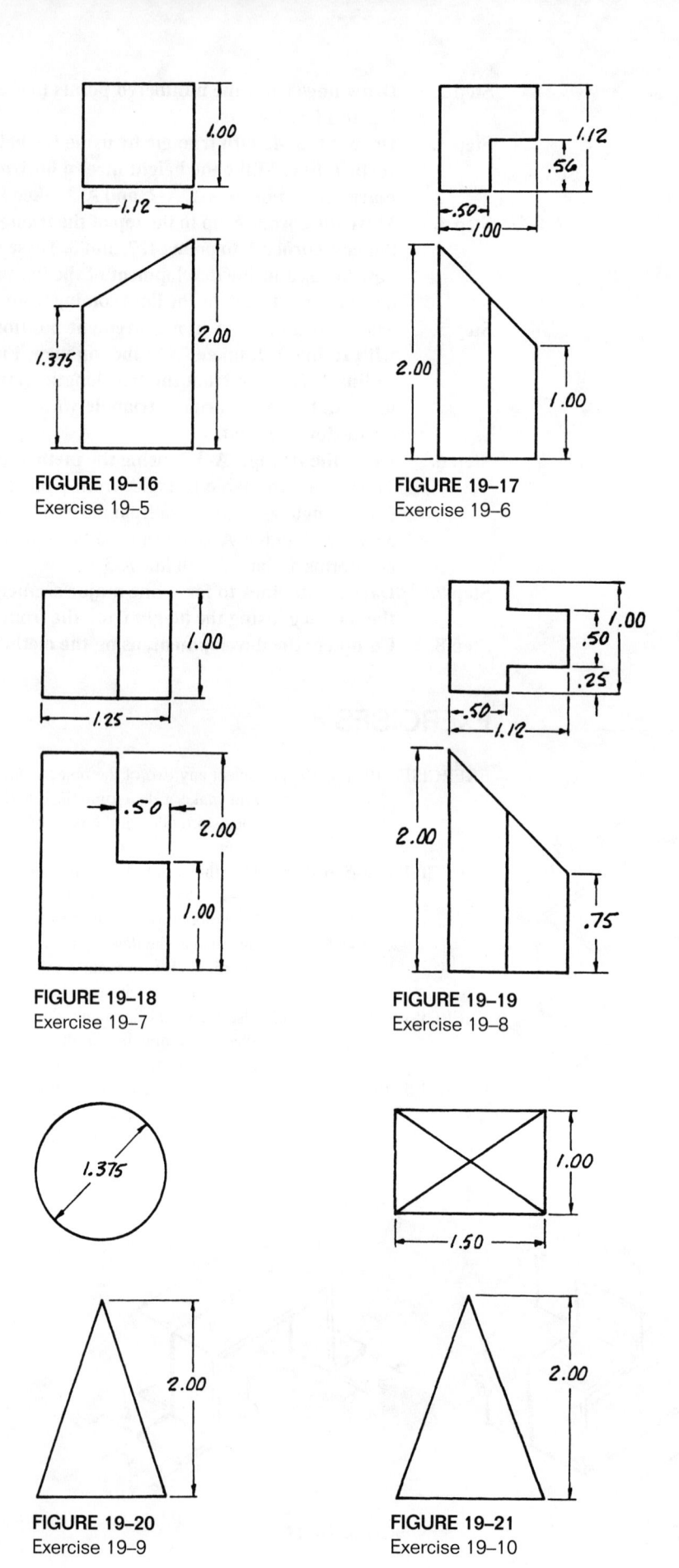

FIGURE 19–16
Exercise 19–5

FIGURE 19–17
Exercise 19–6

FIGURE 19–18
Exercise 19–7

FIGURE 19–19
Exercise 19–8

FIGURE 19–20
Exercise 19–9

FIGURE 19–21
Exercise 19–10

Chapter 19

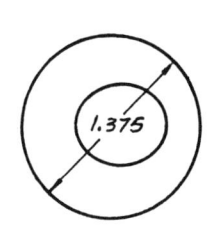

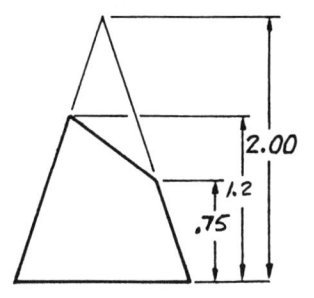

FIGURE 19-22
Exercise 19-11

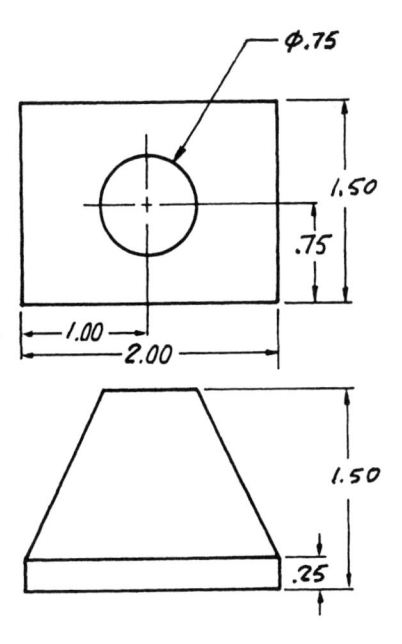

FIGURE 19-23
Exercise 19-12

REVIEW QUESTIONS

Circle the best answer.

1. The most complex method of development is
   a. Parallel line
   b. Radial line
   c. Triangulation
   d. All are equally complex.
2. A parallel line development stretch-out line is used to mark off
   a. Lengths
   b. Heights
   c. Foreshortened measurements
   d. Lengthened measurements
3. The method used to develop a shortened cone is
   a. Parallel line
   b. Radial line
   c. Triangulation
   d. Any one will work.
4. A regular prism is developed using
   a. Parallel line
   b. Radial line
   c. Triangulation
   d. Any one will work.
5. A cylinder can be developed using
   a. Parallel line
   b. Radial line
   c. Triangulation
   d. Any one will work.
6. A truncated cylinder can be developed using
   a. Parallel line
   b. Radial line
   c. Triangulation
   d. Any one will work.

Sketching Development Drawings

7. A pyramid can be developed using
   a. Parallel line
   b. Radial line
   c. Triangulation
   d. Any one will work.
8. A cone can be developed using
   a. Parallel line
   b. Radial line
   c. Triangulation
   d. Any one will work.
9. A truncated pyramid can be developed using
   a. Parallel line
   b. Radial line
   c. Triangulation
   d. Any one will work.
10. A transition piece connecting a round duct to a square duct can be developed using
    a. Parallel line
    b. Radial line
    c. Triangulation
    d. Any one will work.

20 AutoCAD Developments with an Introduction to AutoCAD 3D

OBJECTIVES

When you have completed this chapter, you will be able to

- Draw some of the standard AutoCAD 3D shapes: Box, Cone, Cylinder.
- Draw an extruded shape.
- Draw a parallel line development and fold it in AutoCAD to make a 3D shape.
- Draw a 3D face on the parallel line development
- Draw a radial line development and fold it in AutoCAD to make a 3D shape.

INTRODUCTION

Drawing developments in AutoCAD involves precise two-dimensional (2D) drawings that must be carefully drawn and checked. This chapter uses some of the 2D drawings and also introduces you to the three-dimensional (3D) environment of AutoCAD. It is not a thorough treatment of the subject of sheet metal development, which involves allowances for the bend radius of metal and the many other details necessary for manufacturing. Your first exercise allows you to draw in 3D several of the shapes you drew as 2D developments in Chapter 19.

Drawing and Viewing 3D Shapes in AutoCAD

In Exercise 20–1 you will draw six 3D shapes, view them from a point above and to the right of the shapes, and insert a title block around the shapes. The commands will be described as you use them.

EXERCISE 20–1
Drawing 3D Shapes Using Standard 3D Commands

Your final drawing will look like the drawing in Figure 20–1.

Step 1.   **To begin Exercise 20–1, turn on the computer and start AutoCAD or AutoCAD LT.**
Step 2.   **Open drawing EX20-1 supplied on the disk that came with your book.**
Step 3.   **Use Zoom-All to view the entire drawing area.**

ELEV

The ELEV command allows you to set elevation and thickness for any object you draw after you have made that setting.

FIGURE 20-1
Exercise 20-1 Completed

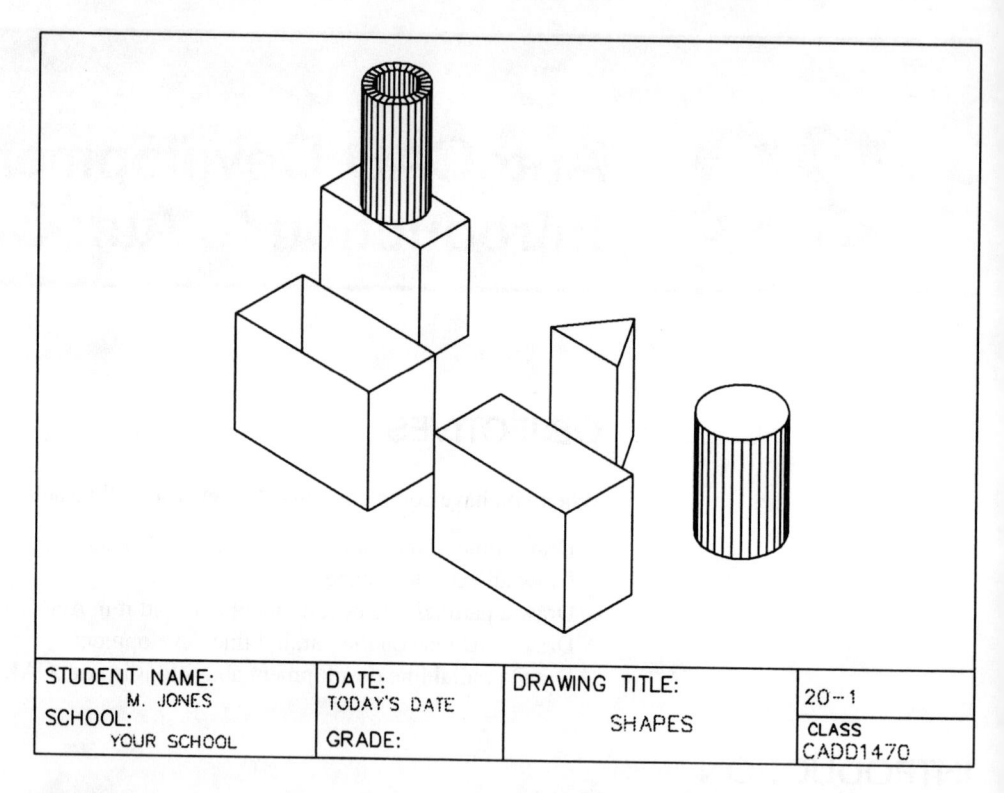

Elevation

The elevation of an object is the location of base or bottom of the model. An object with an elevation of 0 is sitting on the grid. An object with an elevation of 2 is sitting 2″ above the grid.

Thickness

The thickness of an object is its 3D height. An object with an elevation of 0 and a thickness of 2 will sit on the grid and will be 2″ tall. An object with an elevation of 2 and a thickness of 3 will sit 2″ above the grid and will be 3″ tall.

Draw the First 3D Object

Step 4. Use the Solid command to draw a 3D box.

Prompt	Response
Command:	Type: **ELEV**↵
Specify new default elevation <0.0000>:	Type: **0**↵
Specify new default thickness<0.0000>:	Type: **1.5**↵
Command:	Type: **SO**↵
Specify first point:	Type: **2.5,5**↵
Specify second point:	Type: **@1.5<0**↵
Specify third point:	Type: **@-1.5,-.75**↵
Specify fourth point or <exit>:	Type: **@1.5<0**↵
Specify third point:	↵

Although the screen shows only the 2D shape of the box you have just drawn, you will now view this shape in three dimensions. Split the screen into two areas called viewports and make the right viewport one that shows the 3D shapes of the objects you will draw.

Divide the Screen into Two Vertical Viewports and View the 3D Object from a Point Above and to the Right

Step 5. Use the VPORTS command to split the screen into two viewports.

Prompt	Response
Command:	Type: **-VPORTS**⏎
[Save/Restore/Delete/Join/SIngle/?/2<3>/4]:	Type: **2**⏎
Enter a configuration option [Horizontal/Vertical]<Vertical>:	⏎

Your screen is now divided into two vertical viewports. Both viewports contain the same model, so anything you do to one of the models, you do to both of them.

VPORTS

The VPORTS command allows you to divide the screen into as many as 64 areas for many monitors. It also allows you to save a viewport configuration and then restore it or to return the display to a single viewport or to join viewports. You must be aware that the model is the same in all viewports, so if you do anything to the model, the change will be reflected in all viewports.

Step 6. With the right viewport active, use the VPOINT command to view the model in 3D.
Click: any point in the right viewport to make it the active one so the cursor appears as shown in Figure 20–2.

Prompt	Response
Command:	Type: **VPOINT**⏎
Specify a view point or [Rotate] <display compass and tripod>:	Type: **1,-1,1**⏎

VPOINT

The VPOINT command allows you to view the 3D model from any angle. You will use the one typed above for 3D views in this chapter. This view is similar to the isometric drawing you made in an earlier chapter.

Vpoint Coordinates

The coordinates you have typed move your viewpoint 1 unit in the X direction or to the right, −1 unit in the Y direction or in front of the model, and 1 unit in the Z direction or above the object, so the shape appears as shown in Figure 20–3.

FIGURE 20–2
Activating the Right Viewport

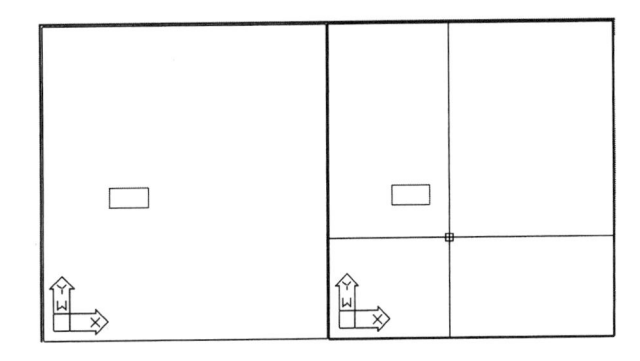

AutoCAD Developments with an Introduction to AutoCAD 3D

FIGURE 20–3
The Viewpoint 1,-1,1

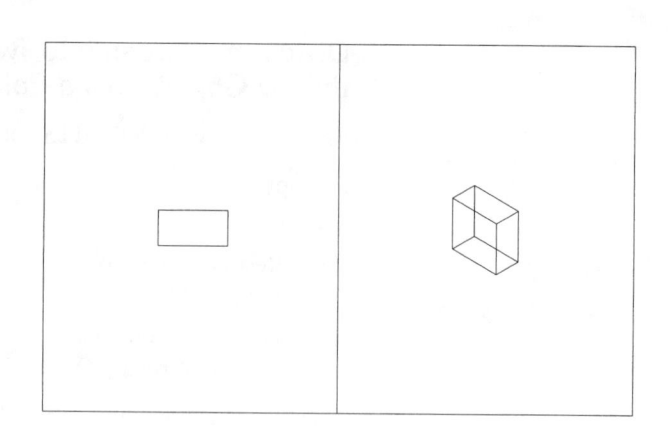

The viewpoint in the left viewport <0,0,1> places your viewpoint directly above the object to give you a plan view.

Draw the Other 3D Shapes

Step 7. Zoom out so you can view the other shapes as they are drawn.

Prompt	Response
Command:	Type: **Z**↵
[All/Center/Extents/Previous/Scale/Window]<real time>:	Type: **.75**↵

Step 8. Draw a prism (Figure 20–4).

Prompt	Response
Command:	Type: **SO**↵
Specify first point:	Type: **6,4.25**↵
Specify second point:	Type: **@1<0**↵
Specify third point:	Type: **@-.5,1**↵
Specify fourth point or <exit>:	↵
Specify third point:	↵

Step 9. Use the ELEV command to set elevation and thickness, and draw a circle with an elevation of 0 and a thickness of 1.5 (Figure 20–5).

Prompt	Response
Command:	Type: **ELEV**↵

FIGURE 20–4
Draw the Prism

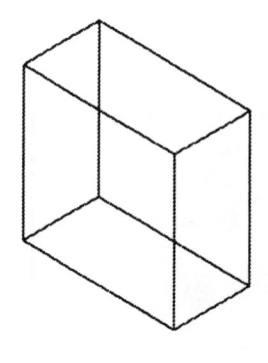

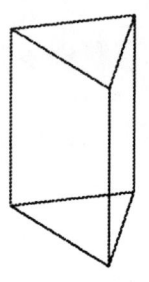

Chapter 20

FIGURE 20–5
Draw a Circle with an Elevation of 0 and a Thickness of 1.5

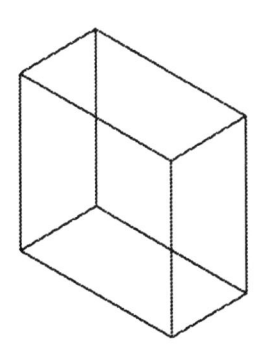

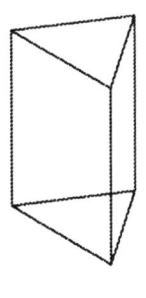

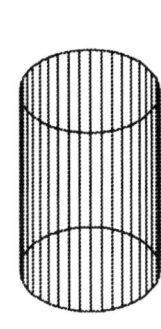

Specify new default elevation<0.0000>:	↵
Specify new default thickness <1.500>:	Type: **1.5**↵
Command:	Type: **C**↵
Specify center point for circle or [3P/2P/Ttr/<tan tan radius>]:	Type: **8.5,4.625**↵
Specify radius of circle or [Diameter]:	Type: **.5**↵

Step 10. With elevation set at 0 and thickness set at 1.5, as you did in step 9, use the Polyline command to draw an open box.

Prompt	Response
Command:	Type: **PL**↵
Specify start point:	Type: **3.5,2**↵
Specify next point or [Arc/Halfwidth/Length/Undo/Width]:	Type: **@2<0**↵
Specify next point or [Arc/Halfwidth/Length/Undo/Width]:	Type: **@1<90**↵
Specify next point or [Arc/Halfwidth/Length/Undo/Width]:	Type: **@2<180**↵
Specify next point or [Arc/Halfwidth/Length/Undo/Width]:	Type: **C**↵

Step 11. With elevation set at 0 and thickness set at 1.5, as you did in step 9, use the Polyline command to draw a solid box by giving the polyline a width of 1″.

Prompt	Response
Command:	Type: **PL**↵ (or just ↵ if you haven't entered another command.)
Specify start point:	Type: **6.5,2.5**↵
Specify next point or [Arc/Halfwidth/Length/Undo/Width]:	Type: **W**↵
Specify starting width <0.0000>:	Type: **1**↵
Specify ending width <1.0000>:	↵

AutoCAD Developments with an Introduction to AutoCAD 3D

FIGURE 20–6
Use the Hide Command to Remove Hidden Lines

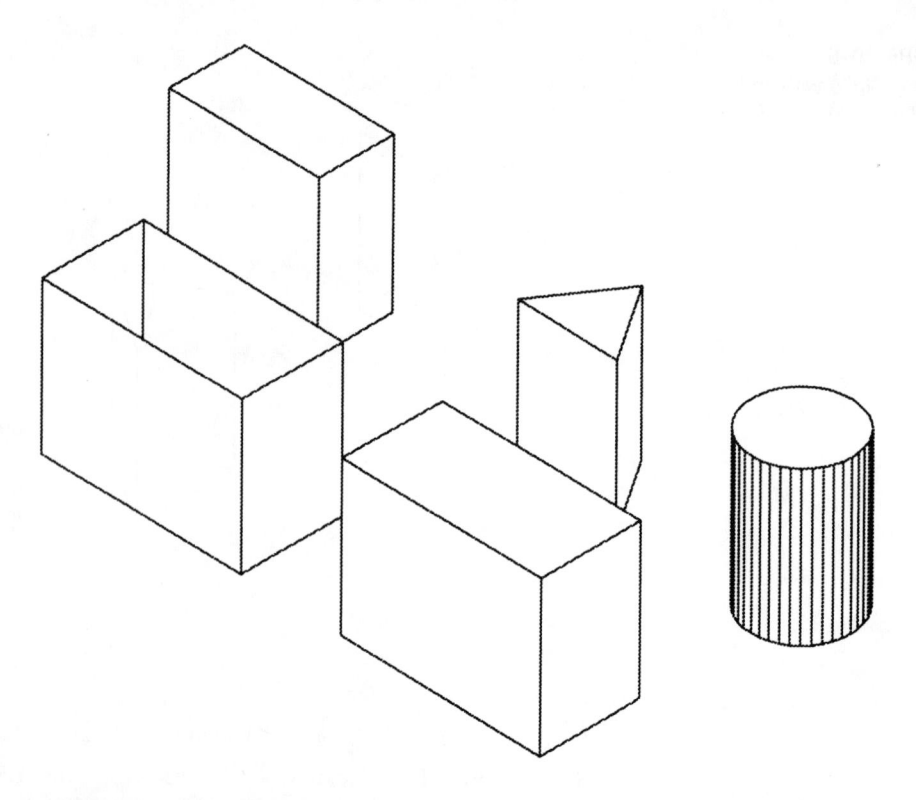

Specify next point or [Arc/Halfwidth/ Length/Undo/Width]:	Type: **@2<0**↵
Specify next point or [Arc/Halfwidth/ Length/Undo/Width]:	↵

Step 12. Use the HIDE command to view what you have drawn (Figure 20–6).

Click: **The right viewport to make it active.**

Prompt	Response
Command:	Type: **HI**↵

HIDE

The Hide command removes the lines describing surfaces that are hidden by surfaces in front of them so that solid objects appear solid.

Step 13. Use the Donut command to draw a tube or pipe sitting on top of the first box you drew (Figure 20–7). Set elevation to 1.5 and thickness to 1.5.

Prompt	Response
Command:	Type: **ELEV**↵
Specify new default elevation<0.0000>:	Type: **1.5**↵
Specify new default thickness <1.5000>:	↵
Command:	Type: **DO**↵
Specify inside diameter of donut <0.5000>:	Type: **.5**↵ (or just ↵ if the default is 0.5000)
Specify outside diameter of donut <1.0000>:	Type: **.75**↵
Specify center of donut or <exit>:	With Snap ON, Click: **a point in the center of the first box in the left viewport**
Center of donut or <exit>:	↵
Command:	Click: **the right viewport** and Type: **HI**↵

302   Chapter 20

FIGURE 20-7
Draw a Donut with an Elevation of 1.5 and a Thickness of 1.5

Note: If your donut is on elevation Ø, move it up using the move command:
Type M↵
Click: The Donut
Click: Any point
Type:@ Ø, Ø, 1.5
(Ø in the X direction;
Ø in the y direction;
1.5 in the z direction)

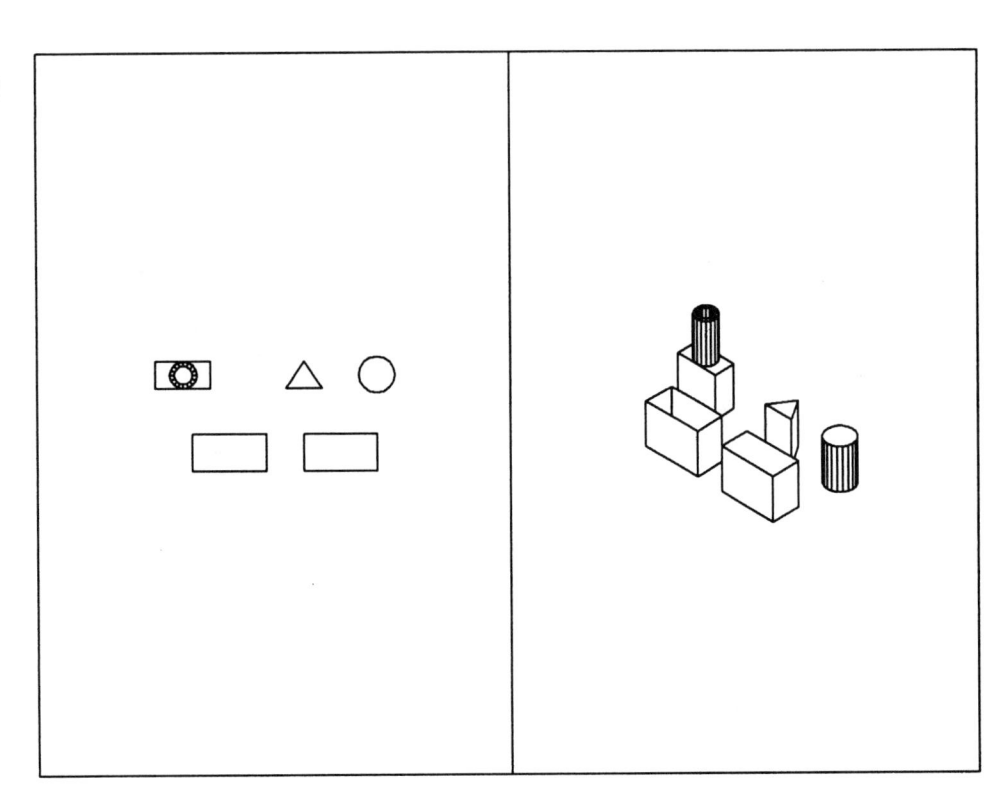

Insert a Title Block around Your Drawing and Complete Exercise 20-1

Step 14. Return the display to a single viewport (Figure 20-8).

Click: **The right viewport to make it active.**

Prompt	Response
Command:	Type: **-VPORTS**↵
[Save/Restore/Delete/Join/ SIngle/?/2/3/4] <3>:	Type: **SI**↵

Step 15. Set UCS to View and insert a title block.

Prompt	Response
Command:	Type: **UCS**↵
Enter an option [New/Move/orthoGraphic/ Prev/Restore/Save/Del/Apply/?/World]:	Type: **V**↵

UCS-View

The View option of the UCS command allows you to draw or insert drawings that are parallel to the screen. This is the only type of situation in which the View option is appropriate. If you draw 3D objects with UCS set to View, they will be oriented in a strange manner and you will never get them rotated correctly.

AutoCAD Developments with an Introduction to AutoCAD 3D

FIGURE 20–8
Return to a Single Viewport

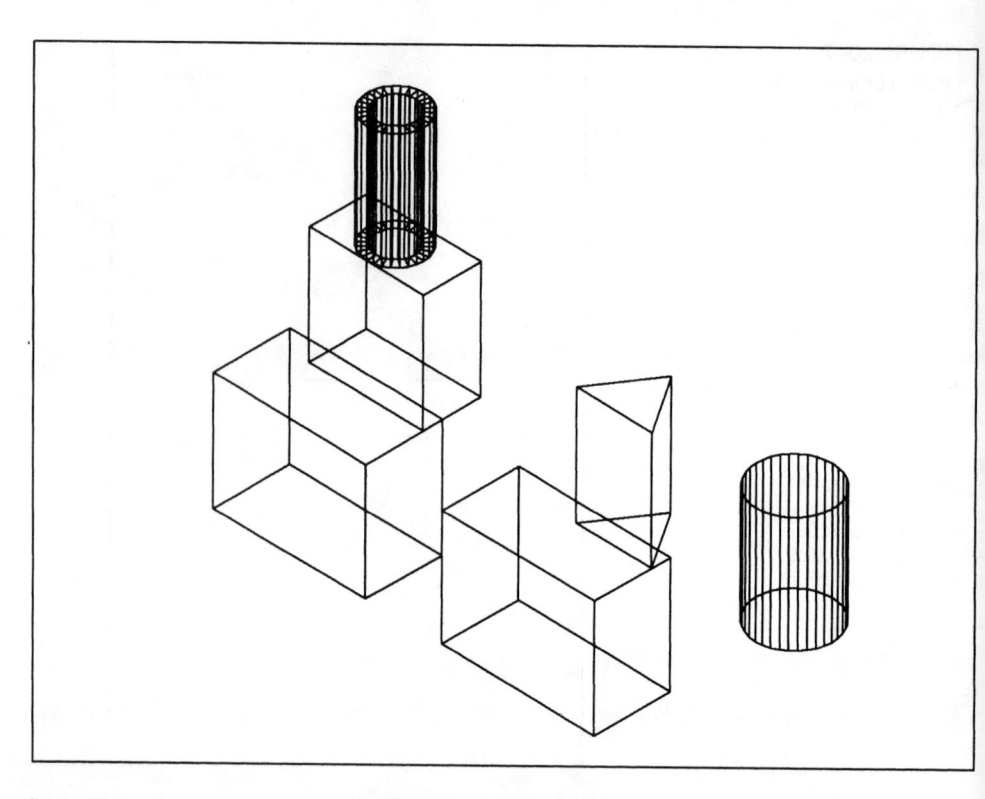

Step 16. Insert a predefined block with attributes named TB (Figure 20–9).

Prompt	Response
Command:	Type: **-I**↵
Enter block name (or ?):	Type: **TB**↵
Specify insertion point or [Scale...PRotate]:	Click: **a point so the objects are centered in the format as shown in Figure 20–9**

FIGURE 20–9
Insert the Block TB Around Your Drawing

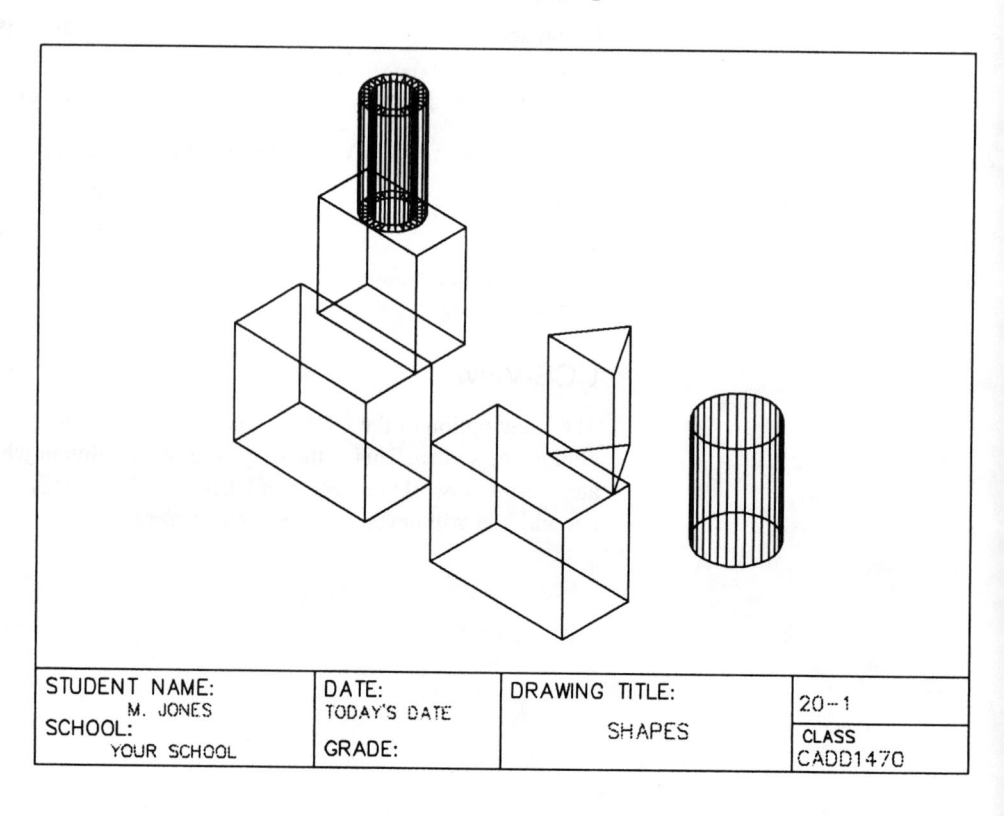

Enter X scale factor, specify opposite corner or [Corner/XYZ] <1>:	↵
Enter Y scale factor <use X scale factor>:	↵
Specify rotation angle <0>:	↵
TYPE THE NAME OF YOUR SCHOOL <XX>:	Type: **YOUR SCHOOL NAME**↵
TYPE YOUR FIRST INITIAL AND LAST NAME <XX>:	Type: **YOUR FIRST INITIAL AND LAST NAME**↵
TYPE TODAY'S DATE <XX>:	Type: **TODAY'S DATE**↵
TYPE THE LETTERS AND NUMBER OF YOUR CLASS NAME <XX>:	Type: **THE CLASS IDENTIFICATION**↵
TYPE THE EXERCISE NUMBER <XX>:	Type: **20-1**↵
TYPE THE TITLE OF YOUR DRAWING <XX>:	Type: **SHAPES**↵

Step 17. Use the SAVEAS command to save your drawing as EX20-1(your initials) on a floppy disk and again on the hard drive of your computer.

Step 18. Plot or print your drawing full size on an 11" × 8½" sheet. Be sure to place a check in the Hide Lines box in the Plot dialog box.

Make a 3D Shape from an Existing 2D Parallel Line Development Drawing (Some AutoCAD LT users do not have the commands to complete this exercise.)

In Exercise 20–2 you will take an existing 2D development drawing, fold it into a 3D shape, and draw 3D faces on all its edges to form a 3D shape that has the appearance of a solid.

EXERCISE 20–2
Making a 3D Shape from an Existing 2D Parallel Line Development Drawing

Your final drawing will look like the drawing in Figure 20–10.

Step 1. To begin Exercise 20–2, turn on the computer and start AutoCAD. AutoCAD LT users will not be able to complete Exercises 20–2 and 20–3.

Step 2. Open drawing EX20-2 supplied on the disk that came with your book.

Step 3. Use the VPOINT command to change the viewpoint to a point above and to the right of the drawing (Figure 20–11).

Prompt	Response
Command:	Type: **VPOINT**↵
Specify a view point or [Rotate] <display compass and tripod>:	Type: **1,-1,1**↵

Step 4. Use ROTATE3D to rotate the drawing 90° about the X axis.

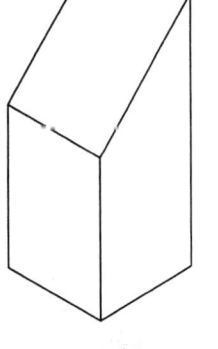

FIGURE 20–10
Exercise 20–2 Completed

Prompt	Response
Command:	Type: **ROTATE3D**↵
Select objects:	Type: **ALL**↵ (or **select all lines with a crossing window**)
Select objects:	↵
Specify first point on axis or define axis by [Object/Last/View/Xaxis/Yaxis/Zaxis/2points]:	Type: **X**↵

AutoCAD Developments with an Introduction to AutoCAD 3D

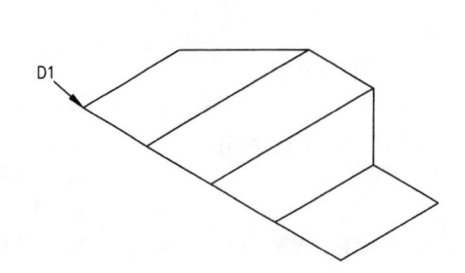

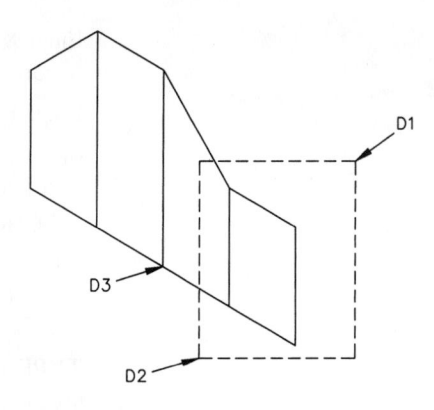

FIGURE 20–11
Changing the Viewpoint and Picking
a Point on the X Axis to Rotate

FIGURE 20–12
Rotating Right and Front Sides
90° Clockwise about the Z-Axis

Prompt	Response
Specify a point on the X axis <0,0,0>:	Type: **END**↵
of	Click: **D1** (Figure 20–11)
Specify Rotation angle or [Reference]	Type: **90**↵
Command:	Type: **Z**↵
[All/Center/Dynamic/Extents/Previous/Scale/Window]<real time>:	Type: **A**↵

Step 5. Use ROTATE3D to rotate the lines forming the two right sections of the drawing 90° in a clockwise direction about the Z axis.

Prompt	Response
Command:	Type: **ROTATE3D**↵
Select objects:	Click: **D1** (Figure 20–12)
Other corner:	Click: **D2**
Select objects:	↵
Specify first point on axis or define by [Object/Last/View/Xaxis/Yaxis/Zaxis/2points]:	Type: **Z**↵
Specify a point on the Z axis <0,0,0>:	Type: **END**↵
of	Click: **D3** (Figure 20–12)
Specify rotation angle or [Reference]:	Type: **-90**↵

Step 6. Use ROTATE3D to rotate the lines forming the front face of the drawing 90° in a clockwise direction about the Z axis.

Prompt	Response
Command:	↵
Select objects:	Click: **D1, D2, D3** (Figure 20–13)
Select objects:	↵
Specify first point on axis or define axis by [Object/Last/View/Xaxis/Yaxis/Zaxis/2points]:	Type: **Z**↵
Specify a point on the Z axis <0,0,0>:	Type: **END**↵
of	Click: **D4** (Figure 20–13)
Specify rotation angle or [Reference]:	Type: **-90**↵

FIGURE 20–13
Rotating the Front View Clockwise 90° More about the Z Axis

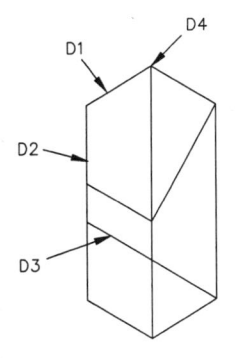

FIGURE 20–14
Rotate the Right Side 90° Clockwise about the Z Axis

Step 7. Use ROTATE3D to rotate the lines forming the left side of the drawing 90° about the Z axis. (Release 12 users will have to use the AXROT command.)

Prompt	Response
Command:	↵
Select objects:	Click: **D1, D2, D3** (Figure 20–14)
Select objects:	↵
Specify first point axis or define axis by [Object/Last/View/Xaxis/Yaxis/Zaxis/2points]:	Type: **Z**↵
Specify a point on the Z axis <0,0,0>: of	Type: **END**↵ Click: **D4** (Figure 20–14)
Specify rotation angle or [Reference]:	Type: **90**↵

Step 8. Use the HIDE command to see that there are no surfaces on this model. This is called a wire frame (Figure 20–15).

Prompt	Response
Command:	Type: **HI**↵

Step 9. Set a running Osnap mode of Endpoint and use the 3DFACE command to put surfaces on all visible faces of the drawing.

Prompt	Response
Command:	Type: **-OSNAP**↵
Enter list of object snap modes:	Type: **END**↵

Now any point you click on a line will automatically go to the endpoint of it.

Prompt	Response
Command:	Type: **3DFACE**↵
Specify first point or [Invisible]:	**D1** (Figure 20–16)
Specify second point or [Invisible]:	**D2**
Specify third point or [Invisible] <exit>:	**D3**
Specify fourth point or [Invisible] <create three-sided face>:	**D4**
Specify third point or [Invisible] <exit>:	↵
Command:	↵
Specify first point or [Invisible]:	**D4**
Specify second point or [Invisible]:	**D5**
Specify third point or [Invisible] <exit>:	**D6**
Specify fourth point or [Invisible] <create three-sided face>:	**D3**
Specify third point or [Invisible] <exit>:	↵
Command:	↵
Specify first point or [Invisible]:	**D1**
Specify second point or [Invisible]:	**D4**
Specify third point or [Invisible] <exit>:	**D5**
Specify fourth point or [Invisible] <create three-sided face>:	**D7**
Specify third point or [Invisible] <exit>:	↵

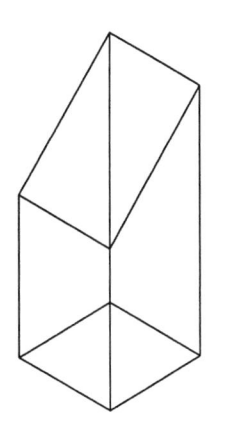

FIGURE 20–15
The Wire Frame

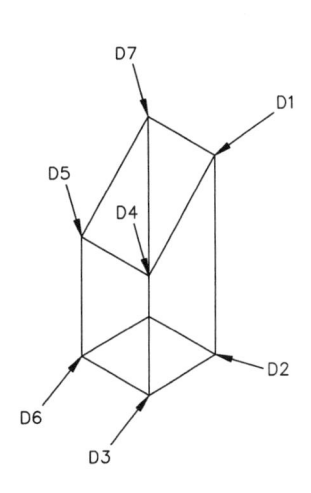

FIGURE 20–16
Drawing 3D Faces

Step 10. Use the HIDE command to remove hidden lines (Figure 20–17).

AutoCAD Developments with an Introduction to AutoCAD 3D

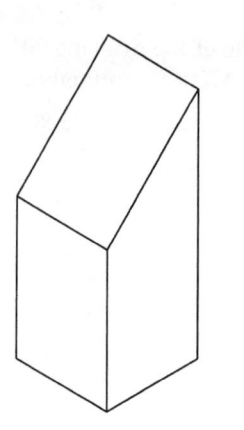

FIGURE 20–17
The Model after the Hide Command Has Been Used

Prompt	Response
Command:	Type: **HI**↵

Step 11. Use the SAVEAS command to save your drawing as EX20-2(your initials) on a floppy disk and again on the hard drive of your computer.

Step 12. Plot or print your drawing full size on an $11'' \times 8\frac{1}{2}''$ sheet. Be sure to place a check in the Hide Lines box in the Plot dialog box.

You will not use the title block on this drawing. In Exercise 20–3 you will insert this drawing into the drawing for Exercise 20–3 and place a title block around the two drawings.

Make a 3D Shape from an Existing 2D Radial Line Development Drawing and Insert EX20-2 and a Title Block into the Drawing (Some AutoCAD LT users do not have the commands to complete this exercise.)

In Exercise 20–3 you will take an existing 2D development drawing, fold it into a 3D shape, and draw 3D faces on all its edges to form a 3D shape that has the appearance of a solid. You will then insert EX20-2 and a title block into the drawing.

EXERCISE 20–3
Making a 3D Shape from an Existing 2D Radial Line Development Drawing and Inserting EX20-2 and a Title Block into This Drawing

Your final radial line development drawing will look like the drawing in Figure 20–18.

Step 1. To begin Exercise 20–3, turn on the computer and start AutoCAD. AutoCAD LT users will not be able to complete Exercises 20–2 and 20–3.

Step 2. Open drawing EX20-3 supplied on the disk that came with your book.

Step 3. Use the Line command to draw an extension line, and then dimension the angle that will determine how much the sides will have to be rotated to form the 3D shape (Figure 20–19).

Step 4. Use the List command to determine the length of the base of the triangle. Then erase all but one side of the development and draw three 2" lines to complete the base (Figure 20–20).

Step 5. Use the VPOINT command to change the viewpoint to a point above and to the right of the drawing (Figure 20–21).

Prompt	Response
Command:	Type: **VPOINT**↵

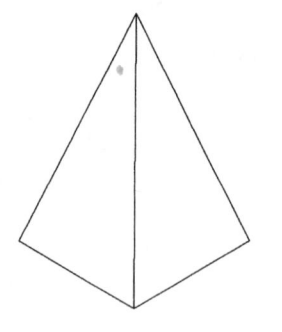

FIGURE 20–18
Radial Line Development Made into a 3D Shape

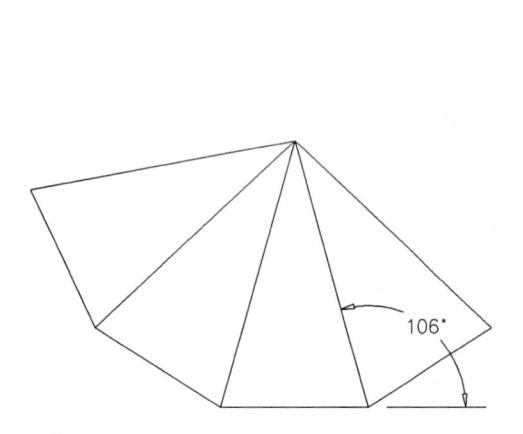

FIGURE 20–19
Rotation Angle about the X-Axis

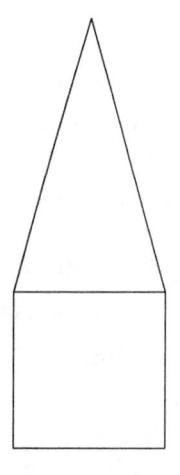

FIGURE 20–20
Drawing the Model Base

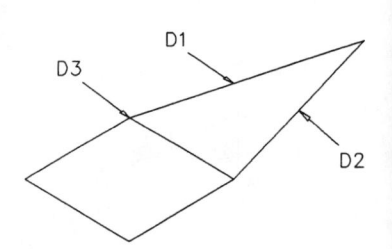

FIGURE 20–21
Rotating the Triangle about the X-Axis

308　　Chapter 20

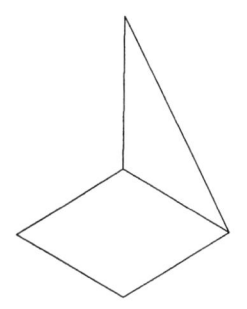

FIGURE 20-22
The Triangle Rotated

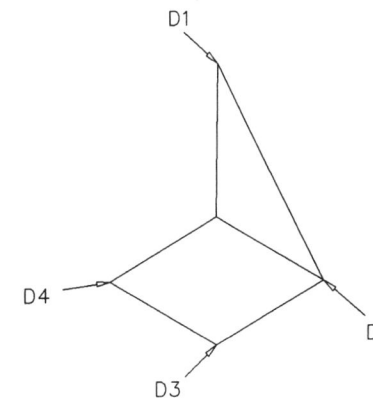

FIGURE 20-23
Drawing 3D Faces

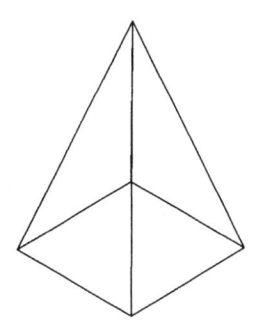

FIGURE 20-24
The Completed Model

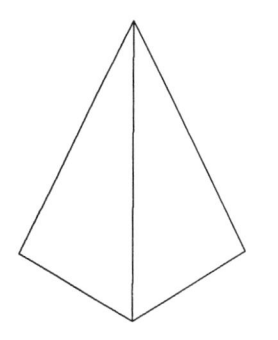

FIGURE 20-25
The Model after the Hide Command Has Been Used

Prompt	Response
Specify a view point or [Rotate] <display compass and tripod>:	Type: **1,-1,1**↵

Step 6. Use ROTATE3D to rotate the two angular lines forming the triangle 90° about the X axis (Figures 20–21 and 20–22).

Prompt	Response
Command:	Type: **ROTATE3D**↵
Select objects:	Click: **D1, D2** (Figure 20–21)
Select objects:	↵
Specify first point on axis or define axis by [Object/Last/View/Xaxis/Yaxis/Zaxis/2points]:	Type: **X**↵
Specify a point on the X axis <0,0,0>:	Type: **END**↵
of	Click: **D3**
Specify rotation angle or [Reference]:	Type: **106**↵
Command:	Type: **Z**↵
All/Center/Dynamic/Extents/Previous/Scale/Window] <real time>:	Type: **A**↵

Step 7. Set a running Osnap mode of Endpoint and use the 3DFACE command to put surfaces on all visible faces of the drawing (Figures 20–23 and 20–24).

Prompt	Response
Command:	Type: **-OSNAP**↵
Enter list of object snap modes:	Type: **END**↵

Now any point you pick on a line will automatically go to the endpoint of it.

Prompt	Response
Command:	Type: **3DFACE**↵
Specify first point or [Invisible]:	**D1** (Figure 20–23)
Specify second point or [Invisible]:	**D2**
Specify third point or [Invisible] <exit>:	**D3**
Specify fourth point or [Invisible] <create three-sided face>:	↵
Command:	↵
Specify first point or [Invisible]:	**D3**
Specify second point or [Invisible]:	**D4**
Specify third point or [Invisible] <exit>:	**D1**
Specify fourth point or [Invisible] <create three-sided face>:	↵

Step 8. Use the HIDE command to remove hidden lines (Figure 20–25).

Prompt	Response
Command:	Type: **HI**↵

Step 9. Insert drawing EX20-2 into this drawing.

Place the floppy disk containing drawing EX20-2 into the A drive:

Prompt	Response
Command:	Type: **-I**↵
Enter block name (or ?):	Type: **A:EX20-2(your initials)**↵
Specify insertion point or [Scale...PRotate]:	Click: any point in the approximate location shown in Figure 20–26.

AutoCAD Developments with an Introduction to AutoCAD 3D

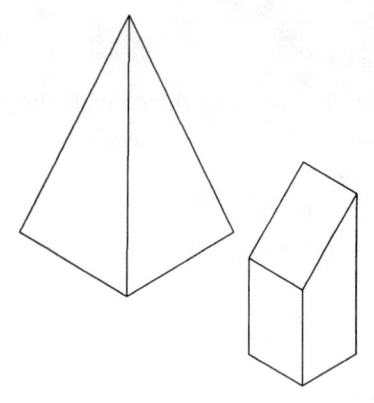

FIGURE 20–26
Inserting Drawing EX20-2

Prompt	Response
Enter X scale factor, specify opposite corner or [Corner/XYZ]:	↵
Enter Y scale factor <use X scale factor>:	↵
Specify rotation angle <0>:	↵

Step 10. Set UCS to View and insert the predefined block TB containing attributes into the drawing.

Prompt	Response
Command:	Type: **UCS**↵
Enter an option [New/Move/orthoGraphic/Prev/Restore/Save/Del/?/Apply/World]:	Type: **V**↵
Command:	Type: **-I**↵
Enter block name (or ?) <EX20-2>:	Type: **TB**↵
Specify insertion point or [Scale...PRotate]:	Click: **any point in the lower left corner of your display**
Enter X scale factor, specify opposite corner or [Corner/XYZ] <1>:	↵
Enter Y scale factor <use X scale factor>:	↵
Specify rotation angle <0>:	↵
TYPE THE NAME OF YOUR SCHOOL <XX>:	Type: **YOUR SCHOOL NAME**↵
TYPE YOUR FIRST INITIAL AND LAST NAME <XX>:	Type: **YOUR FIRST INITIAL AND LAST NAME**↵
TYPE TODAY'S DATE <XX>:	Type: **TODAY'S DATE**↵
TYPE THE LETTERS AND NUMBER OF YOUR CLASS NAME <XX>:	Type: **THE CLASS IDENTIFICATION**↵
TYPE THE EXERCISE NUMBER <XX>:	Type: **20-3**↵
TYPE THE TITLE OF YOUR DRAWING <XX>:	Type: **DEVELOPMENT SHAPES**

Step 11. On your own:
1. Zoom-All so you can see the entire drawing.
2. Use the MOVE command to move the shapes to the approximate locations shown in Figure 20–27.
3. Use the HIDE command to make sure all figures are correct.
4. Use the UCS command to return the drawing to the World UCS just in case you have to change something on the models.

Step 12. Use the SAVEAS command to save your drawing as EX20-3(your initials) on a floppy disk and again on the hard drive of your computer.

Step 13. Plot or print your drawing full size on an 11" × 8½" sheet. Be sure to place a check in the Hide Lines box in the Plot dialog box.

FIGURE 20–27
Exercise 20–3 Completed

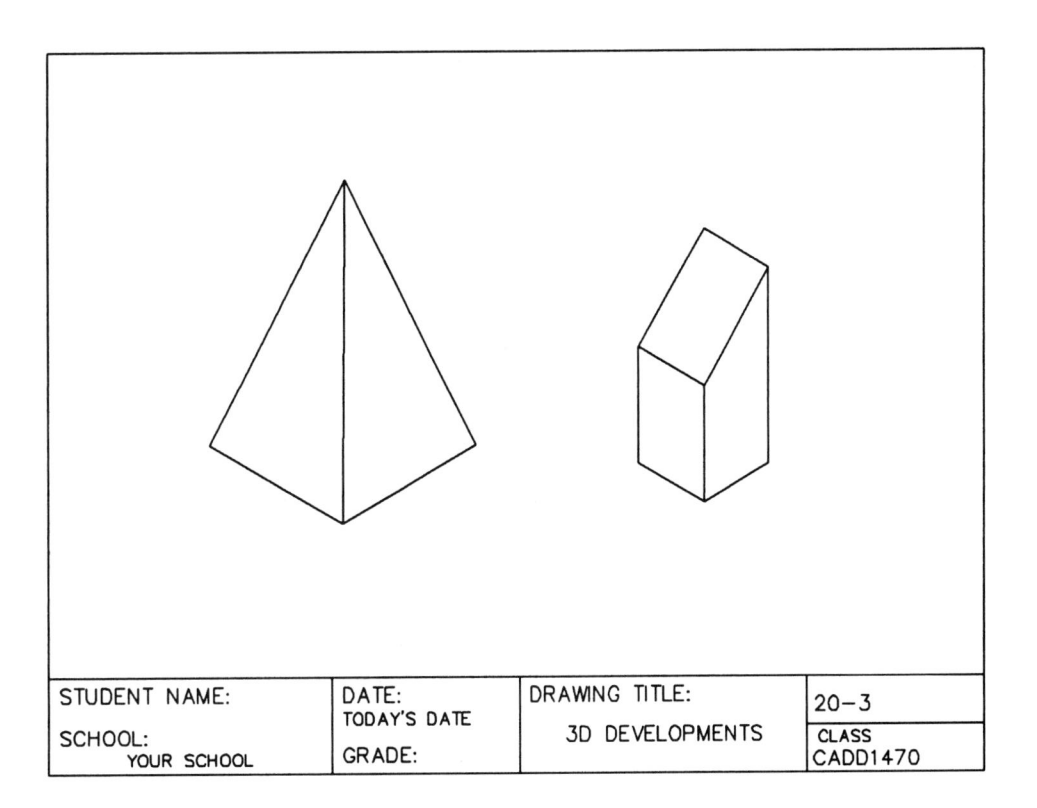

EXERCISES

EXERCISE 20–1   Complete Exercise 20–1 using steps 1 through 18 described in this chapter.
EXERCISE 20–2   Complete Exercise 20–2 using steps 1 through 12 described in this chapter.
EXERCISE 20–3   Complete Exercise 20–3 using steps 1 through 13 described in this chapter.

REVIEW QUESTIONS

1. Which of the following is not one of the prompts for the Solid command?
    a. First point:
    b. Second point:
    c. Third point:
    d. Fourth point:
    e. Fifth point
2. Which of the following commands splits the screen into two viewports?
    a. VPOINT
    b. ID
    c. VPORTS
    d. UCS
    e. ELEV
3. Which of the following commands allows you to view a 3D object from any angle?
    a. VPOINT
    b. ID
    c. VPORTS
    d. UCS
    e. ELEV
4. Which of the following viewpoints gives you a view similar to an isometric drawing?
    a. 1,0,0
    b. 0,0,1
    c. 1,−1,−1
    d. 1,−1,1
    e. −1,1,1

AutoCAD Developments with an Introduction to AutoCAD 3D

5. Which of the following commands allows you to set elevation and thickness?
   a. INSERT
   b. THICKNESS
   c. VPORTS
   d. UCS
   e. ELEV
6. For an object to be drawn so that it is 2" above the grid and 5" tall which are the correct settings?
   a. Elevation 0, Thickness 5
   b. Elevation 2, Thickness 5
   c. Elevation 5, Thickness 2
   d. Elevation 7, Thickness 0
   e. Elevation 0, Thickness 7
7. To draw a solid box that is 1" wide using the PLINE command which of the following polyline options is correct?
   a. Beginning width 1", ending width 1"
   b. Beginning width 0", ending width 1"
   c. Beginning width 0", ending width 1"
   d. Halfwidth 1"
   e. Halfwidth 2"
8. Which of the following is used to remove the lines describing surfaces that are hidden by surfaces in front of them?
   a. HIDE
   b. HIDDEN
   c. VPORTS
   d. UCS
   e. 3DFACE
9. Which of the following allows you to rotate an object about X, Y, or Z axes?
   a. ROTATE
   b. MOVE
   c. ROTATE3D
   d. 3DFACE
   e. VPOINT
10. Which of the following allows you to draw a surface on a wire frame?
    a. HIDE
    b. HIDDEN
    c. VPORTS
    d. UCS
    e. 3DFACE

21 The Drawing System

OBJECTIVES

When you have completed this chapter, you will be able to

- Identify and label drawings within a drawing system, going from top assembly to detail drawing, and provide cross references.
- Identify and describe the functions of parts of a drawing, such as the title block, revision block, list of materials, next assembly number, and notes.
- Correctly label a set of drawings with parts lists, title block information, and drawing numbers.

INTRODUCTION

All drawing systems have some means of reducing drawings of complex equipment or structures into less complex drawings that can be more easily understood. The most common system involves the use of three different types of drawings: top assembly, subassembly, and detail (Figure 21–1).

Top Assembly Drawings

A *top assembly drawing* is an overall drawing that shows all parts of a machine or structure as they are finally assembled. It is used as a means of identifying subassemblies or separate parts, depending on the complexity of the machine. Figure 21–2 is a simplified sketch of a top assembly drawing. Notice that the item numbers pointing to separate subassemblies are identified by part number on the parts list (also called list of materials or bill of material).

FIGURE 21–1
Drawing Relationships

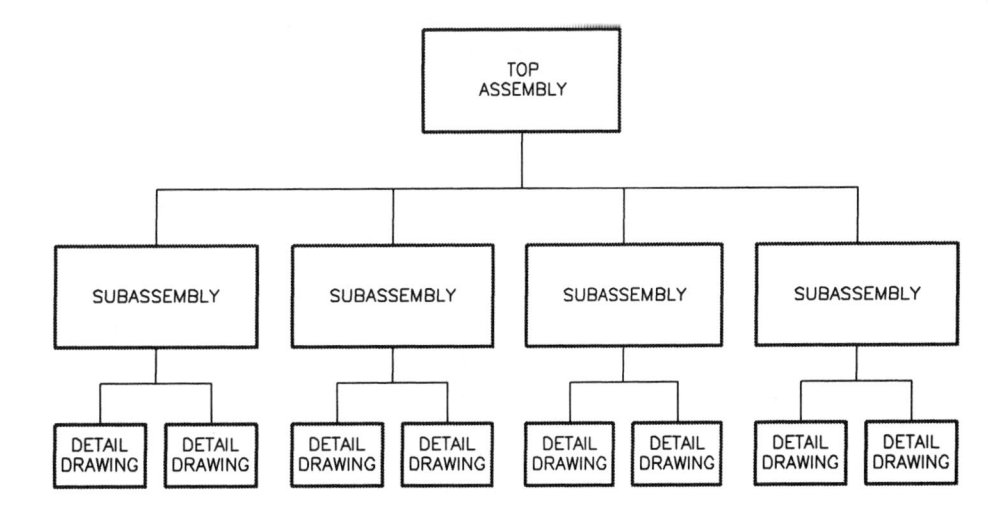

313

FIGURE 21–2
Top Assembly Drawing

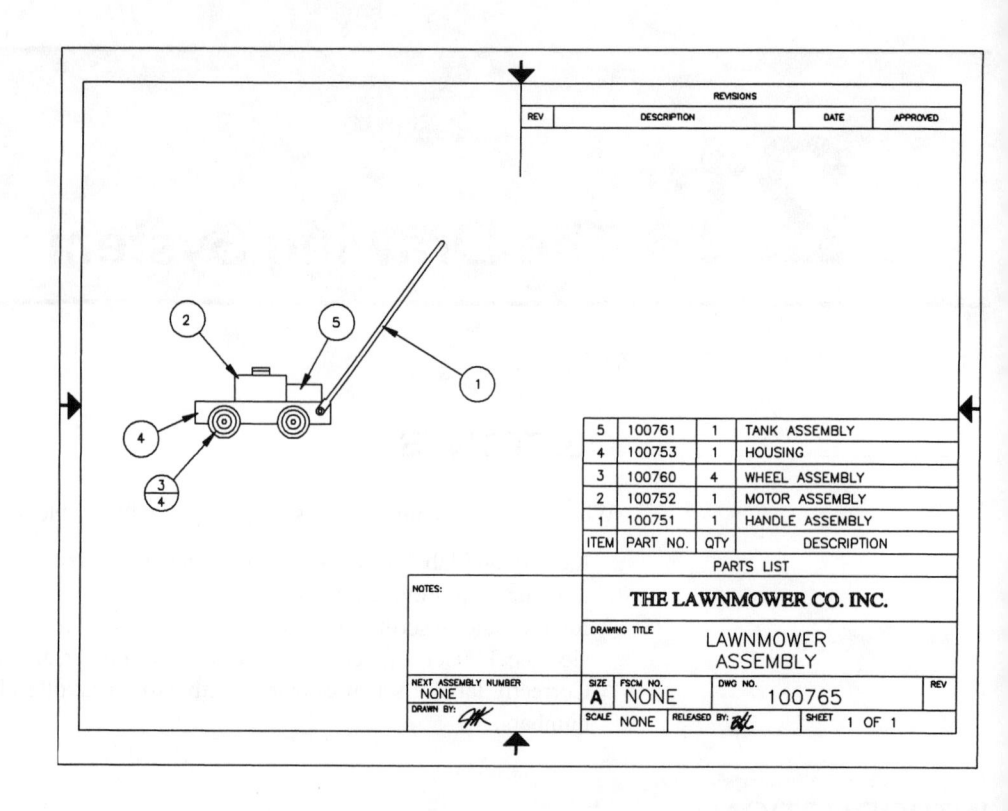

The part number of a particular part or subassembly is usually the same as its drawing number, so the parts list on the top assembly drawing is a means of referring to other drawings that give more information about the parts. The parts list also gives a description of the parts and tells how many of each part are used on the assembly. The top assembly drawing of the lawnmower in Figure 21–2, for example, shows the complete lawnmower and points out the wheel subassembly (Item 3, four each), the motor subassembly (Item 2, one each), the handle subassembly (Item 1, one each), and so forth.

Subassembly Drawings

A *subassembly drawing* contains the same elements as a top assembly drawing except that there are one or more higher assemblies in the drawing system to which the subassembly must refer. A space is provided in the title block of the subassembly drawing for the next assembly number, meaning the next higher assembly in the drawing chain.

The wheel subassembly of the lawnmower drawing set, for example, refers to the complete lawnmower drawing, number 100765, in its Next Assembly Number block (Figure 21–3). Notice that the parts list on this drawing refers to specific parts, not to subassemblies. If this were more complex equipment, the parts of the subassembly could refer to lesser subassemblies rather than to detail parts drawings.

Detail Drawings

A *detail drawing* has only one part on it. The purpose of a detail drawing is to describe in detail how to manufacture a particular part. The detail drawing contains all dimensions and manufacturing information needed.

Because it is often necessary to locate other parts associated with the particular detail part, the next assembly number is very useful in locating (1) the assembly or subassembly to which the part belongs and (2) from the assembly or subassembly drawing, other parts associated with the particular part.

Notice that the Next Assembly Number (100760) on the detail drawing of the bushing (100821) shown in Figure 21–4 refers to the wheel assembly drawing shown in Figure 21–3. The bushing in Figure 21–4 is thus a detail part of the wheel subassembly in Figure 21–3, which, in turn, is a subassembly of the lawnmower assembly in Figure 21–2.

FIGURE 21-3
Subassembly Drawing

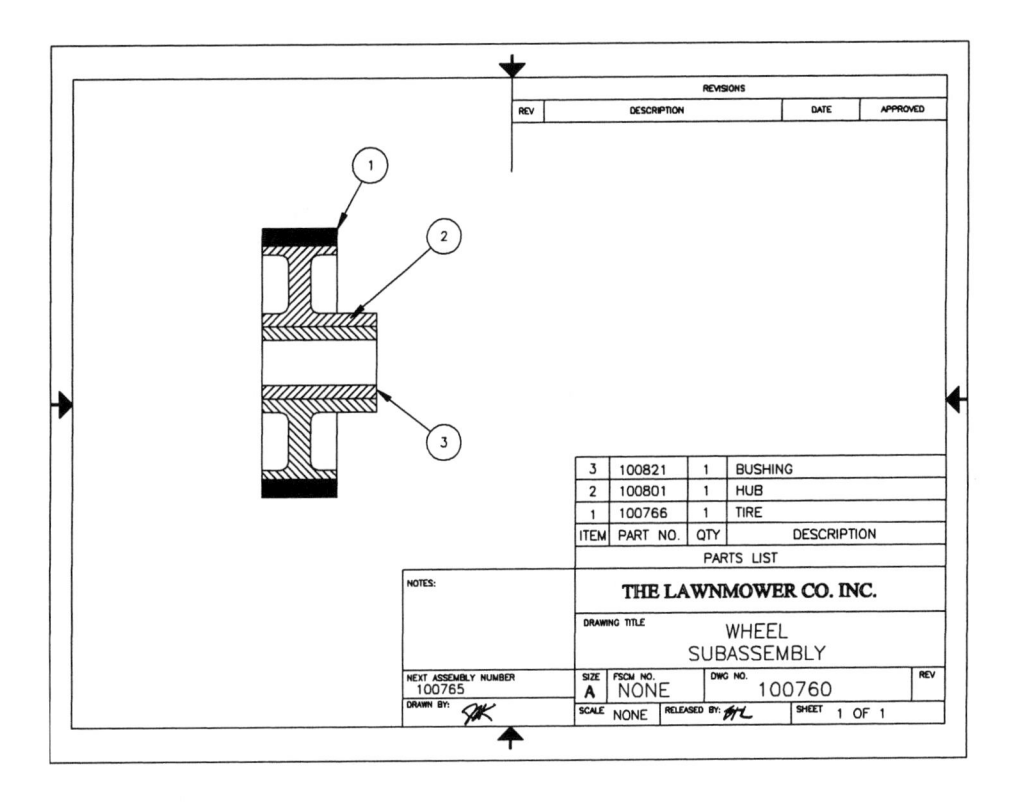

Parts of a Drawing

There are five major areas or items on a drawing: the field of the drawing, the title block, the revision block, the parts list, and notes. All these items may appear on any drawing. Many complex assemblies have separate parts lists, and many detail drawings have no parts list because the detail drawing often describes how to manufacture a single part.

FIGURE 21-4
Detail Drawing

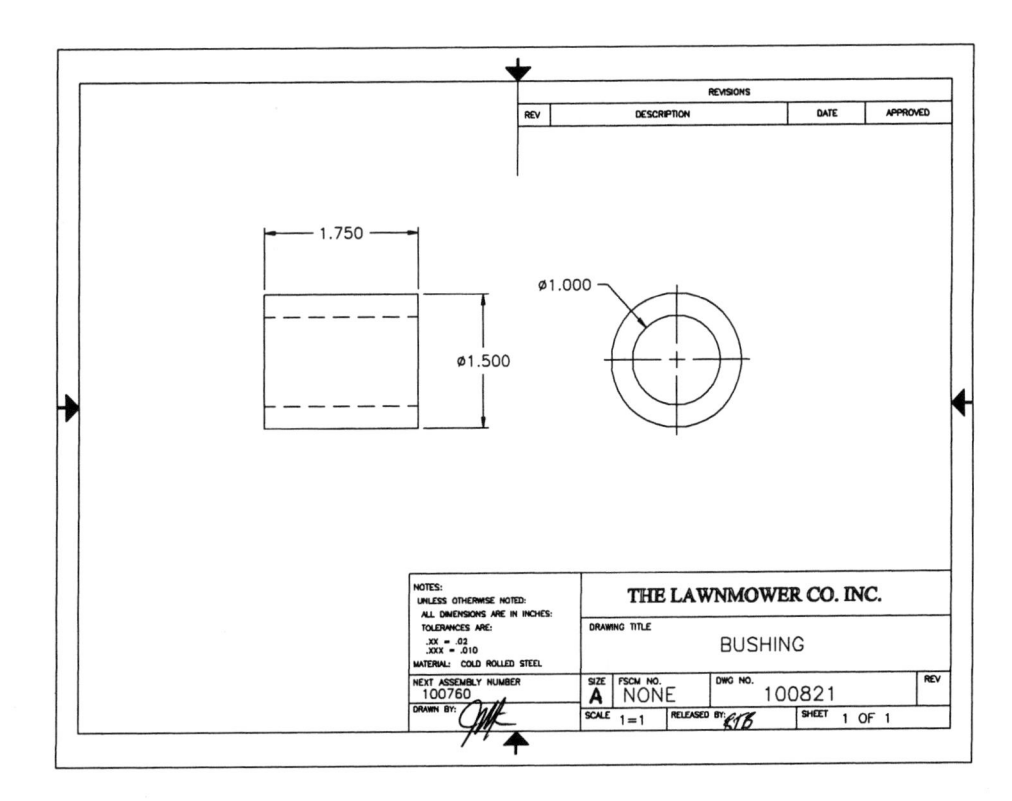

The Drawing System

315

Field of the Drawing

The field of the drawing is the area inside the border containing the drawing, exclusive of the title block and the revision block.

Title Block

The title block is usually placed in the lower right corner or across the bottom of the drawing. Figure 21–5 shows a commonly used type of title block. The main items in the title block are as follows:

The drawing title   This is the name of the part or assembly and is either selected from a standard list of titles or is assigned by the designer.

The next assembly number   This number identifies the next higher assembly of which this drawing is a part. A gear drawing in an automobile set, for example, could have the transmission subassembly as its next higher assembly, the transmission subassembly would have the drive train assembly drawing as its next assembly number, and so forth. The top assembly drawing has no next higher assembly number because there is no higher assembly than the top assembly.

The drawing number   This number is usually placed in at least two places on a drawing so it can be read and identified easily. The drawing number is often assigned from a central log of drawing numbers. There is no attempt to keep all drawings from the same set in numerical order; for example, the top assembly drawing number could be 10064 and a detail drawing from the same set could be 10001. The assignment of number is determined by whoever asks for a number first. This system eliminates the need to reshuffle numbers when changes are made.

Size   The letter in this block refers to the size of the drawing. The standard sheet sizes are as follows:

	Mechanical	Architectural
A:	$8\frac{1}{2}'' \times 11''$	$9'' \times 12''$
B:	$11'' \times 17''$	$12'' \times 18''$
C:	$17'' \times 22''$	$18'' \times 24''$
D:	$22'' \times 34''$	$24'' \times 36''$
E:	$34'' \times 44''$	$36'' \times 48''$

FIGURE 21–5
Title and Revision Block

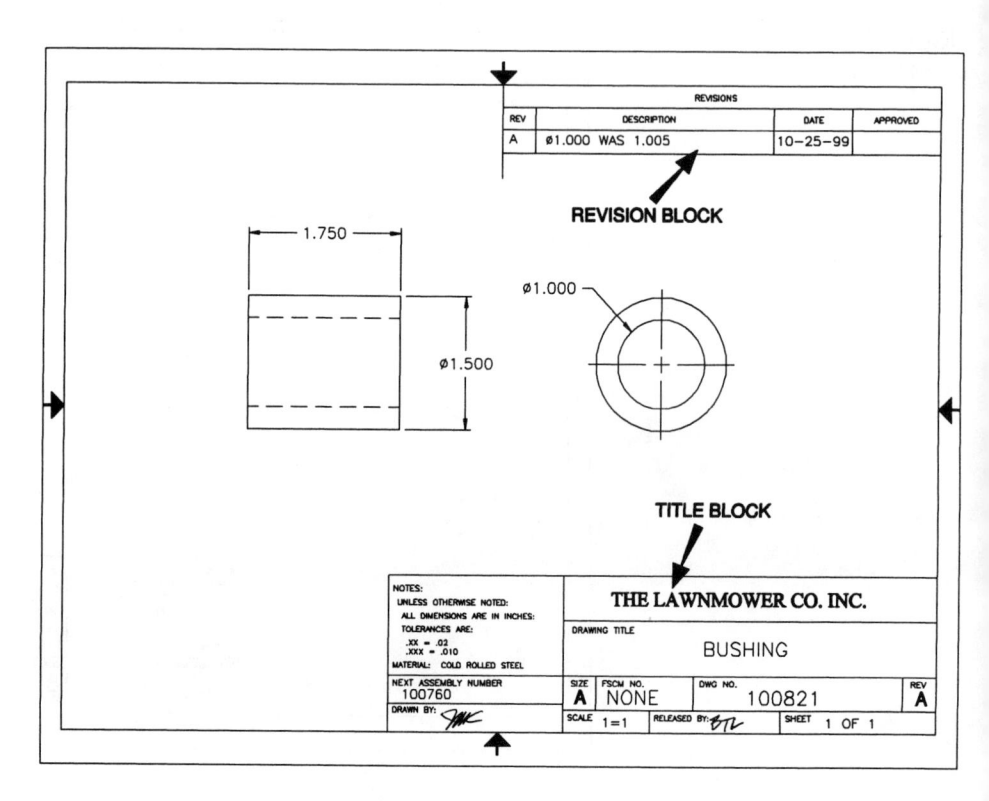

FSCM number This is a number that is assigned to manufacturers by the federal government. Every company that supplies goods to the government has its own FSCM number.

Scale This is the scale at which the drawing was made. Full size is noted as 1=1 or FULL, half scale is noted as 1=2, an architectural scale of $\frac{1''}{4}=1'$ is noted as $\frac{1''}{4}=1'$. If the scale is unimportant or if some part of the drawing is not to scale, NONE is entered in the Scale block.

Signature blocks The initials or the full names of the person who drew the drawing and the person who approved its release are required so any questions regarding the drawing can be directed to those persons.

Revision Block

When changes are made to a part or assembly, it is essential that these changes be recorded so that all versions of the part or assembly can be identified (Figure 21–5). If you order a part for your car and it doesn't fit, it may be that a change to the part was not properly recorded. These changes are called *revisions*. The revision is usually identified by a letter (first revision is A, second is B ....) in the revision block, stating, for example, "This dimension changed from 4.000 to 4.500" or simply, "was 4.000." The engineering change order number is often recorded as well, and the change is dated and initialed.

Parts List

The parts list lists every part or subassembly used to make the assembly or subassembly described by the drawing. Every screw, washer, and piece of wire, down to the most insignificant part, is listed in such a way that it can be easily identified. The parts list is used to order parts so they are in stock when it is time to assembly them into a product that can be sold. The parts list is also used to determine the price of the product, so it must be accurate. The parts list contains the item number (or index number), the part number, the quantity, and a description.

Item number The item numbers are listed on the parts list in order and appear on the field of the drawing to point out each item, as was shown in Figures 21–2 and 21–3. Notice that item 3 in Figure 21–2 shows the quantity beneath the item number.

Part number This is usually the same number as the drawing number of the detail or subassembly called out. For a part that is purchased from another manufacturer, it is often necessary to list the manufacturer and the part number assigned by that manufacturer as well.

Quantity The quantity noted for each item is the number of that item necessary to manufacture the assembly.

Description The description of the item is the name that appears in the title block of the drawing for that item. If there is no drawing for the item, as is sometimes the case for screws, bolts, washers, and the like, a clear and accurate description must appear in the parts list. A typical entry for a screw might be: "Hex Head Cap Screw, .250-10UNC-2A, × .750 long." In many drawing systems all parts have part numbers even if there are no drawings for them.

Notes

Most drawings have several standard notes. The most common one describes dimensions and their tolerances. Other notes describe how smoothly the part is to be finished, how it is to be painted, what material is to be used to build the part, and any other information needed to build the part exactly as it was designed.

EXERCISE

EXERCISE 21–1 Drawings 100353, 100303, 100351, 100352, and 100356 are five simplified sketches representing part of a set of drawings of a semitrailer. You are to remove these drawings from your book and fill in the missing information on these drawings according to the instructions given.

A. Use your best lettering WITH ALL CAPITAL LETTERS to complete the title block and make a parts list for drawing 100353.

Title the drawing CITY DELIVERY TRAILER.

The descriptions for each of the items called out for a parts list on the drawing are as follows:

ITEM NO.	DESCRIPTION
1	WHEEL ASSEMBLY (just wheels, not the complete tandem)
2	LANDING GEAR ASSEMBLY
3	ROOF ASSEMBLY
4	LEFT HAND DOOR ASSEMBLY
5	RIGHT HAND DOOR ASSEMBLY
6	LEFT HARDWARE ASSEMBLY
7	RIGHT HARDWARE ASSEMBLY
8	BUMPER

Make up six-digit part numbers for all but the left hardware assembly; use the part number 100303 for this item.

Show the Scale of this drawing as NONE.

There are no notes on this drawing.

B. Use your best lettering WITH ALL CAPITAL LETTERS to complete the title block and make a parts list for drawing 100303.

Title the drawing LEFT HARDWARE ASSEMBLY.

Be sure you fill in the correct Next Assembly Number.

The descriptions for each of the items called out on the drawing are:

ITEM NO.	DESCRIPTION	PART NUMBER
1	CATCH	100356
2	DOG	XXXXXX
3	ROD RETAINING PLATE	100352
4	ROD	XXXXXX
5	WASHER	XXXXXX
6	GUARD	XXXXXX
7	HANDLE	XXXXXX
8	RETAINER	XXXXXX
9	HANDLE PLATE	100351

Make up six-digit part numbers for all those shown as XXXXXX.

Show the Scale of this drawing as NONE.

There are no notes on this drawing.

C. Use your best lettering WITH ALL CAPITAL LETTERS to complete the title blocks for drawings 100351, 100352, and 100356.

Title drawing 100351, HANDLE PLATE.
Title drawing 100352, ROD RETAINING PLATE.
Title drawing 100356, CATCH.

Be sure you fill in the correct Next Assembly Number on each drawing.

Measure each drawing, and dimension it completely using the principles described in Chapter 15. Use the following specifications:

Make all dimensions 2-place decimals except:

All hole diameters and dimensions between holes must be 3-place decimals. Be sure to use the diameter symbol from Chapters 15 and 16.

The leader on the smaller hole on 100352 should show the diameter on the first line and 2 PL on the second line to show that the diameter is the same on both small holes.

REVIEW QUESTIONS

1. Which of the following is not part of a complete set of drawings?
   a. Top assembly
   b. Subassembly
   c. Detail
   d. Parts list
   e. All are part of a set of drawings.
2. Which of the following would most likely be a top assembly drawing?
   a. Lawnmower
   b. Lawnmower engine
   c. Blade
   d. Housing
   e. Washer
3. Which of the following would most likely be a subassembly drawing?
   a. Lawnmower
   b. Lawnmower engine
   c. Blade
   d. Housing
   e. Washer
4. Which of the following would most likely be a detail drawing?
   a. Lawnmower
   b. Lawnmower engine
   c. Blade
   d. Wheel
   e. Starter
5. Which of the following is not included in a parts list?
   a. Item number
   b. Quantity
   c. Part number
   d. Description
   e. All are included in a parts list.
6. The next assembly number refers to
   a. The next number on a parts list
   b. The next number on a drawing log
   c. The next higher assembly
   d. The next lower assembly
   e. The next part to fit on this one
7. The FSCM number refers to
   a. A code number given to the part
   b. A code number given to the metal from which the part is made
   c. A code number classifying the part into a general group
   d. The federal identification number of the manufacturer
   e. The identification number of a particular group of manufacturers
8. Which of the following is not a reason for making sure that the parts list is complete and accurate?
   a. So that the correct quantity of parts can be kept in inventory
   b. So that the purchasing agent knows what to buy
   c. So that the cost of the assembly can be computed accurately
   d. So that the correct parts can be sent to the assembly area
   e. All are reasons for a complete and accurate parts list.
9. Every item listed on a parts list must have an individual number.
   a. True
   b. False
10. The field of the drawing does not contain which of the following?
    a. The revision block
    b. The drawing of the part
    c. Dimensions on the part
    d. Detail views
    e. Assembly views

Appendix

Decimal and Millimeter Equivalents

4ths	8ths	16ths	32nds	64ths	To 4 Places	To 3 Places	To 2 Places	Milli-meters	4ths	8ths	16ths	32nds	64ths	To 4 Places	To 3 Places	To 2 Places	Milli-meters
				1/64	.0156	.016	.02	.397					33/64	.5156	.516	.52	13.097
			1/32		.0312	.031	.03	.794				17/32		.5312	.531	.53	13.494
				3/64	.0469	.047	.05	1.191					35/64	.5469	.547	.55	13.891
		1/16			.0625	.062	.06	1.588			9/16			.5625	.562	.56	14.288
				5/64	.0781	.078	.08	1.984					37/64	.5781	.578	.58	14.684
			3/32		.0938	.094	.09	2.381				19/32		.5938	.594	.59	15.081
				7/64	.1094	.109	.11	2.778					39/64	.6094	.609	.61	15.478
	1/8				.1250	.125	.12	3.175		5/8				.6250	.625	.62	15.875
				9/64	.1406	.141	.14	3.572					41/64	.6406	.641	.64	16.272
			5/32		.1562	.156	.16	3.969				21/32		.6562	.656	.66	16.669
				11/64	.1719	.172	.17	4.366					43/64	.6719	.672	.67	17.066
		3/16			.1875	.188	.19	4.762			11/16			.6875	.688	.69	17.462
				13/64	.2031	.203	.20	5.159					45/64	.7031	.703	.70	17.859
			7/32		.2188	.219	.22	5.556				23/32		.7188	.719	.72	18.256
				15/64	.2344	.234	.23	5.953					47/64	.7344	.734	.73	18.653
1/4					.2500	.250	.25	6.350	3/4					.7500	.750	.75	19.050
				17/64	.2656	.266	.27	6.747					49/64	.7656	.766	.77	19.447
			9/32		.2812	.281	.28	7.144				25/32		.7812	.781	.78	19.844
				19/64	.2969	.297	.30	7.541					51/64	.7969	.797	.80	20.241
		5/16			.3125	.312	.31	7.938			13/16			.8125	.812	.81	20.638
				21/64	.3281	.328	.33	8.334					53/64	.8281	.828	.83	21.034
			11/32		.3438	.344	.34	8.731				27/32		.8438	.844	.84	21.431
				23/64	.3594	.359	.36	9.128					55/64	.8594	.859	.86	21.828
	3/8				.3750	.375	.38	9.525		7/8				.8750	.875	.88	22.225
				25/64	.3906	.391	.39	9.922					57/64	.8906	.891	.89	22.622
			13/32		.4062	.406	.41	10.319				29/32		.9062	.906	.91	23.019
				27/64	.4219	.422	.42	10.716					59/64	.9219	.922	.92	23.416
		7/16			.4375	.438	.44	11.112			15/16			.9375	.938	.94	23.812
				29/64	.4531	.453	.45	11.509					61/64	.9531	.953	.95	24.209
			15/32		.4688	.469	.47	11.906				31/32		.9688	.969	.97	24.606
				31/64	.4844	.484	.48	12.303					63/64	.9844	.984	.98	25.003
					.5000	.500	.50	12.700						1.0000	1.000	1.00	25.400

American National Standard Unified Threads

Nominal Diameter	Coarse (UNC) Thds. per Inch	Fine (UNF) Thds. per Inch	Extra Fine (UNEF) Thds. per Inch
0 (.060)		80	
1 (.073)	64	72	....
2 (.086)	56	64	....
3 (.099)	48	56	....
4 (.112)	40	48	....
5 (.125)	40	44	....
6 (.138)	32	40	....
8 (.164)	32	36	....
10 (.190)	24	32	....
12 (.216)	24	28	32
$\frac{1}{4}$	20	28	32
$\frac{5}{16}$	18	24	32
$\frac{3}{8}$	16	24	32
$\frac{7}{16}$	14	20	28
$\frac{1}{2}$	13	20	28
$\frac{9}{16}$	12	18	24
$\frac{5}{8}$	11	18	24
$\frac{11}{16}$	....	....	24
$\frac{3}{4}$	10	16	20
$\frac{13}{16}$	....	....	20
$\frac{7}{8}$	9	14	20
$\frac{15}{16}$	....	....	20
1	8	12	20
$1\frac{1}{16}$	....	....	18
$1\frac{1}{8}$	7	12	18
$1\frac{3}{16}$	....	....	18
$1\frac{1}{4}$	7	12	18
$1\frac{5}{16}$	....	....	18
$1\frac{3}{8}$	6	12	18
$1\frac{7}{16}$	....	....	18
$1\frac{1}{2}$	6	12	18
$1\frac{9}{16}$	....	....	18
$1\frac{5}{8}$	....	....	18
$1\frac{11}{16}$	....	....	18
$1\frac{3}{4}$	5	....	....
2	$4\frac{1}{2}$	....	....
$2\frac{1}{4}$	$4\frac{1}{2}$	....	....
$2\frac{1}{2}$	4	....	....
$2\frac{3}{4}$	4	....	....
3	4	....	....
$3\frac{1}{4}$	4	....	....
$3\frac{1}{2}$	4	....	....
$3\frac{3}{4}$	4	....	....
4	4	....	....

Appendix

American National Standard Square and Hexagon Bolts and Nuts and Hexagon Head Cap Screws

Nominal Size Body Diameter of Bolt	Regular Bolts Width Across Flats Sq.	Regular Bolts Width Across Flats Hex.	Regular Bolts Height H Sq.	Regular Bolts Height H Hex.	Regular Bolts Height H Hex. Cap Scr.	Heavy Bolts Width Across Flats W	Heavy Bolts Height H Hex.	Heavy Bolts Height H Hex. Screw
$\frac{1}{4}$ 0.2500	$\frac{3}{8}$	$\frac{7}{16}$	$\frac{11}{64}$	$\frac{11}{64}$	$\frac{5}{32}$	....	....	....
$\frac{5}{16}$ 0.3125	$\frac{1}{2}$	$\frac{1}{2}$	$\frac{13}{64}$	$\frac{7}{32}$	$\frac{13}{64}$	....	....	....
$\frac{3}{8}$ 0.3750	$\frac{9}{16}$	$\frac{9}{16}$	$\frac{1}{4}$	$\frac{1}{4}$	$\frac{15}{64}$	....	....	....
$\frac{7}{16}$ 0.4375	$\frac{5}{8}$	$\frac{5}{8}$	$\frac{19}{64}$	$\frac{19}{64}$	$\frac{9}{32}$	....	....	....
$\frac{1}{2}$ 0.5000	$\frac{3}{4}$	$\frac{3}{4}$	$\frac{21}{64}$	$\frac{11}{32}$	$\frac{5}{16}$	$\frac{7}{8}$	$\frac{11}{32}$	$\frac{5}{16}$
$\frac{9}{16}$ 0.5625		$\frac{13}{16}$	....	....	$\frac{23}{64}$	....	....	....
$\frac{5}{8}$ 0.6250	$\frac{15}{16}$	$\frac{15}{16}$	$\frac{27}{64}$	$\frac{27}{64}$	$\frac{25}{64}$	$1\frac{1}{16}$	$\frac{27}{64}$	$\frac{25}{64}$
$\frac{3}{4}$ 0.7500	$1\frac{1}{8}$	$1\frac{1}{8}$	$\frac{1}{2}$	$\frac{1}{2}$	$\frac{15}{32}$	$1\frac{1}{4}$	$\frac{1}{2}$	$\frac{15}{32}$
$\frac{7}{8}$ 0.8750	$1\frac{5}{16}$	$1\frac{5}{16}$	$\frac{19}{32}$	$\frac{37}{64}$	$\frac{35}{64}$	$1\frac{7}{16}$	$\frac{37}{64}$	$\frac{35}{64}$
1 1.000	$1\frac{1}{2}$	$1\frac{1}{2}$	$\frac{21}{32}$	$\frac{43}{64}$	$\frac{39}{64}$	$1\frac{5}{8}$	$\frac{43}{64}$	$\frac{39}{64}$
$1\frac{1}{8}$ 1.1250	$1\frac{11}{16}$	$1\frac{11}{16}$	$\frac{3}{4}$	$\frac{3}{4}$	$\frac{11}{16}$	$1\frac{13}{16}$	$\frac{3}{4}$	$\frac{11}{16}$
$1\frac{1}{4}$ 1.2500	$1\frac{7}{8}$	$1\frac{7}{8}$	$\frac{27}{32}$	$\frac{27}{32}$	$\frac{25}{32}$	2	$\frac{27}{32}$	$\frac{25}{32}$
$1\frac{3}{8}$ 1.3750	$2\frac{1}{16}$	$2\frac{1}{16}$	$\frac{29}{32}$	$\frac{29}{32}$	$\frac{27}{32}$	$2\frac{3}{16}$	$\frac{29}{32}$	$\frac{27}{32}$
$1\frac{1}{2}$ 1.5000	$2\frac{1}{4}$	$2\frac{1}{4}$	1	1	$1\frac{5}{64}$	$2\frac{3}{8}$	1	$\frac{15}{16}$
$1\frac{3}{4}$ 1.7500	....	$2\frac{5}{8}$	....	$1\frac{5}{32}$	$1\frac{3}{32}$	$2\frac{3}{4}$	$1\frac{5}{32}$	$1\frac{3}{32}$
2 2.0000	....	3	....	$1\frac{11}{32}$	$1\frac{7}{32}$	$3\frac{1}{8}$	$1\frac{11}{32}$	$1\frac{7}{32}$
$2\frac{1}{4}$ 2.2500	....	$3\frac{3}{8}$	....	$1\frac{1}{2}$	$1\frac{3}{8}$	$3\frac{1}{2}$	$1\frac{1}{2}$	$1\frac{3}{8}$
$2\frac{1}{2}$ 2.5000	....	$3\frac{3}{4}$	....	$1\frac{21}{32}$	$1\frac{17}{32}$	$3\frac{7}{8}$	$1\frac{21}{32}$	$1\frac{17}{32}$
$2\frac{3}{4}$ 2.7500	....	$4\frac{1}{8}$	....	$1\frac{13}{16}$	$1\frac{11}{16}$	$4\frac{1}{4}$	$1\frac{13}{16}$	$1\frac{11}{16}$
3 3.0000	....	$4\frac{1}{2}$	....	2	$1\frac{7}{8}$	$4\frac{5}{8}$	2	$1\frac{7}{8}$
$3\frac{1}{4}$ 3.2500	....	$4\frac{7}{8}$	....	$2\frac{3}{16}$	....	....	....	....
$3\frac{1}{2}$ 3.5000	....	$5\frac{1}{4}$	....	$2\frac{5}{16}$	....	....	....	....
$3\frac{3}{4}$ 3.7500	....	$5\frac{5}{8}$	....	$2\frac{1}{2}$	....	....	....	....
4 4.0000	....	6	....	$2\frac{11}{16}$	....	....	....	....

American National Standard Square and Hexagon Bolts and Nuts and Hexagon Head Cap Screws *(continued)*

Nominal Size Body Diameter of Bolt	Regular Nuts					Heavy Nuts			
	Width Across Flats		Thickness T			Width Across Flats	Thickness T		
	Sq.	Hex.	Sq.	Hex. Flat	Hex.	W	Sq.	Hex. Flat	Hex.
$\frac{1}{4}$ 0.2500	$\frac{7}{16}$	$\frac{7}{16}$	$\frac{7}{32}$	$\frac{7}{32}$	$\frac{7}{32}$	$\frac{1}{2}$	$\frac{1}{4}$	$\frac{15}{64}$	$\frac{15}{64}$
$\frac{5}{16}$ 0.3125	$\frac{9}{16}$	$\frac{1}{2}$	$\frac{17}{64}$	$\frac{17}{64}$	$\frac{17}{64}$	$\frac{9}{16}$	$\frac{5}{16}$	$\frac{19}{64}$	$\frac{19}{64}$
$\frac{3}{8}$ 0.3750	$\frac{5}{8}$	$\frac{9}{16}$	$\frac{21}{64}$	$\frac{21}{64}$	$\frac{21}{64}$	$\frac{11}{16}$	$\frac{3}{8}$	$\frac{23}{64}$	$\frac{23}{64}$
$\frac{7}{16}$ 0.4375	$\frac{3}{4}$	$\frac{11}{16}$	$\frac{3}{8}$	$\frac{3}{8}$	$\frac{3}{8}$	$\frac{3}{4}$	$\frac{7}{16}$	$\frac{27}{64}$	$\frac{27}{64}$
$\frac{1}{2}$ 0.5000	$\frac{13}{16}$	$\frac{3}{4}$	$\frac{7}{16}$	$\frac{7}{16}$	$\frac{7}{16}$	$\frac{7}{8}$	$\frac{1}{2}$	$\frac{31}{64}$	$\frac{31}{64}$
$\frac{9}{16}$ 0.5625	....	$\frac{7}{8}$	....	$\frac{31}{64}$	$\frac{31}{64}$	$\frac{15}{16}$	....	$\frac{35}{64}$	$\frac{35}{64}$
$\frac{5}{8}$ 0.6250	1	$\frac{15}{16}$	$\frac{35}{64}$	$\frac{35}{64}$	$\frac{35}{64}$	$1\frac{1}{16}$	$\frac{5}{8}$	$\frac{39}{64}$	$\frac{39}{64}$
$\frac{3}{4}$ 0.7500	$1\frac{1}{8}$	$1\frac{1}{8}$	$\frac{21}{32}$	$\frac{41}{64}$	$\frac{41}{64}$	$1\frac{1}{4}$	$\frac{3}{4}$	$\frac{47}{64}$	$\frac{47}{64}$
$\frac{7}{8}$ 0.8750	$1\frac{5}{16}$	$1\frac{5}{16}$	$\frac{49}{64}$	$\frac{3}{4}$	$\frac{3}{4}$	$1\frac{7}{16}$	$\frac{7}{8}$	$\frac{55}{64}$	$\frac{55}{64}$
1 1.0000	$1\frac{1}{2}$	$1\frac{1}{2}$	$\frac{7}{8}$	$\frac{55}{64}$	$\frac{55}{64}$	$1\frac{5}{8}$	1	$\frac{63}{64}$	$\frac{63}{64}$
$1\frac{1}{8}$ 1.1250	$1\frac{11}{16}$	$1\frac{11}{16}$	1	1	$\frac{31}{32}$	$1\frac{13}{16}$	$1\frac{1}{8}$	$1\frac{1}{8}$	$1\frac{7}{64}$
$1\frac{1}{4}$ 1.2500	$1\frac{7}{8}$	$1\frac{7}{8}$	$1\frac{3}{32}$	$1\frac{3}{32}$	$1\frac{1}{16}$	2	$1\frac{1}{4}$	$1\frac{1}{4}$	$1\frac{7}{32}$
$1\frac{3}{8}$ 1.3750	$2\frac{1}{16}$	$2\frac{1}{16}$	$1\frac{13}{64}$	$1\frac{13}{64}$	$1\frac{11}{64}$	$2\frac{3}{16}$	$1\frac{3}{8}$	$1\frac{3}{8}$	$1\frac{11}{32}$
$1\frac{1}{2}$ 1.5000	$2\frac{1}{4}$	$2\frac{1}{4}$	$1\frac{5}{16}$	$1\frac{5}{16}$	$1\frac{9}{32}$	$2\frac{3}{8}$	$1\frac{1}{2}$	$1\frac{1}{2}$	$1\frac{15}{32}$
$1\frac{5}{8}$ 1.6250	....	....	....	....	....	$2\frac{9}{16}$	....	....	$1\frac{19}{32}$
$1\frac{3}{4}$ 1.7500	....	....	....	....	....	$2\frac{3}{4}$	....	$1\frac{3}{4}$	$1\frac{23}{32}$
$1\frac{7}{8}$ 1.8750	....	....	....	....	....	$2\frac{15}{16}$	....	....	$1\frac{27}{32}$
2 2.0000	....	....	....	....	....	$3\frac{1}{8}$	....	2	$1\frac{31}{32}$
$2\frac{1}{4}$ 2.2500	....	....	....	....	....	$3\frac{1}{2}$	....	$2\frac{1}{4}$	$2\frac{13}{64}$
$2\frac{1}{2}$ 2.5000	....	....	....	....	....	$3\frac{7}{8}$	....	$2\frac{1}{2}$	$2\frac{29}{64}$
$2\frac{3}{4}$ 2.7500	....	....	....	....	....	$4\frac{1}{4}$	....	$2\frac{3}{4}$	$2\frac{45}{64}$
3 3.0000	....	....	....	....	....	$4\frac{5}{8}$	....	3	$2\frac{61}{64}$
$3\frac{1}{4}$ 3.2500	....	....	....	....	....	5	....	$3\frac{1}{4}$	$3\frac{3}{16}$
$3\frac{1}{2}$ 3.5000	....	....	....	....	....	$5\frac{3}{8}$	....	$3\frac{1}{2}$	$3\frac{7}{16}$
$3\frac{3}{4}$ 3.7500	....	....	....	....	....	$5\frac{3}{4}$	....	$3\frac{3}{4}$	$3\frac{11}{16}$
4 4.0000	....	....	....	....	....	$6\frac{1}{8}$	....	4	$3\frac{15}{16}$

Exercises to Chapters

← START ALL LINES FROM THIS VERTICAL GRID

ARCHITECT'S SCALE

4'–3"   @3/4"=1'
6'–2"   @3/4"=1'
15'–6"   @1/8"=1'
7'–10"   @1/2"=1'
5–3/4"   @3/4"=1"
6–3/4"   @3/4"=1"
20–1/2"   @1/8"=1"
8–1/4"   @1/2"=1"

CIVIL ENGINEER'S SCALE

84'   @1"=20'
68'   @1"=20'
1650'   @1"=400'
214 MILES   @1"=50 MILES
5.5"   @1/2"=1"
6.2"   @1/2"=1"
17.5"   @1/4"=1"
11.4"   @1/4"=1"

NAME:
SCHOOL:
DATE:
GRADE:
DRAWING TITLE:
CLASS
EXERCISE 2-1 SH 1 OF 2

← START ALL LINES FROM THIS VERTICAL GRID

METRIC RULER

86mm FULL SCALE
35mm FULL SCALE
122mm FULL SCALE
155mm FULL SCALE

STEEL RULE

1.55" FULL SCALE
2.875" FULL SCALE
4.65" FULL SCALE
5.735" FULL SCALE
1-3/8" FULL SCALE
4-1/8" FULL SCALE
3-3/64" FULL SCALE
2-5/32" FULL SCALE

NAME:
SCHOOL:
DATE:
GRADE:
DRAWING TITLE:
CLASS
EXERCISE 2-2

Kirkpatrick, *Basic Drafting Using Pencil Sketches and AutoCAD*, © 2003 by Pearson Education, Inc.

NAME:
SCHOOL:
DATE:
GRADE:
DRAWING TITLE:
CLASS
EXERCISE 2-3

Kirkpatrick, *Basic Drafting Using Pencil Sketches and AutoCAD*, © 2003 by Pearson Education, Inc.

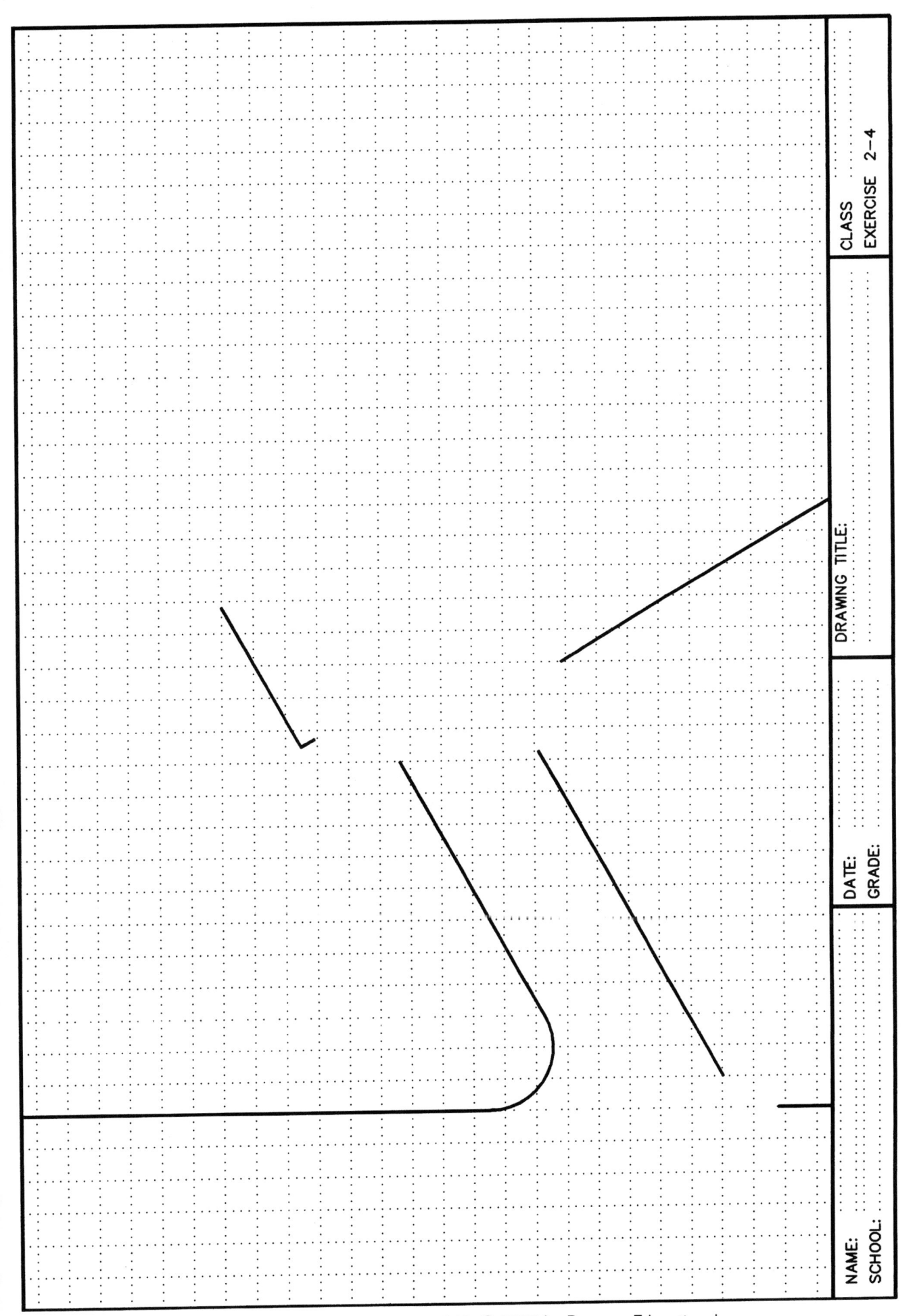

NAME:	DATE:	DRAWING TITLE:	CLASS
SCHOOL:	GRADE:		EXERCISE 2-4

Kirkpatrick, *Basic Drafting Using Pencil Sketches and AutoCAD,* © 2003 by Pearson Education, Inc.

DRAW LETTERS AND NUMBERS 7 TIMES

A						B
C						D
E						F
G						H
I						J
K						L
M						N
O						P
Q						R
S						T
U						V
W						X
Y						Z
1						2
3						4
5						6
7						8
9						0

$1\frac{1}{2}$					$2\frac{3}{4}$

    5 TIMES				 5 TIMES

GOOD LETTERING MAKES A GOOD SKETCH BETTER
G
G
G
G

                    5 TIMES

NAME:	DATE:	DRAWING TITLE:	CLASS.
SCHOOL:	GRADE:		EXERCISE 4-1

Kirkpatrick, *Basic Drafting Using Pencil Sketches and AutoCAD*, © 2003 by Pearson Education, Inc.

DRAW LETTERS AND NUMBERS 7 TIMES

A	B
C	D
E	F
G	H
I	J
K	L
M	N
O	P
Q	R
S	T
U	V
W	X
Y	Z
1	2
3	4
5	6
7	8
9	0
$1\frac{1}{2}$	$2\frac{3}{4}$
5 TIMES	5 TIMES

GOOD LETTERING MAKES A GOOD SKETCH BETTER
G
G
G
G

5 TIMES

NAME: DATE: DRAWING TITLE: CLASS.
SCHOOL: GRADE: EXERCISE 4-2

Kirkpatrick, *Basic Drafting Using Pencil Sketches and AutoCAD,* © 2003 by Pearson Education, Inc.

NAME:	DATE:	DRAWING TITLE:	CLASS
SCHOOL:	GRADE:		EXERCISE 5-1

		NAME:	DATE:	DRAWING TITLE:	CLASS
		SCHOOL:	GRADE:		EXERCISE 5-2

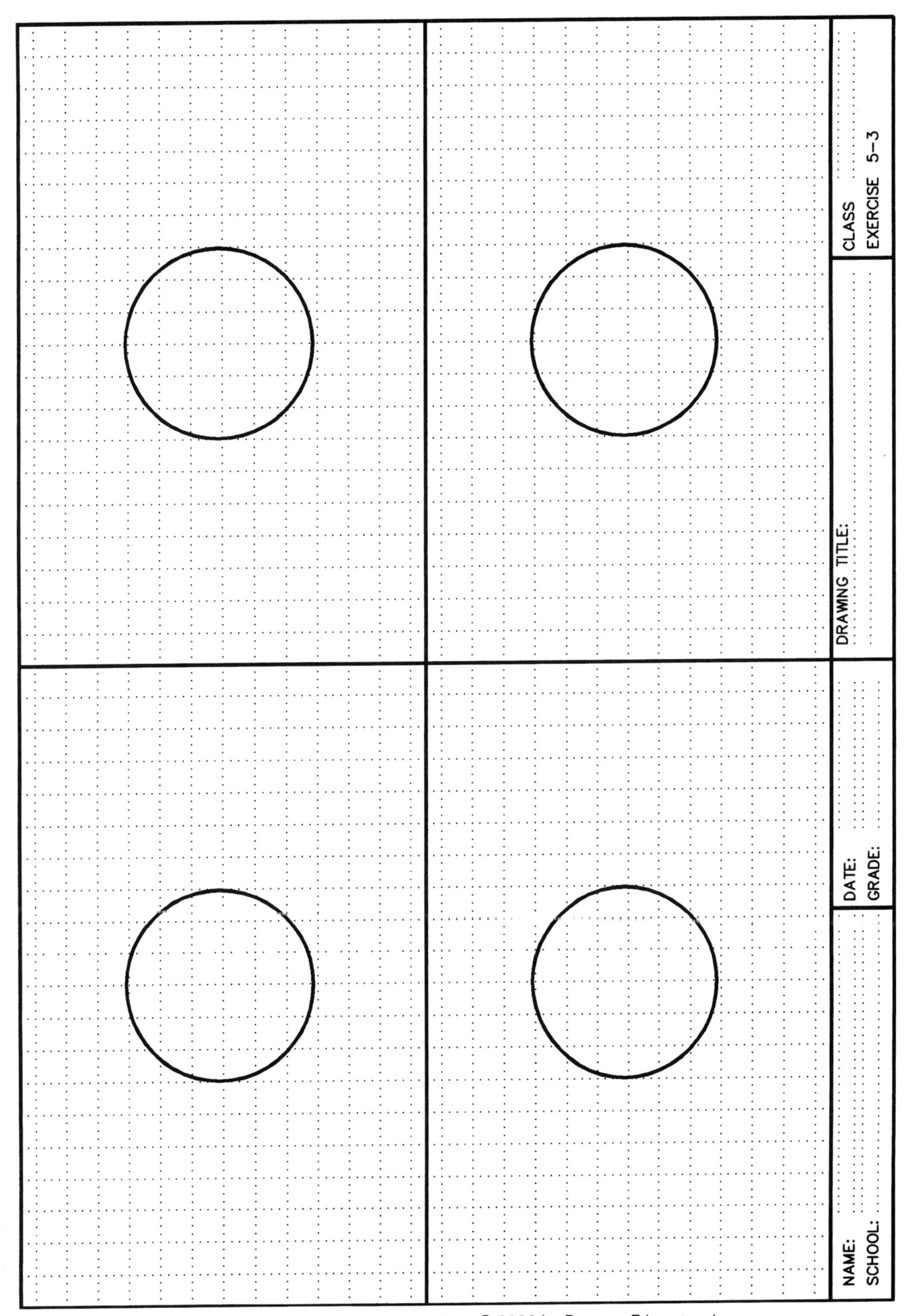

NAME:
SCHOOL:
DATE:
GRADE:
DRAWING TITLE:
CLASS
EXERCISE 5-3

Kirkpatrick, *Basic Drafting Using Pencil Sketches and AutoCAD*, © 2003 by Pearson Education, Inc.

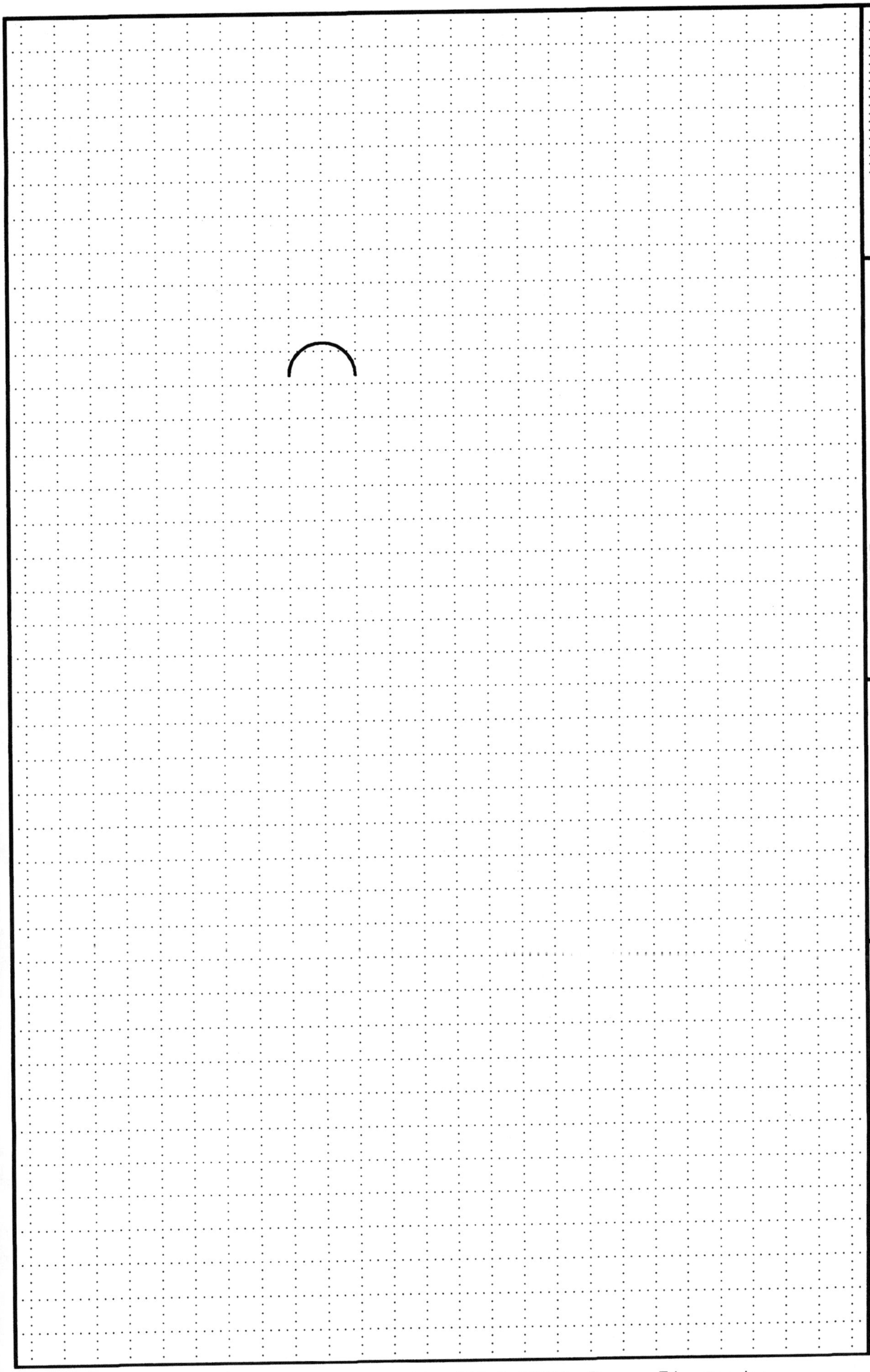

NAME:
SCHOOL:
DATE:
GRADE:
DRAWING TITLE:
CLASS
EXERCISE 5–4

NAME:	DATE:	DRAWING TITLE:	CLASS
SCHOOL:	GRADE:		EXERCISE 5-5

Kirkpatrick, *Basic Drafting Using Pencil Sketches and AutoCAD,* © 2003 by Pearson Education, Inc.

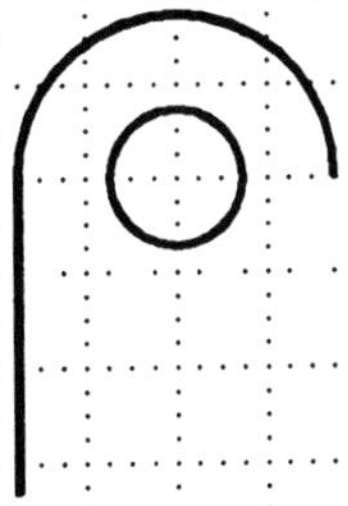

NAME:
SCHOOL:
DATE:
GRADE:
DRAWING TITLE:
CLASS
EXERCISE 5–6

CLASS
EXERCISE 7-1

DRAWING TITLE:

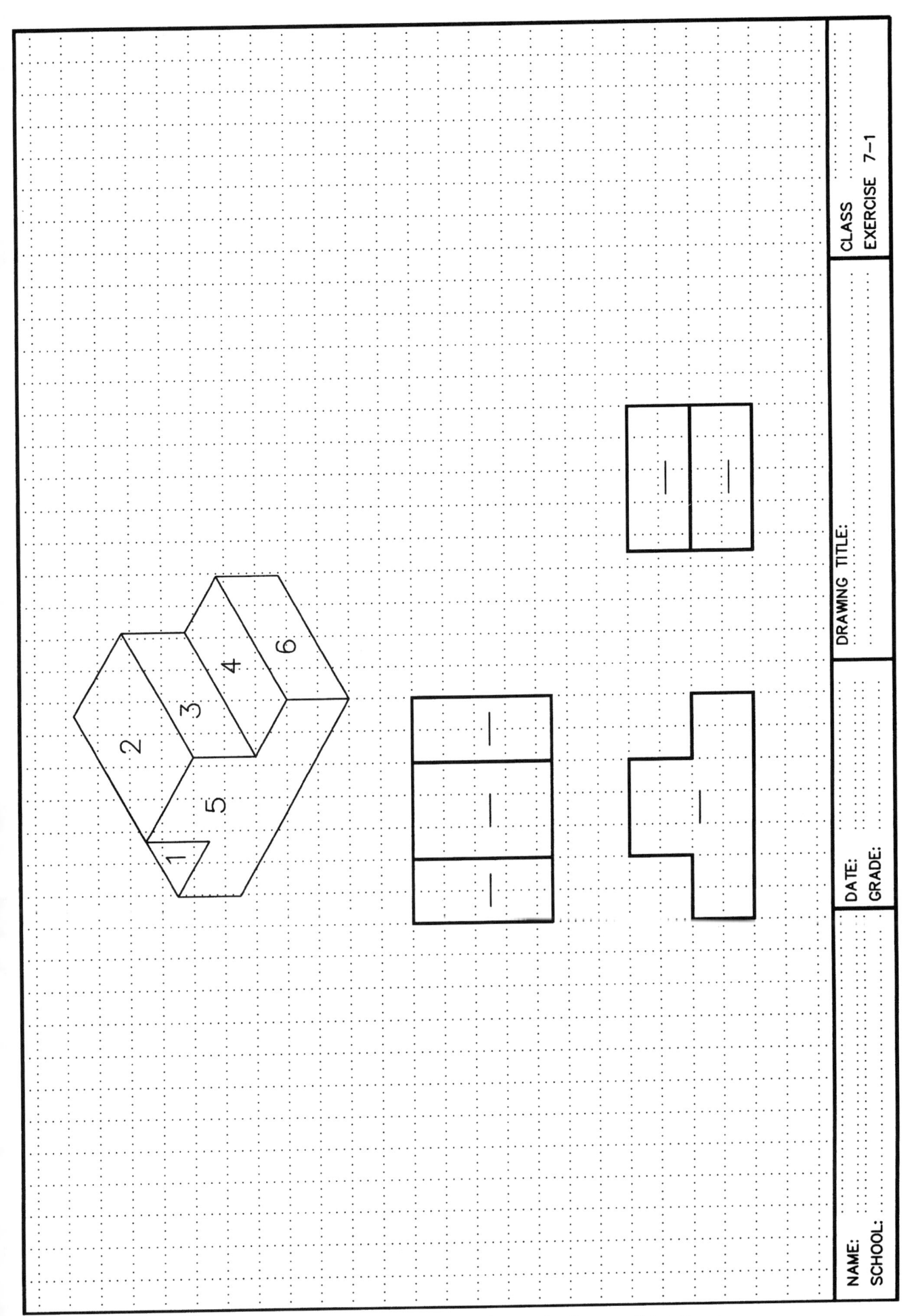

DATE:
GRADE:

NAME:
SCHOOL:

Kirkpatrick, *Basic Drafting Using Pencil Sketches and AutoCAD,* © 2003 by Pearson Education, Inc.

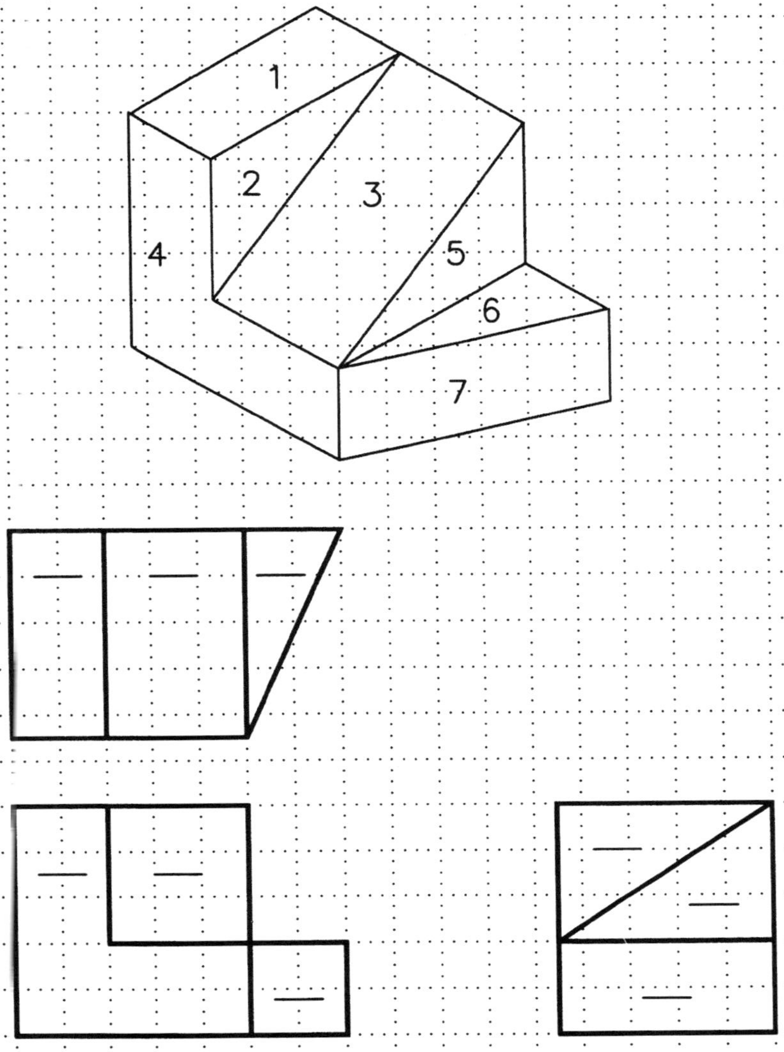

NAME: SCHOOL:	DATE: GRADE:	DRAWING TITLE:	CLASS EXERCISE 7–2

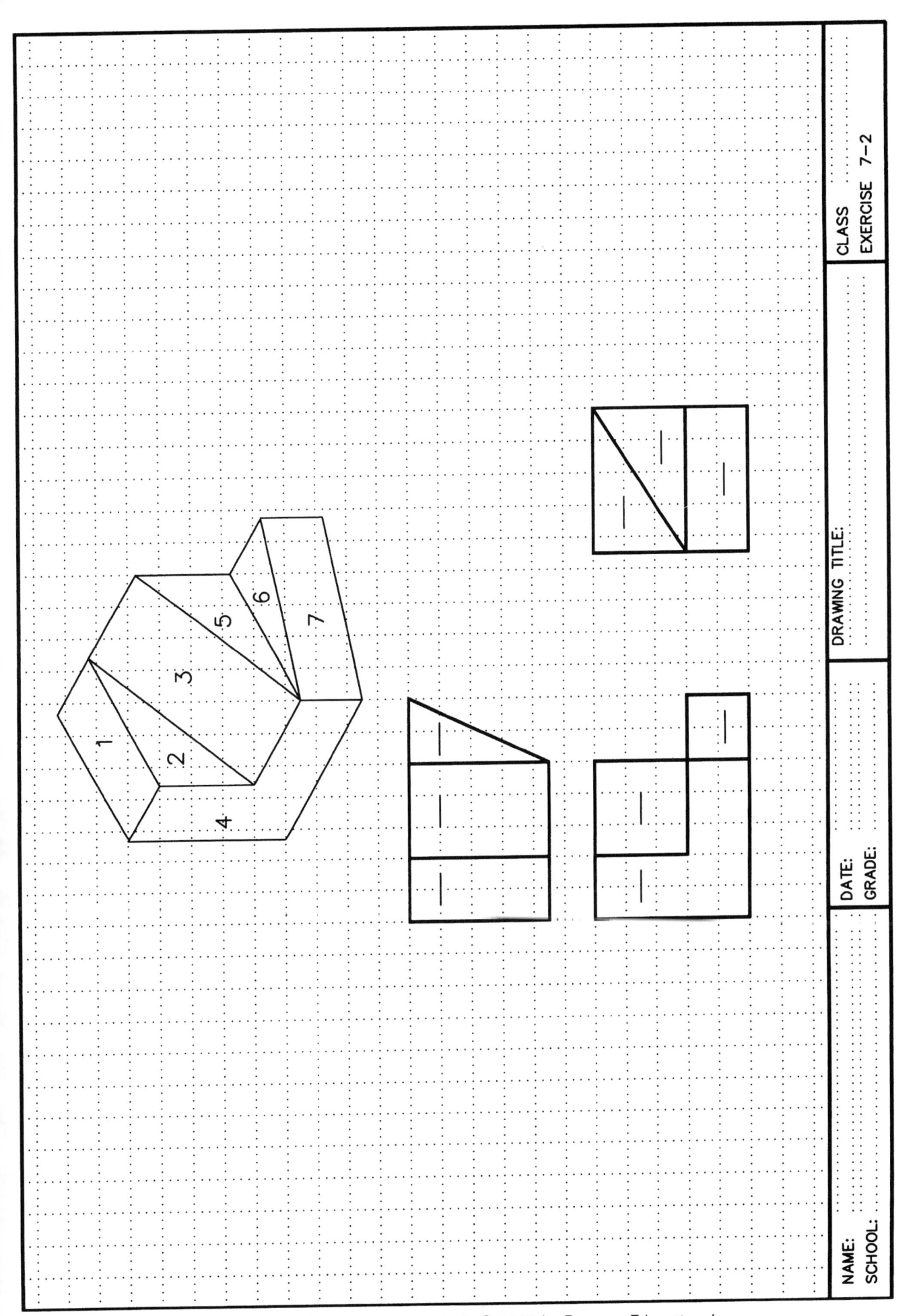

EXERCISE 7-3

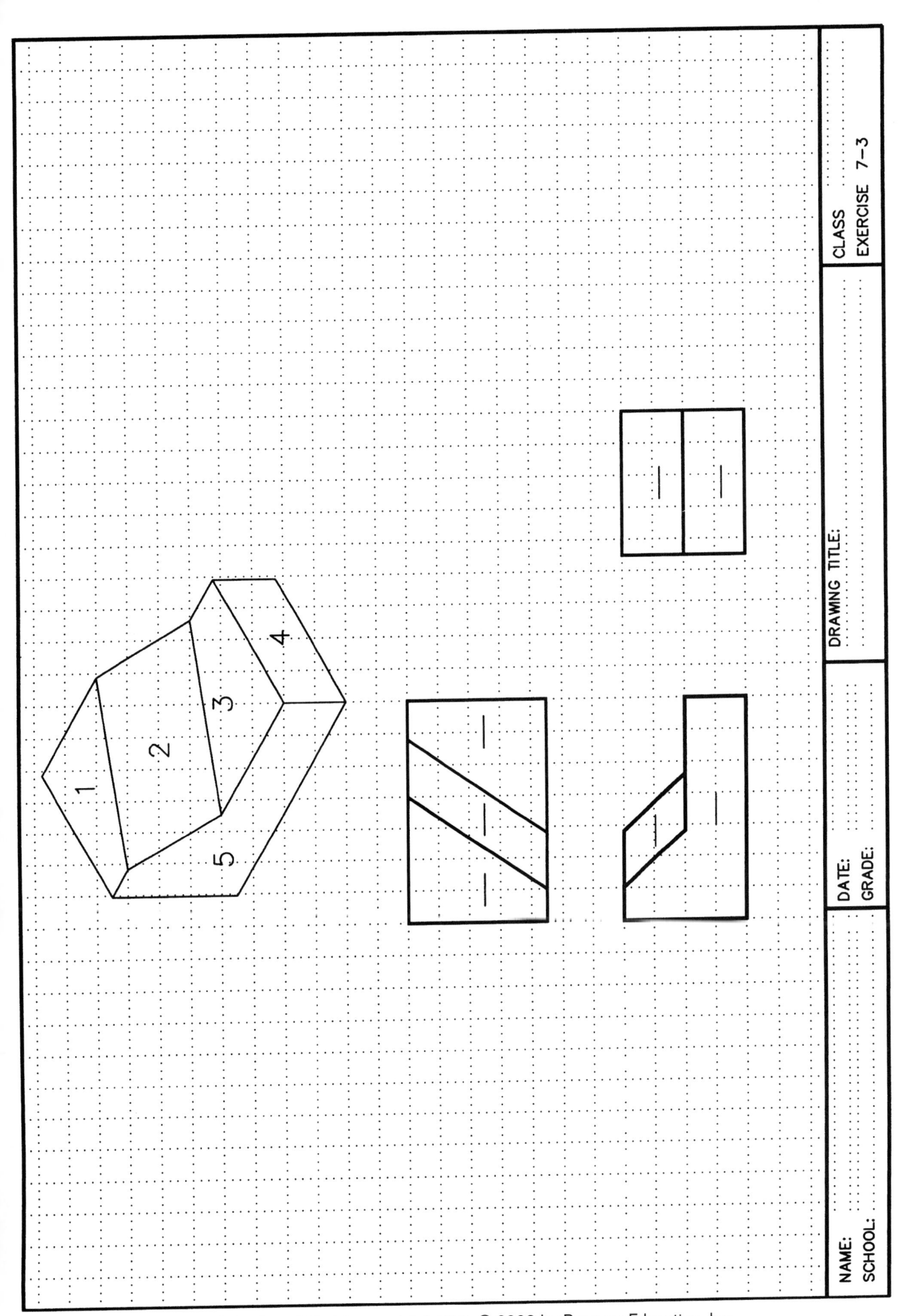

11 (BACK)
12 (LEFT SIDE)
10 (BOTTOM)

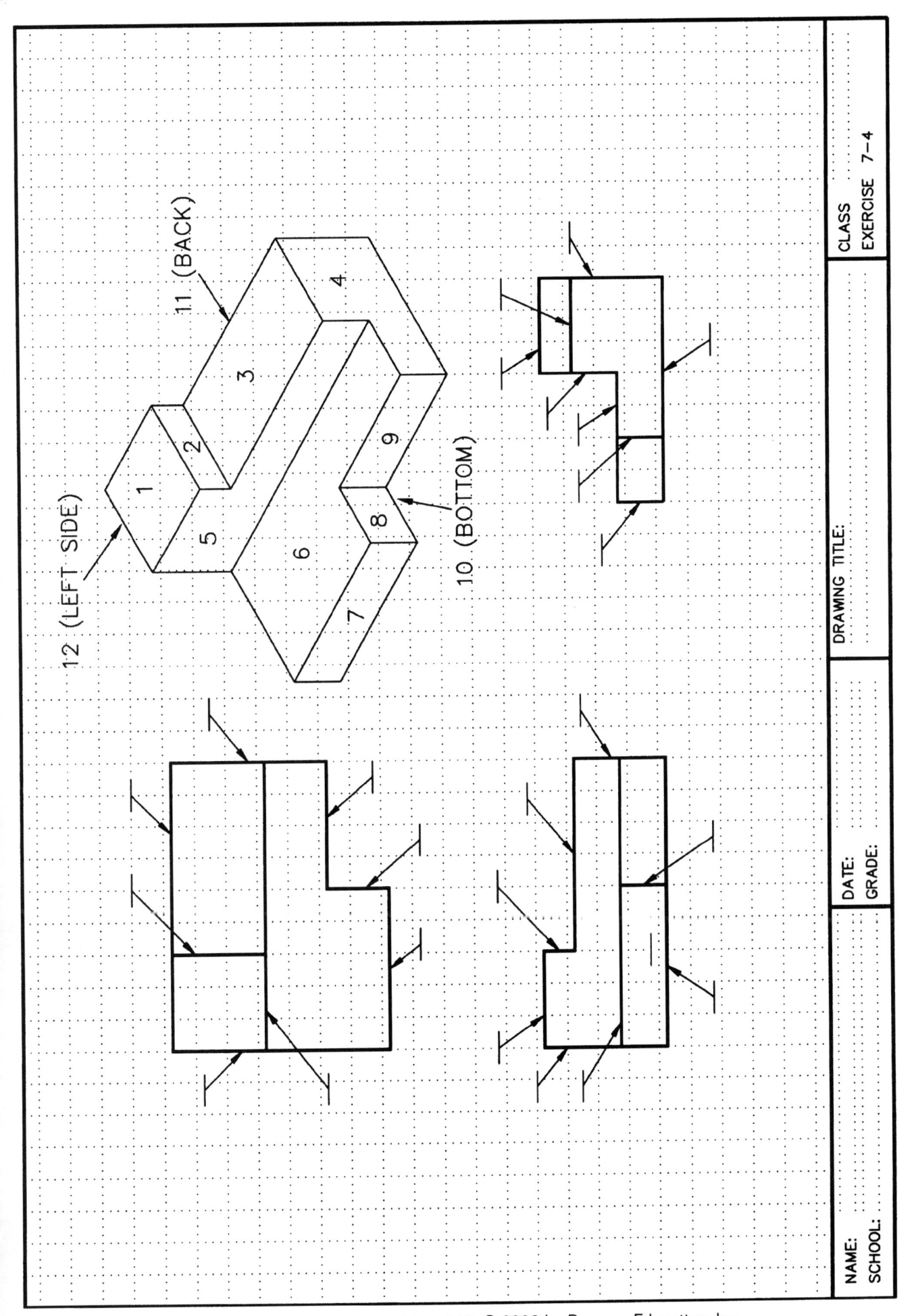

NAME:
SCHOOL:
DATE:
GRADE:
DRAWING TITLE:
CLASS
EXERCISE 7-4

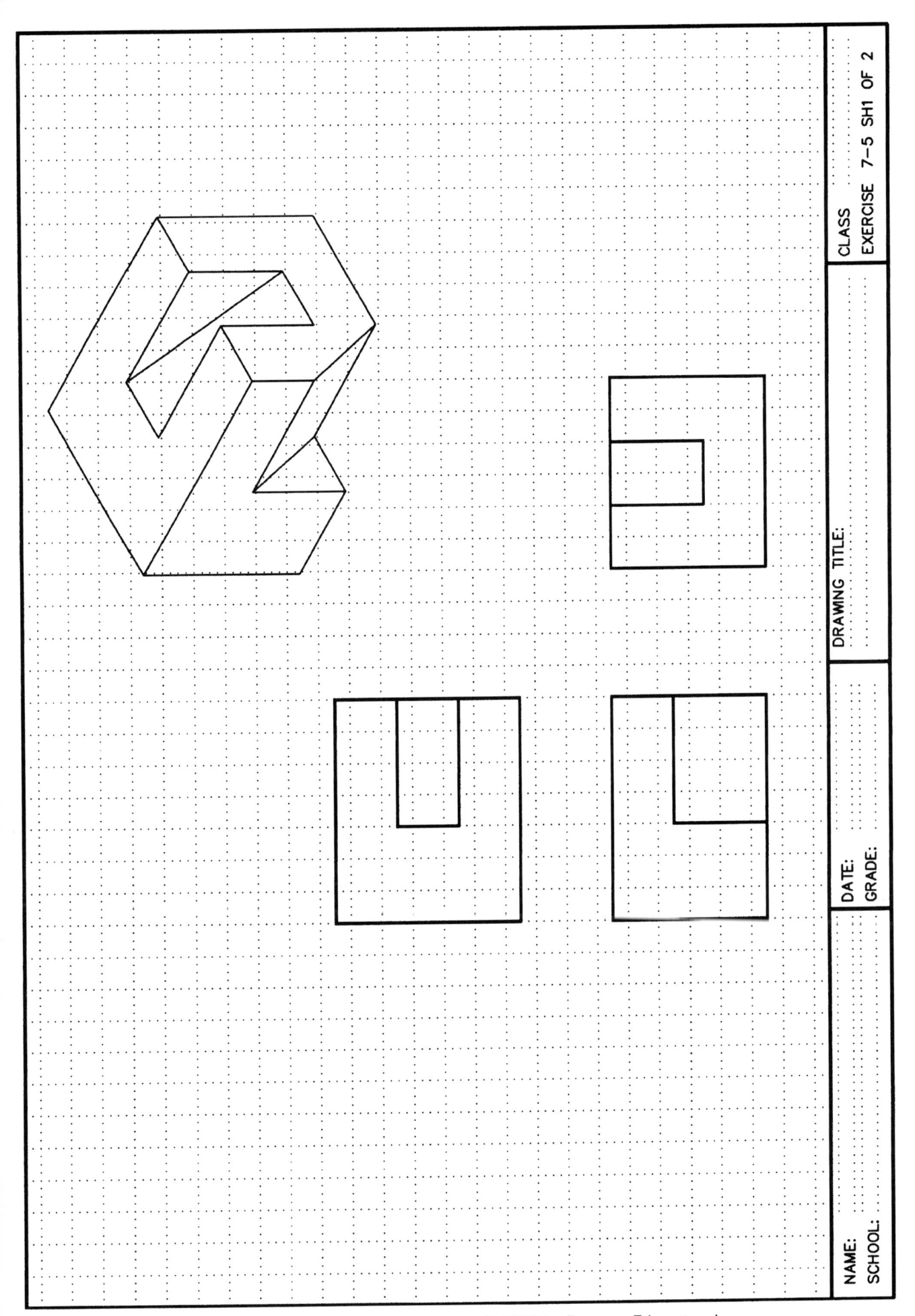

Kirkpatrick, *Basic Drafting Using Pencil Sketches and AutoCAD,* © 2003 by Pearson Education, Inc.

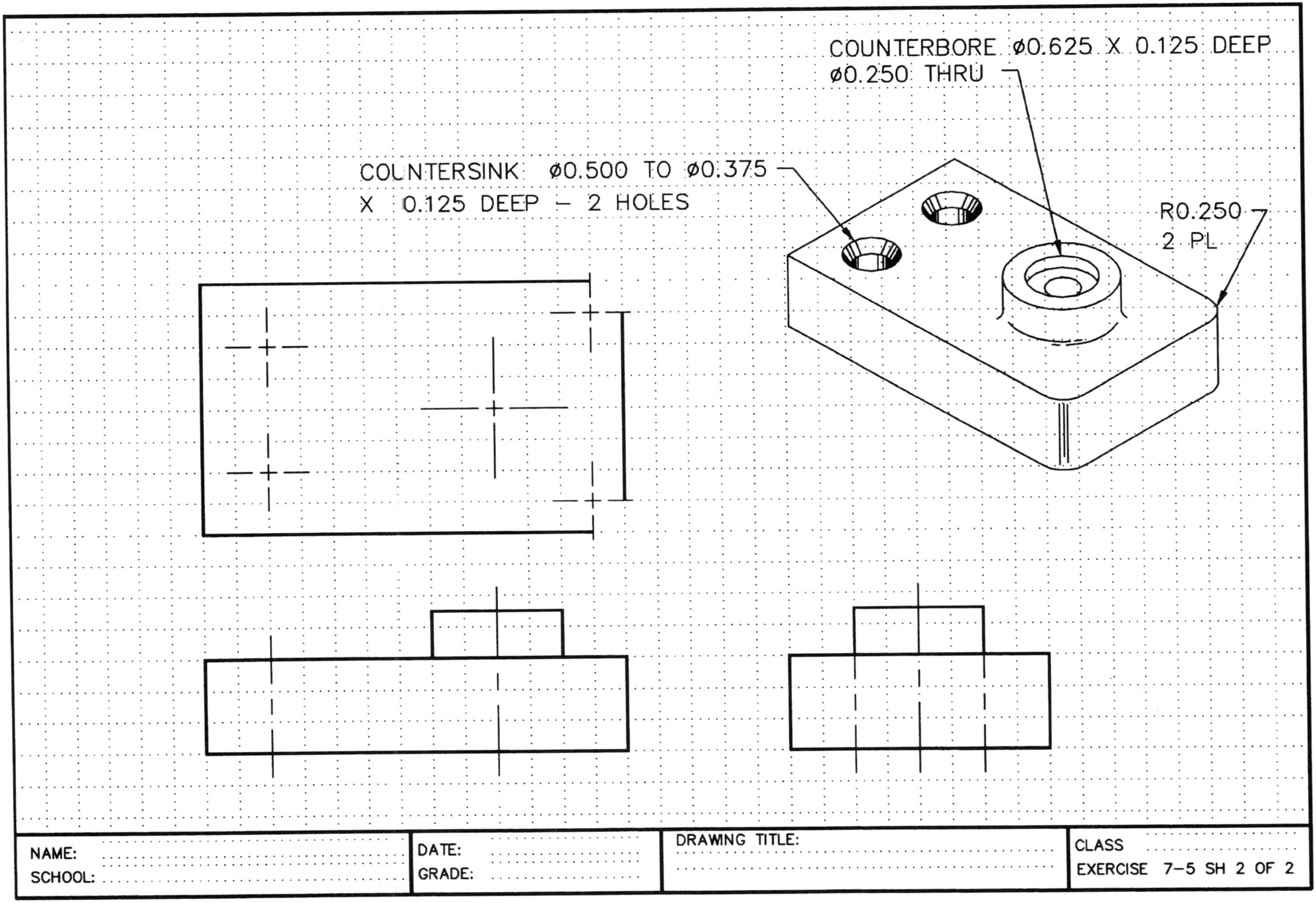

COUNTERBORE ⌀0.625 X 0.125 DEEP
⌀0.250 THRU

R0.250
2 PL

COUNTERSINK ⌀0.500 TO ⌀0.375
X 0.125 DEEP – 2 HOLES

NAME:
SCHOOL:

DATE:
GRADE:

DRAWING TITLE:

CLASS
EXERCISE 7-5 SH 2 OF 2

Kirkpatrick, *Basic Drafting Using Pencil Sketches and AutoCAD*, © 2003 by Pearson Education, Inc.

EXERCISE 7-6

EXERCISE 7-7

CLASS
EXERCISE 7-8

DRAWING TITLE:

DATE:
GRADE:

NAME:
SCHOOL:

Kirkpatrick, *Basic Drafting Using Pencil Sketches and AutoCAD,* © 2003 by Pearson Education, Inc.

DATE:	DRAWING TITLE:	CLASS
GRADE:		EXERCISE 7-9
NAME:		
SCHOOL:		

Kirkpatrick, *Basic Drafting Using Pencil Sketches and AutoCAD,* © 2003 by Pearson Education, Inc.

NAME:
SCHOOL:
DATE:
GRADE:
DRAWING TITLE:
CLASS
EXERCISE 7-10

Kirkpatrick, *Basic Drafting Using Pencil Sketches and AutoCAD,* © 2003 by Pearson Education, Inc.

| NAME: | DATE: | DRAWING TITLE: | CLASS |
| SCHOOL: | GRADE: | | EXERCISE 7-11 |

Kirkpatrick, *Basic Drafting Using Pencil Sketches and AutoCAD,* © 2003 by Pearson Education, Inc.

NAME:
SCHOOL:
DATE:
GRADE:
DRAWING TITLE:
CLASS
EXERCISE 7–13

EXERCISE 7-14

CLASS	EXERCISE 9–1
DRAWING TITLE:	
DATE: GRADE:	
NAME: SCHOOL:	

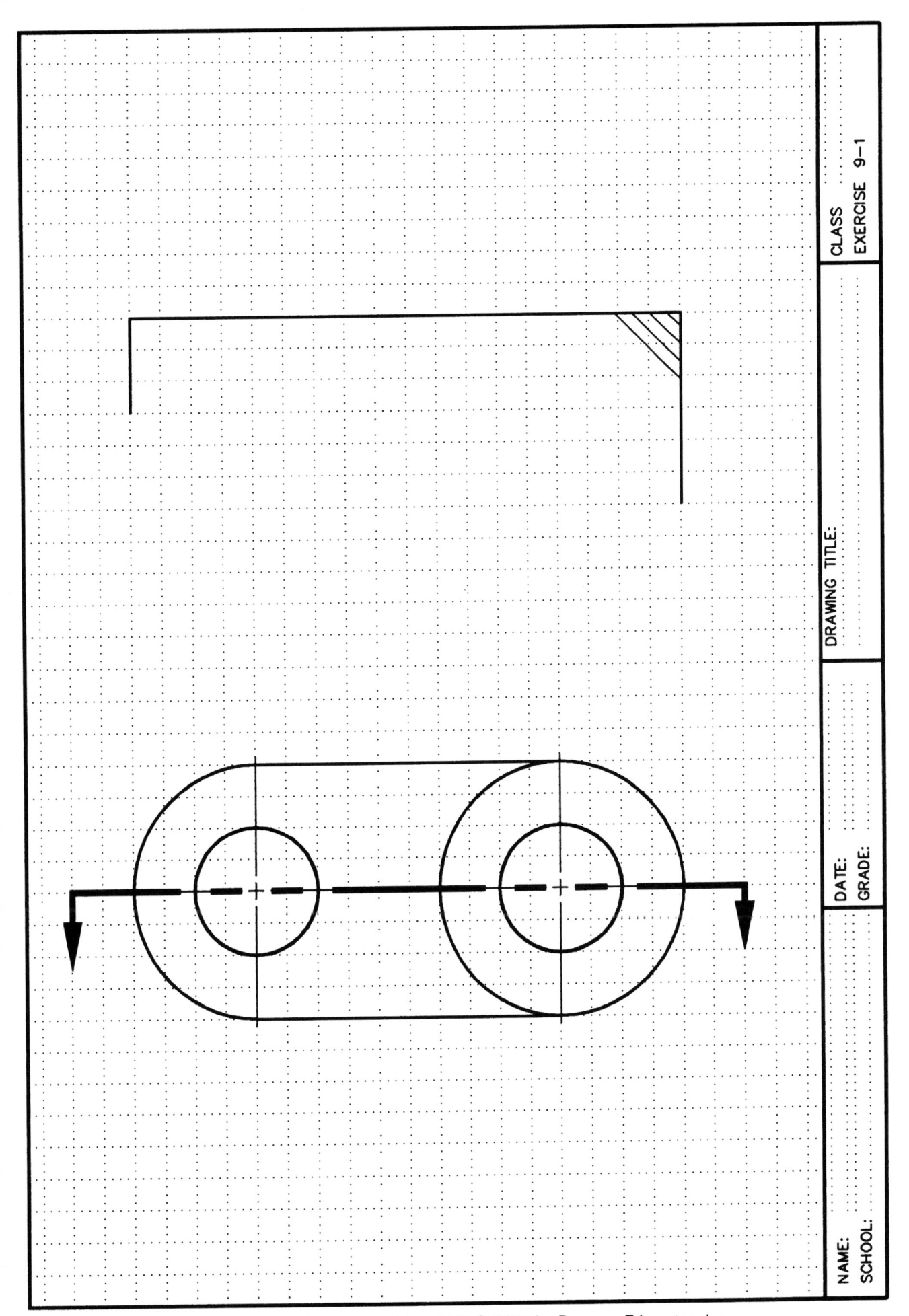

Kirkpatrick, *Basic Drafting Using Pencil Sketches and AutoCAD,* © 2003 by Pearson Education, Inc.

CLASS	EXERCISE 9-2
DRAWING TITLE:	
DATE:	
GRADE:	
NAME:	
SCHOOL:	

Kirkpatrick, *Basic Drafting Using Pencil Sketches and AutoCAD*, © 2003 by Pearson Education, Inc.

NAME:	DATE:	DRAWING TITLE:	CLASS
SCHOOL:	GRADE:		EXERCISE 9-3

CLASS	EXERCISE 9–4
DRAWING TITLE:	
DATE:	GRADE:
NAME:	SCHOOL:

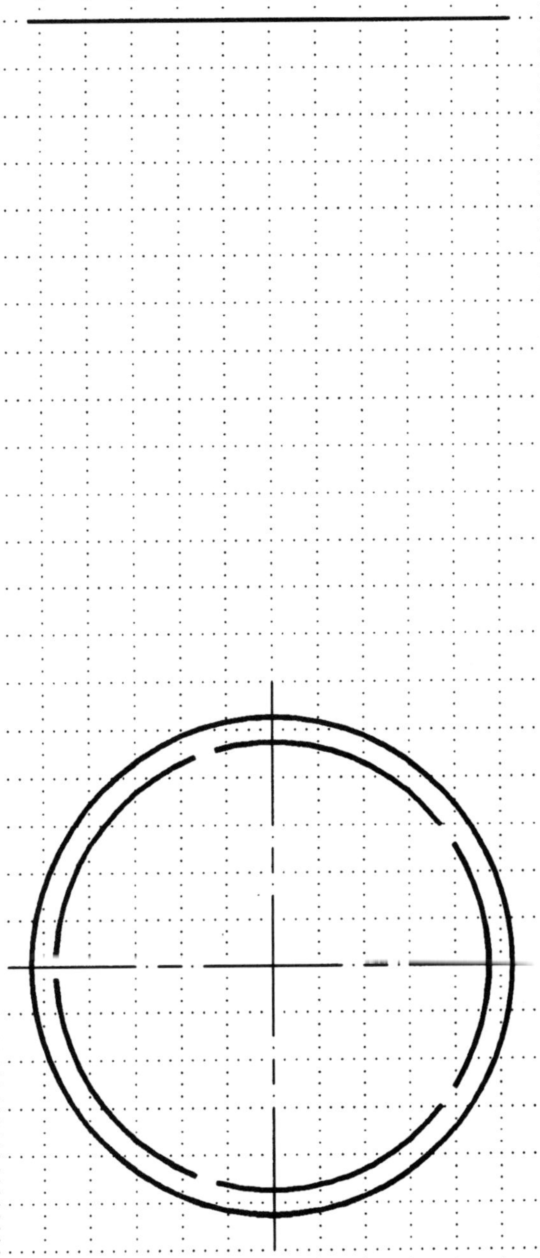

CLASS
EXERCISE 9–5

DRAWING TITLE:

DATE:
GRADE:

NAME:
SCHOOL:

CLASS	EXERCISE 9-6
DRAWING TITLE:	
DATE:	
GRADE:	
NAME:	
SCHOOL:	

Kirkpatrick, *Basic Drafting Using Pencil Sketches and AutoCAD*, © 2003 by Pearson Education, Inc.

NAME:
SCHOOL:
DATE:
GRADE:
DRAWING TITLE:
CLASS
EXERCISE 9–7

Kirkpatrick, *Basic Drafting Using Pencil Sketches and AutoCAD,* © 2003 by Pearson Education, Inc.

CLASS:	EXERCISE 11-1
DRAWING TITLE:	
DATE:	
GRADE:	
NAME:	
SCHOOL:	

Kirkpatrick, *Basic Drafting Using Pencil Sketches and AutoCAD,* © 2003 by Pearson Education, Inc.

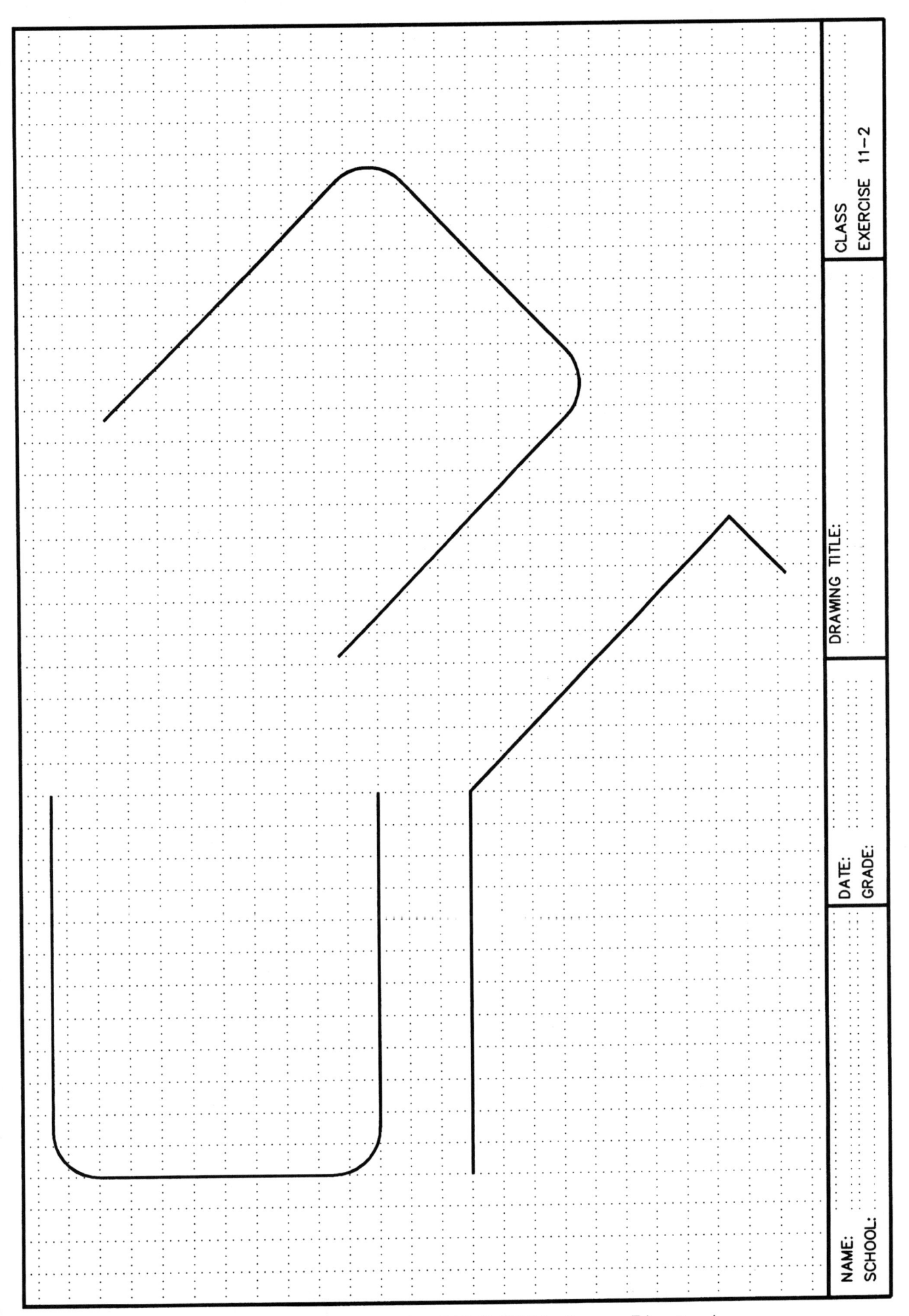

NAME:
SCHOOL:
DATE:
GRADE:
DRAWING TITLE:
CLASS
EXERCISE 11–2

Kirkpatrick, *Basic Drafting Using Pencil Sketches and AutoCAD*, © 2003 by Pearson Education, Inc.

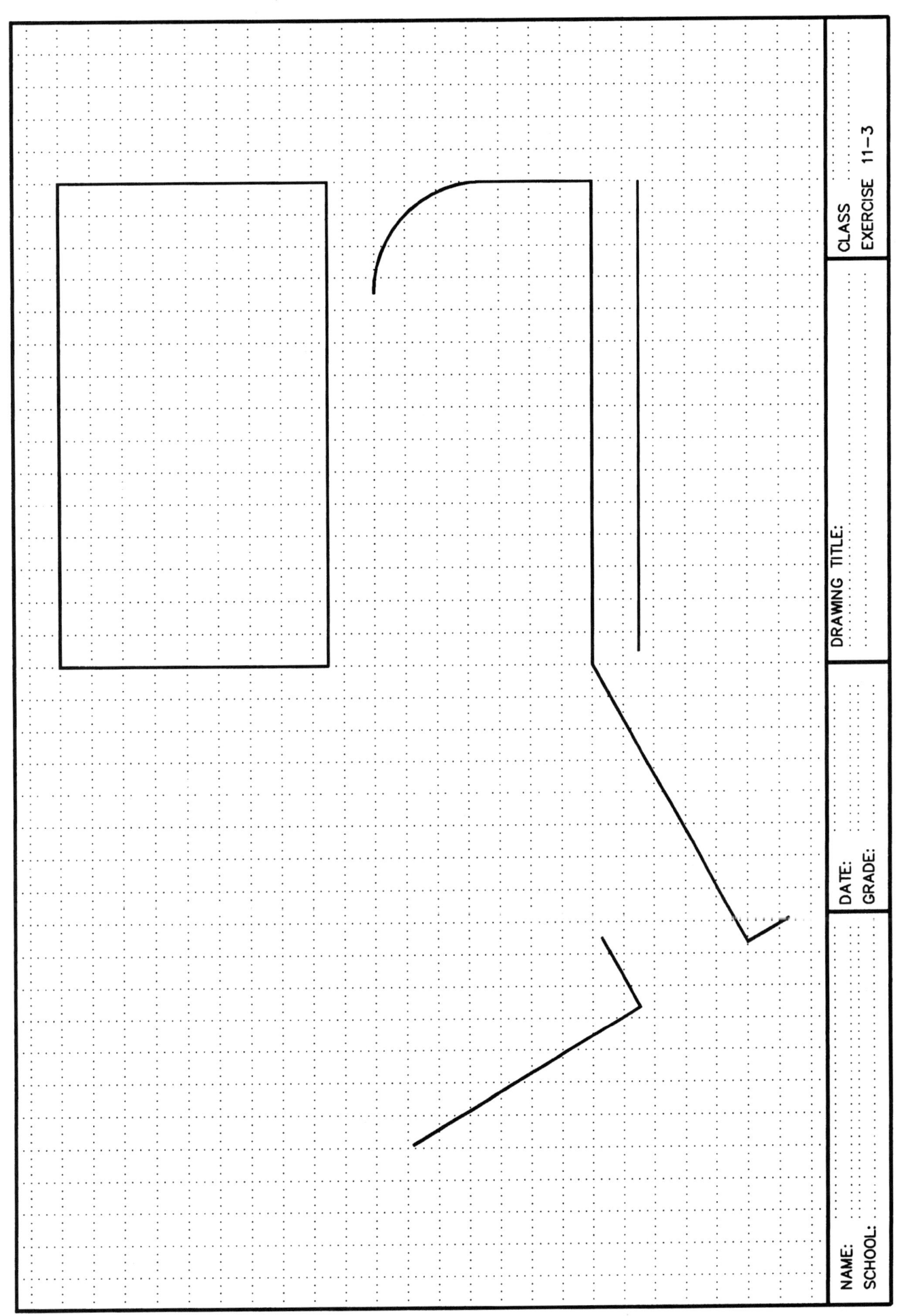

CLASS	EXERCISE 11-3

DRAWING TITLE:

DATE:
GRADE:

NAME:
SCHOOL:

Kirkpatrick, *Basic Drafting Using Pencil Sketches and AutoCAD,* © 2003 by Pearson Education, Inc.

Kirkpatrick, *Basic Drafting Using Pencil Sketches and AutoCAD*, © 2003 by Pearson Education, Inc.

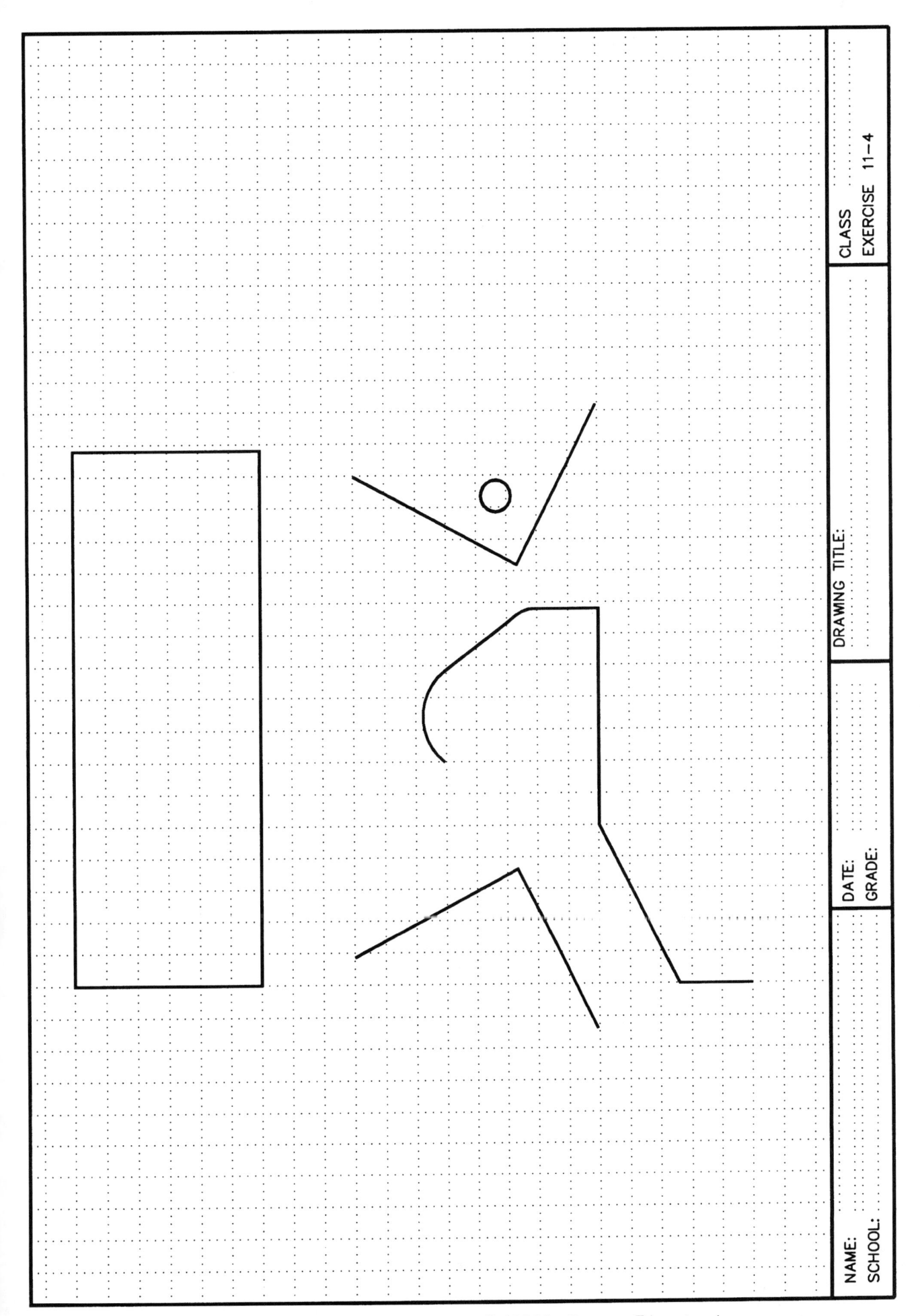

NAME:
SCHOOL:
DATE:
GRADE:
DRAWING TITLE:
CLASS
EXERCISE 11–5

Kirkpatrick, *Basic Drafting Using Pencil Sketches and AutoCAD,* © 2003 by Pearson Education, Inc.

Kirkpatrick, *Basic Drafting Using Pencil Sketches and AutoCAD,* © 2003 by Pearson Education, Inc.

CLASS
EXERCISE 13–2

DRAWING TITLE:

DATE:
GRADE:

NAME:
SCHOOL:

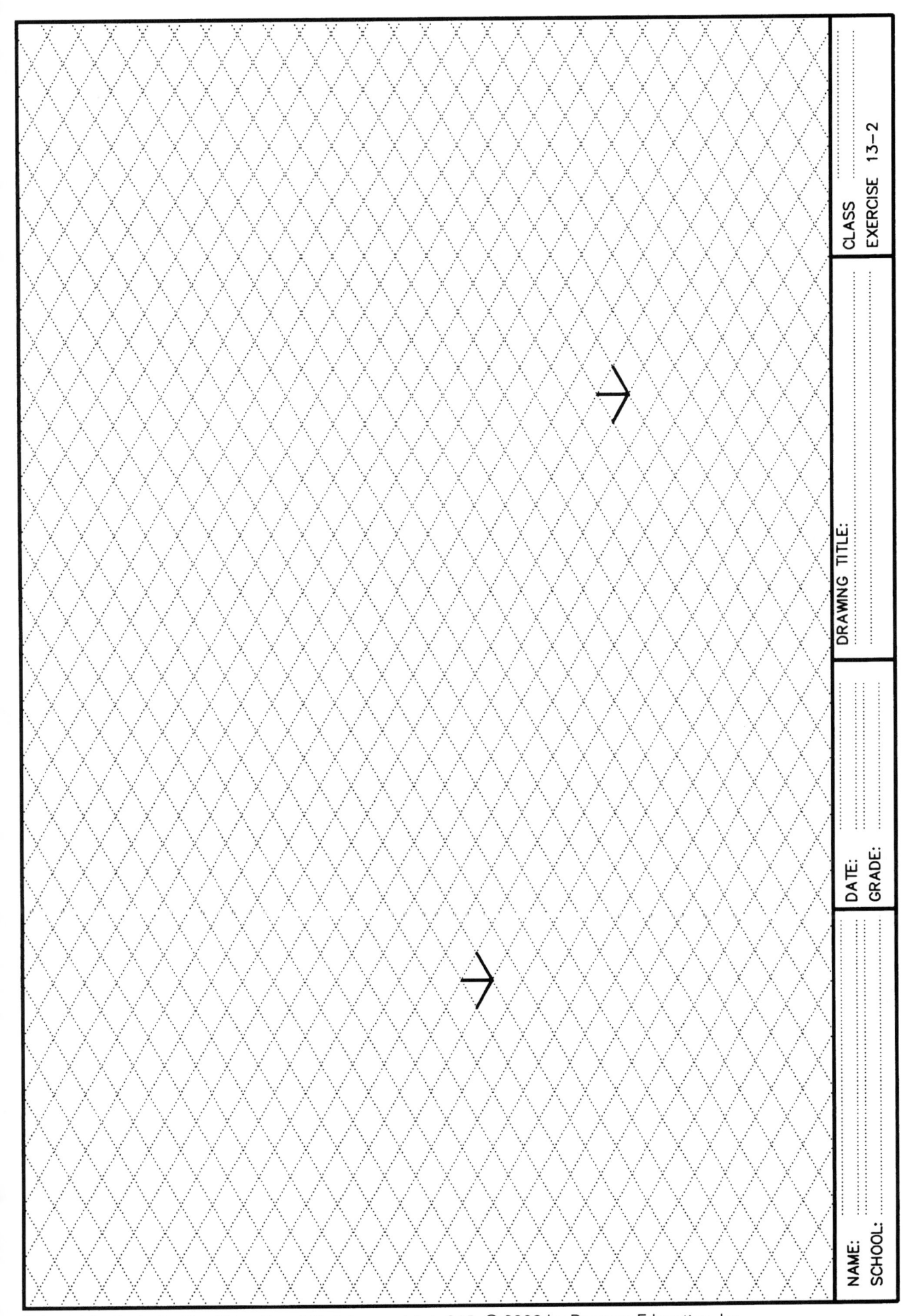

Kirkpatrick, *Basic Drafting Using Pencil Sketches and AutoCAD,* © 2003 by Pearson Education, Inc.

NAME: SCHOOL: DATE: GRADE: DRAWING TITLE: CLASS EXERCISE 13–3

Kirkpatrick, *Basic Drafting Using Pencil Sketches and AutoCAD,* © 2003 by Pearson Education, Inc.

NAME:
SCHOOL:
DATE:
GRADE:
DRAWING TITLE:
CLASS
EXERCISE 13-4

Kirkpatrick, *Basic Drafting Using Pencil Sketches and AutoCAD,* © 2003 by Pearson Education, Inc.

NAME:	DATE:	DRAWING TITLE:	CLASS
SCHOOL:	GRADE:		EXERCISE 13-5

Kirkpatrick, *Basic Drafting Using Pencil Sketches and AutoCAD,* © 2003 by Pearson Education, Inc.

Kirkpatrick, *Basic Drafting Using Pencil Sketches and AutoCAD,* © 2003 by Pearson Education, Inc.

NAME:
SCHOOL:
DATE:
GRADE:
DRAWING TITLE:
CLASS
EXERCISE 13–6

NAME:	DATE:	DRAWING TITLE:	CLASS
SCHOOL:	GRADE:		EXERCISE 13–7

Kirkpatrick, *Basic Drafting Using Pencil Sketches and AutoCAD,* © 2003 by Pearson Education, Inc.

NAME: SCHOOL: DATE: GRADE: DRAWING TITLE: CLASS: EXERCISE 13-8

Kirkpatrick, *Basic Drafting Using Pencil Sketches and AutoCAD,* © 2003 by Pearson Education, Inc.

NAME:
SCHOOL:
DATE:
GRADE:
DRAWING TITLE:
CLASS
EXERCISE 13–9

Kirkpatrick, *Basic Drafting Using Pencil Sketches and AutoCAD,* © 2003 by Pearson Education, Inc.

Kirkpatrick, *Basic Drafting Using Pencil Sketches and AutoCAD,* © 2003 by Pearson Education, Inc.

NAME:	DATE:	DRAWING TITLE:	CLASS
SCHOOL:	GRADE:		EXERCISE 13–11

Kirkpatrick, *Basic Drafting Using Pencil Sketches and AutoCAD*, © 2003 by Pearson Education, Inc.

Kirkpatrick, *Basic Drafting Using Pencil Sketches and AutoCAD,* © 2003 by Pearson Education, Inc.

NAME:	DATE:	DRAWING TITLE:	CLASS
SCHOOL:	GRADE:		EXERCISE 15–2

Kirkpatrick, *Basic Drafting Using Pencil Sketches and AutoCAD,* © 2003 by Pearson Education, Inc.

NAME:	DATE:
SCHOOL:	GRADE:

DRAWING TITLE:

CLASS
EXERCISE 15-3

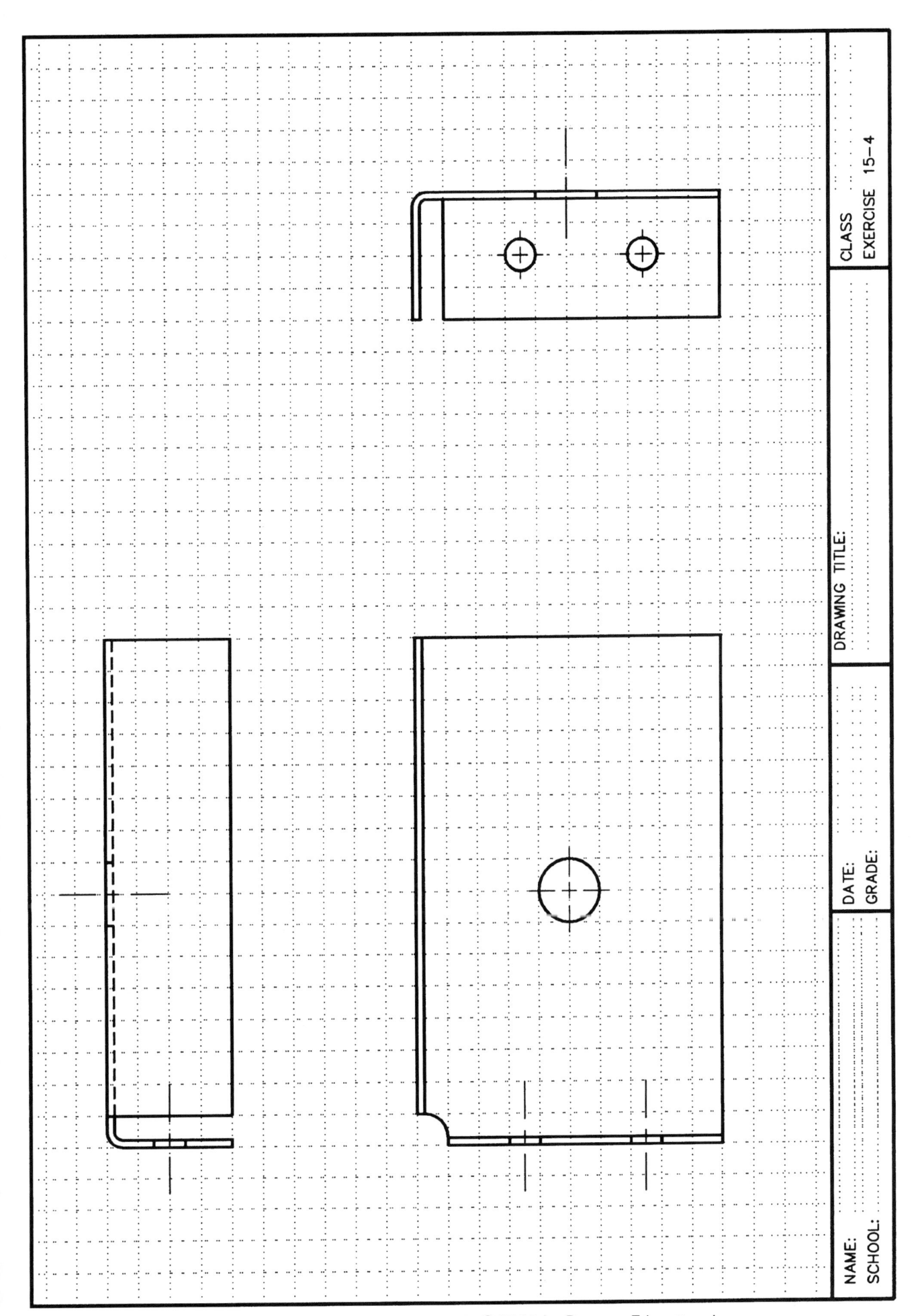

Kirkpatrick, *Basic Drafting Using Pencil Sketches and AutoCAD,* © 2003 by Pearson Education, Inc.

CLASS
EXERCISE 15-4
DRAWING TITLE:
DATE:
GRADE:
NAME:
SCHOOL:

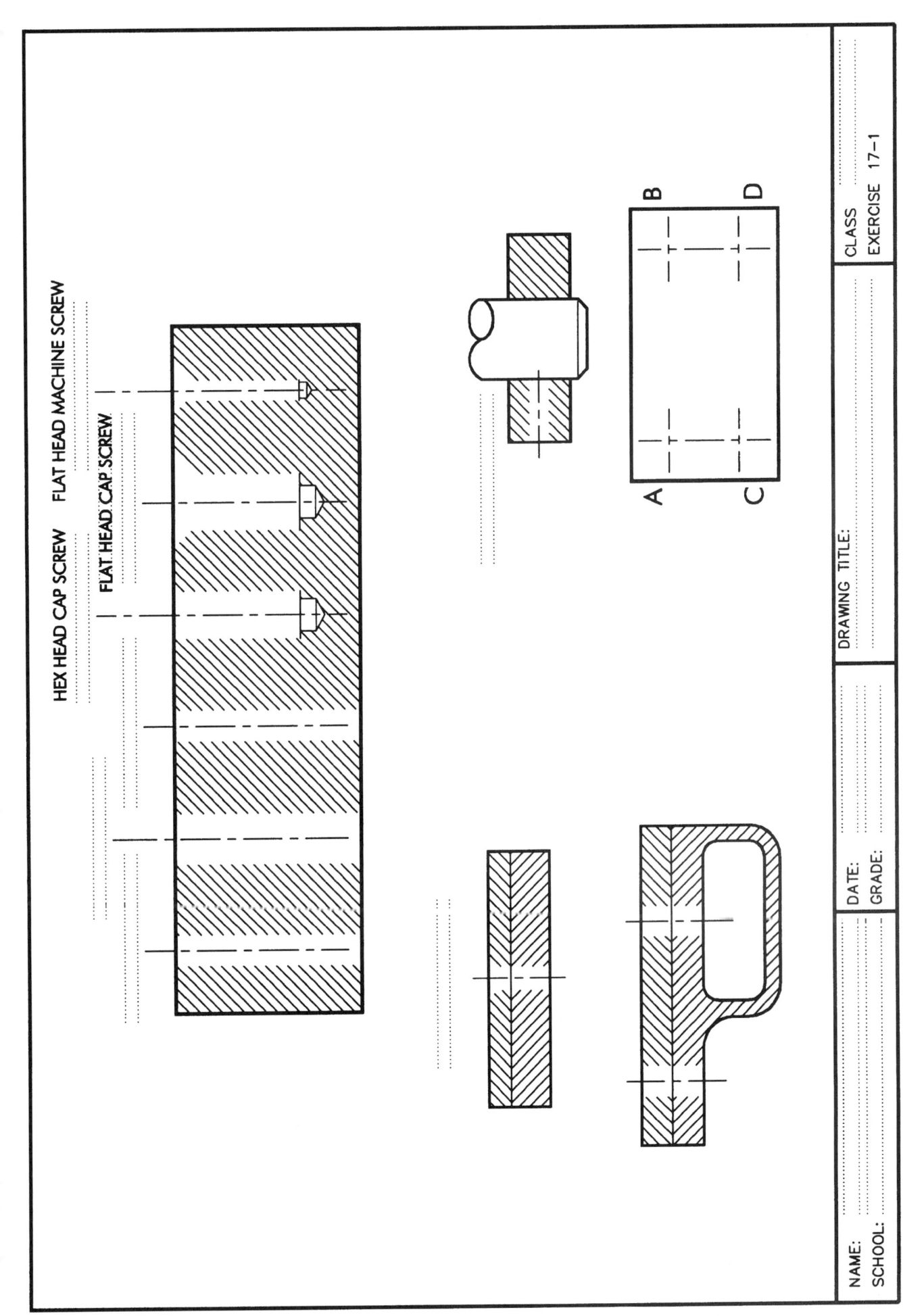

HEX HEAD CAP SCREW   FLAT HEAD MACHINE SCREW
FLAT HEAD CAP SCREW

B D
A C

NAME:
SCHOOL:
DATE:
GRADE:
DRAWING TITLE:
CLASS
EXERCISE 17-1

| NAME: | DATE: | DRAWING TITLE: | CLASS |
| SCHOOL: | GRADE: | | EXERCISE 19-1 |

NAME:
SCHOOL:
DATE:
GRADE:
DRAWING TITLE:
CLASS
EXERCISE 19-2

Kirkpatrick, *Basic Drafting Using Pencil Sketches and AutoCAD,* © 2003 by Pearson Education, Inc.

NAME:
SCHOOL:
DATE:
GRADE:
DRAWING TITLE:
CLASS
EXERCISE 19–3

Kirkpatrick, *Basic Drafting Using Pencil Sketches and AutoCAD,* © 2003 by Pearson Education, Inc.

NAME:	DATE:	DRAWING TITLE:	CLASS
SCHOOL:	GRADE:		EXERCISE 19–4

Kirkpatrick, *Basic Drafting Using Pencil Sketches and AutoCAD,* © 2003 by Pearson Education, Inc.

NAME:
SCHOOL:
DATE:
GRADE:
DRAWING TITLE:
CLASS
EXERCISE 19-5

Kirkpatrick, *Basic Drafting Using Pencil Sketches and AutoCAD,* © 2003 by Pearson Education, Inc.

Kirkpatrick, *Basic Drafting Using Pencil Sketches and AutoCAD,* © 2003 by Pearson Education, Inc.

Kirkpatrick, *Basic Drafting Using Pencil Sketches and AutoCAD,* © 2003 by Pearson Education, Inc.

NAME:	DATE:	DRAWING TITLE:	CLASS
SCHOOL:	GRADE:		EXERCISE 19-8

Kirkpatrick, *Basic Drafting Using Pencil Sketches and AutoCAD,* © 2003 by Pearson Education, Inc.

NAME:	DATE:	DRAWING TITLE:	CLASS
SCHOOL:	GRADE:		EXERCISE 19–9

Kirkpatrick, *Basic Drafting Using Pencil Sketches and AutoCAD,* © 2003 by Pearson Education, Inc.

NAME:	DATE:	DRAWING TITLE:	CLASS
SCHOOL:	GRADE:		EXERCISE 19–10

Kirkpatrick, *Basic Drafting Using Pencil Sketches and AutoCAD,* © 2003 by Pearson Education, Inc.

NAME:
SCHOOL:

DATE:
GRADE:

DRAWING TITLE:

CLASS
EXERCISE 19–11

Kirkpatrick, *Basic Drafting Using Pencil Sketches and AutoCAD,* © 2003 by Pearson Education, Inc.

NAME:	DATE:	DRAWING TITLE:	CLASS
SCHOOL:	GRADE:		EXERCISE 19–12

Kirkpatrick, *Basic Drafting Using Pencil Sketches and AutoCAD,* © 2003 by Pearson Education, Inc.

REVISIONS		
REV	DESCRIPTION	DATE	APPROVED

PARTS LIST

ITEM	PART NO.	QTY	DESCRIPTION
1			
2			
3			
4			
5			
6			
7			
8			

THE TRAILER CO. INC.

DRAWING TITLE

SIZE	FSCM NO. NONE	DWG NO. 100353	REV
SCALE NONE	RELEASED BY:		SHEET 1 OF 1

NOTES:
UNLESS OTHERWISE NOTED:
ALL DIMENSIONS ARE IN INCHES:
TOLERANCES ARE:
 .XX = .02
 .XXX = .010
MATERIAL: COLD ROLLED STEEL
NEXT ASSEMBLY NUMBER
DRAWN BY:

REVISIONS			
REV	DESCRIPTION	DATE	APPROVED

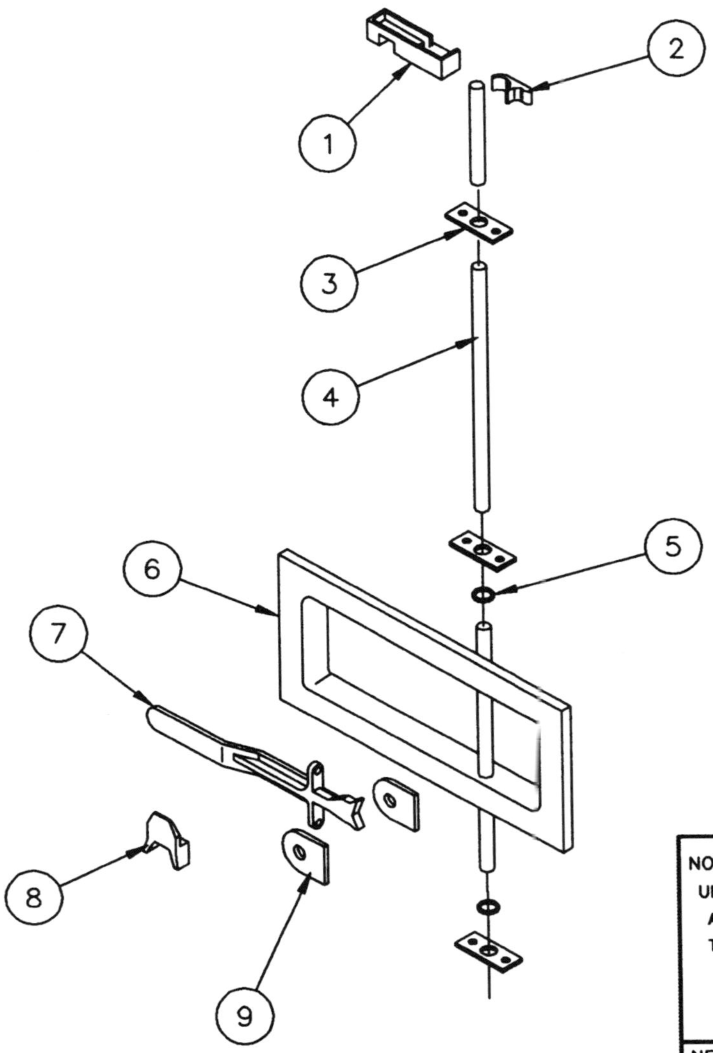

ITEM	PART NO.	QTY	DESCRIPTION
9			
8			
7			
6			
5			
4			
3			
2			
1			

PARTS LIST

NOTES:
UNLESS OTHERWISE NOTED:
ALL DIMENSIONS ARE IN INCHES:
TOLERANCES ARE:
.XX = .02
.XXX = .010

THE TRAILER CO. INC.

DRAWING TITLE

NEXT ASSEMBLY NUMBER	SIZE	FSCM NO. NONE	DWG NO. 100303	REV
DRAWN BY:	SCALE NONE	RELEASED BY:	SHEET 1 OF 1	

REV	DESCRIPTION	DATE	APPROVED

REVISIONS

THE TRAILER CO. INC.

DRAWING TITLE

SIZE	FSCM NO.	DWG NO.	REV
	NONE	100351	
SCALE 1=1	RELEASED BY:	SHEET 1 OF 1	

NOTES:
UNLESS OTHERWISE NOTED:
ALL DIMENSIONS ARE IN INCHES:
TOLERANCES ARE:
 .XX = .02
 .XXX = .010
MATERIAL: COLD ROLLED STEEL
NEXT ASSEMBLY NUMBER
DRAWN BY:

Kirkpatrick, *Basic Drafting Using Pencil Sketches and AutoCAD,* © 2003 by Pearson Education, Inc.

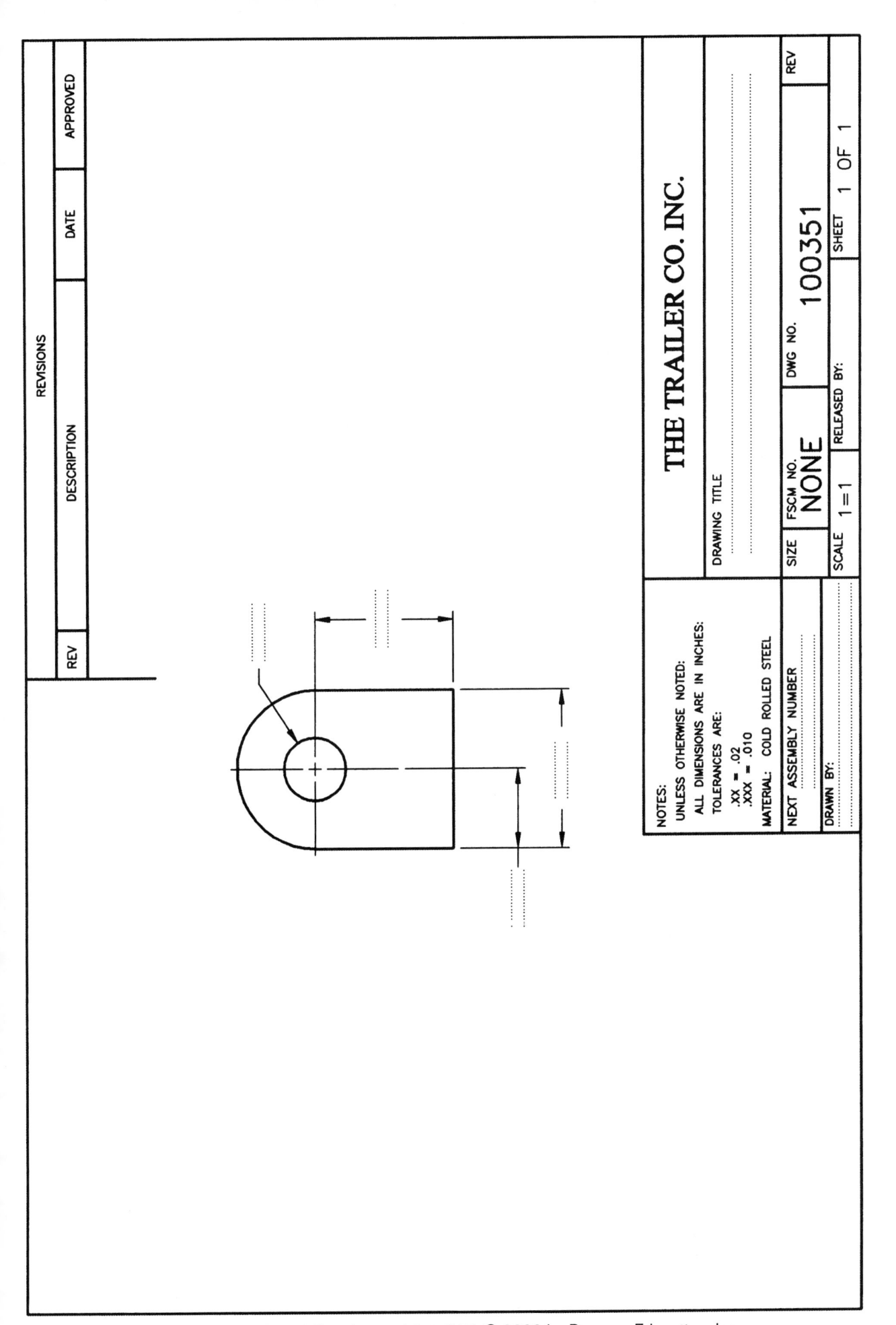

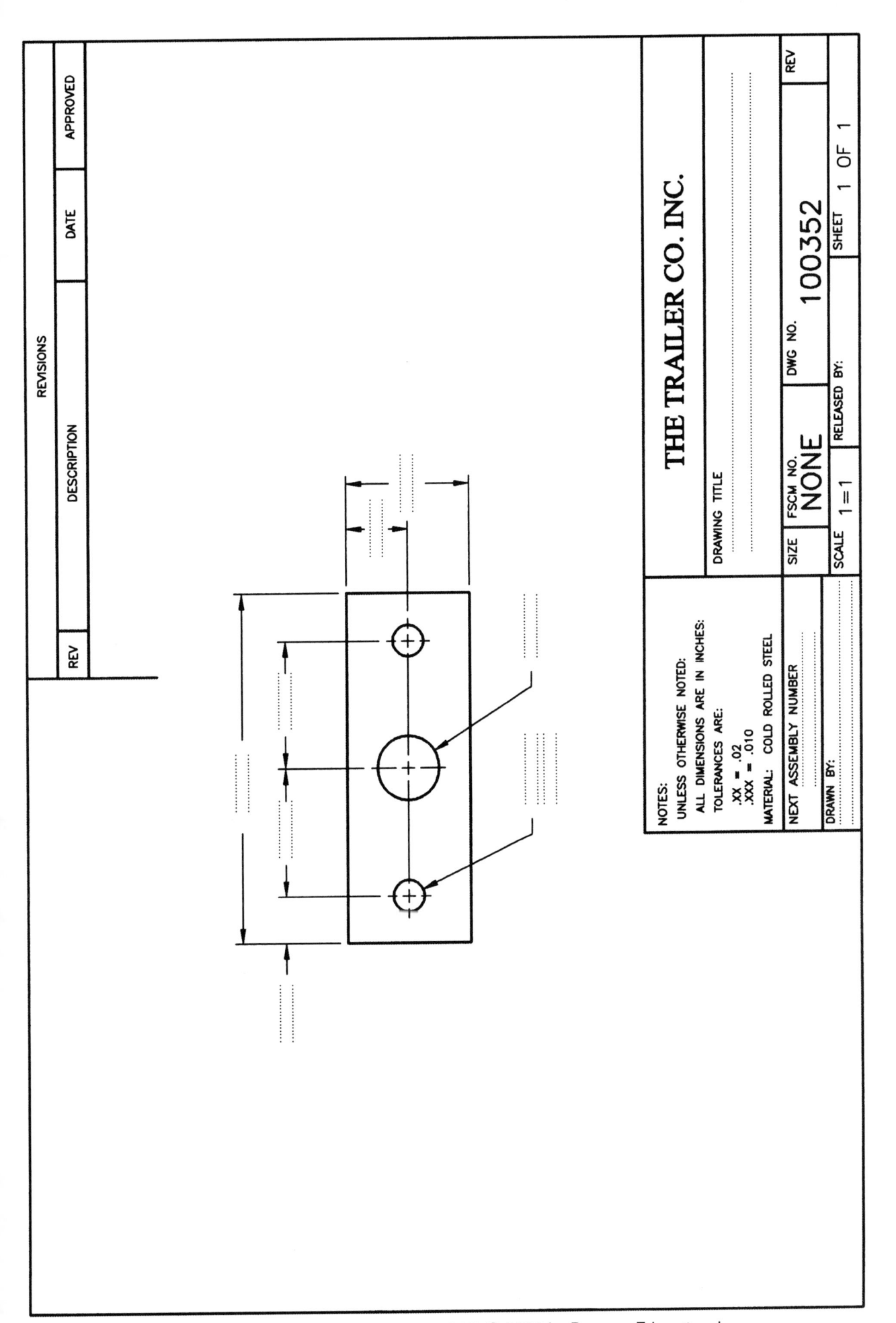

| REV | DESCRIPTION | DATE | APPROVED |
|---|---|---|---|//REVISIONS

THE TRAILER CO. INC.

DRAWING TITLE

SIZE	FSCM NO. NONE	DWG NO. 100356	REV
SCALE 1=1	RELEASED BY:	SHEET 1 OF 1	

NOTES:
UNLESS OTHERWISE NOTED:
ALL DIMENSIONS ARE IN INCHES:
TOLERANCES ARE:
 .XX = .02
 .XXX = .010
MATERIAL: COLD ROLLED STEEL
NEXT ASSEMBLY NUMBER
DRAWN BY:

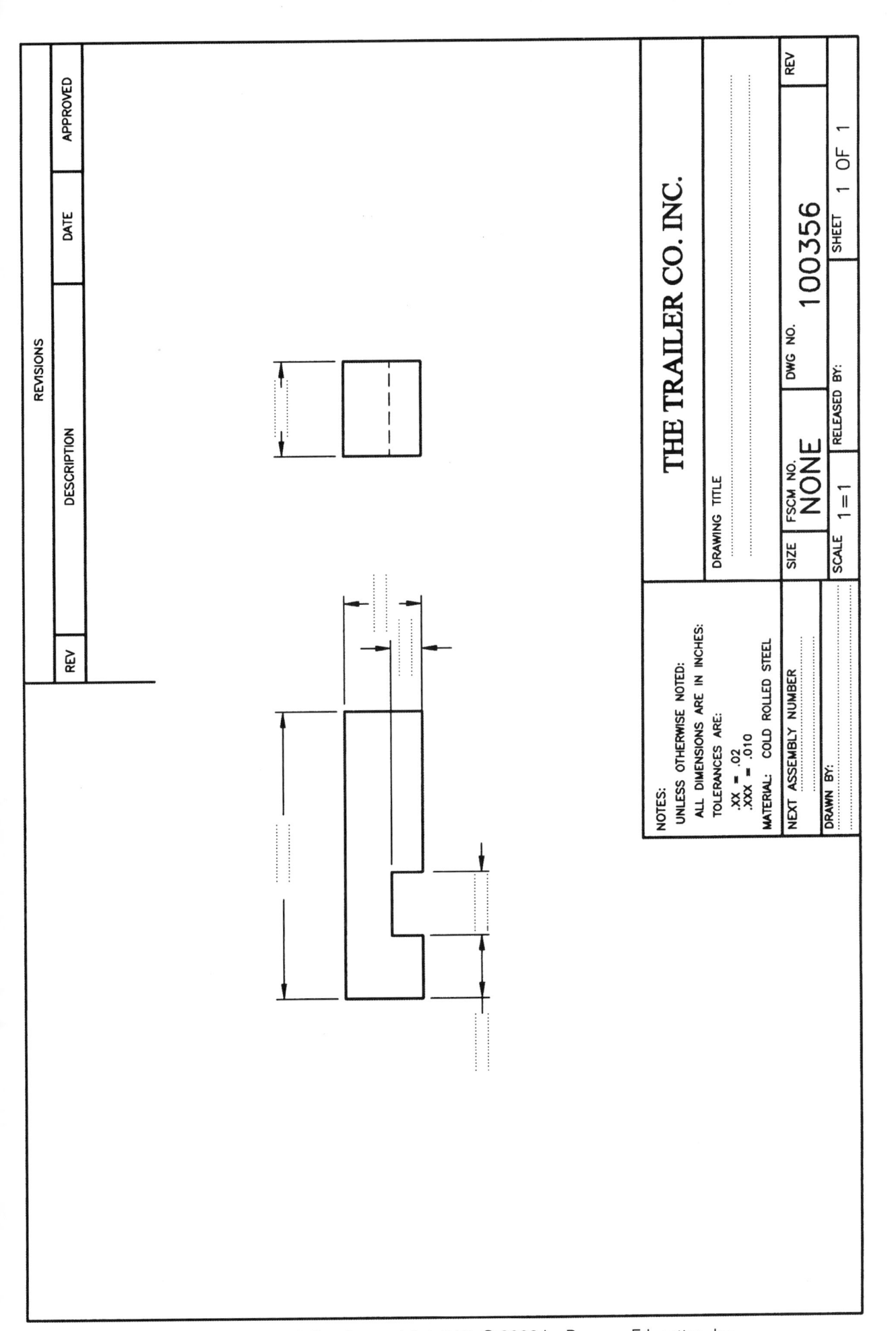

Index

3D, AutoCAD, 297–310
3DFACE, 307
3D objects, 298

Absolute coordinates, 70
Across corners, 52
Across flats, 52
Acute angle, 52
Aligned dimensioning, 238
Aligning views, 102
Alphabet for sketching, 44
Alphabet of lines, 49
Angles in isometric sketching, 205
Architect's scale, 20
Arcs, sketching, 102
Assembly drawings, 9
Attributes, 268
AutoCAD, 15
AutoCAD fundamentals, 32
AutoCAD LT, 15
AutoCAD 3D, 297–310
Auxiliary, primary, 178, 179
Auxiliary, secondary, 179–81
Auxiliary view, partial, 177, 178
Auxiliary views, 5, 6, 177–81
Auxiliary views in AutoCAD, 186–201
Axonometric drawing, 203

Backspace key, 35
Bisected, 51
Block, 191, 268, 270
Blocks with attributes, 272, 273
Bolts, 260, 322, 323
Break, 120

Cap screws, 259, 260, 322, 323
Centerline, 50, 234
Centerlines, drawing, 75, 167
Chamfer command, 142
Change properties, 135
Character keys, 35
Chprop command, 135, 195, 196
Circle command, 71
Circles, sketching, 101
Circles in isometric sketching, 206
Circle template, 19
Circumference, 51
Civil engineer's scale, 21
Compact disk drive, 14
Computer, 12
Construction lines, 50
Coordinates:
 absolute, 70
 polar, 70
 relative, 70
Copy command, 120, 121
Counterbore, 100, 171

Countersink, 100
Curves in isometric sketching, 206
Cutaways:
 features not to cut, 212
 general guidelines for, 211, 212
 isometric, 209–12
Cut cylinders, 97
Cutting-plane line, 153
Cylinders, cut, 97

Datum dimensioning, 242
Decimal equivalents, 320
Detail drawing, 9, 314
Development:
 parallel line, 282–86, 305
 radial line, 287–90, 308
 triangulation, 291–93
Development drawings, 9, 282–93
Developments with AutoCAD, 297–310
Dialog box, components of, 34
Diameter, 51
DIMCEN, 246
Dimension lines, 50, 234
Dimensioning:
 aligned, 238
 AutoCAD, 246–52
 datum, 242
 overdimensioning, 237
 tabular, 242
 unidirectional, 238
Dimensioning features, 238–40
Dimensioning notes, 240, 241
Dimensioning practices, 233
Dimensioning symbols and abbreviations, 235
Dimensions:
 horizontal, 247
 placement of, 236
 size and location, 235, 236
 sketching, 233–42
 vertical, 247
DIMUNIT, 250
Directories button, 39
Dividing lines and angles, 61–63
Drafting fundamentals, 1
Drafting powder, 19
Drafting Settings dialog box,
Drawing:
 detail, 314
 parts of, 315, 316
 subassembly, 314
 top assembly, 313
Drawing Aids dialog box, 34
Drawing limits, 33
Drawing precision, 33
Drawing size, 316
Drawing system, 313–18

Drives button, 39
Dtext command, 76

Elevation, 298
Ellipses, 103
 isometric, 226, 227
Erase command, 72
Erasers, 18
Exit AutoCAD, 38–40
Extend command, 119
Extension line, 50, 234

Fasteners, 259–64
Fasteners, AutoCAD drawing of, 268–79
Field of drawing, 316
File name input button, 39
Fillet, 52
Fillet command, 71, 81, 118
First-angle orthographic projection, 91–92
Floor plan, 115–17
Floppy disk, 13, 14
Floppy disk drive, 13
Fractions, 44
FSCM number, 317

Good lettering style guidelines, 46
Grid paper, 18
Guidelines for lettering, 45

Hard disk drive, 14
Hardware, 12
Hatch command, 163–75
Hatching, 163–75
Hatch lines, 50
Hexagon, 52
Hidden features, 98
Hidden lines, 50
Hidden lines, drawing, 74, 127
Hidden lines in sectional views, 133
Hide command, 302
Horizontal dimensions, 247
How to use this book, 2–4

ID command, 70, 122
Identifying edges, 96
Inclined surfaces, 94, 95
Insert command, 191, 269, 270, 273, 274
Intersection, 51
Isometric:
 angles in, 205
 circles in, 206
 curves in, 206
 cutaways, 209–12
 spheres in, 207
Isometric drawing in AutoCAD, 223–30
Isometric drawings, 7
Isometric ellipses, 226, 227

471

Isometric sketching, 203–16
Isometric snap, 223
Isoplanes, 223, 224
   toggling to, 224

Keyboard, 14
Keys, 262

Layer, set current, 193
Layers, 35–38
Leader, 234
Lead hardness, 17
Lead holder, using, 50
Left arrow, 35
Lettering, developing a good style, 46
Lettering for pencil sketches, 44–47
Lettering guidelines, 46
Lettering slant, 45
Limits, 33
Line:
   center, 234
   dimension, 234
   extension, 234
Line command, 69
Lines:
   alphabet of, 49
   sketching, 101
   types used in sketching, 50
Lines for sketches, 49
Linetypes, AutoCAD, 69
Line weights and drawing constructions, 4

Machine screw, 261
Mechanical pencil, using, 50
Metric equivalents, 320
Metric scale, 22
Mirror, isometric, 226
Mirror command, 71, 79, 124
Mouse, 14

Next assembly number, 316
Normal cylinders, 99
Normal surfaces, 94
Notes, 317
Nuts, 261, 322, 323

Oblique drawing, 203, 204
Oblique surfaces, 95
Obtuse angle, 52
Offset command, 71, 118
Offset cutting-plane line, 154
OK button, 39
Order of sketching, 107
Orthographic drawings, 4, 114–47
Orthographic projection, 91–106
Orthographic views from an isometric sketch, 136–39, 140–44
Osnap, 70
   running, 309

Paper, grid, 18
Parallel, 51
Parallel line development, 282–86, 305
Parallel lines, sketching, 53
Parts list, 317
Pencils, 17, 18
Perpendicular lines, sketching, 54
Perpendicular osnap, 51
Personal characteristics, 1

Perspective drawing, 203
Phantom lines, 50
   drawing, 169
Pictorial views, 203–16
Pins, 262
Placing views on a drawing, 106
Pline command, 69
Plotter, 15
Polar coordinates, 70
Polygon command, 52
Polygons, sketching, 63, 64
Polyline command, 69, 115, 116
Primary auxiliary view, 178, 179
Printer, 15
Prism, 300
Proportional, 52
Purpose, 2

Radial line development, 287–90, 308
Radius, 51
Redo command, 41
Relative coordinates, 70
Right angle, 52
Right arrow, 35
Rivets, 262
Rotate command, 279
Rotate3d command, 305
Round, 52
Rule, steel, 22–24
Running osnap, 126
Running osnap mode, removing, 173
Running osnap modes, 71
Runouts, 97

Save Drawing As dialog box, 39
Save file as type button, 39
Save the drawing, 38–40
Scale command, 317
Scales, 19–22
Screw:
   cap, 322, 323
   machine, 261
   set, 261
Screws, cap, 259, 260
Secondary auxiliary, 179–81
Sectional drawings, 5
Sectional view:
   assembly, 151, 156
   broken-out, 156
   constructing, 152
   full, 155
   half, 155
   wall, 151
Sectional views, 150–60
   AutoCAD, 163–75
   hatching, 163–75
   hidden lines in, 153
   uses of, 150, 151
Selecting views, 103
Set layer current, 193
Set screw, 261
Site plan, 117–22
Sketching, 17
   order of, 107
Sketching orthographic views, 101–6
Sketching parallel lines, 53
Sketching perpendicular lines, 54
Sketching polygons, 63, 64
Sketching tangents, 55–61

Sketching tools, 17
Slanted surfaces, 95
Snap, 34
Software, 15
Solid command, 299
Spheres in isometric sketching, 207
Springs, 262
Starting a new drawing, 32
Steel rule, 22–24
Studs, 260
Subassembly, drawing, 314

Tabular dimensioning, 242
Tangent, 51
Tangents, sketching, 55–61
TEDIT, 248, 249
Template, circle, 19
Thickness, 298
Third-angle orthographic projection, 91–92
Thread class, 258
Thread forms, 254
Threads, 254–59
   American National Standard, 321
   drawing, 262–65
Threads and fasteners, 9
Thread series, 257
Thread specification:
   English units, 254, 255
   metric units, 258
Thread symbols, 259
Thread terms, 254
Tips for AutoCAD users, 41
Title block, 316
Toggling to isoplanes, 224
Toleranced drawings, 7
Tools, sketching, 17
Top assembly drawing, 313
Triangles, 18
Triangulation development, 291–93
Trim command, 71, 79, 116
Types of drawings, 4–12

UCS, view, 304
Undo, 41
Unidirectional dimensioning, 238
Units, 33
Uppercase letters, 44

Vertical dimensions, 247
Video monitor, 14
Viewports, 299
Views:
   aligning, 102
   auxiliary, 177–81
   pictorial, 203–16
   sectional, 150–60
   selecting, 103
Views for orthographic projection, 91, 92
Vpoint command, 299
VPORTS command, 299

Washers, 261
Wblock command, 268, 271
Wood pencil, using, 50

Zoom, window, 172
Zoom command, 72